UPSTREAM

Salmon and Society in the Pacific Northwest

Committee on Protection and Management of
Pacific Northwest Anadromous Salmonids

Board on Environmental Studies and Toxicology

Commission on Life Sciences

NATIONAL ACADEMY PRESS
Washington, D.C. 1996

NATIONAL ACADEMY PRESS • 2101 Constitution Ave., N.W. • Washington, D.C. 20418

NOTICE: The project that is the subject of this report was approved by the Governing Board of the National Research Council, whose members are drawn from the councils of the National Academy of Sciences, the National Academy of Engineering, and the Institute of Medicine. The members of the committee responsible for the report were chosen for their special competences and with regard for appropriate balance.

This report has been reviewed by a group other than the authors according to procedures approved by a Report Review Committee consisting of members of the National Academy of Sciences, the National Academy of Engineering, and the Institute of Medicine.

The project was supported by Department of Commerce, National Oceanic and Atmospheric Administration under grant #NA26FH0208.

Library of Congress Cataloging-in-Publication Data

Upstream : salmon and society in the Pacific Northwest / Committee on Protection and Management of Pacific Northwest Anadromous Salmonids, Board on Environmental Studies and Toxicology, Commission on Life Sciences.

p. cm.

Includes bibliographical references (p.) and index.

ISBN 0-309-05325-0

1. Pacific salmon—Northwest, Pacific. 2. Fishery conservation—Social aspects—Northwest, Pacific. 3. Pacific salmon fisheries—Pacific Northwest—Management. 4. Pacific salmon—Effect of habitat modification on—Northwest, Pacific. I. National Research Council (U.S.). Committee on Protection and Management of Pacific Northwest Anadromous Salmonids.

QL638.S2U67 1996

597′.55—dc20 95-51662

CIP

Cover art: *Midnight Run* by Ray Troll, Ketchikan, Alaska

Printed in the United States of America

COMMITTEE ON PROTECTION AND MANAGEMENT OF PACIFIC NORTHWEST ANADROMOUS SALMONIDS

John J. Magnuson (*Chair*), University of Wisconsin, Madison, Wisconsin
Fred W. Allendorf, University of Montana, Missoula, Montana
Robert L. Beschta, Oregon State University, Corvallis, Oregon
Peter A. Bisson, Weyerhaeuser Company, Tacoma, Washington
Hampton L. Carson, University of Hawaii at Manoa, Honolulu, Hawaii
Donald W. Chapman, Don Chapman Consultants, Inc., Boise, Idaho
Susan S. Hanna, Oregon State University, Corvallis, Oregon
Anne R. Kapuscinski, University of Minnesota, St. Paul, Minnesota
Kai N. Lee, Williams College, Williamstown, Massachusetts
Dennis P. Lettenmaier, University of Washington, Seattle, Washington
Bonnie J. McCay, Rutgers, The State University, New Brunswick, New Jersey
Gordon M. MacNabb, independent consultant, Vernon, British Columbia, Canada
Thomas P. Quinn, University of Washington, Seattle, Washington
Brian E. Riddell, Department of Fisheries and Oceans, Pacific Biological Station, Nanaimo, British Columbia, Canada
Earl E. Werner, University of Michigan, Ann Arbor, Michigan

Staff

David J. Policansky, Project Director
Tania Williams, Research Associate
Norman Grossblatt, Editor
Adriénne Davis, Senior Project Assistant

Sponsor

Department of Commerce

BOARD ON ENVIRONMENTAL STUDIES AND TOXICOLOGY

OTHER RECENT REPORTS OF THE BOARD ON ENVIRONMENTAL STUDIES AND TOXICOLOGY

Carcinogens and Anticarcinogens in the Human Diet: A Comparison of Naturally Occurring and Synthetic Substances (1996)
Science and the Endangered Species Act (1995)
Wetlands: Characteristics and Boundaries (1995)
Biologic Markers in Urinary Toxicology (1995)
Review of EPA's Environmental Monitoring and Assessment Program (three reports, 1994-1995)
Science and Judgment in Risk Assessment (1994)
Ranking Hazardous Sites for Remedial Action (1994)
Pesticides in the Diets of Infants and Children (1993)
Issues in Risk Assessment (1993)
Setting Priorities for Land Conservation (1993)
Protecting Visibility in National Parks and Wilderness Areas (1993)
Biologic Markers in Immunotoxicology (1992)
Dolphins and the Tuna Industry (1992)
Environmental Neurotoxicology (1992)
Hazardous Materials on the Public Lands (1992)
Science and the National Parks (1992)
Animals as Sentinels of Environmental Health Hazards (1991)
Assessment of the U.S. Outer Continental Shelf Environmental Studies Program, Volumes I-IV (1991-1993)
Human Exposure Assessment for Airborne Pollutants (1991)
Monitoring Human Tissues for Toxic Substances (1991)
Rethinking the Ozone Problem in Urban and Regional Air Pollution (1991)
Decline of the Sea Turtles (1990)
Tracking Toxic Substances at Industrial Facilities (1990)
Biologic Markers in Pulmonary Toxicology (1989)
Biologic Markers in Reproductive Toxicology (1989)

Copies of these reports may be ordered from
the National Academy Press
(800) 624-6242
(202) 334-3313

Preface

The Committee on Protection and Management of Pacific Northwest Anadromous Salmon was formed in 1992 under the auspices of the National Research Council's Board on Environmental Studies and Toxicology (BEST). The committee was formally charged as follows:

> *The committee will review information concerning the seven species of the genus Oncorhynchus in the Pacific Northwest. The review will focus on the population status, habitat, and environmental requirements of the stocks. It will include analysis of information about their genetics, history, management, and production by hatcheries, as well as federal, state, tribal, and other management regimes. The committee will evaluate options for improving the prospects for long-term sustainability of the stocks, and will consider economic and social implications of such changes. The committee will undertake the following general tasks:*
>
> *Assess the state of the stocks. This assessment will include a consideration of the nature of local adaptations of stocks to environmental conditions. More specific questions will probably include the degree to which the adaptations are due to phenotypic plasticity versus genetic differences, the nature and appropriate definition of a stock, the value of a diversity of stocks for preserving evolutionary potential, and which stocks are in danger of extinction or of becoming irretrievably mixed with other stocks.*
>
> *Analyze the causes of decline. This analysis will consider all stages of*

the life history of the seven species: spawning and nursery habitat, downstream migration, life at sea, and the return upriver to spawn.

Analyze options for intervention (management). Again, this will involve all stages of the life history. It will include a consideration of previous activities, such as the Pacific Northwest Power Planning Council's program in adaptive management; federal, state, and local regulations and enhancements; and other activities. The analysis will include some consideration of socioeconomic costs and benefits.

Composition of this "salmon" committee was especially broad; it had to be to encompass the geographic, scientific, and human breadth of the issues in the continuing interactions in the Pacific Northwest between salmon and humans. Members included experts with general and specific knowledge of genetics, fish ecology, stream ecology, fish biology, inland waters and ocean science, anthropology, social science, political science, international fisheries and transboundary issues, habitat and habitat rehabilitation, hydrology, hatcheries, dams, fishery management, and fishery science. The committee's biographies are in Appendix A. I especially appreciated the degree to which the committee was able to frame and synthesize the many facets of the "salmon problem." I continue to be amazed by the high quality of the National Research Council staff; on behalf of the committee I especially thank David Policansky, Tania Williams, and Adriénne Davis.

The committee met seven times from December 1992 to June 1994; two meetings were held in Portland, Oregon, and two in Seattle, Washington, to allow persons from the Pacific Northwest with information relevant to the issues easy access to the committee. In addition, a writing subgroup met and individual committee members met in pairs and trios on specific issues. The dates and meeting locations are in Appendix B.

The committee thanks the many persons who provided information and points of view to us in open sessions and public hearings. These people further broadened the scope of the committee, and the specific information we received greatly helped develop our own perspectives on the issues. These individuals and organizations are identified in Appendix C. Also, I thank Courtland L. Smith for his many contributions during the period he served on the committee, both to the process of the committee and the text of our report.

The "salmon problem" in the Pacific Northwest is one that can be dealt with only if the diverse participants work together on the many issues that unfold over the long meandering path laid out by the salmon over their lives—from clear cold streams where they hatch, to the ocean where they grow, and back again to their natal streams to spawn. Owing to the wide distribution of Pacific anadromous salmon, the region over which the salmon-human interactions play out cuts across local, state, and international boundaries; across the responsibilities of many human institutions designed to deal with parts of the problem; and across the lands and waters used in many different human endeavors. The interactions with human activity cannot be solved by the action of single groups or by focusing on single issues and single causes. In a sense, *Upstream: Salmon and Society in the Pacific Northwest* challenges society to deal with this large issue—involving environmental, resource, and human considerations—at the time and space scales necessary to prevent further declines or perhaps even rehabilitate the human-salmon system.

John J. Magnuson
Chair

The National Academy of Sciences is a private, nonprofit, self-perpetuating society of distinguished scholars engaged in scientific and engineering research, dedicated to the furtherance of science and technology and to their use for the general welfare. Upon the authority of the charter granted to it by the Congress in 1863, the Academy has a mandate that requires it to advise the federal government on scientific and technical matters. Dr. Bruce Alberts is president of the National Academy of Sciences.

The National Academy of Engineering was established in 1964, under the charter of the National Academy of Sciences, as a parallel organization of outstanding engineers. It is autonomous in its administration and in the selection of its members, sharing with the National Academy of Sciences the responsibility for advising the federal government. The National Academy of Engineering also sponsors engineering programs aimed at meeting national needs, encourages education and research, and recognizes the superior achievements of engineers. Dr. Harold Liebowitz is president of the National Academy of Engineering.

The Institute of Medicine was established in 1970 by the National Academy of Sciences to secure the services of eminent members of appropriate professions in the examination of policy matters pertaining to the health of the public. The Institute acts under the responsibility given to the National Academy of Sciences by its congressional charter to be an adviser to the federal government and, upon its own initiative, to identify issues of medical care, research, and education. Dr. Kenneth I. Shine is president of the Institute of Medicine.

The National Research Council was organized by the National Academy of Sciences in 1916 to associate the broad community of science and technology with the Academy's purposes of furthering knowledge and advising the federal government. Functioning in accordance with general policies determined by the Academy, the Council has become the principal operating agency of both the National Academy of Sciences and the National Academy of Engineering in providing services to the government, the public, and the scientific and engineering communities. The Council is administered jointly by both Academies and the Institute of Medicine. Dr. Bruce Alberts and Dr. Harold Liebowitz are chairman and vice chairman, respectively, of the National Research Council.

Contents

APPENDIXES

Facing page: Image of the Haida Dog Salmon (Saagi) by Haida artist Bill Reid. Courtesy of Bill Reid and the Buschlen Mowatt Gallery, 111-1445 W. Georgia St., Vancouver, BC, Canada V6G 2T3.

UPSTREAM

Salmon and Society in the Pacific Northwest

Executive Summary

Pacific salmon have disappeared from about 40% of their historical breeding ranges in Washington, Oregon, Idaho, and California over the last century, and many remaining populations are severely reduced. Most runs that appear plentiful today are largely composed of fish produced in hatcheries. Recreational and commercial fishing for several salmon species has been restricted or even prohibited from the coastal waters of the region to the headwaters of many streams, and tribal fishing has been much reduced. Petitions have been filed to list several populations as endangered or threatened under the Endangered Species Act; a few have been listed, and more could be soon.

Salmon have great cultural, economic, recreational, and symbolic importance in the Pacific Northwest. As a result, their declines—which have numerous interacting causes—have resulted in much concern. The often expensive efforts to reverse the declines have been controversial and unsuccessful in many cases. Faced with the possibility of dozens or perhaps even hundreds of listings of Pacific salmon under the Endangered Species Act, and faced with controversies over the effectiveness of proposed actions to slow, halt, or reverse the salmon declines, Congress requested advice from the National Research Council (NRC). In response, the NRC's Board on Environmental Studies and Toxicology assembled the expert Committee on Protection and Management of Pacific Northwest Anadromous Salmonids to review information concerning the seven species of anadromous salmonids[1] in the Pacific Northwest.

---

[1]This report deals with anadromous forms of the seven species of the genus *Oncorhynchus*. They are chinook, chum, coho, pink, and sockeye salmon and the anadromous forms of rainbow and cutthroat trout: steelhead and sea-run cutthroat. In this report, the general term *salmon* refers to all seven species.

The committee was asked to "evaluate options for improving the prospects for long-term sustainability of the stocks, and [to] consider economic and social implications of such changes" (statement of task; see Preface). It was asked to perform the following tasks:

- Assess the status of the salmon stocks.
- Analyze the causes of declines.
- Analyze options for intervention.

The committee was asked to consider all stages of salmon life histories, including the ocean phase, and to consider the appropriate roles of hatcheries. Congress did not request advice on whether society *should* make the investments needed to halt and reverse salmon declines. However, the committee's analysis of options for intervention and their likely effectiveness should help to inform that policy decision.

STATUS OF SALMON POPULATIONS

The status of many specific salmon populations in the Pacific Northwest is uncertain, and there are exceptions to most generalizations with regard to overall status. Nevertheless, a general examination of the evidence of population declines over broad areas is helpful for understanding the current status of species with different life cycle characteristics and geographical distributions, and with some caution, the following generalizations are justified:

- *Pacific salmon have disappeared from about 40% of their historical breeding ranges in Washington, Oregon, Idaho, and California over the last century, and many remaining populations are severely depressed in areas where they were formerly abundant.* If the areas in which salmon are threatened or endangered are added to the areas where they are now extinct, the total area with losses is two-thirds of their previous range in the four states. Although the overall situation is not as serious in southwestern British Columbia, some populations there also are in a state of decline, and all populations have been completely cut off from access to the upper Columbia River in eastern British Columbia. Even if the estimate of population losses of about 40% is only a rough approximation, the status of naturally spawning salmon populations gives cause for pessimism.
- *Coastal populations tend to be somewhat better off than populations inhabiting interior drainages.* Species with populations that occurred in inland subbasins of large river systems (such as the Sacramento, Klamath, and Columbia rivers)—spring/summer chinook, summer steelhead, and sockeye—are extinct over a greater percentage of their range than species limited primarily to coastal rivers. Salmon whose populations are stable over the greatest percentages

of their range (fall chinook, chum, pink, and winter steelhead) chiefly inhabit rivers and streams in coastal zones.

- *Populations near the southern boundary of species' ranges tend to be at greater risk than northern populations.* In general, proportionately fewer healthy populations exist in California and Oregon than in Washington and British Columbia. The reasons for this trend are complex and appear to be related to both ocean conditions and human activities.
- *Species with extended freshwater rearing (up to a year)—such as spring/summer chinook, coho, sockeye, sea-run cutthroat, and steelhead—are generally extinct, endangered, or threatened over a greater percentage of their ranges than species with abbreviated freshwater residence, such as fall chinook, chum, and pink salmon.*
- *In many cases, populations that are not smaller than they used to be are now composed largely or entirely of hatchery fish.* An overall estimate of the proportion of hatchery fish is not available, but several regional estimates make clear that many runs depend mainly or entirely on hatcheries.

Chapter 4 discusses some of the difficulties in evaluating the status of wild populations and how these difficulties have been addressed in recently published status reports. Regional trends are summarized, and the overall conditions of the species are presented.

THE SALMON PROBLEM

The salmon problem is the decline of wild salmon runs and the reductions in abundance of salmon even after massive investments in hatcheries. The declines—largely a result of human impacts on the environment caused by activities such as forestry, agriculture, grazing, industrial activities, urbanization, dams, hatcheries, and fishing—are widespread, although not universal. They have a variety of causes, and they are exacerbated by the unusual life cycle of Pacific anadromous salmon, which spawn in freshwater, migrate to sea to grow and mature, and return to their natal streams to reproduce. Salmon thus require high-quality environments from mountain streams, through major rivers, to the ocean. Economic development and population growth have created widespread declines in anadromous salmon abundance in the Pacific Northwest. Variations in ocean conditions—especially in water temperature and currents and the associated biological communities—also contribute to the rise and fall of salmon abundance, often thwarting the interpretation of events in freshwater and the surrounding terrestrial systems.

GENERAL CONCLUSION

To achieve long-term protection for a diversity and abundance of salmon in the Pacific Northwest, two general goals must be achieved:

- The long-term survival of salmon depends crucially on a diverse and rich store of genetic variation. Because of their homing behavior and the distribution of their populations and their riverine habitats, salmon populations are unusually susceptible to local extinctions and are dependent on diversity in their genetic makeup and population structure (Chapter 6). Therefore, management must recognize and protect the *genetic diversity* within each salmon species, and it must recognize and work with local breeding populations and their habitats. It is not enough to focus only on the abundance of salmon.
- The social structures and institutions that have been operating in the Pacific Northwest have proved incapable of ensuring a long-term future for salmon, in large part because they do not operate at the right time and space scales. As described in Chapter 13, differences among watersheds mean that different approaches are likely to be appropriate and effective in different watersheds, even where the goals are the same. This means that institutions must be able to operate at the scale of watersheds; in addition, a coordinating function is needed to make sure that larger perspectives are considered.

As a framework in which to approach its deliberations, the committee chose to focus on *rehabilitation*—a pragmatic approach that relies on natural regenerative processes in the long term and the selected use of technology and human effort in the short term—rather than on attempts to restore the landscape to some pristine former state and rather than on a primary reliance on substitution, i.e., the use of technologies and energy inputs, such as hatcheries, artificial transportation, and modification of stream channels. Rehabilitation would protect what remains in an ecosystem and encourage natural regenerative processes.

The solutions will not be easy or inexpensive to implement; even a holding action to prevent further declines will require large commitments of time and money from many people in many segments of society in the Pacific Northwest. Therefore, broad-based societal decisions are needed to successfully provide a long-term future for natural salmon populations.

ENVIRONMENTAL FACTORS

Natural and human-caused environmental changes affect all aspects of salmon life histories. Although humans can do little in the short term to control or even predict large-scale changes in environmental conditions, salmon-management programs must expect such changes and take them into account. Managers must also recognize that the natural variability in environmental conditions and people's desires for large and stable catches of salmon are often not compatible. Natural changes in environmental conditions in the ocean, in fresh water, and on land occur continually; sometimes they can lead to increased salmon productivity in an area; at other times they can lead to decreased productivity.

The emerging understanding of interdecadal changes in the ocean climate and the related mechanisms that affect salmon at sea have implications that are both exciting and disconcerting to scientists thinking about resource management. Humans are beginning to understand what happens to salmon during the majority of their lives—the portion spent at sea. Although we know little of the details, the new insights already demonstrate that variations in salmon abundance are linked to phenomena on spatial and temporal scales that humans and human institutions do not ordinarily take into account. Consider that the apparent effectiveness of hatcheries might have resulted from favorable ocean and climatic conditions in the era when the hatcheries were built; what looked like human manipulation of the total number of salmon might have been only a reapportionment among different populations. Or consider that the decline of some populations might be a direct result of introducing new hatchery populations into an ocean pasture of limited capacity.

The scale of human endeavor often has been incommensurate with the scale of salmon ecology. Some of our current policies are based on deep ignorance: it is not reasonable to assume that ocean conditions vary in ways that are generally uniform and random in their impacts on populations of salmon. Interdecadal variations and the importance of the ocean phase should be incorporated into human thought, planning, and actions in response to the effects of and attempts to repair damage that occurred during the freshwater phases of the salmon lives. The possible overriding effects of interdecadal changes in ocean conditions on salmon, the results of freshwater salmon management, and the overwhelming focus of human attention on the more-visible freshwater phases of the salmon history combine to provide the key ingredients for surprises in future.

Recently, natural environmental conditions in the Pacific Northwest appear to have been unfavorable to salmon production. As changes continue to occur, environmental conditions will probably favor salmon and lead to larger runs in some areas for a time, even without human intervention. If such changes do occur, they should be regarded as providing time to develop better strategies for rehabilitation of salmon populations. They should *not* be used as reasons for abandoning efforts to rehabilitate salmon, for they will surely be followed by other natural changes. Inappropriate short-term responses to large-scale environmental changes at sea or on land should be avoided, because there can be long lags between causes and effects.

LIMITS ON SALMON PRODUCTION

The salmon production cycle has three principal components that determine abundance: reproductive potential of adults returning from the sea to spawn, which is affected by their growth at sea; production of offspring from natural reproduction in streams and artificial propagation in hatcheries; and sources of mortality (including natural mortality, fishing mortality, dam-caused mortality, mortality from habitat alterations and changes in environmental conditions, and so on). All three components are affected by changes in environmental conditions as well as by human activities. Variation in the three components and their interactions ultimately determine the ability to sustain salmon populations and their production. These limitations cannot be easily overcome through technology. Although it has been widely assumed that a loss of natural salmon production can be compensated by enhancement (e.g., by increasing hatchery production), chapters 6, 11, and 12 show that such an assumption is untenable by explaining the need to conserve sufficient genetic variation in natural populations to support the evolutionary and ecological processes needed for sustained salmon production. Compensating for salmon loss from any source over the long term therefore requires reducing other losses. Furthermore, an increasing appreciation of the marine environment and its effects on the above components is emerging as an essential consideration in salmon management.

VALUES

The salmon problem, like many other environmental issues, has been addressed through choices made within economic, political, and individual ethical frameworks. Values and ethical positions held by people involved in and affected by the salmon problem encompass a pluralistic, pragmatic, and evolutionary approach to natural resource management. Recognizing and articulating that pluralism is important because problems in managing and protecting fish populations are due in part to the failure to articulate divergent interests, goals, and values and to address them explicitly. Chapter 5 describes how the widely varied ways that humans intervene in salmon populations are linked to socially validated values.

From a policy perspective, the salmon problem is one of long-standing and serious conflict in fact, interest, and values. People often invoke widely held values to protect particular interests, but values are genuine sources of conflict in themselves. Value conflict stems from different assessments of the desirable goals of public action. From a scientific perspective, wild salmon populations are an example of an ecosystem's natural capital. Our greatest success has been in designing ways to use human-food benefits from wild salmon. Our corresponding failure has been in protecting indirect and nonhuman benefits.

One way to present the salmon problem is to say that the value of the Pacific Northwest's salmon-capital asset has depreciated over time as its productivity has declined. A major problem is that the market does not account for the full range of costs and benefits of salmon. That is called a market distortion. When such market distortions exist, some resources are underpriced and overused, and others overpriced and underused. Many nonmarket values of salmon are underrepresented and are not easy to measure or compare. Thus management decisions often do not adequately reflect the importance of salmon to society and decisions about resource use may not achieve societal goals. To correct the discrepancy between social values and resource use, attempts can be made to design policies that reflect the full range of resource values.

Full value is a public, not a private, question. Consequently, public choices are central to the salmon problem. Public choices have to take into account many owners with multiple preferences, attributes that are not fully observable and sometimes unknown, and prices that reflect only part of the resources' full value to society. The concept of full value points to the problem of "externalities"—the problem that some costs and benefits are beyond the accounting of the decision-making unit.

Environmental variability creates economic uncertainty, which causes people to discount the future more heavily, and this leads to pressures to increase rates of immediate, direct use. Environmental variability also creates scientific uncertainty about biological processes, which can be perceived to call for a cautious approach and lead to pressures to lower rates of immediate, direct use. The resulting tension between economic and scientific responses to uncertainty adds complexity to decisions about appropriate rates of resource use. That tension is widespread in decisions concerning the salmon problem.

Problems like these emphasize the need to develop more appropriate interdisciplinary approaches. The idea of rebuilding the salmon runs of an industrialized ecosystem is heroically optimistic—a hope that might not have occurred to anyone except those who had rehabilitated the Willamette River Basin in Oregon or Lake Washington near Seattle. Those environmental successes came through the disciplined execution of the planning paradigm that has been fitfully applied to the much larger Columbia River Basin. The extension of those experiences to the multijurisdictional, multifunctional situations of the Pacific Northwest would require coordinated action and learning on a new, larger scale—a scale on which planning and action have been tried but have not been successful. A more explicit appreciation of the values, interests, and institutions involved in this undertaking is required. Chapter 13 explores this further and urges constructive change in institutions that include cooperative management, bioregional governance, and adaptive management.

GENETICS AND CONSERVATION

Pacific salmon reproduce in freshwater streams. Their progeny migrate to the sea to grow and mature, and then return to freshwater streams to reproduce and (nearly always) die. This pattern of freshwater reproduction and growth at sea is called *anadromy*. Most of the adults actually return to the streams where they hatched. This behavior—called *homing*—is an essential part of salmon biology and makes their genetics and conservation unusual. There is a great deal of environmental variation among the various streams and lakes where salmon spawn and in the rivers through which they migrate. Because of their anadromous life cycles and homing behaviors and the variety of environments they occupy, each salmon species tends to differentiate into local breeding populations—called *demes*—that are in general reproductively isolated from other populations and adapted to each stream. To sustain productive natural populations of salmon, it is crucially important to maintain this genetic variation and local adaptation. Chapter 6 describes examples of such local adaptation.

However, more is involved than only local adaptation to various streams. Natural environmental fluctuations, including major disruptions caused by geological activity, can cause the extinction of local populations. Because homing is not perfect, fish that stray from nearby streams can replenish those populations. Strays are more likely to re-establish a population if the environment in the new stream is similar to that in the stream where they hatched. Thus, strays into tributaries in the same major river system or into nearby streams are more likely to succeed than those that stray into very different environments. This network of local populations (known as a *metapopulation*) provides a balance between local adaptation and the evolutionary flexibility that results from exchange of genetic material among local populations (Chapter 6). It likely also explains why artificial attempts to re-establish populations from a captive broodstock have often failed—too often, the gene pool of the broodstock has had reduced variation or has been derived from a population adapted to a different environment (Chapter 12). The metapopulation structure provides a balance between local adaptation and evolutionary flexibility; therefore, maintaining a metapopulation structure with good geographic distribution should be a top management priority to sustain salmon populations over the long term. Many of the committee's recommendations are based on this crucial conclusion.

There is no "correct" answer to the question of precisely how much biological diversity and population structure should be maintained or can be lost to provide a long-term future for salmon. Scientific estimates—including uncertainties associated with them—are only part of the argument. Society must decide what degree of biological security would be desirable and affordable if it could be achieved, i.e., the desired probability of survival or extinction of natural populations, over what time and what area, and at what cost. Nonetheless, biological diversity and the structure of salmon populations are being lost at a substantial rate, and this loss threatens the sustainability of naturally reproducing salmon populations in the Pacific Northwest.

HABITAT LOSS AND REHABILITATION

The main habitat requirements of salmon in freshwater include a stream or lake, the adjacent border of vegetation (riparian zone) that serves as the interface between aquatic and terrestrial ecosystems, and the quality and quantity of water (Chapter 7). The water must be clean enough and cool enough to support returning adults, for eggs to hatch, and for young to survive and grow until they migrate to sea. There must be enough water in the rivers at crucial times to make migration possible, to allow fish to escape predators, and to allow fish to find adequate food. Well-aerated streambed gravels are important for spawning. Streamside vegetation provides shade, which keeps the water cool; it provides a buffer against soil erosion, which maintains water quality; it provides living space for various animals that provide food and nutrients for streams; and it provides a source of large woody debris, which plays a key role in the formation of physical habitat and storage of sediment and organic matter and provides habitat complexity in stream channels, thus improving the stream environment for salmon. These requirements for environmental conditions in streams and adjacent riparian zones depend on the condition of the entire watershed in which they occur.

Many human activities—such as forestry; agriculture; grazing; industrial uses; commercial, residential, and recreational development; and flood control—have a variety of adverse effects on salmon habitats. For example, they can increase soil erosion, reduce the amount of woody debris in streams, raise the water temperature, add contaminants to the water, affect water flow, and reduce the amount of water available, with resultant loss or degradation of riverine and adjacent riparian and near-river habitat. Therefore, protection and rehabilitation of riverine and riparian habitats and associated watershed processes will be an integral part of rehabilitating salmon populations, although it is a major and difficult undertaking (Chapter 8). In the past few years, genuine improvements in protecting forested streams have been initiated. Nonetheless, for real progress to occur, habitat protection must be coordinated at landscape scales appropriate to salmon life histories, and they must be more consistent across different types of land use (chapters 8 and 13).

DAMS

Hundreds of dams have been built on rivers of the Pacific Northwest. They range from small irrigation dams with a hydraulic head of only a few feet to massive dams at Grand Coulee, Dworshak, and Hells Canyon on the Columbia and Snake rivers that are several hundred feet high and completely block upstream and downstream passage of anadromous fish. Dams on various rivers—some of them impassable—have greatly reduced wild runs. Even smaller dams (e.g., those associated with many hatchery operations and irrigation-diversion dams) can block salmon runs. In addition to their effects on migration, large storage dams affect the quantity and timing of water flow in the river as well as flow velocities, water chemistry, and water temperatures. Reservoirs behind dams can also inundate extensive areas of spawning and rearing habitat, although in some cases the reservoirs provide new (but different) rearing habitat. Many water diversions for irrigation lack protective fish screens of modern design; installing such screens would reduce mortality of smolts as they migrate downstream.

Even when fish ladders provide passage for adult salmon, many young salmon (smolts) migrating downriver die at dams. Although as many as 90% of young salmon might survive passage over, around, and through any single major project on the Columbia-Snake mainstem, the cumulative reduction in survival caused by passing many projects has adversely affected salmon populations. To counteract these effects, it is essential to improve the survival of smolts migrating through hydropower projects, especially in the Columbia and Snake rivers. Serious consideration needs to be given to all available alternatives for doing so; even a small improvement in survival would be helpful if it were repeated at several dams.

Controversy surrounds the effects of dams and how best to mitigate them. Alternatives include removal of dams, modification of turbines and other structural aspects of dams to improve fish survival during passage, drawdown of the water during the seaward migration of smolts to restore the river's profile to its pre-dam (river-grade) configuration to increase the flow rate and diminish the smolts' travel time, drawdown of the river to some level above river grade, augmentation of water flows during smolt migration to speed their passage downriver, transportation of smolts around dams by truck or by barge, control of predators in reservoirs and below dams, and spilling of water over dams instead of through the turbines. However, there is a dearth of good scientific information on which to base evaluations of the alternatives, some of which would be very expensive and would cause large losses of hydropower revenues.

Dam removal and drawdown of those rivers to river grade would be enormously expensive, would take many years, and probably would have long-term adverse impacts on the rivers. However, because the many dams on the Columbia River and its tributaries cumulatively have large effects on salmon survival, the addition of any new major dams in undammed reaches in the system (e.g., the Hanford Reach of the Columbia River) would make the situation worse; existing dams should have adequate fish-passage facilities where feasible and appropriate before being relicensed. The committee is unaware of any scientific data that unequivocally support drawdown to a level above river grade as the best available dam-mitigation option for the Columbia River or the Snake River. Based on limited information, transportation appears to be the most biologically effective and cost-effective approach for moving smolts downstream. It should be continued on an adaptive basis (i.e., in such a way that additional information can be obtained about its effectiveness). Additional information is needed on effects of transportation on survival to the adult return stage, on homing, on success of natural spawning, and on genetic diversity of returning adults. Because any action that could jeopardize all of the fish in a stream must be avoided, not all the fish in any stream should be transported.

Research is needed on the effects of various options on the survival of both smolt and adult migration through dam and reservoir systems. Any management option should be applied on an adaptive (experimental) basis. The committee is not recommending that the salmon be "studied to death," a criticism often leveled at those who urge further studies. Indeed, enough is known now to take some actions. In recommending "adaptive" actions, the committee is recommending that any mitigative actions be taken in a way that allows their effects and effectiveness to be measured and assessed objectively. For example, if some fish in a stream are transported downstream, the action should be designed so its effectiveness can be assessed and compared with other alternatives. Despite the paucity of information, it is clear that no single approach would eliminate the adverse effects of dams on salmon.

HATCHERIES

Hatcheries have been used for more than 100 years in attempts to mitigate the effects of human activities on salmon and to replace declining and lost natural populations. As a result, a major proportion of salmon populations in the Pacific Northwest now consist largely of hatchery fish. These hatchery fish appear to have had substantial adverse effects on native fish populations.

For many years, people did not recognize the potential for hatchery fish to affect wild fish and did not believe that there was any limit to the ocean's capacity to provide food for growing salmon. It therefore seemed that producing more juveniles would result in more returning adults. The difficulties and shortcomings of hatchery production did not become apparent until fishing pressure and habitat-related mortality increased and marking technologies became available. As a result, hatcheries were not part of an adaptive-management program; that is, they were not considered as scientific experiments—they were not even adequately monitored—so many of their effects were not well known.

It is now clear from synthesis of experience and from consideration of well-established biological knowledge that hatcheries have had demographic, ecological, and genetic impacts on wild salmon populations and have caused problems related to the behavior, health, and physiology of hatchery fish. They have resulted (among other effects) in reduced genetic diversity within and between salmon populations, increased the effects of mixed-population fisheries on depleted natural populations, altered behavior of fish, caused ecological problems by eliminating the nutritive contributions of carcasses of spawning salmon from streams, and probably displaced the remnants of wild runs (Chapter 12). Hatchery fish have at times exceeded the capacity of streams and are increasingly being associated with reduced marine growth and survival in wild salmon populations (Chapter 12).

Many of the problems stem from purposes to which hatcheries have been put—mainly to provide substitutes for natural populations lost or displaced because of human development activities. Because of their deleterious impacts, however, hatcheries should no longer be viewed solely as factories for producing fish. Hatcheries should also be thought of as laboratories that can provide controlled environments for studying juvenile fish and for testing treatments to improve our understanding of what happens to juveniles after they leave spawning areas. Seen in that light, hatcheries can be a powerful tool for learning about salmon.

Hatchery planning, management, and operations should be changed so that their goals are to assist recovery of wild populations and to increase knowledge about salmon. As described above and in many parts of this report, especially chapters 6, 11, and 12, precautions must be taken to protect the genetic diversity and ecological productivity of naturally spawning populations of salmon. Those precautions will include an overall decrease in hatchery-fish production and—over the short term—in fishing opportunities. The basic guideline is to ensure that any hatchery production for fishing is not detrimental to natural populations. Because adaptive-management experiments should be tailored to the circumstances in different watersheds of the Pacific Northwest, decisions about use of hatcheries will differ across these watersheds. Therefore, decisions about uses of hatcheries should include a focus on the whole watershed and its linkage to the region and the ocean pasture, rather than only on the fish.

FISHING

Fishing for salmon is important in the Pacific Northwest. It includes commercial, recreational, and treaty fishing at sea and in rivers and is an important source of mortality, especially for adults returning to spawn. Salmon mortality caused by other human activities and structures such as dams, habitat loss or degradation, pollution, and water diversion and by natural factors such as predators, disease, and environmental variability together usually exceed fishing mortality. Those causes of mortality have a major effect on the production of adult fish and thus influence the rate of fishing that can be sustained. However, fishing is the easiest mortality factor to control. Control of fishing has rehabilitated marine and anadromous fish populations in various parts of the United States.

Managing salmon fisheries is more difficult than managing many other fisheries because of the geographic distribution of salmon, their metapopulation structure, and the fact that most adult fish spawn only once and then die. In the jargon of Pacific salmon fisheries, managers refer to groups of salmon populations that are identifiable for management as *stocks*. Frequently, *stock* refers to a geographic aggregate of populations that includes many local breeding populations of varied size and productivity; this is too large a unit for conservation of genetic diversity and rehabilitation of salmon production. Managing at the stock level obscures critical biological complexity. But even managing such large units is difficult because of the complex relationships, responsibilities, and obligations among a large number of institutional entities in the region (including nations, states, provinces, federal agencies, tribes, interest groups, and other organizations), the mandates of the Endangered Species Act and other laws, and the diverse array of interests and values in the region.

For rehabilitation of salmon populations, the aim for fishery management—as for other management efforts—should be to achieve long-term sustainability based on maintaining diversity of gene pools and population structures. Therefore, a successful fishery-management component for protecting natural salmon runs in the Pacific Northwest should explicitly recognize the need to maintain and rehabilitate the genetic diversity of salmon and recognize the interdependence of genetic diversity, habitat, and salmon production. It must also account for the uncertainty in scientific predictions and the inherent variability of biotic and abiotic environmental factors.

In general, the aim should be to assure adequate escapements for depleted populations. To achieve long-term sustainability, which requires sufficient genetic diversity, fishing should occur only where the identity (i.e., the originating population) of the salmon is known, when total fishing mortality is consistent with productivity of the fish, and when the catching technology ensures minimal mortality in depleted demes. This will require fishing methods that allow different degrees of fishing effort on various salmon populations and that allow identification of fish taken from depleted demes so that they can be avoided or released alive. Two methods of achieving these goals (but not the only ones) are terminal fisheries and live-catch fisheries.

In general, the serious declines of wild salmon populations show that not enough fish are being allowed to return to spawn. The number of fish returning to spawn (escapements) must be substantially increased to conserve genetic diversity within and between demes, use available habitats, rehabilitate ecological processes (including the return of nutrients to aquatic ecosystems), and increase the sustainable production of salmon. Increasing escapements will disrupt fisheries, industry, and communities, but it is necessary for restoring production. As salmon abundance increases and fisheries begin operating at lower, but sustainable, catch *rates*, actual catches will gradually increase, although probably not to the sizes of some historical catches, because those were based on excessive catch rates. Implementing this recommendation will initially require low fishing effort in many areas, especially in the ocean, and it will require cooperation from British Columbia and Alaska, because many salmon that originate in the Pacific Northwest are caught at sea off British Columbia and southeastern Alaska (chapters 10 and 11).

A more holistic management approach must recognize the connections between the genetic resource base, habitat, and the resulting salmon production; it must also account for the uncertainty in our scientific advice and for inherent environmental variability. The committee has outlined a process intended to improve the potential sustainability of salmon in the Pacific Northwest. Furthermore, the committee does not believe that the sustainability of Pacific Northwest salmon can be achieved without limiting the interceptions of U.S. salmon in Canada and obtaining the cooperation of Alaska. An effective and cooperative Pacific Salmon Treaty is necessary. The committee does not provide specific recommendations about altering specific fisheries, because there are numerous options and interactions between fisheries. Achieving agreement on changes in fisheries will be difficult and necessitates an effective institutional process.

INSTITUTIONAL CHANGE

The long and serious decline of salmon in the Pacific Northwest has been promoted—often unwittingly—by human institutions; effective remedies, if they are to be found, will have to involve changes in those institutions. Growth in human populations and economic activity threatens the continued existence of salmon in the Pacific Northwest. Institutions developed in different times for diverse purposes have been asked to do things foreign to their original objectives and capabilities. Political changes have hindered attempts to take a long-term perspective. There has been fragmentation of effort and responsibility.

Changing institutional structures is notoriously difficult, but it is possible. Because the problems facing salmon have many aspects, a multidisciplinary approach to their solution is essential. Indeed, if the money that has been spent to date on salmon research had been spent with a more unified, regional vision, greater progress would have been made in maintaining viable salmon populations (Chapter 14). Unless agencies cooperate more effectively, salmon populations are unlikely to recover.

One problem is that current institutions and the boundaries of their jurisdictions usually do not match the spatial, temporal, or functional scales of the salmon problem. In addition, current institutional structures lack both a fine-grained aspect to respond to local concerns and variations and a coarse-grained aspect to integrate across small regions and to make sure that the interests of a few small areas do not jeopardize larger regional interests.

Because we often do not know what the effects of a management option will be, management must be undertaken with an experimental, adaptive point of view. Flexibility must be built into institutional structures to allow for changes in management practices based on experience. Institutions must allow and encourage refocusing the energies of salmon management to recognize the importance of demes in maintaining genetic processes and to maintain and expand their diversity. The goal of management should be to achieve a biologically sound escapement (instead of focusing on a "sustainable" or permissible catch) for each metapopulation and an explicit adoption of time scales for management and planning that are commensurate with the multiyear scale of salmon life cycles.

Beyond those facilitating changes, the formal institutions that manage salmon need to be restructured or refocused to reflect three important institutional principles. First, decision-making authority should be shared among all legitimate interests (cooperative management); legitimate interests that are excluded from decision-making are likely to block desirable changes. Second, the organizational structures and decision-making processes should allow for local conditions and variations and the management strategies should vary accordingly. Third, systematic learning using appropriate experimental designs (adaptive management) should be an essential goal.

As a first step, the relevant agencies in the Pacific Northwest, including the National Marine Fisheries Service, should agree on a process to permit the formulation of salmon recovery plans *in advance* of listings under the Endangered Species Act, and the Pacific Northwest states, acting individually and through the Northwest Power Planning Council, should provide technical and financial assistance to watershed-level organizations to prepare and implement these preemptive recovery plans.

A SCIENTIFIC ADVISORY BOARD TO ADDRESS SALMON PROBLEMS

A great deal is known about salmon and their difficulties, but a great deal remains unknown or controversial despite the expenditure of large amounts of money and time on research. Part of the reason for the lack of knowledge is that people have not agreed on what information is needed, have duplicated each other's work, and have been unwilling to fund needed research. An independent, multidisciplinary, standing scientific advisory board should be established to ensure that the limited money available for research is spent most productively to answer the most critical questions in a timely manner. A standing scientific advisory board would also help to ensure that when urgently needed actions are taken, they are designed so that their effects and effectiveness can be properly assessed. The board's reports should be public.

AN APPROACH TO SOLVING THE SALMON PROBLEM

The salmon problem took many years to develop, and its solution will require the commitment of considerable time, money, and effort. The committee's analyses of the problems and potential solutions lead to the conclusion that there is no "magic bullet." Therefore, like the problem itself, solutions will be complex and often hard to agree on; to be successful, they will need to be based on scientific information, including information provided by social and economic sciences. In addition, to be successful, consensus will be needed about the size of the investments to be made in solving the problem and how the costs should be allocated. This means that solutions will have to be regionally based, just as the salmon problem has regional variations (see Chapter 13).

The committee recommends the following general approach. *For each major watershed or river basin*, the following should be assessed:

- All causes of salmon mortality, including their estimated magnitude and the uncertainties associated with the estimates. Factors known to decrease natural production should also be listed.
- Ways of reducing those sources of mortality or compensating for them, their probable effectiveness, and their drawbacks.
- The probable costs of each method of reducing mortality. To be most useful, the estimates should include both market and nonmarket costs. To the degree possible, it is important to identify what societal groups would bear the major portion of the costs of each method and significant uncertainties in the estimates. (For example, reductions in catch rates would primarily affect fishers and tourists; changes in water use could affect agricultural interests or ratepayers; changes in riparian management could affect forest-products industries or private landowners.)

All the estimates would include substantial uncertainties, due both to lack of knowledge and to fundamental environmental, socioeconomic, and biological uncertainties. Nonetheless, such a process of assessment and evaluation is essential for rational decision making. They will provide a basis for evaluating options—for weighing benefits and costs—and for identifying areas where research is critical. *All the recommendations in this report should be viewed in this context: they need to be considered on a regional basis (i.e., major watersheds) and in a comprehensive framework that includes an analysis of their costs, probable effectiveness, and the ability and willingness of various sectors to bear the costs.*

This will be challenging for several reasons. First, in many cases, the desired information has not been collated or does not exist. Second, considerable time and resources will be needed to perform such analyses even for one watershed. But the most important reason is that estimates of costs and how they might be distributed will require intimate knowledge of each watershed and of people's preferences and habits. These essential estimates should be made with input from the people involved. The committee believes this approach will lead to improved effectiveness and—if not reduced costs—at least increased cost-effectiveness and reduced controversy.

THE FUTURE

The best approach to establishing a sustainable future for salmon in the Pacific Northwest is to use currently available information to develop workable, comprehensive programs rather than reacting to crises. This report has analyzed many parts of the salmon problem and assessed many options for intervention. However, if current trends continue, the Pacific Northwest will continue to see the effects of more people, more resource consumption, changing economic demands and technologies, and changing societal values. Because the success of programs to improve the long-term prospects for salmon in the Pacific Northwest will depend on the societal and environmental contexts, it is important to develop ways for improving our ability to identify changing contexts and to respond to them. As long as human populations and economic activities continue to increase, so will the challenge of successfully solving the salmon problem.

1

Introduction

THE SALMON PROBLEM

The "salmon problem" is easy to state, hard to analyze, and even more difficult to solve. Salmon have lived together with people in the Pacific Northwest for at least 10,000 years. Before European and American explorers and settlers arrived in the Pacific Northwest of North America in the nineteenth century, salmon were so abundant there that an American Indian economy had been founded on them; yet human exploitation rates and disturbances of the fishes' environment were small enough that the salmon populations did not diminish over the long run.

Wild salmon, which once numbered more than 8-10 million returning adults in the Columbia River basin alone, have declined to less than one-tenth that number up and down the coast of the Pacific Northwest. Most of the fish that now return began their lives in hatcheries. Although we refer to the decline in the numbers of salmon as "the salmon problem," it is a problem for people—those who make their living from the fish, such as Indians in subsistence lifestyles and trollers who catch fish commercially; those charged with regulating fisheries in the public interest; and the 9 million human inhabitants of the Pacific Northwest, for whom salmon are a symbol of their geography, a way of life, a delicacy, and an indicator of ecosystem condition.

Because salmon are important across the society of the Northwest and in the United States, they are protected under the Endangered Species Act, the Northwest Power Act, the Fishery Conservation and Management Act, and other federal statutes. The web of rights and obligations that form salmon policy draws into the salmon problem a remarkable variety of interests and actors. Salmon migrate down and up rivers, and some of the rivers have dams, so dam operators are involved. Salmon rely on high-quality water, so pollution control—including regulation of agriculture, mining, ranching, and suburban runoff—is important. Salmon originating in the Pacific Northwest are caught by fishers in several U.S. states and western Canada and by members of Indian tribes under treaties to which the United States is a party. In addition, salmon are produced in hatcheries, and the wild and hatchery fish interact biologically. Thus, the salmon problem is entangled with so many arms of human society that it is a policy problem affecting the economy.

The decline in salmon numbers has been observed and lamented for at least a century and a half, as the human population has grown and economic activity has increased. As early as 1848, the Oregon Territory's constitution prohibited dams across salmon streams; the Washington Territory created management regulations as early as 1859, although many of them were more for allocation than for conservation (Wendler 1966). Hatcheries were seen as important to maintain the salmon runs on the Columbia River as early as the 1870s. By 1919, regulations in Oregon and Washington called for closed seasons, restrictions on gear, and prohibition of purse seines; but they were not well enforced (F.J. Smith 1979). More recently, petitions have been filed to list some species or populations as endangered or threatened under the Endangered Species Act (ESA); lawsuits have been filed, meetings have been held, federal laws have been passed, and more than $1 billion has been spent over the last 10 years alone to improve salmon runs in the Pacific Northwest.

Despite extensive recent efforts and activities to improve conditions for salmon, their overall populations in the region continue to decline. (The decline is not universal: some populations or stocks in some streams are not declining, and some in the northern part of the range are even increasing; see Chapter 4 for details.) In response to concern over the continuing decline of salmon stocks in the Pacific Northwest and the controversies about how to arrest or reverse the decline, Congress requested advice from the National Research Council (Senate Report 102-106, Appropriations Bill for Departments of Commerce, Justice, and State, the Judiciary, and Related Agencies, 1992). The National Research Council thereupon established the Committee on Protection and Management of Pacific Northwest Anadromous Salmon to evaluate information on the status of various stocks in the Pacific Northwest, identify causes of their decline, and suggest a comprehensive approach to protecting and managing them (the complete charge to the committee is in Appendix A).

COMPONENTS OF THE SYSTEM

The Region

The Pacific Northwest includes Washington, Oregon, and Idaho; northern California and southeastern Alaska are often considered to be part of the region. The southwestern part of British Columbia is in every physical and biological aspect part of the Pacific Northwest, but it is part of Canada. The committee focused on the region shown in Figure 1-1, although it considered conditions in adjacent areas to the north and as far south as Monterey Bay ir Northern California. Chapter 2 reviews the region in detail.

The Salmon

In the Pacific Northwest, seven species of Pacific salmon (members of the genus *Oncorhynchus*, including cutthroat and steelhead trout; see Table 1-1) spawn in freshwater, migrate to sea, and return to freshwater to spawn and complete their life cycle; fish with this complex life cycle are called *anadromous* (Figure 1-2). Some of the species have populations that are not anadromous. Two of the species—steelhead trout (the anadromous form of rainbow trout) and cutthroat trout—have recently been reclassified from *Salmo* to the Pacific salmon genus *Oncorhynchus* (Smith and Stearley 1989).

This report considers only anadromous salmon species and populations. Two other salmonids[1] that have anadromous populations, Dolly Varden (*Salvelinus malma*) and bull trout (*S. confluentus*), are not discussed, because their anadromous forms are scarce and not well known in the Pacific Northwest.

Freshwater habitats seem to provide optimal spawning and rearing areas for these fishes, but the ocean provides better environments for growth. Some species migrate thousands of miles at sea, where they grow quite large; some populations of some species never go to the ocean. The species vary considerably in the time they spend in freshwater or at sea. Salmon return to spawn, or "home" with various degrees of fidelity to the same place in the same stream where they were spawned. The homing habit results in local populations being adapted to particular streams: The genetic distinctions of local populations, or demes, make the genetic structure of salmon complex and different from that of most fish species. Chapter 2 contains more information on the biology and ecology of the Pacific salmon and Chapter 4 discusses the status of wild salmon populations, including difficulties in evaluation of status.

[1]Members of the family Salmonidae, which includes freshwater and anadromous fishes in three major subgroups: the whitefishes and ciscoes (subfamily Coregoninae); graylings (Thymallinae); and salmon, trout, and chars (Salmoninae).

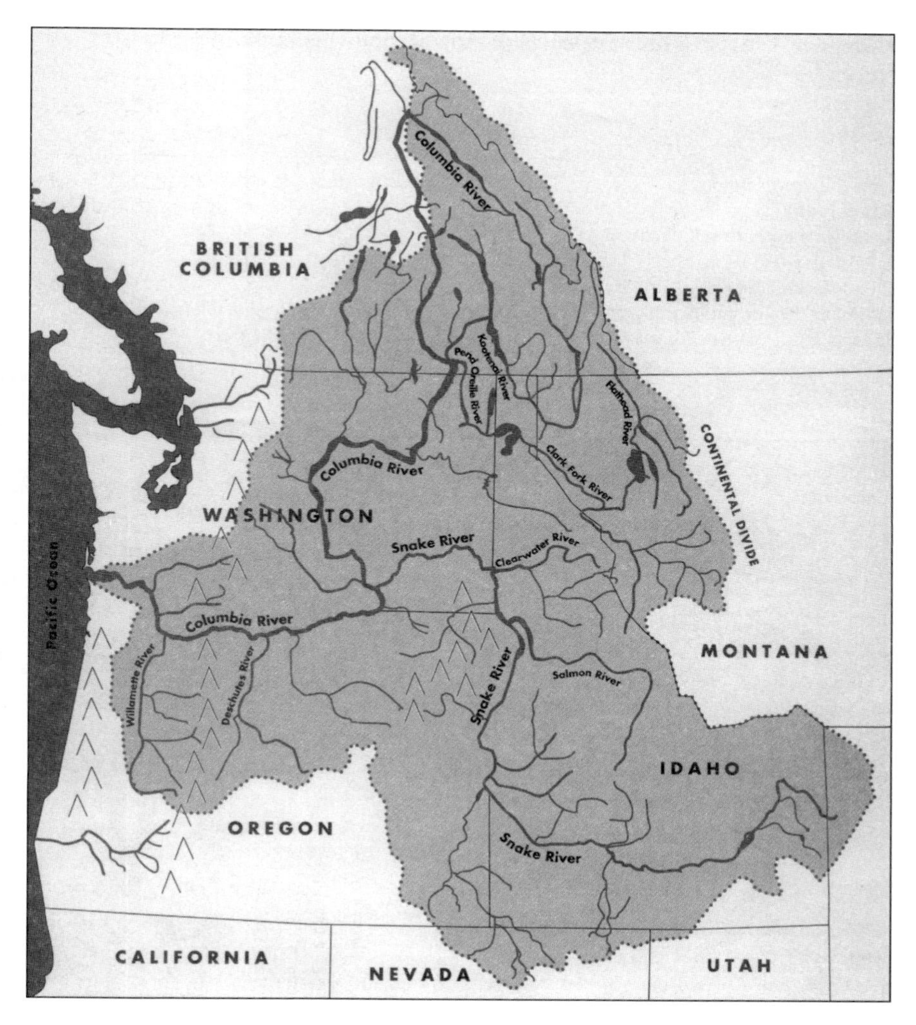

FIGURE 1-1 The Pacific Northwest and the Columbia River Basin. The committee focused on the area up to the U.S.-Canada border and southward to Monterey Bay, California (not shown). Note how much of the region is encompassed by the Columbia River Basin (dotted line). Source: Columbia River System Operation Review 1993.

Evolutionary, Genetic, Ecological, and Spatial Units of Concern

Pacific salmon are widely distributed and occur in diverse habitats; they are correspondingly biologically diverse. Each species is divided genetically and evolutionarily into subspecies (races) and into narrower and narrower local breeding populations (demes). These are described in detail in Chapter 6. The Endan-

TABLE 1-1 Seven Native Species of Anadromous Salmon in Pacific Northwest

Common Name(s)	Scientific Name
Pink (humpy) salmon	*Oncorhynchus gorbuscha*
Chum (dog) salmon	*O. keta*
Sockeye (red, blueback) salmon	*O. nerka*
Coho (silver) salmon	*O. kisutch*
Chinook (king, spring) salmon	*O. tshawytscha*
Steelhead (anadromous rainbow) trout	*O. mykiss* (formerly *Salmo gairdneri*)
Sea-run cutthroat trout	*O. clarki* (formerly *Salmo clarki*)

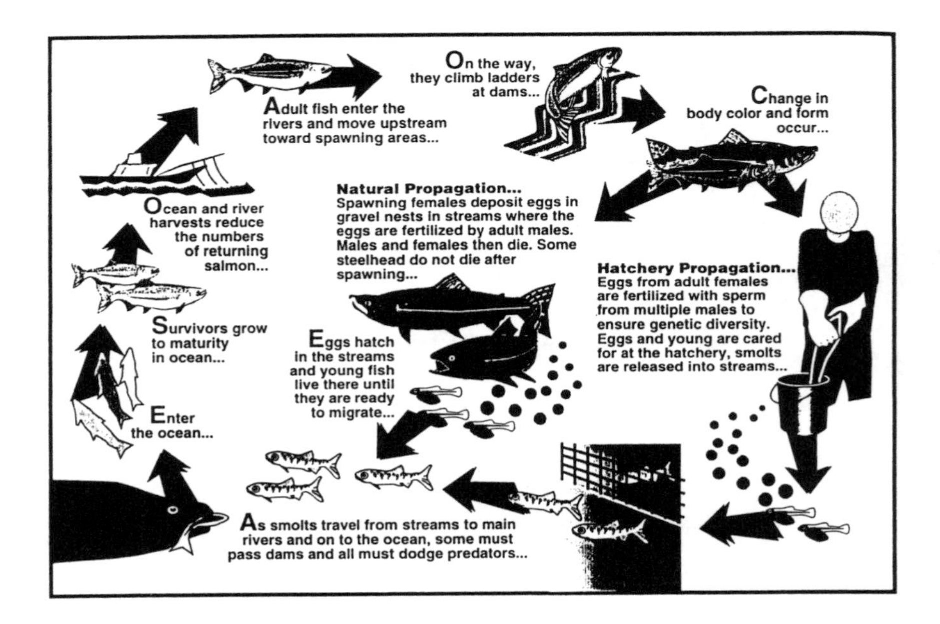

FIGURE 1-2 General life cycle of salmon and steelhead. Source: NMFS 1995.

gered Species Act protects not only species but, for vertebrates, "subspecies, varieties, and distinct population segments." However, the act does not define "distinct population segments." It would be important to understand what levels of evolutionary (or taxonomic) and genetic variation are important biologically (see NRC 1995b for a discussion of identifying distinct population segments). What are the biological consequences of preserving or not preserving spatially distributed salmon populations, such as all the salmon in a watershed, river, or stream? Spatial units of concern include both freshwater and ocean habitats. Biological relationships of salmon to freshwater are well known because a river has an easily studied upland watershed and a lake has a clear shoreline and is without abyssal depths. Considerably less is known about the vast and widely dispersed ocean habitats of salmon.

Salmon spend some of their lives in freshwater and some in the sea and are subject to management in both. Fishery management in the two systems is quite different (Magnuson 1991). In general, the difference is characterized by the use of manipulations in freshwater systems, sometimes as drastic as poisoning a lake or stream and restocking it. Hatchery supplementation of freshwater systems and management of their species composition, including introduction of exotic species, is common. One major reason that such an approach can work is that freshwater systems have well-defined boundaries and are usually fairly small. In contrast, the management of marine fisheries historically has depended only or mainly on attempts to assess or predict the size of the stock and regulate fishing effort. Enhancement and other manipulations of the marine environment are comparatively rare.

Natural environmental variations and humans' activities in freshwater affect the salmon stocks at sea, and vice versa; this adds complexity to the salmon problem and to attempts to solve it.

The People

As the glaciers receded about 10,000 years ago, the Pacific Northwest was populated by the ancestors of the American Indians who live there today. They benefited from and depended on the natural resources of the region, including salmon. Salmon became an important part not only of the native peoples' sustenance, but also of their culture. As salmon became an integral component of American Indians' culture, the effect of people on salmon increased as well. There is evidence that native peoples caught large amounts of salmon in some areas (Hewes 1947, Walker 1967).

When Euro-Americans first arrived in the region in the eighteenth century, they too took advantage of Nature's bounty, including salmon. By the 1870s, salmon populations had begun to decrease (C.L. Smith 1979). But salmon continue to have important commercial and cultural values for inhabitants of the Pacific Northwest. The diversity of cultures and institutions, the diversity of values and goals, and the complexity of physical and social infrastructures in the region all contribute to the salmon problem. All must be considered in any proposed approach to a solution.

COMPONENTS OF THE PROBLEM

Limits to Biological Production

Human populations and institutions have reduced the numbers of salmon not only by fishing, but also by modifying the land and waters of the Pacific Northwest. When people noticed declines in salmon runs, they tried to compensate for them by increasing salmon production with hatcheries or technological modifications to the fishes' environment, such as fish ladders and mechanical transportation schemes and instream projects that attempted to improve habitat. Salmon fishing also was restricted. Those approaches have not been successful, at least on a regional scale, inasmuch as salmon populations continue to decline.

The salmon production cycle has three principal components that determine abundance: reproductive potential of adults returning from the sea to spawn, which is affected by their growth at sea; production of offspring from natural reproduction in streams and artificial propagation in hatcheries; and sources of mortality (including natural mortality, fishing mortality, dam-caused mortality, mortality from habitat alterations and changes in environmental conditions, and so on). All three components are affected by changes in environmental conditions as well as by human activities. Production from salmon returning to freshwater to spawn is affected by interactions between natural and hatchery stocks. Habitat sets an upper limit on salmon production, which cannot be exceeded easily even with technological assistance. The number of salmon removed from the system by all agents (e.g., fishing, dams, habitat loss, and consumptive water use) has steadily increased. The growing losses have been easy to overlook because it has not occurred to everyone that a dam that reduces a salmon run by 100 adult fish takes just as much as a person who catches 100 adult salmon. Thus, a dam or an irrigation system reduces the amount of fish available for fishing just as much as ocean trolling reduces the amount of fish available for tribal set-nets or recreational fishing upriver. Solving the salmon problem requires that people understand those equalities.

Institutions

Many state, federal, local, regional, tribal, international, and private institutions have jurisdiction over or influence some aspect of the salmon problem. As described in Chapter 13, institutions have complex behaviors and commonly change slowly; understanding and managing them is not easy. Yet because they are essential to the conduct of human lives in modern society, understanding them and perhaps adapting them are essential to solving the salmon problem. Current institutional arrangements in the Pacific Northwest have contributed to the salmon problem and probably will need modification if the problem is to be solved.

Knowledge

A great deal is known about salmon, rivers, climate, the oceans, human behavior, institutions, how to manage fisheries, and other relevant matters. Yet much remains unknown, and some gaps are crucial to a long-term, stable solution to the salmon problem. As the rest of this report makes clear, enough is known now to improve the prospects for salmon in the short term if knowledge is applied wisely and quickly, but not enough is known to warrant confidence in a long-term regional plan for salmon. In addition, the components of the salmon problem and the pertinent disciplines are so diverse that no person can know all that needs to be known for a comprehensive solution. Thus, the salmon problem is in a sense a cognitive problem whose solution will depend on close cooperation and collaboration of people with many kinds of experience and expertise. This report attempts to provide the background for such a solution.

APPROACHES

Goals

Finding cures to the salmon problem promises to be even harder than recognizing its character and scope. People do not agree on how or even whether they should modify the region's human population size, their behavior, and their economic contributions to halt or reverse the decline in salmon, even though they do agree that such a reversal should be achieved. The salmon problem, like so many other controversies in natural-resource management (Ludwig et al. 1993), reflects conflicting goals and values, many of which are unarticulated and so cannot be squarely addressed.

Part of the need to articulate and discuss diverse goals and values is practical: actions undertaken without clear articulation of goals are unlikely to achieve anything useful. Consideration of goals and values has been essential to this committee in addressing how to approach the salmon problem, and has led it to frame its conclusions and recommendations in terms of rehabilitation of salmon stocks—a pragmatic approach to improving the situation that relies on natural regenerative processes in the long term and the selected use of technology and human effort in the short term—rather than on attempts to restore the ecosystem to some pristine former state and rather than on a primary reliance on substitution, i.e., the use of technologies and energy inputs, such as hatcheries, artificial transportation, and modification of stream channels. The concepts of habitat restoration, rehabilitation, and substitution are discussed in detail in Chapter 8.

Framework

In considering options for solving the salmon problem, the committee first had to consider a range of frameworks or contexts into which the solutions would fit. For example, the framework in which some people operate is the desire to increase catch levels; a framework for others is the desire to restore the salmon's environment to some pristine state; yet others might prefer to focus on economic development of the various natural resources in the region. The committee considered four general categories of framework: continued degradation of the salmon's environment, restoration of the natural environment to some pristine condition, substitution, and rehabilitation.

- **Continued Degradation.** Unless a substantial coordinated effort is made to reverse the trajectory of environmental decline, it will continue, and the number of salmon runs and their abundance will continue to decline along with it. Although there might be worthy reasons for allowing the decline to continue, the committee's charge (see Preface) did not include this option, and the committee did not consider it.
- **Restoration.** Restoration of Pacific Northwest ecosystems is no longer possible in many portions of the region. Many ecosystems have been so altered that it is difficult to decide what the pristine condition was. The process by which the environment arrived at its current condition is not totally reversible. Genetic variability has been lost; evolution has occurred; exotic species have been introduced; human populations in the region have increased, and people have developed dependence on a variety of modern technologies, cultures, and economic systems; and other natural and anthropogenic environmental changes have changed the range of biophysical and socioeconomic possibilities for future states of the system. In brief, the past provides opportunities for the future but also constrains it. Because of anthropogenic impacts and because of the dynamic nature of most stream systems, it is not possible to return many streams and rivers of the Pacific Northwest to a specific pristine condition. However, there are major opportunities for restoring riparian and aquatic ecosystem functions and processes to many streams and rivers throughout the region, particularly those in public ownership. It will be a major challenge to current and future generations to see how many, and to what extent, riparian and aquatic functions and habitats can be restored throughout the range of anadromous salmon.
- **Substitution.** By *substitution* the committee means investing substantial energy, time, and money on a continuing basis to replace natural ecosystem processes that have been destroyed or degraded. Examples of substitution could include production hatcheries, barging of young salmon around dams, construction of fish ladders and stream channels, altering instream flow by drawdown or changing storage, and other activities that can reasonably be expected to require continuing human input. Substituting for natural processes to maintain salmon runs is possible in at least some portions of most streams, but it would be expensive, and the cost in human and financial resources is expected to increase rather than to decrease or stabilize in the future. This approach has several things to

recommend it, but there is no compelling evidence that it will reverse salmon declines or protect against unwelcome surprises in the future status of various stocks. Such activities also will limit other uses of natural ecosystems and their natural functioning. Finally, as the ability of human actions to make up for natural processes lags behind expectations, the danger is that either more and more drastic interventions will be undertaken or the whole effort will be abandoned and the salmon will be lost. Therefore, the committee has focused most of its attention on rehabilitation as a goal.

- **Rehabilitation.** By *rehabilitation* the committee means a process of human intervention to modify degraded ecosystems and habitats to make it possible for natural processes of reproduction and production to take place. Rehabilitation would protect what remains in an ecosystem context and regenerate natural processes where cost-effective opportunities exist. It might be necessary to use the technologies and techniques suggested in the preceding paragraph to maintain the essential ecosystem components in the short term, but the ultimate goal is to modify (i.e., rehabilitate) the systems to the point where human input is substantially reduced or even stopped altogether. Substantial local opportunities for ecosystem rehabilitation exist throughout the region and they should be taken advantage of. Although this framework implies reduced management costs over the long term, it requires a long-term commitment to achieve positive results.

2

Salmon Geography and Ecology

INTRODUCTION

Salmon live and travel over a large geographic area. In North America, they begin their lives in streams as far south as California, as far north as the Arctic Circle, as far west as the Pacific shore, and as far east as the Continental Divide. Particular populations migrate from central Idaho more than 1,500 km to the sea; they might travel many times that distance during the ocean phase of their lives and return to spawn and die in the same freshwater areas where they hatched. Thus, during their lives, salmon cross many geographic and human boundaries. Salmon are parts of ecosystems that cover areas much larger than marked properties and government jurisdictions. Salmon generation times are usually 2-6 years, and interdecadal changes in climate influence salmon abundance and production. Human decisions can be made quickly, but they can have consequences for irrigation, land-clearing, urbanization, and dams that last for much longer periods. Conflicts between human wants and the needs of fish occur in a context of mismatches between time and space scales. To make the mismatches explicit, this chapter describes the geographic and ecological scales of salmon lives.

The chapter also concerns salmon ecology, the interaction of salmon with their physical and biological environment. It addresses the state of knowledge on salmon ecology and raises cautionary notes relevant to plans and expectations for human interventions. Two aspects are emphasized: ecological interactions that occur in the river basins and those that occur in the ocean. Salmon are most visible in river basins, where changes to their environment—human-caused and other—are most apparent. But they spend most of their life in the ocean. The ocean can seem distant and constant, but processes going on in the ocean are probably neither constant nor unimportant with respect to the salmon problem. Attempts to restore, rehabilitate, or otherwise manage salmon populations will affect and be affected by abiotic and biotic interactions in both river and ocean ecosystems.

SALMON LIFE HISTORY AND DISTRIBUTION

The life histories of Pacific anadromous salmon set the geographic stage for solutions to the problem of their decline. Their life histories have three characteristics—anadromy, homing, and semelparity—that are key to considering trends in abundance and factors responsible for salmon declines; these characteristics differ among species and individual runs. (See Figure 1-2.)

Anadromy

Anadromous fishes begin their lives in freshwater, migrate to the sea after some variable period of freshwater residence, feed and attain most of their size at sea, and then migrate back to freshwater to spawn. Anadromy makes salmon dependent on a complex of freshwater and marine environments. The large marine pasture environment supports a much greater biomass of salmon than the small freshwater environments could possibly support. Freshwater spawning sites provide a relatively safe haven for egg and larval development.

Homing

Individuals that survive their marine residence and return to spawn almost invariably spawn in the stream where they began their lives years earlier (Quinn and Dittman 1992). This "homing" reduces genetic exchange among salmon populations. Combined with the action of natural selection in diverse habitats, homing results in differentiation of populations into distinct units that are reproductively isolated.

Semelparity

Most anadromous salmon die after their first spawning season and thus spawn only once in their lifetime—they are semelparous. Exceptions are cutthroat and steelhead, some of which migrate back to sea after spawning, and then return to spawn again after one or more seasons. Semelparity is thought to evolve in species with a high growth rate and high mortality during long and stressful migrations; i.e., these species reach a large body size by time of first spawning, and their probability of surviving more than one migration to the spawning area is low. A consequence of this life cycle of growing at sea and dying in freshwater is that the carcasses transport nutrients into comparatively impoverished freshwater environments, and these nutrients are used by various terrestrial animals (Cederholm et al. 1989) or cycled through aquatic pathways (Kline et al. 1990, 1993).

Generalized Life Cycle

The salmon life cycle integrates anadromy, homing, and semelparity (Figure 1-2). Generally, in fall or spring, females select sites for spawning, dig crater-shaped depressions in the stream gravel, and there deposit from a few hundred to several thousand pea-sized eggs. Males compete to fertilize the eggs, which females then bury and guard. After about 2 weeks, adults of both sexes die near the spawning site. The eggs develop in the gravel and hatch several months later. The larvae, termed alevins, remain buried until their yolk sac is absorbed; they then emerge as free-swimming fry that either begin feeding in the stream or migrate from it. Pinks migrate directly to sea, chum do so after a few days or weeks, chinook after a few months to a year, and coho generally after a year. Most sockeye populations migrate to a lake for a year before continuing to the ocean, but some rear in rivers and others migrate to the ocean shortly after emergence. But some sockeye populations, called kokanee, never migrate to the ocean but remain in the lakes. Young steelhead and cutthroat usually migrate to sea after about 2 years in freshwater; some never migrate, but form freshwater populations. Young salmon (parr), which are adapted to freshwater, become prepared to migrate seaward and live in saltwater by a complex developmental transformation (known as *smoltification*) that involves physiological, biochemical, morphological, and behavioral changes.

Anadromous salmon display four patterns of marine distribution that differ in how far the salmon move from the river of origin. First, juvenile pink, chum, and sockeye (smolts) enter the ocean and migrate north along the continental shelf, reaching the northern Gulf of Alaska in late fall (Hartt and Dell 1986); they then migrate south into the open ocean, where they remain until they mature and return to coastal waters, usually making landfall north of their river of origin. Second, coho and chinook usually rear in coastal waters, although some migrate to the open ocean as well. Third, steelhead migrate to the open ocean, but—unlike pink, chum, and sockeye—appear to migrate directly to open ocean as smolts and return directly to the vicinity of their river of origin, rather than migrating along the coast (Burgner et al. 1992). Fourth, cutthroat make relatively little use of the ocean for feeding, generally spending only the summer at sea near their rivers of origin.

The duration of marine residence differs among populations—even populations of the same species—and among years. Some populations tend to mature at a later age than others, but ocean conditions that are poor for growth for any population can delay maturation and return to freshwater. Pink, chum, and coho generally return to freshwater in the fall shortly before their population-specific spawning time. Sockeye might return in fall or several months in advance of spawning. Time differences between return and spawning are highly varied for chinook and steelhead.

The differences in life history among chinook and steelhead populations are reflected, in part, by designations that indicate the timing of adult or juvenile migrations (Healey 1991, Taylor 1990). Chinook are classified by the season when adults return to spawn. They also are distinguished by their juvenile residence in freshwater: stream-type chinook spend a year or two in freshwater and migrate to the sea in the spring of their second or third year of life, whereas ocean-type chinook migrate to the sea during their first year of life, often after a few months in the river. Stream-type chinook tend to migrate to open-ocean waters, whereas ocean-type chinook have a more coastal marine distribution. Ocean-type chinook also spend much more time in estuaries than do stream-type chinook. In the Pacific Northwest, stream-type juveniles are often spring-run adults, and ocean-type juveniles are most often fall-run adults. Fall-run chinook predominate in the lower reaches of large rivers and in smaller coastal rivers, spring/summer chinook tend to occur farther inland. Stream-type chinook are more prevalent in the northern part of the range. Common designations for steelhead are "summer" and "winter." Summer steelhead return to freshwater during late summer and fall, spend the winter in rivers, and spawn the following spring; they typically occur in upper reaches of rivers. Winter steelhead, which are more prevalent in the lower reaches of rivers, enter freshwater in winter and spawn soon thereafter.

Individual Species Distributions

Pink Salmon

Pink salmon, the most abundant anadromous salmon in North America, are relatively scarce in the Pacific Northwest (Heard 1991). Although they spawn in Puget Sound streams, they are apparently absent in coastal Washington, Oregon, and California. They generally are fished in nearshore areas as they approach maturity, which is always reached in their second year. In the southern part of their range, runs are much larger during odd-numbered years; in the northern part of their North American range, run sizes tend to be greater in even-numbered years. Virtually all pink salmon that return to Puget Sound and the Fraser River do so in odd years.

Sockeye Salmon

Sockeye are the second most abundant anadromous salmon and predominate in Alaska and British Columbia (Burgner 1991). They are abundant in the Fraser

River, and significant populations existed in the Columbia River before dam construction (Chapman 1986). The Columbia River has self-sustaining sockeye populations in Lake Wenatchee and Lake Osoyoos. One population, which has been listed as endangered, migrates nearly 1,600 km upstream to Redfish Lake, Idaho. Populations exist in Puget Sound (in Baker Lake and Lake Washington) and on the Washington coast in (Lake Quinault, Ozette Lake, and Lake Pleasant). No anadromous populations seem to exist south of the Columbia River. Sockeye are fished on their homeward migrations in coastal waters.

Chum Salmon

Chum are distributed more broadly both to the north and to the south than pink or sockeye. Although today they are relatively scarce south of the Columbia River (Salo 1991), they were abundant in the Columbia River and in Oregon coastal streams south to Tillamook Bay and also in California as far south as Monterey Bay. Chum tend to make greater use of estuaries than pink or sockeye (Healey 1982), but like those species, they are distributed in offshore waters and are taken almost exclusively by commercial, rather than recreational, fisheries. Although chum can grow to be relatively large, about 5 kg in Washington (Salo 1991), they are not highly valued, because they are often in an advanced state of maturity and are considered unpalatable by some when caught in coastal waters (some American Indians do consider chum caught in coastal water palatable and their roe is highly valued).

Coho Salmon

Coho are native to coastal and interior rivers from Alaska to Monterey Bay, California. They tend to spawn in small streams. Most coho in the Pacific Northwest mature at the age of 3 years. Some populations migrate to offshore waters, but many remain relatively near the coast during their marine residence (Sandercock 1991). Coho are caught in commercial and recreational fisheries in waters off the coast and in Puget Sound and the Strait of Georgia.

Chinook Salmon

Chinook, like coho, occur from Alaska to the Sacramento River system but are generally less abundant than coho. They attain the largest size of any Pacific salmon (over 40 kg, although 10 kg is typical) and are relatively long-lived, commonly maturing at the age of 4-5 years. Chinook tend to spawn in large rivers near the coast and in the interior. As mentioned above, different populations have markedly different life-history patterns. Stream-type chinook are fished primarily on their return to freshwater from their wide distribution in open-ocean waters. Ocean-type chinook, however, are vulnerable to recreational and commercial fisheries much of their lives because they tend to stay near the coast.

Steelhead Trout

Steelhead occur from Alaska to central California. They migrate directly to open-ocean waters with little estuarine residence time and usually spend 1 or 2 years at sea (Burgner et al. 1992). Juvenile steelhead commonly co-occur with coho in streams but are usually less abundant than coho. They are caught generally in commercial fisheries (mostly by American Indians, although bycatch in other fisheries can be significant) and recreational fisheries in rivers.

Cutthroat Trout

Cutthroat have similar habitat requirements to steelhead (Bisson et al. 1988). They are common as anadromous and resident populations in coastal and interior streams and are native to drainage systems of the continental interior. Cutthroat are smaller and often less numerous than steelhead. They have a relatively complex life cycle (Trotter, 1989; Behnke, 1992) generally spending only a summer at sea. They are caught by recreational fishers but generally are underemphasized in research and management, as evidenced by the inability of scientists to define cutthroat populations in a meaningful way (Nehlsen et al. 1991) or even to consider them in stock-status reports (WDF et al. 1993).

PACIFIC NORTHWEST SALMON AREAS

The region's political divisions do not correlate closely with the life history and distribution of anadromous Pacific salmon (see Chapter 4 for details of populations). As a political region, the Pacific Northwest is usually considered to encompass Idaho, Oregon, and Washington (Figure 1-1). The Columbia River Basin, however, extends into Canada and includes portions of Montana, Wyoming, Utah, and Nevada. The U.S.-Canada border cuts not only across the Columbia River Basin, but also across the migratory paths of salmon moving north and south along the Pacific coast. Oregon and Washington share 522 km of the Columbia River as a common boundary. The Snake River serves as a portion of the boundary of Idaho and Oregon. Klamath River headwaters are east of the Cascades in Oregon and empty into the Pacific Ocean in northern California. Forest planners consider the area to the west of the Cascades one unit (FEMAT 1993) and the area to the east another (Bormann et al. 1993).

The committee focused on five biogeographical areas of the Pacific Northwest and on relevant parts of the Pacific Ocean and adjacent Canadian coastal areas and rivers that differentiate important aspects of the salmon problem (Figure 1-1). Three of the biogeographic areas—Puget Sound, the Columbia River, and the Sacramento River—are drainage basins. The two others divide the portion of the U.S. Pacific Coast used by salmon into a north and south region at Humbug Mountain, 20 km south of Cape Blanco, Oregon. Appendix D discusses the major landforms and their rivers.

The economies in the five areas are diverse. The Sacramento River, Columbia River, and Puget Sound drainage basins have heavily urbanized regions. Sacramento and San Francisco are the major urban centers in the Sacramento drainage. The Columbia River Basin differs from the Sacramento River Basin in that the largest urban area, Portland, is farther upstream, more than 185 km from the river mouth. Two smaller urban areas are in the upper reaches of the Columbia River Basin, on tributaries of the Columbia—Spokane and Boise. The Seattle-Tacoma urban area covers much of the east side of Puget Sound. In contrast, the northern and southern coastal areas do not have major urban centers, and their economies depend primarily on forestry, fishing, and tourism. Here, salmon are an important part of the total economy. Although Seattle is a major fishing port, the Seattle-Tacoma area has a diverse economy in which fishing plays only a small part. Fishing is least important to the economies of Portland, Sacramento, Spokane, and Boise. The decline of salmon has produced significant impact to what was an important commercial fishing industry, not only in San Francisco, but also along the entire northern California coast; hundreds of fishing boats have been idled and a whole way of life has been changed. The cultural effects have been and continue to be important. As a tourist attraction, fishing still is important to the San Francisco waterfront economy. In the interior of the Sacramento River, Columbia River, and Puget Sound basins, agriculture, ranching, hydropower generation, forestry, and tourism are the dominant economic activities.

River Basins

The fundamental freshwater geographic unit for salmon is the river basin. A river basin is the drainage network of a river and its tributaries. It ranges in size from small, unnamed coastal streams that drain less than a few square kilometers to large rivers, such as the Columbia and Sacramento, with drainage areas of thousands of square kilometers. River basins can have subbasins of major tributaries, such as the Salmon River of Idaho, the Yakima River of Washington, and the Willamette River of Oregon, which are all part of the Columbia River basin. The term watershed is applied to small and large river systems but usually is taken to mean intermediate-size drainages of 20-500 km^2 (FEMAT 1993, Washington Forest Practices Board 1993).

River basins originate in numerous steep, headwater streams that often are fed by glaciers, snowmelt, or groundwater springs. These small, fishless streams make up 70% or more of the total length of the drainage network in the Pacific Northwest (Naiman et al. 1992). Although not inhabited by salmon, they are important for salmon because they carry cool water, nutrients, and organic matter downstream to areas used for spawning and as nurseries. The tiny headwater streams come together to form slightly larger streams with gradients low enough for some fishes, but not usually salmon. Farther downstream, gradients lessen and channels widen; one or two salmon species reach these streams to spawn. Confluences of small and medium streams eventually form large streams and rivers that include lake systems and well-developed flood plains that support salmon. Before flowing into the ocean, a river enters an estuary, where freshwater and salt water mingle. Estuaries vary considerably in size and in the extent to which freshwater and salt water intermingle.

Ocean

The North Pacific Ocean forms the second major geographic unit for salmon. The continental shelf along the Pacific Northwest underlies coastal waters to a depth of about 200 m and ranges in width from about 7 km off Cape Blanco, Oregon, to about 30 km off Grays Harbor, Washington (Good 1993). At the edge of the continental shelf, the slope drops 2,000 m to the sea floor.

Off the Pacific Northwest coast during summer, the south-flowing California Current along the outer shelf and open ocean is underlain by the north-flowing California Undercurrent, which, when combined with offshore winds, results in a movement of nearshore surface water offshore and upwelling of cold, nutrient-rich waters from the deep ocean. In some years, warm surface waters from the tropics move northward along the coast and inhibit upwelling; this results in increased sea-surface temperatures and depletion of nutrients (El Niño conditions). During winter, southerly winds result in the north-flowing Davidson Current along the continental shelf, an onshore flow of surface waters, and downwelling.

As discussed previously, some salmon do not remain in nearshore waters but travel well out to sea; some tagged salmon of U.S. origin are recovered in the western Pacific Ocean off the Russian coast. Most salmon migrate along the North American coast, either north or south, to reach feeding areas, where they then follow major ocean currents while foraging (Pearcy 1992).

SALMON ECOLOGY IN RIVER BASINS

To understand the ecology of anadromous Pacific salmon in freshwater, one must consider species interactions, complex adaptations of juveniles to stream habitat, and the influence of returning spawners on stream ecosystems.

Species Interactions

Salmon interact with other species in complex ways. The dynamics of ecological communities are shaped by many direct and indirect interactions among species. Traditionally, the study of direct pairwise interactions, such as the interaction between a predator and a prey species, has dominated approaches to ecological systems, but recent work has demonstrated the importance of indirect effects that cascade through the web of other species and the physicochemical components of the system (Carpenter 1988, Carpenter and Kitchell 1993). Competitors can have positive effects on each other in more complex ecological systems; removing a predator can negatively affect its prey through interactions that its prey will then have with other species. The physicochemical environment not only affects but is affected by biology.

More than 100 freshwater and andaromous fishes inhabit the Pacific Northwest, of which about one-third are introduced (Scott and Crossman 1973, Moyle 1976, Wydoski and Whitney 1979, McPhail and Lindsey 1986). The Columbia River contains 52 native species, including 13 endemic species (species found in the basin and not elsewhere). The Chehalis River contains 34 native species, most of which also occur in the Columbia River. However, except in such large river basins as the Columbia, Chehalis, and Sacramento, the region's freshwater communities tend to contain few fish species. Typical Puget Sound and Pacific Coast streams might have two or three salmons (family Salmonidae), two sculpins (family Cottidae), one or two minnows (family Cyprinidae), and a lamprey (family Petromyzontidae).

Large rivers—such as the Columbia, Chehalis, and Sacramento—have more-diverse native fish communities (25-50 fishes versus 1-6), more-complex food webs, and less-variable streamflow and temperature regimes than small streams. Large rivers also contain more introduced fishes (Li et al. 1987). Introduced fishes in this region—such as sunfish (*Lepomis* spp.), black bass (*Micropterus* spp.), perch (*Perca flavescens)*, walleye *(Stizostedion vitreum*), striped bass (*Morone saxatilis*), and catfish (*Ictalurus* spp.)—prefer or tolerate higher temperatures and use more standing-water habitats than do juvenile salmons; thus, they are often suited to live in reservoirs behind dams. Introduced fishes often are large and piscivorous, inasmuch as usually they have been introduced to provide recreational fishing opportunities. Their potential interactions with salmon (competition or predation), have seldom been evaluated.

Perhaps the most intensively studied ecological interaction between salmon and nonsalmonid fishes in a large river has been the predation on salmon smolts in the Columbia River. Nonnative species, such as walleye and smallmouth bass, eat smolts, but the most important predator is the native northern squawfish, *Ptychocheilus oregonensis* (Rieman et al. 1991). Squawfish removals have been initiated as a management measure, but experience with other aquatic ecosystems suggests that any benefits to salmon could be confounded by other species interactions (e.g., squawfish might control other salmon predators) or interactions between life-history stages (e.g., reduction in predation by large squawfish on smaller squawfish might result in rapid population rebound and more intense predation on salmon).

Juvenile Adaptability

Juvenile salmon in streams and rivers tend to consume mostly aquatic and terrestrial invertebrates carried along by the flowing water (Mundie 1969), but they also eat small fish, salmon eggs, and occasionally the carcasses of adult salmon (Kline et al. 1990). Diet varies among species and seasons. In small watersheds with dense forest canopies, much of the organic matter in streams originates in the surrounding forest, and the invertebrate communities are dominated by organisms specialized for processing wood and leaves (Gregory 1983). In larger watersheds, where streams are too wide to be completely shaded, or in small streams where the canopy has been removed, the food base of the aquatic community shifts to algae produced photosynthetically in the stream itself (Bilby and Bisson 1992). This shift in primary production from mostly terrestrially produced to aquatically produced organic matter as streams become larger is of fundamental importance for the composition of the invertebrate community (Vannote et al. 1980). In spite of the shift in the food web, juvenile salmon generally are able to capitalize on whatever is abundant (Chapman and Bjornn 1969).

Juvenile salmon move within watersheds to take advantage of seasonal environmental changes (Bustard and Narver 1975), but the importance of winter feeding areas has become recognized only recently. Despite temperatures well below the optimum for growth, some species show impressive weight gains in late winter and early spring. Juvenile salmon in Pacific Coast streams often overwinter in areas sheltered from high flows, including off-channel ponds (Peterson 1982), sloughs, wetlands, and even estuaries (Tschaplinski and Hartman 1983). In the interior, salmon overwinter in areas free of ice accumulation. Thus, salmon display remarkable adaptations to the availability of food and refuge in different parts of their range. However, human activities can alter the physical characteristics of the habitat and the types of food available for salmon (e.g., Hawkins et al. 1983).

Spawners' Effects on Streams

One way in which salmon are connected to their ecosystems that has attracted some attention is what happens to them after they die. Carcasses of postspawning adults are seldom washed out to sea. Usually, they are removed from the stream and eaten by a wide variety of terrestrial animals (Cederholm et al. 1989); are deposited by floods in the riparian zone, where they are eaten or decompose; or decompose within the stream. Although terrestrial animals must persist through the year on other sources of food, the salmon can be seasonally important, particularly for animals fattening for the onset of winter. Migratory animals also can take advantage of salmon carcasses, as exemplified by the large population of bald eagles that winter on the lower Skagit River, Washington. The absence of salmon carcasses can reduce wildlife abundance (Spencer et al. 1991). Decomposition in the riparian zone also can contribute a large amount of nutrients to vegetation (Bilby and Bisson, unpublished data).

Marine-derived nutrients (carbon, nitrogen, and phosphorus) imported into freshwater systems by salmon are important to the trophic support system for salmon and other components of the aquatic and riparian communities (Donaldson 1967, Kline et al. 1990, Kline et al. 1993), as evidenced in concentrations of dissolved organic nutrients and stable-isotope ratios. When not eaten by wildlife, carcasses can be used through four pathways: autotrophic fixation by aquatic plants and subsequent transfer through periphyton-based food webs, heterotrophic fixation by bacteria and fungi and transfer through decomposer-based food webs, abiotic uptake by sorption within the stream substrate, and direct consumption of carcass tissues and remaining eggs by fish.

Before they die, however, adult salmon can exert another important influence on the stream ecosystem by their digging activities. Digging by females dramatically (but only temporarily) decreases the density of benthic invertebrates (Field-Dodgson 1987). In some instances, nest-digging can affect the dimensions and stability of the stream. At moderate to high fish densities—such as often occur in healthy pink, chum, and sockeye populations—salmon can widen a stream by digging along its margins, filling in its pools, and coarsening the streambed's substrate. Thus, contrary to the general belief that high densities of spawners are undesirable (Hunter 1959, McNeil 1964), high salmon density can be associated with high survival. If a flood, as might result when warm rain melts snow accumulated on a deforested slope, coincided with a poor run of salmon to a stream, the resulting increase in streambed scour could reduce recruitment. This could force the population into a cycle of smaller and smaller runs that could affect the physical structure of the stream and indirectly affect other inhabitants, such as fish, amphibians, and invertebrates.

Cautions

The complexity of salmon life cycles and the communities in which they exist should engender caution in proposing simplistic solutions to the salmon problem in the Pacific Northwest, especially in light of our limited understanding of salmon ecosystems. A weakness of natural-resource management has been the tendency to emphasize single-species management by manipulating either the "valued" species or the physical factors thought to influence its abundance or, at most, to consider some of the obvious interactions between pairs of species. Salmon restoration and rehabilitation must be approached in an ecosystem context, bearing in mind that habitat changes might not bring equivalent improvements for all species and could even be detrimental to some. Research on Pacific salmon has not kept pace with the theoretical and empirical literature on complex multispecies interactions in other systems. Caution must be practiced in the application of relatively simplistic salmon-management approaches, such as altering habitat, increasing hatchery production of juveniles, controlling predators, and introducing or eliminating nonnative species, because we have little ability to predict their ecological consequences.

SALMON ECOLOGY IN THE OCEAN

What happens at sea is important to the conservation and management of Pacific salmon (Pearcy 1992). Pacific salmon spend most of their lives in the ocean, where they grow to maturity after leaving freshwater streams as smolts. Two aspects are critical to understand the salmon problem and select appropriate management systems: interdecadal changes in the ocean environment and the effects of density on growth and survival. Interdecadal changes in the ocean environment—especially in water temperature and currents and associated biological communities—influence growth and survival rates and thus the return of adults (Francis and Sibley 1991, Beamish and Bouillon 1993, Francis and Hare 1994). Density-dependent effects—how salmon at sea are affected by the number of young salmon entering the ocean from their natal streams or hatcheries—might be experienced during the ocean phase and could thwart attempts to supplement populations by stocking during periods of poor ocean conditions (R. Francis and R. Brodeur, Production and management of coho salmon: A simulation model incorporating environmental variability, pers. comm., 1991; Beamish and Bouillon 1993).

Ocean factors are important not only because they directly influence the number of adults that return to their spawning streams but also because a poor understanding of their effects can result in misinterpreting both population status and the potential for an acceptable level of fishing. Management can overestimate or underestimate the abundance of returning adults if ocean effects are not realistically acknowledged. For example, Lawson (1993) has pointed out that the success of freshwater salmon-habitat improvements can be masked by changes in the ocean condition. Likewise, favorable ocean conditions could mask the effects of habitat degradation. Ocean effects are logistically difficult to observe because they occur over such large spatial and temporal scales that they are not easy to observe directly, in contrast with the more local effects of land use, fishing, and impoundments. Although hatchery stocking rates and high-seas fishing mortality can be altered, purposeful manipulation of the ocean climate to counter natural variations is currently beyond human capability. It is possible to respond appropriately to long-term, interdecadal environmental variation but this is difficult in a management and conservation milieu of many short-term crises and large interyear variability. A fishery and conservation management strategy can be developed that is better attuned to the large spatial scale and long-term variability in oceanic conditions relevant to Northwest Pacific salmon.

Interdecadal Variation in Ocean Climate

Evidence of wide-scale interdecadal variation in the Northeast Pacific and the Gulf of Alaska is increasingly convincing (UNESCO 1992). Over the period of dependable records (from 1946 to the present), the greatest anomalies in mean decadal sea surface temperatures occurred in the decade 1977-1986. During that period, lower sea-surface temperatures (more than 0.5°C below average) occurred in the central North Pacific, and higher sea-surface temperatures (more than 1.5°C above average) occurred along the West Coast of North America. Those changes were associated with changes in many other climatic variables. For example, the average annual North Pacific sea-level pressures from November through March (Trenberth 1990) exhibited a step decrease in about 1976. Stream flows in many rivers increased in coastal Alaska but decreased in the Pacific Northwest indicating that interdecadal climatic variations might influence not only the ocean environment, but also inland conditions affecting salmon habitat (Cayan et al. 1991). The Pacific Northwest Index, a composite index of climatic variation, incorporates the average annual coastal temperature, the average annual basin temperature, and snow depth in March. Variation in the index from 1901 to 1985 in the Puget Sound basin (Ebbesmeyer and Coomes 1989) not only shows warming in the late 1970s but demonstrates that interdecadal variations are regular features of the region during the twentieth century. Recognition of the interdecadal patterns enhances our knowledge of seasonal and interannual variations in the ocean environment (Wooster 1992).

Interdecadal changes have influenced the migration routes of adult salmon returning to spawning streams and, more important, the physics and biology of the ocean (UNESCO 1992) related to salmon production. Groot and Quinn (1987) documented a change in the percentage of salmon approaching the Fraser River from the northern passage around Vancouver Island versus the southern passage. Only about 17% took the northern passage from 1953 to 1977, but about 40% did from 1977 to 1984. Changes in migratory route are accompanied by differences in the timing of migrations as well (reviewed by Quinn 1990).

Predation of salmon at sea has been studied almost exclusively in nearshore areas as smolts have been leaving and adults returning. A wide variety of fishes and other animals eat salmon, and major predation by spiny dogfish (*Squalus acanthias*) (Beamish et al. 1992) and pinnipeds, such as harbor seals (*Phoca vitulina*) (Olesiuk 1993) and California sea lions (reviewed by Palmisano et al. 1993; see Box 2-1), can occur in some situations. Predation has apparently eliminated the smolt production from a hatchery in southern Vancouver Island in sets of years when warmer waters allowed mackerel (*Scomber japonicus*) to move north and prey on the smolts entering the sea (B. Riddell, pers. comm., June 1994).

Salmon at sea eat fish, crustaceans, and soft-bodied organisms in the upper 1-200 m; variation in diet among species (e.g., Beacham 1986) and years (Brodeur

1992) has been noted. The abundance of those prey resources varies seasonally, from year to year, over decadal time scales, and across broad regions of the North Pacific Ocean (Frost 1983, Brodeur and Ware 1992). In analyses for the North Pacific Ocean from 1965 to 1980 (Beamish and Bouillon 1993), 25% of the variation in the abundance of copepods was statistically associated with the Aleutian Low Pressure Index; from 1976 to 1980, copepods were 1.5 times more abundant than from 1965 to 1975. Analyses by Brodeur and Ware (1992) suggest that zooplankton biomass in the Alaska Gyre during the 1980s was more than twice that in the late 1950s and 1960s. Because copepods are an important food of salmon and many of their prey, their abundance is expected to affect the welfare of salmon on the high seas. The parallel trends in the catches of pink, chum, and sockeye salmon across the entire North Pacific Ocean suggest that common events across wide areas influence salmon production (Beamish and Bouillon 1993). The pattern in the total catches of those species by Canada, Japan, Russia, and the United States corresponds with the Aleutian Low Pressure Index (see Figure 7 in Beamish and Bouillon 1993); this suggests that the fish are fluctuating in response to long-term ocean variations (Beamish and Bouillon 1993). Beamish and Bouillon (1993) suggested (on the basin of meager data) that the index and the catches of Pacific salmon were low at the turn of the century—further evidence is needed to support their conclusion.

The relation of both salmon catches and zooplankton abundance in the North Pacific Ocean to the Aleutian Low Pressure Index suggests that the interdecadal changes in salmon abundance are mediated via the food web. Beamish and Bouillon (1993) presented a rationale for relating the Aleutian Low Pressure Index to salmon abundance via changes in survival upwelling, nutrient availability and phytoplankton production, and trophic relationships to copepod abundance. There is evidence that the increase in marine survival of chum and pink salmon in the 1970s corresponded to an increase in the Aleutian Low Pressure Index; Japan's salmon catch during the 1970s indicated that marine survival might have increased by 200%–300% (see Figure 9 in Beamish and Bouillon 1993).

The interdecadal changes in salmon abundance that are linked to climate appear to reflect increases and decreases in survival. Year-class strength seems to be set early in the ocean phases of salmon life history (Pearcy 1992). However, the actual source of mortality and the mechanism that causes survival to change are not known. It has been speculated that they are linked to north-south changes in the distribution of other marine fishes, such as hake (*Merluccius productus*) or other taxa that prey on young salmon (Holtby et al. 1990).

Although salmon catches across the North Pacific Ocean appear to vary together in time, there is evidence that salmon catches along the Alaska coasts and the Washington-Oregon-California coasts are out of phase with each other (Francis and Sibley 1991). The reason for this phase difference in salmon catches is probably that low-frequency changes in the California Current and the Alaska

Box 2-1 Marine Mammals and Salmon

The effect on salmon of predation by marine mammals has been controversial at least since the nineteenth century (Merriam 1901). Fish predation by California sea lions (*Zalophus californianus*) has received attention again recently largely because of its high visibility in one location: the Hiram M. Chittenden locks in Ballard (a section of Seattle). The Chittenden (or Ballard) locks were completed in 1917 as part of a project of the U.S. Army Corps of Engineers to allow ship traffic between Puget Sound and Lake Washington (Willingham 1992). The current drainage into Puget Sound is through the Lake Washington Ship Canal, Lake Union, and the Ballard locks. Steelhead trout migrated through Lake Washington to spawn in its tributaries before 1917; they still do today. The runs consist of wild and hatchery fish, but the hatchery program was recently discontinued. Although some people believed that the current "wild" runs were derived from hatchery fish, genetic analysis indicates that they are probably descendants of the original wild runs (Fraker 1994).

Lake Washington steelhead are winter-run fish, returning to the fresh water to spawn from December to April. As the fish enter the ship canal below the locks, some of them are captured and eaten by California sea lions. The first observation of such predation was made in 1980; by the middle 1980s, as many as 60 sea lions were in the area around the locks, and more than 50% of the returning steelhead were being eaten. The number of fish in the run that escaped to spawn, which ranged from 474 to 2,575 fish from 1980-1981 to 1985-1986, declined to 184 fish in 1992-1993 (Fraker 1994) and to 70 in 1993-1994 (NMFS and WDFW 1995). Harbor seals and fish-eating birds appear also to be eating large numbers of steelhead smolts leaving the lake (Fraker 1994).

It is clear that sea lions have affected the Lake Washington steelhead run, but

Current are out of phase, as well. Francis and Hare (1994) based an explanation for the phenomenon on suggestions by Hollowed and Wooster (1992). They proposed that

> 1. there are two mean states of winter atmospheric circulation in the North Pacific which relate to the intensity and location of the winter mean Aleutian Low . . .;
>
> 2. oceanic flow in the Subarctic Current and the resultant bifurcation at its eastern boundary into the California and Alaska Currents is fundamentally different in these two states;
>
> 3. the patterns in Alaska salmon production tend to indicate long interdecadal periods of oscillating 'warm' and 'cool' regimes: early 1920s to late 1940s/early 1950s (warm), early 1950s to mid 1970s (cool), mid 1970s to present (warm);
>
> 4. the hypothesized inverse behavior of the long-term production dynamics

they are not entirely responsible for its recent decline. Cooper and Johnson (1992) found that steelhead had declined generally since 1985. The major possible contributing factors included were competition for food with other salmon, in particular 8 billion hatchery salmon released since the late 1980s; authorized and unauthorized drift-net fisheries (probably not a factor today); predation by birds and mammals; and large-scale environmental changes.

Predation by marine mammals is probably not a major factor in the current decline of salmon in general. Anadromous salmon and marine mammals coexisted for thousands of years before the current declines in salmon, and California and Steller sea lions were much more abundant in the first half of the nineteenth century—a time when salmon were also abundant—than later. For example, Bonnot (1928) counted only 1,429 sea lions in his census of California haulouts and rookeries in 1927-1928. And marine mammals do not normally specialize on salmon; they eat a wide variety of prey items, determined by what is available and how easy it is to catch (Gearin et al. 1988, Fraker 1994, Olesiuk 1993).

However, many marine-mammal populations are increasing, at least partly because of the protections of the Marine Mammal Protection Act (MMPA). California sea lions now number more than 100,000 (Fraker 1994). Other human activities, combined with increased marine-mammal populations, could cause increasing problems, especially in areas where the fish congregate and have few refuges, as in the case of the steelhead at Ballard. If the Lake Washington steelhead were listed as endangered or threatened under the Endangered Species Act (ESA), the conflict would be brought into sharper focus by the requirements of the ESA and the MMPA; a petition for such listing has been filed. Indeed, the MMPA was amended in 1994 to allow the killing of marine mammals under particular circumstances, and Washington state has filed a petition to remove sea lions from the Ballard locks and kill them (NMFS and WDFW 1995).

> of the Alaska Current salmonids . . . and zooplankton . . . is related to effects of these two states of winter atmospheric circulation on the dynamics of . . . physical oceanographic systems . . . and, subsequently, on biological processes at the base of the food chain.

The impact of interdecadal oscillations apparently occurs in a salmon's first year in the ocean (Francis and Hare 1994, Hare and Francis 1995). That conclusion is based on time-series intervention analyses that identify years when "step changes" occur in a long-term record of annual values. Step changes are sudden increases and decreases in a data record that persist for a period of years and are apparent in spite of large year-to-year variation. The time series analyzed were of western-Alaska (Bering Sea) sockeye-salmon catches from 1920 to 1990, northern Gulf of Alaska pink-salmon catches from 1935 to 1990, winter air temperatures at Kodiak Island, and the North Pacific index of the intensity of the Aleutian Low.

Step changes were apparent in all four time series; in the air-temperature data, a cooling step began in 1947 and a warming one in 1977. There was a lag of 1 year between the step change in the climatic time series and that in the catch of pink salmon, which migrate directly to the ocean after emerging from the gravel. In comparison, a lag of 2 years was observed for the sockeye salmon, which spend 2-3 years in freshwater before entering the ocean. Thus, the effects of interdecadal climate changes are reflected in the fish catches and match the time lags expected on the basis of when the salmon first enter the ocean. The analyses also suggest that, at least for the coastal streams of Alaska that have not been extensively degraded, the fish respond to changes in the ocean conditions rather than to changes in conditions of the natal streams.

Density-Dependent Effects

A corollary to the influence of interdecadal changes in ocean conditions on salmon survival during their first year at sea is how they are density-dependent. Whether salmon at sea experience density-dependent growth or survival is controversial. But it appears that salmon growth and therefore the age structure of returning spawners can be affected by high densities at sea. That is exemplified by systems with high natural abundance, such as sockeye salmon from Bristol Bay, Alaska (Rogers and Ruggerone 1993), and regions with extensive artificial propagation, such as of chum salmon from Japan (Ishida et al. 1993). It also appears that density-dependent growth reductions occur only in the early and final periods of marine residence, when salmon are in coastal waters. Salmon from a given region are thus competing primarily with each other; there is little evidence of competitive interaction among population complexes from distant regions, such as Japanese chum salmon and Bristol Bay sockeye (Rogers and Ruggerone 1993).

Beamish and Bouillon (1993) suggested that with the exception of Japanese chum salmon, hatcheries were not important contributors to the increase in salmon catches in the North Pacific during the 1970s and 1980s. In the Gulf of Alaska, increased hatchery production of pink salmon contributed to the increase in catches from the early to middle 1980s, but natural reproduction increased at the same time by a factor of 3 to 4. The increases in Alaskan catches of chum and sockeye were so great in the late 1970s that hatcheries could have contributed to them only slightly. For Canadian fisheries, some increases can be attributed to enhancement programs for sockeye and chum salmon, but these increases in enhancement could not account for the marked regional increase in all northern salmon populations, because Canada's contribution to the total catch of Canada, Japan, Russia, and the United States was relatively small. Russian chum-salmon catches increased from 1976 to 1990 even though releases from hatcheries were reduced.

Beamish and Bouillon (1993) suggested that hatchery production strategies should take into account the interdecadal changes occurring in ocean quality relative to salmon production. In general, hatchery production tended to track wild production in the 1970s and 1980s but did not play a large role in the increase in production. However, Beamish and Bouillon (1993) speculated that

> it may not be an appropriate strategy to continue to release large numbers of artificially reared smolts during a period of decreasing marine survival of salmon. . . . Declines in marine survival should be managed differently from declines resulting from fishing mortality that is too high.

That is an interesting idea of great potential importance, but Beamish and Bouillon provided no evidence that hatchery releases influence marine survival of wild fish when ocean conditions are either good or poor. Nevertheless, density-dependent effects, even between or among hatchery populations, may result in decreased returns, smaller sizes, and decreased value.

Larger Spatial and Temporal Scales

The emerging understanding of interdecadal changes in the ocean climate and the related mechanisms that affect salmon at sea have implications that are both exciting and disconcerting to scientists thinking about resource management. Humans are beginning to understand what happens to salmon during the majority of their lives—the portion spent at sea. Although we know little of the details, the new insights already demonstrate that variations in salmon abundance are linked to phenomena on spatial and temporal scales that humans and human institutions do not ordinarily take into account. Consider that the apparent effectiveness of hatcheries might have resulted from favorable ocean and climatic conditions in the era when the hatcheries were built; what looked like human manipulation of the total number of salmon might have been only a reapportionment among different populations. Or consider that the decline of some populations might be a direct result of introducing new hatchery populations into an ocean pasture of limited capacity.

The scale of human endeavor often has been incommensurate with the scale of salmon ecology. Some of our current policies are based on deep ignorance: it is not reasonable to assume that ocean conditions vary in ways that are generally uniform and random in their impacts on populations of salmon. Interdecadal variations and the importance of the ocean phase should be incorporated into human thought, planning, and actions in response to the effects of and attempts to repair damage that occurred during the freshwater phases of the salmon lives. The possible overriding effects of interdecadal changes in ocean conditions on salmon, the results of freshwater salmon management, and the overwhelming focus of human attention on the more-visible freshwater phases of the salmon history combine to provide the key ingredients for surprises in the future.

3

Human History and Influences

The general decline of salmon in the Pacific Northwest has initiated a wide range of technical, social, and political debates concerning what can and should be done to maintain or restore native populations. The situation is very difficult because of the complexity of the species' life cycles and the diversity of human activities and land uses that affect them. Throughout the various environments that make up the Pacific Northwest, the life history of salmon is intertwined with human history.

HISTORICAL SETTING

The American Indian settlements in the Pacific Northwest constituted "one of the most densely populated nonagricultural regions of the world" (Boyd 1990:135). There were perhaps 100,000 living in the Pacific Northwest in 1770 when Euro-Americans began to interact with them with some frequency (Boyd 1990:136). The Indians were very successful in using salmon to meet their own needs.

In the late 1700s, events in eastern North America set the stage for the changes that were about to commence in the Pacific Northwest. After the signing of the Declaration of Independence in 1776, a newly formed nation of states along the eastern seaboard looked westward. In the early 1800s, Meriwether Lewis and William Clark led a party across the recently acquired Louisiana Purchase and continued into the largely unknown Oregon Country to the mouth of the Columbia River. Lewis wrote remarkably detailed and accurate descriptions of Pacific salmon long before they were given formal taxonomic recognition. Descriptions of the region's flora, fauna, landforms, and climate by Lewis and Clark and others indicated that the Northwest was a special place. For example, accounts of plentiful beaver and muskrat populations helped to initiate a rush of trappers to the Northwest. Early reports of vast and valuable natural resources prompted a westward migration. Immigrants were aided by Lieutenant John Charles Fremont's 1842 expedition that examined the Platte River-South Pass route into Oregon. His well-publicized exploits made Americans more aware of the Oregon Country.

With the discovery of gold on the American River in 1848, a flood of prospectors headed west into the future California. In addition, tens of thousands, desiring opportunities to develop, use, and control the natural resources of the West, journeyed along the Oregon Trail via horseback and wagon during the 1840s and 1850s. By now, major impacts on American Indian cultures were well under way. The capture of Chief Joseph and his people during their flight toward Canada in 1877 was one of many events that marked the uneasy truce between the rights and needs of American Indians and the surging immigration of Euro-Americans. Even before the immigrating Euro-Americans arrived in large numbers, their diseases had a substantial impact: by the late 1850s, the American Indian population had decreased by 80–90%, and some tribal groups had disappeared.

By the mid-1800s, Euro-Americans along the West Coast had become sufficiently numerous that statehood was reasonable; California obtained statehood in 1850, Oregon in 1859, Washington in 1889, and Idaho in 1890. The Euro-American population had reached 100,000 by about 1870; the American Indian population had declined to under 10,000. By 1900, the combined population of Idaho, Oregon, and Washington had reached nearly 1 million. By 1990, the total population of the region was 8.7 million; 133,000 identified themselves as American Indians. From 1940-1990, the population grew at an annual rate of 1.9%, mostly as a result of inmigration (Figure 3-1).

The immigration of Euro-Americans into the Pacific Northwest, with their accompanying cultural and industrial perspectives, transformed the region in ways that were previously unimaginable. The bountiful natural resources and the desire to use them for a growing economy were precursors to the widespread use of forests, water, salmon, and other resources of the region. Unless the current population growth rate slows dramatically, which appears unlikely, these transformations will continue.

CULTURES AND TREATIES

The Euro-American settlers that migrated to the region in large numbers after 1800 were farmers. To address the conflicts between American Indian and non-Indian ways, the U.S. government negotiated treaties in the 1850s with many of the Indian groups. Those of a Euro-American background wanted formal

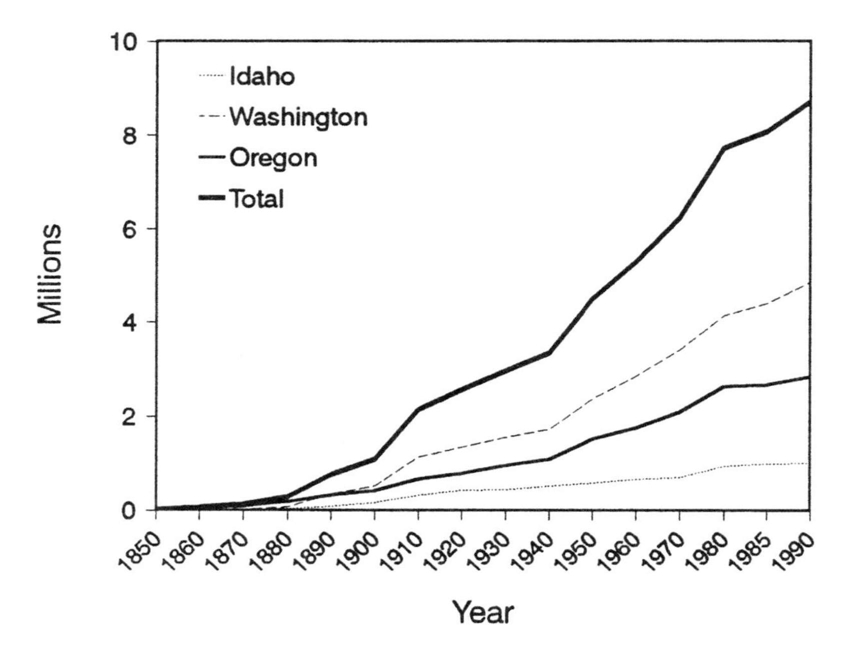

FIGURE 3-1 Population growth in the Pacific Northwest states from 1850-1990. Source: Data from Statistical Abstracts, USDC Bureau of the Census 1975, 1995.

treaties and required the signing of agreements to assign land ownership, sovereignty, and rules for fishing and hunting. The treaty-making process created treaty and nontreaty tribes.

Treaty-making in the Northwest began with the Medicine Creek Treaty of 1854. Over the next year, eight additional treaties tried to establish and settle relations between Indians and Euro-Americans. Each treaty envisioned tribal peoples becoming family farmers, each family with its own independent piece of land. The transcript of the 1855 Treaty at Walla Walla gives insight into the cultural differences. Isaac Stevens, governor of Washington, commented about American Indian land ownership: "On these tracts the land was all in common: there were one or more larger fields for the tribes but no man has his special field" (U.S. Superintendent of Indian Affairs, Oregon Territory, 1855).

The treaties signified radical changes in property rights. They were primarily about land division and private land ownership, and they marked a formal transition from a culture coevolved with salmon and their landscapes toward a cultural assemblage that substituted intervention, engineering, markets, and mitigations—all undertaken on time scales shorter than a single human generation—as ways to mediate humans' needs and nature's capacities.

Provisions of the treaties have been taken to the U.S. Supreme Court for interpretation eight times (Cohen 1986). Two major decisions advanced the treaty rights to fishing: the Belloni decision in 1969 and the 1974 Boldt decision. As a result, the treaties now serve as a critical legal basis for the contemporary salmon problem. Among other things, the treaties guarantee signatory tribes a right of access to salmon and other resources, implicitly signaling the importance of the natural world to the Indian cultures.

DECLINE OF THE BEAVER

Beaver (*Castor canadensis*) had a key role in creating and maintaining conditions of many headwater streams, wetlands, and riparian systems that were fundamentally important to the rearing of many salmon. Not only did they provide an important disturbance regime that helped to maintain environmental heterogeneity (see Chapter 7 for a discussion of the ecological role of disturbance), but their dams and ponds created storage locations for water, sediment, and nutrients. Many riparian plants and aquatic organisms' life cycles required change in the water-table depth; beaver dams and ponds caused such depth alterations (Naiman et al. 1992). Beaver ponds were of particular importance in the more arid regions but also had important implications for coastal streams, where they also provided rearing habitat for salmon.

The regional decline of the beaver was an early example of the capacity of Euro-American exploitation to deplete resources. Even in Oregon, which ultimately adopted the beaver as its state animal, current beaver populations are diminished greatly from their former extent and numbers; persistent trapping pressure over the decades has continued to keep beaver populations relatively small throughout much of the state and the region. The general decline of beaver and their associated habitats constituted perhaps the first major impact on salmon populations from the influx of Euro-Americans.

FISHING PRESSURES

The size of salmon and steelhead runs in the Columbia River before significant non-Indian presence has been estimated at 10-16 million fish per year (NPPC 1986) and 7-8 million fish per year (PFMC 1978, Chapman 1986). The first salmon cannery along the West Coast was established in 1864 along the Sacramento River of northern California (Hittell 1882, Goode and others, 1884-1887). However, sediment from hydraulic mining so devastated the runs that the cannery was soon shut down and moved to the lower Columbia River in 1866. In the first year on the Columbia, the company packed 4,000 cases of salmon, or about 240,000 lb. Salmon canning spread from the Columbia River to Puget Sound, British Columbia, and Alaska; soon southeastern Alaska and Bristol Bay dominated. Columbia River packers sought to create and retain the advantage of high-quality, prime spring-caught chinook salmon (Cobb 1930, Craig and Hacker

1940, Smith 1979). Historically, the Columbia River runs were huge by any standard. Catches in the late 1800s reached 43 million lb. Peak catches might have been 3-4 million fish of all species (Chapman 1986). From one cannery and two gill-net boats in 1866, the Columbia River fishery grew to 40 canneries in the early 1900s (Smith 1979, Netboy 1980). After the 1870s, the river catch of spring chinook began a steady decline; canners extended their season to include fall chinook runs and broadened the species they caught to include sockeye, steelhead, and coho. The last cannery on the Columbia closed in 1975 (Figure 3-2). By the early 1990s, the run size dropped to about 2.5 million fish (NPPC 1994).

Although much of the decline in the Columbia River fishery has been attributed to increases in inriver and ocean fishing, other factors, such as dam construction and modifications of freshwater spawning and rearing habitats in the Columbia basin, are important contributors as well (Simenstad et al. 1992, Bottom 1994).

PROPAGATING FISH

Early fish propagators had confidence that the salmon-producing environment could be made better. The belief that humans could tinker with one part of nature—reproduction of salmon—and obtain expected results has turned out to be simplistic. Today, freshwater and ocean ecosystems are understood to exhibit complex, often unpredictable interactions and feedbacks among their countless parts. And the appropriate role of artificial propagation in salmon management has become a much more complex question than was conceived by the early fish propagators (see Chapter 12).

The federal government and canners supported artificial propagation. In 1877, concerns about overfishing led the Oregon and Washington Fish Propagating Company to construct a salmon-breeding station on the Clackamas River (Whale and Smith 1979 as cited by Columbia Basin Fish and Wildlife Authority 1990). Its problems presaged today's salmon problem. At first, the station produced many eggs, but "gradually mills and dams, timber-cutting on the upper waters of the Clackamas, and logging in the river, together with other adverse influences, so crippled its efficiency that it was given up in 1888" (Stone 1897:218). Federal hatcheries were built in the 1890s. State hatcheries also increased in number: Oregon had 12 hatcheries releasing 27 million salmon fry by 1907 (Lichatowich and Nicholas in press).

Even in the first half-century of artificial propagation (1877-1930), Pacific salmon abundance did not always increase in response to increased releases of hatchery fish (Smith 1979, Lichatowich and Nicholas in press). In the early 1900s, tens of millions of fry were released annually from Columbia River and Oregon coastal hatcheries, but declining catches discouraged the hatchery propagators and, starting in 1910, stimulated them to rear fish until they were larger

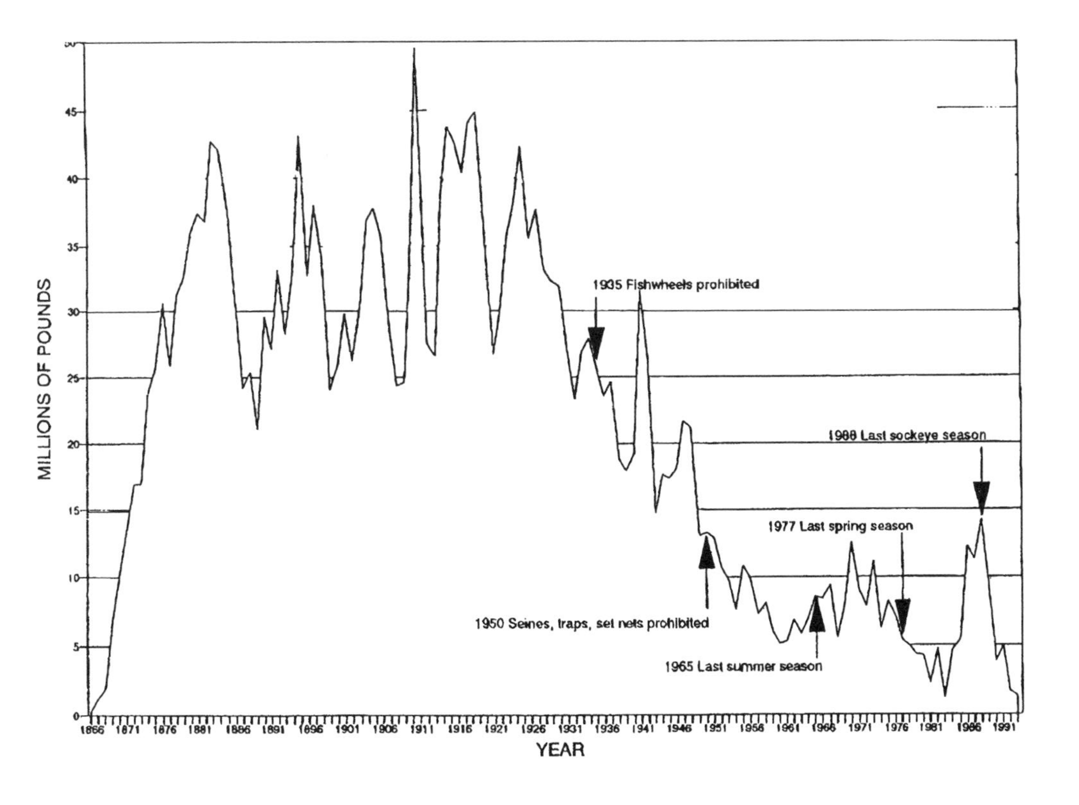

FIGURE 3-2 Commercial landings (pounds) of salmon and steelhead in the Columbia River, 1866-1993. Source: WDFW and ODFW 1994.

before releasing them. The hope was that larger hatchery fish would survive better in the wild.

When the catch did increase, in 1914, fishery managers were quick to conclude a cause-effect relationship between the first releases of larger hatchery fish and the improved catch. The Oregon Fish and Game Commission (1919, as cited by Lichatowich and Nicholas in press) boldly stated its conviction:

> This new method has now passed the experimental stage, and . . . the Columbia River as a salmon producer has "come back." By following the present system, and adding to the capacity of our hatcheries, thereby increasing the output of young fish, there is no reason to doubt . . . that the annual pack in time can be built up to greater numbers than ever before known in the history of the industry.

In retrospect, it is impossible to rule out the possibility that if there had been no releases of larger hatchery fish, catches nevertheless might have increased by around 1914 in response to changes in freshwater or ocean conditions. Interviews with residents who witnessed the 1914 salmon runs to the Umatilla River, where no hatchery fish had been released at that time, indicate that 1914 was the year of "the largest run of chinook salmon within the memory of white men" (Van Cleve and Ting 1960). If the early releases of larger hatchery fish had been handled as a legitimate field experiment, they would have involved estimating adult returns for both the hatchery population and one or more control, wild populations in carefully matched tributaries not influenced by hatchery fish (Eberhardt and Thomas 1991).

For many artificial-propagation facilities, the lack of long-term monitoring makes it nearly impossible to differentiate impacts of the hatchery program from impacts of other human interventions or of natural environmental trends. And no effort was made to evaluate cumulative effects of releasing large numbers of fish from different programs.

From the 1930s through the early 1950s, support for hatcheries dropped considerably because of poor returns and disease problems (Columbia Basin Fish and Wildlife Authority 1990, F. J. Smith 1979). In the Columbia River Basin, many early facilities were closed; if not for the rapid expansion of dam construction, the use of hatcheries might not have resurged.

With increasing dam-building on the Columbia River from the 1930s through the 1970s, the purpose of hatcheries gradually shifted from improving on nature to merely making up for huge losses of salmon populations and their spawning habitats caused by dams for hydropower, irrigation, and navigation. In the 1960s, the invention of pasteurized and formulated feeds that reduced the incidence of disease brought new expectations that artificial propagation could overcome negative dam effects and even increase salmon abundance (Anonymous 1959).

Eventually, more than 80 hatcheries were built in the Columbia River Basin, with the Mitchell Act of 1938 playing a major role in the development of 39 federally funded facilities (Columbia Basin Fish and Wildlife Authority 1990). Although hatchery construction originally was authorized to mitigate damage from the dams, an agreement (which did not include Indian tribes) put most of the artificial propagation downstream from most of the dams. That arrangement avoided losses from dams and reservoirs and functionally allocated the bulk of the fish to non-Indian fishers (Lee 1993a:26-27). Negative effects of dams on upstream habitat and the down-river placement of hatcheries dramatically shifted the geographical distribution of Columbia River salmon production from mostly the upper river to mostly the lower river. The downstream siting of hatcheries combined with fishery management decisions to favor certain species (primarily coho and fall chinook) led to a major alteration of species composition: comparisons between the period before 1850 and 1977-1981 show more than a doubling of the relative proportion of coho and a virtual elimination of sockeye and chum salmon (Lee 1993a).

Hatchery facilities are widely distributed throughout the five regions of the Pacific Northwest; Figure 3-3 illustrates their distribution within the Columbia River Basin. Coho, chinook, and steelhead are the principal species cultured; chum and pink are cultured to a smaller extent and primarily in Washington; and small-scale captive breeding of sockeye was initiated to help recover the endangered Snake River population in Redfish Lake, Idaho, and to compensate for hydropower-caused losses of sockeye from Osoyoos and Wenatchee lakes.

The abundance of many coastal and Columbia River populations has declined sharply since the mid-1970s and again in the late 1980s. This decline has occurred while reliance on artificial propagation has been at a historic high and hatchery-released fish have dominated the overall composition of anadromous salmon originating in Washington, Idaho, Oregon,and California. Coastwide estimates of relative abundance of hatchery fish are also difficult or impossible to make with existing data. There is little coastwide coordination to mark all hatchery fish physically[1] or to collect and analyze the relevant data. Available data allow for only rough estimates of the proportion of hatchery fish of some species in some locations (see Box 12-1). Even so, Light (1987) and Burgner et al. (1992) estimated that hatchery steelhead adults made up half the 1.6 million steelhead adults that return annually to the Pacific coast of North America.

The history of artificial propagation reveals a recurring cycle of technological optimism followed by pessimism. With the increasing reliance on artificial propagation, concerns became greatly heightened that contemporary hatchery programs are having negative effects on the genetic diversity and persistence of wild populations and that increasing releases of hatchery fish cannot override

[1]All hatchery steelhead released from Columbia River hatcheries have the adipose fin removed so that adult wild fish can be identified. A similar mark is applied to all spring-summer chinook smolts released from Snake River hatcheries.

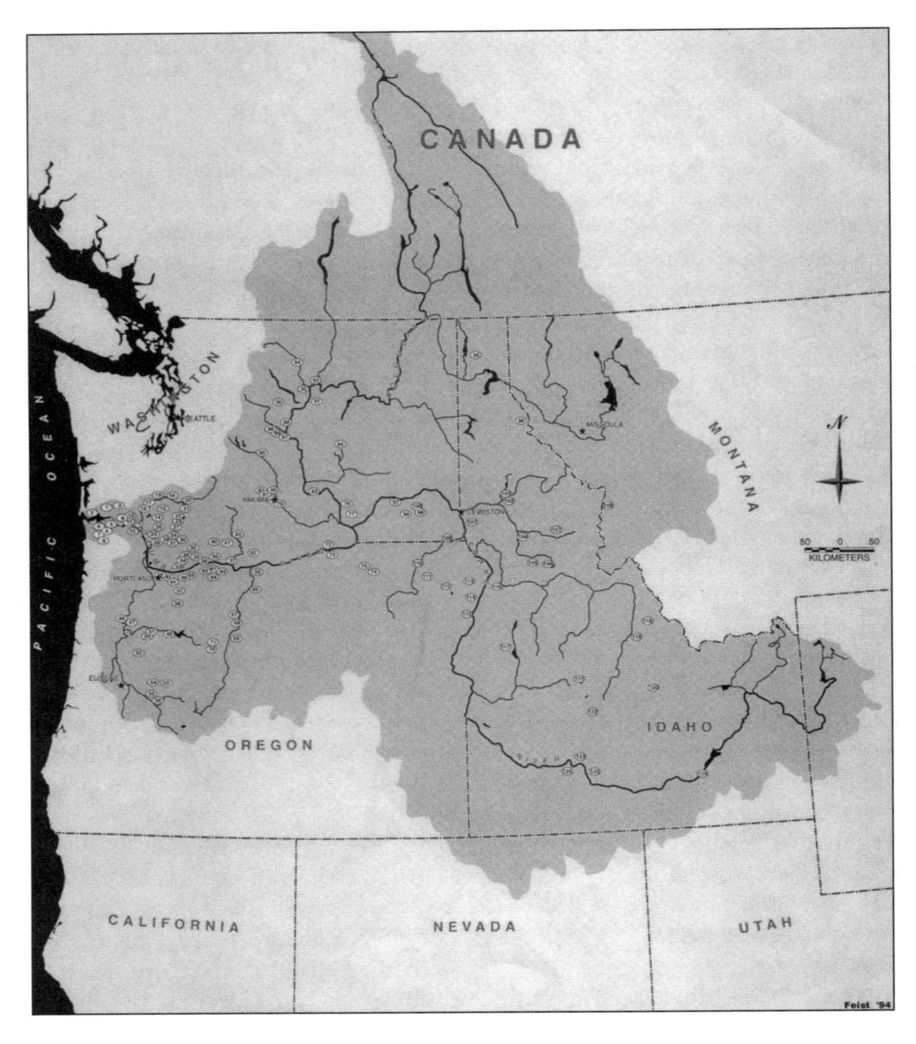

FIGURE 3-3 Approximate geographical location of existing facilities in the Columbia River Basin for artificial propagation of anadromous Pacific salmon. Source: Feist 1994.

other factors contributing to an overall decline of salmon. Some recent trends also call into question the sustainability of a fishery dependent on large-scale hatchery releases. These include decreasing body size at maturity and increasing age at maturity of Japanese chum as total returns have increased, suggesting density-dependent rearing limitations in the oceanic environment (Kaeriyama 1989 cited by Riddell 1993a); reduced catches of chinook salmon in the Strait of Georgia, British Columbia, when hatchery releases exceeded 8.3 million fish per year (Riddell 1993a); and the suggestion that interannual variability in fish abundance might increase as releases of hatchery fish increase (McCarl and Rettig 1983, Fagen and Smoker 1989).

Disappointment has resurfaced whenever sufficient data have accumulated to indicate that hatchery programs had failed either to improve on nature, to circumvent natural fluctuations in ocean conditions, or to make up sufficiently for large, human-induced losses of natural reproduction. Each turn of the cycle formed a larger orbit as the scale of artificial propagation has increased, naturally reproducing populations declined more precipitously, and the number of hatchery critics has increased. Prevention of another repetition of the cycle will require development of more realistic hatchery goals (see Chapter 12), overhaul of hatchery practices, and serious commitment to evaluation of hatcheries in an adaptive-management context.

GRAZING RANGELANDS

Because riparian areas provide a crucial link between aquatic and terrestrial ecosystems, sustained grazing of these areas can substantially affect fish and aquatic habitats. Overgrazing, both inside and outside the topographic boundaries of the Columbia River Basin, has caused sedimentation of spawning gravels, changes in channel structure, loss of shading, high stream temperatures, channel incision, and other deleterious effects (NPPC 1986, Elmore and Beschta 1987). Fish production in grazed streams is much lower than in ungrazed streams (NPPC 1986).

As ranchers and settlers entered the Columbia River system, livestock numbers rapidly increased, and they probably peaked well before the turn of the century (Wilkinson 1992). Although the forage productivity and resilience of this previously ungrazed region was initially able to sustain intense livestock pressure, ultimately the ecological costs of overgrazing western rangelands included increased erosion and surface runoff, loss of shrub and riparian communities along stream systems, extensive alteration of native plant communities, continued decline of beaver populations, widespread channel downcutting, and broad impacts on fish and wildlife habitats.

In 1934, Congress passed the Taylor Grazing Act (TGA), which established the Grazing Service—later to become the Bureau of Land Management (BLM) in the Department of the Interior—to regulate grazing on public lands. High levels of grazing in the previous decades and an extended period of drought had contributed to the widespread degradation of many rangelands by the 1930s.

In 1941, the Grazing Service authorized the highest level of use ever, 22 million animal unit months (an animal unit month represents about one cow and her calf or five sheep grazing for 1 month), on the 258 million acres of public rangelands (USDI Bureau of Land Management Undated-a, NRC 1994). Since then, the extent of livestock grazing on public lands has declined steadily (Figure

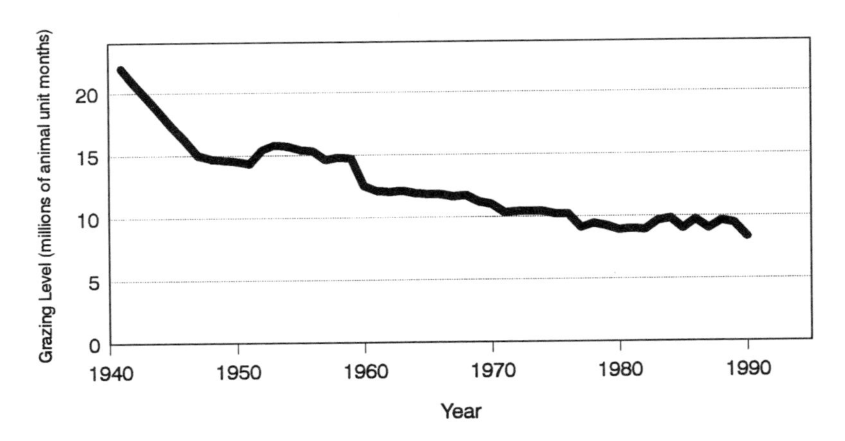

FIGURE 3-4 Levels of livestock grazing on Bureau of Land Management Section 3 public rangelands in the 11 western states in 1941-1991 (Section 3 lands encompass about 90% of the total grazing use of public rangelands). Source: Annual reports of Public Land Statistics, USDI Bureau of Land Management.

3-4). Although the effects of declining use might have translated into some improvement of upland range condition (USDI Bureau of Land Management Undated-b), riparian and aquatic resources remain in poor condition and in urgent need of improvement (General Accounting Office 1988, Chaney et al. 1993).

Other aspects of livestock grazing might also affect fishery resources. For example, in southeastern Oregon, the Vale District of BLM established 28 deep wells and storage tanks, constructed 440 miles of pipeline, and developed 1,000 reservoirs and springs from 1962 to 1973. The effects of those developments on fish habitat are largely unknown. In addition, intensive grazing occurs along many lowland streams and estuaries west of the Cascades. Such practices might have substantial local impacts on riparian resources and fish habitat, but hardly any research has been undertaken to evaluate their magnitude.

HARVESTING THE OLD GROWTH

To the first settlers and loggers, the extensive coniferous forests of the Pacific Northwest appeared vast and endless, and it was difficult to imagine that they could ever run out of trees. However, in much less time than it takes to develop an old-growth forest, much of the forest land was harvested or converted to other uses. The first sawmill was constructed at Vancouver, Washington, in 1827. Lumbering and logging became the leading industry in the Pacific Northwest in the last several decades (Figure 3-5). There have been dramatic declines in harvest rates on some federal lands in recent years, but nonfederal commercial

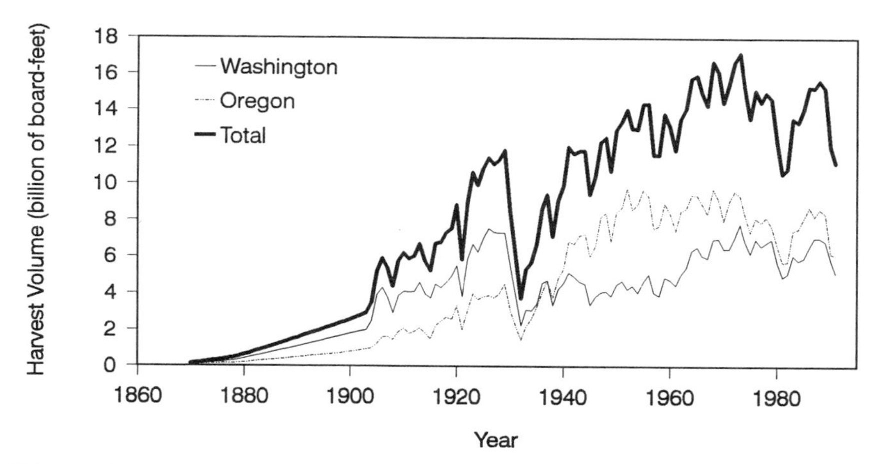

FIGURE 3-5 Timber harvest volumes (Scribner scale) in Oregon and Washington in 1870-1990. Source: Oregon data for 1870-1924 based on lumber-production data from Oregon State Board of Forestry, Forest Resources of Oregon, 1943; Washington data for 1870-1924 obtained from unpublished records provided by D. Larsen, forest economist, Washington Department of Natural Resources; data for 1925-1950 from USDA Forest Service, 1972, Resource Bulletin PNW-42, Log Production in Washington and Oregon; data for 1951-1990 obtained from timber-harvest reports published by state agencies.

forest lands continue to experience high levels of harvest, much of it harvest of relatively young second-growth stands.

Some of the earliest logging operations were along the banks of larger rivers and streams, where logs could be floated downstream for milling. When the timber within easy access of a navigable stream was exhausted, logging operations moved to another location. According to Sedell and Luchessa (1982), "by 1880 the land along the western banks of Puget Sound and all around Hood Canal had been cleared of trees for 2 miles inland and up to 7 miles around the major streams and rivers." The complete harvest of these riparian forests had important implications for the production of anadromous salmon.

In addition to harvesting riparian timber, it was common practice to remove and salvage large wood from coastal streams and major rivers in the late 1800s and early 1900s. Removing snags and downed trees from the streams and rivers was well established by the turn of the century (Sedell and Beschta 1991). Rivers used for navigation were routinely "cleaned" of all large wood and boulders to provide unobstructed passage for log rafts. Salvage logging of timber in rivers and streams, especially western red cedar, had a serious impact on small streams throughout western portions of Washington and Oregon. Loss of large woody debris to salvage logging likely reduced both the size and frequency of pools in these systems, and diminished the amount of cover available to rearing salmon (Bisson et al. 1987).

The use of splash dams to transport logs downstream degraded many miles of important spawning and rearing habitat of salmon. Splash dams consisted of temporary structures that created a sizable backwater pond. Logs were dumped into the pond and the waterway below it. Once filled with logs and water, the dam was collapsed and a torrent of logs and water rushed downstream. The procedure was then repeated. More than 150 major splash dams existed in coastal Washington rivers, and an additional 160 were used on coastal and Columbia River tributaries in Oregon (Figures 3-6 and 3-7). Although splash damming is no longer used to transport logs, its effects are still evident in many streams.

With the increased availability of heavy equipment after World War II (e.g., portable yarders, loaders, bulldozers, and trucks), areas formerly inaccessible

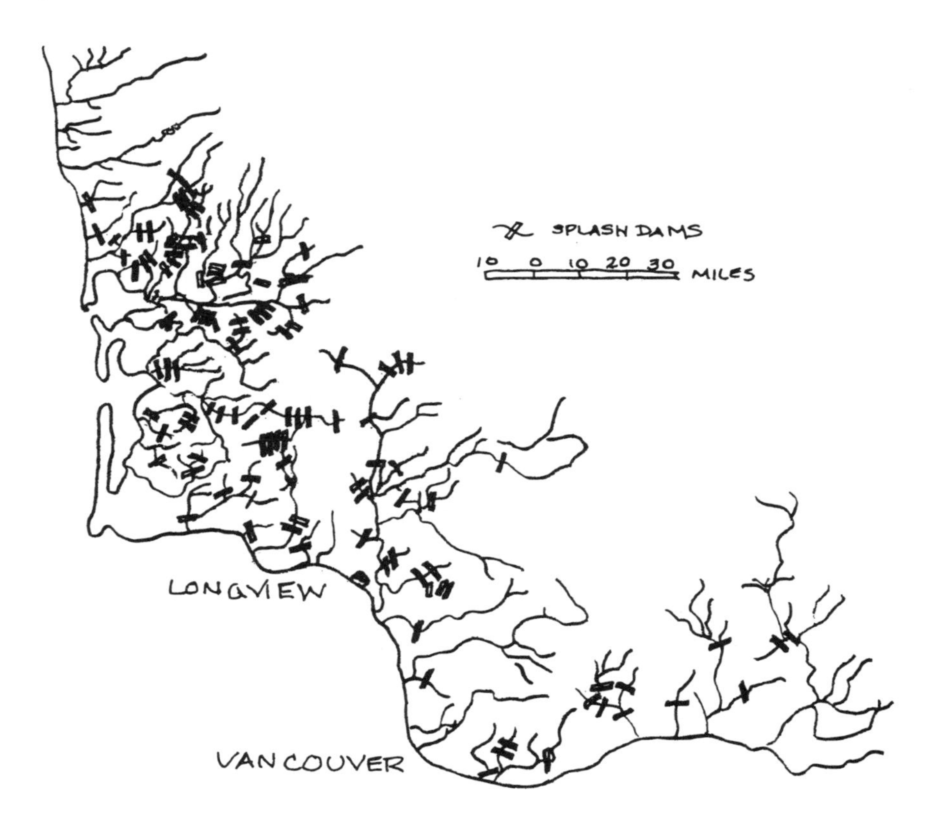

FIGURE 3-6 Splash dams in western Washington rivers in 1880-1910. Source: Sedell and Luchessa 1982.

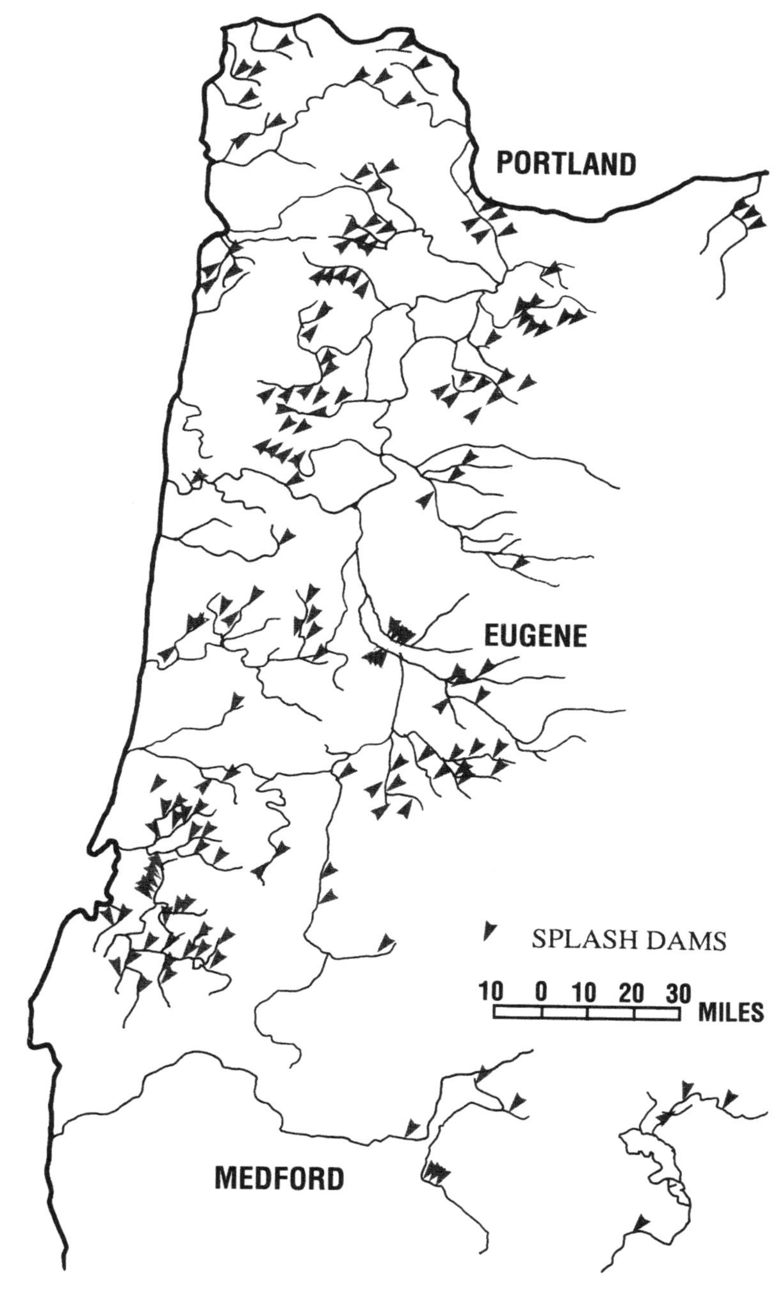

FIGURE 3-7 Splash dams in western Oregon rivers in 1880-1910. Source: Sedell and Luchessa 1982.

because of steep terrain or distance became accessible for commercial logging. Extensive networks of roads were constructed throughout mountainous terrain to provide access. According to The Federal Ecosystem Management Assessment Team (FEMAT) (1993), federal lands in western Oregon and Washington average between 3 and 4 miles of road per square mile of watershed area; road densities of over 12 miles per square mile occur in some commercial forests. The frequency of landslides associated with logging roads constructed in steep terrain during the 1940s, 1950s, and 1960s was up to several hundred times greater than that in unroaded terrain (Swanston and Swanson 1976, Sidle et al. 1985). Even with improvements in road location, design, construction, and maintenance the potential for mass failures, accelerated sedimentation, and resulting effects on water quality and fish habitat remains an important concern.

Forest-harvesting practices used into the early 1970s generally produced large volumes of logging slash and woody debris in many northwestern streams. However, during the 1964-1965 winter, major floods throughout the Pacific Northwest, in conjunction with numerous road and hillslope mass failures, caused extensive damage to streams and riparian zones. Large debris jams were common. Fishery biologists largely viewed the accumulations of large woody debris as barriers to fish migration or as material that might scour channels during later large storms (Sedell and Luchessa 1982). Hence, the removal and salvage logging of woody debris accumulations from streams were encouraged and required. Since the early 1980s, the practice has largely been curtailed.

DAMMING THE NORTHWEST

Of the various human-caused changes in the region, particularly the Columbia River Basin, perhaps none has had greater impact than dams. The potential for dams to affect salmon runs was recognized early in the Pacific Northwest's development. The constitution of the Oregon Territory, drafted in 1848, prohibited dams on any river or stream in which salmon were found, unless the dam were constructed to allow salmon to pass freely upstream and downstream (Stahlberg 1993). At the turn of the century, R.D. Hume (1908) warned of the impacts of dams:

> Hundreds of years ago men of my name were resisting the maintenance of dams and other obstructions in the river Tweed, Berwickshire, Scotland, that prevented the passage of salmon to the spawning grounds, and the lapse of centuries finds me opposing like structures on the Rogue (p. 25).

Dam construction began during the late 1800s when hydroelectric facilities were built on some of the larger Columbia River tributaries, such as the Willamette and Spokane. Dam construction proceeded relatively slowly into the early 1900s but increased thereafter. The congressional authorization of Grand Coulee and Bonneville dams during the early 1930s signaled the start of a period

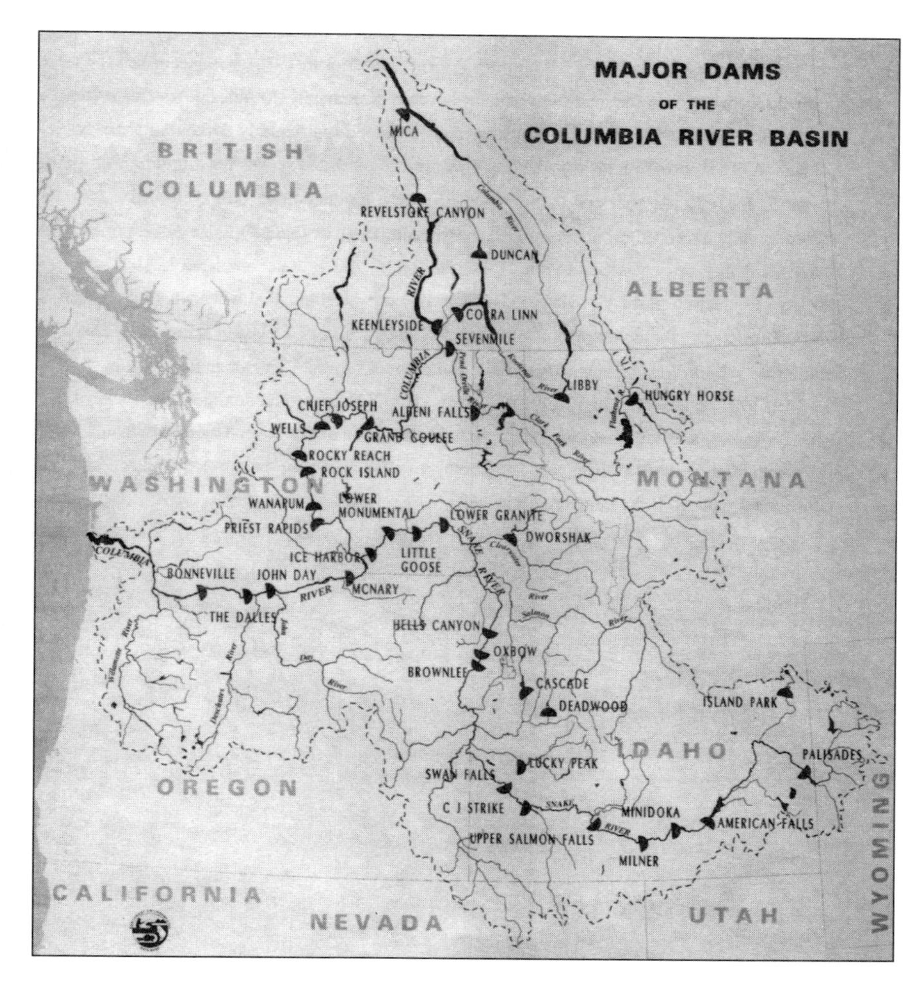

FIGURE 3-8 Location of major dams on the Columbia River and tributaries. Source: NMFS 1995.

of intense dam construction. The emphasis during this period was on "taming the Columbia" through a series of major dams that provided, in addition to hydropower, benefits to navigation, flood control, and irrigation. By 1975, 14 mainstem Columbia River and 13 Snake River dams were completed within the natural range of anadromous fish runs (Figure 3-8, Table 3-1). Within the Columbia basin, 58 dams were constructed exclusively for hydropower and another 78 are classified as multipurpose (NPPC 1986). The Pacific Northwest currently de-

TABLE 3-1 Characteristics of U.S. Mainstem Columbia River and Snake River Dams That Affect Anadromous Fish

Dam	Year Started	Year Completed	Distance from Ocean (km)
Columbia River			
Bonneville	1933	1938	233
The Dalles	1952	1957	307
John Day	1958	1968	348
McNary	1947	1954	470
Priest Rapids	1956	1959	639
Wanapum	1959	1963	668
Rock Island	1930	1933	729
Rocky Reach	1956	1961	763
Wells	1963	1967	832
Chief Joseph[a]	1950	1955	877
Grand Coulee[a]	1934	1941	961
Snake River			
Ice Harbor	1957	1961	538
Lower Monumental	1962	1969	589
Little Goose	1963	1970	636
Lower Granite	1965	1973	716
Hells Canyon[a]	1961	1967	919
Oxbow	1958	1961	961
Brownlee	1955	1958	980
Swan Falls	1906	1910	1,255
C.J. Strike	1950	1952	1,313
Bliss	1948	1949	1,423
Lower Salmon	1910	1910	1,444

[a]Blocks anadromous-fish migration.

Source: Pacific Northwest River Basins Commission 1971.

pends on hydropower for about 90% of its electrical energy (Jackson and Kimerling 1993).

In addition to the major dams, many smaller projects throughout the region provide water for municipal, industrial, irrigation, livestock, and rural uses (Table 3-2). Unlisted dams that are too small to require state or federal safety inspections are also numerous. The Oregon Water Resources Department estimates that additional thousands of smaller dams are not included in the state inventory (B. Norris, Oregon Water Resources Department, pers. comm., 1993). Similarly, other states have numerous unlisted small dams. Most of the small dams have no fish-passage facilities; the extent to which they impede anadromous-salmon migration or affect their spawning and rearing habitats has not been documented.

Trends in the number of dams constructed over time (Figure 3-9) and im-

TABLE 3-2 Dams in Pacific Northwest That Meet Minimum Federal Criteria for Inventory and Inspection

State	Minimum Height (ft)	Minimum Storage (acre-ft)	Number of Dams
California[a]	25	50	674
Idaho	10	50	523
Oregon	10	10	905
Washington	10	10	842

[a]Includes only Amador, Alpine, Sacramento, and Solano counties and counties to the north.
NOTE: Dam inventory data for each state were obtained in different forms. Idaho Department of Water Resources and Oregon Water Resources Department provided dam inventory data in digital format taken directly from their own inventory databases. Inventory information for dams in Washington and northern California, however, was obtained from reports published by regulatory agencies; data were then transferred into digital format by hand.

Quality of data varies. Data from Idaho and Oregon were of poorer quality, with a higher proportion of missing records. Data from Washington and California had been published, so they are of much higher quality, with few missing records. Where there was no information regarding date of construction or normal storage capacity, the record was excluded from analysis. For Oregon and Idaho, this procedure substantially reduced number of dams included in analysis. For example, the database for Oregon contained 3,635 dams, but date of construction or normal storage capacity was known for only 905. In Idaho, 602 dams were included in database, but only 523 were used, because of similar gaps in information.

Source: Department of Water Resources 1988 and Washington State Department of Ecology 1981.

pounded water volumes (Figure 3-10) indicate that many streams and rivers have experienced a rapid and massive change in their hydrology. Even though the rate of increase in storage volume has leveled since the mid-1970s, the total number of dams continues to increase; this suggests that new construction is focused on smaller dams.

Before any water-resources development, over 163,000 mi^2, including several thousand square miles in Canada, of the total 260,000-mi^2 Columbia River Basin was accessible to salmon and steelhead (Figure 3-11). Netboy (1986) estimated that by 1947 "about 40% of the original spawning areas had been lost" because of blockages due to dams. Today, only 73,000 mi^2 of the original area is open to anadromous fish with 7,600 linear miles of stream habitat above and 2,500 linear miles below Bonneville Dam; access to all Canadian habitats has been eliminated by dams. Of the original salmon and steelhead habitat available in the Columbia River Basin, 55% of the area and 31% of the stream miles have been eliminated by dam construction. Habitat availability for salmon and steelhead before 1850 and in 1975 is shown in Table 3-3.

Total blockage of upstream spawning and rearing areas for large portions of

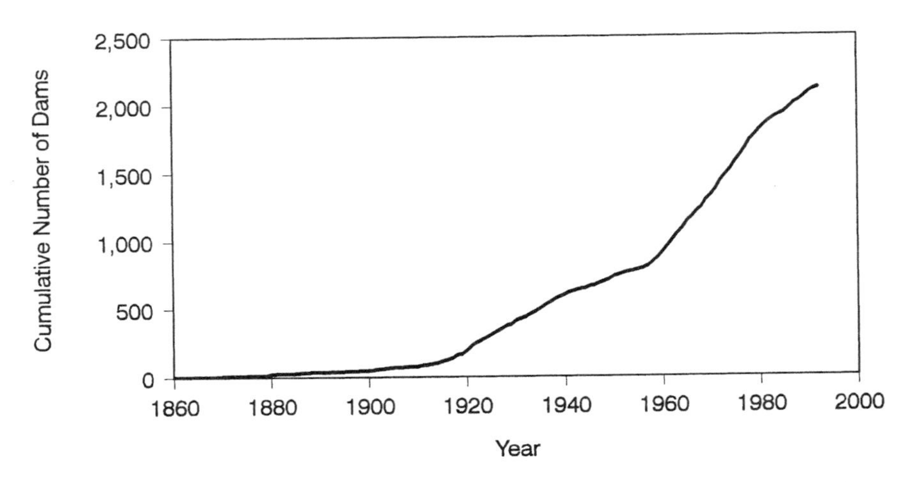

FIGURE 3-9 Cumulative number of federal and nonfederal dams in the Pacific Northwest (Idaho, Oregon, Washington, and northern California) from 1860 to 1990. (Data from individual state water-resources agencies; minimum size of dams varies by state—see Table 3-2.)

the upper Columbia River became a reality with construction of the Grand Coulee Dam (1941) and Chief Joseph Dam (1950) on the mainstem Columbia River. The Hells Canyon Dam (1961) had an equivalent effect on salmon stocks that formerly spawned and reared along some portions of the Snake River and several of its tributaries. Similarly, many dams associated with tributaries of the Columbia River or with coastal streams do not provide for the migration of anadromous fish. In addition, instream barriers prohibiting the upstream migration of adult salmon were built during the construction of many Pacific Northwest fish hatcheries; although these barriers were apparently installed because of concerns about disease in hatchery fish, the structures delineate watershed areas that have become off limits to anadromous fish. Dams of various sizes and functions provide important benefits to human populations and industries, but their ability to eliminate habitat access constitutes a major contribution to the decline of salmon runs in the Pacific Northwest.

Before dam construction, the mainstem Columbia River was an important spawning area for anadromous fish (NPPC 1986). Aerial surveys conducted in 1946, after construction of Grand Coulee had already blocked upriver reaches of the Columbia, showed that chinook salmon used gravels throughout a 210-mi reach between the confluences of the Snake and Okanogan rivers with the Columbia. The only spawning habitat remaining after dam construction in this reach is a 50-mi portion between the McNary Reservoir and the Priest Rapids Dam. One of the most productive populations in the Columbia system is the

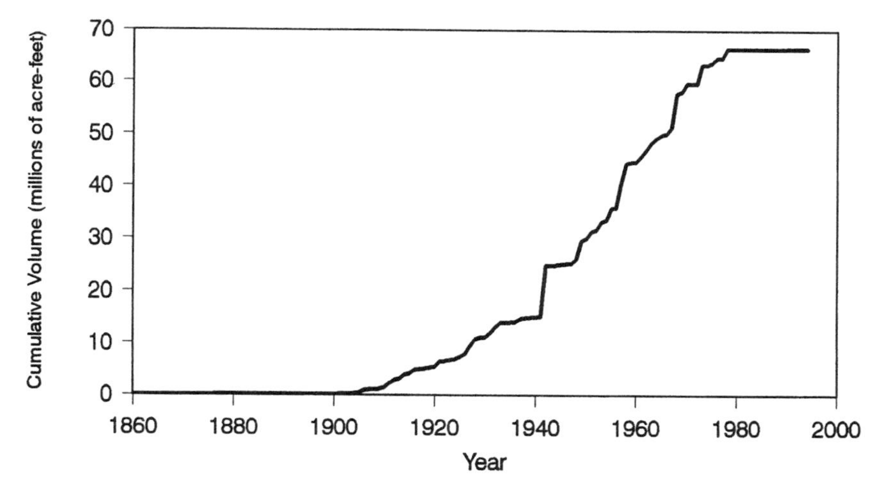

FIGURE 3-10 Cumulative volume of water impounded by federal and nonfederal dams in Pacific Northwest (Idaho, Oregon, Washington, and northern California) from 1860 to 1990. (Data from individual state water-resources agencies; minimum size of dams varies by state—see Table 3-2.)

Hanford Reach chinook, which spawns in the only free-flowing stretch left in the mainstem.

The John Day and McNary pools inundate about 137 mi of river and numerous spawning areas. Before construction of the Chief Joseph Dam in 1950, the Grand Coulee Dam inundated a 103-mi stretch of river that once supported great numbers of chinook salmon that spawned on gravel bars in the main river and near the mouths of tributaries; it also eliminated access to other upriver areas in the United States and Canada (NPPC 1986).

Fish-passage facilities have been installed at many of the mainstem Columbia River dams and other dams in the Pacific Northwest. These, however, can result in delays in upstream migration, increased stress, prespawning mortality, and reduction in success of late spawners. In 1970, mortalities of 13% for migrating adult chinook salmon were reported for Bonneville Dam. In the same year, adult mortalities of 12-25% were reported for the Dalles Dam (NPPC 1986). More recent work indicated average per-project losses of 5% or less (Pratt and Chapman 1989, Stuehrenberg et al. 1994).

In addition, smolts migrating downstream must negotiate reservoirs and the physical barrier of a dam, where they will pass over a spillway, be routed through a bypass facility, or be drawn through the turbines. Although controversy surrounds the question of how many smolts pass through a spillway, bypass facility, or generating system and what mortalities are associated with a given dam during a given year for a given species of salmon, there is a consensus that migration

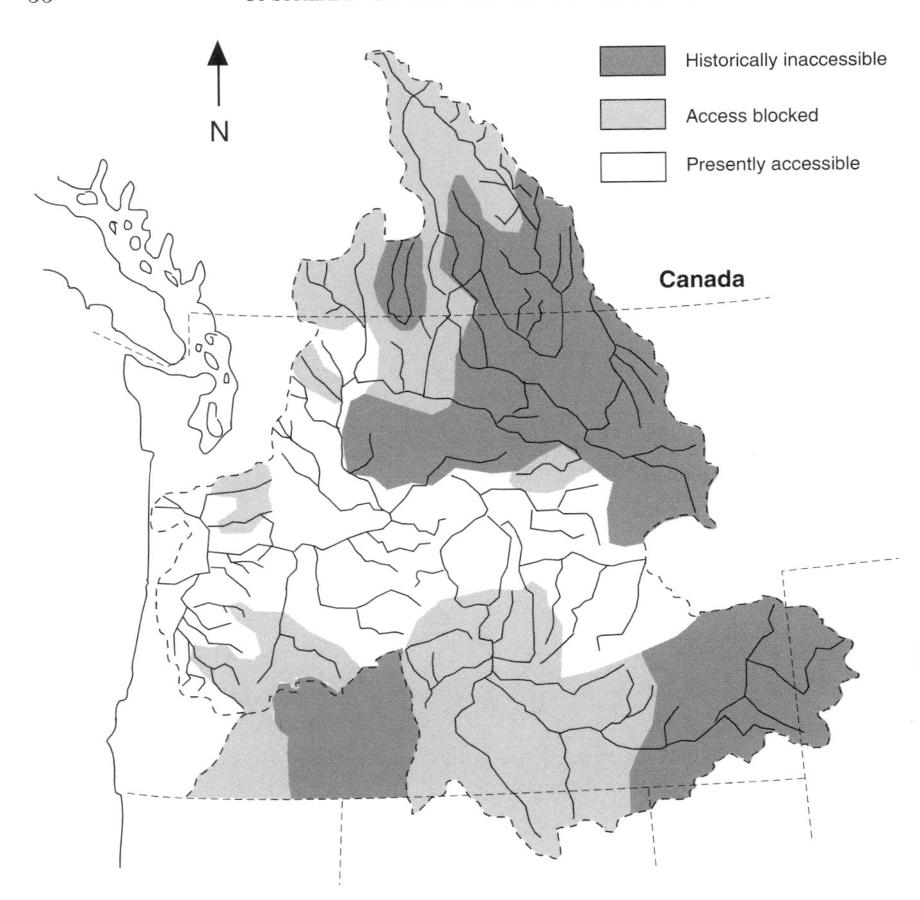

FIGURE 3-11 Columbia River Basin and Oregon Closed Basin, showing areas that were historically accessible to anadromous salmon and areas that have become inaccessible because of dam construction. Source: Kaczynski and Palmisano 1993:310.

hazards (e.g., time of travel, predators, turbine passage) associated with mainstem dams are a leading factor in the mortality of smolts as they migrate downriver (Table 3-4).

WATERING THE LAND

As rainwater or snowmelt flows to lower elevations, it is used for hydropower, irrigation, industrial, and municipal demands. Not surprisingly, of all the water withdrawn from lakes, streams, and rivers, irrigation uses the most (Figure 3-12). The demand for irrigation water is particularly great for the mid-Columbia

TABLE 3-3 Salmon and Steelhead Habitat[a] in Columbia River Basin Before Water Development and in 1975

River Location	Habitat Available (mi of stream)		Change %
	Pre-1850	1975	
Columbia River below Bonneville Dam			
Spring chinook	1,835	1,191	-35
Summer chinook	0	0	0
Fall chinook	861	1,047	+22[b]
Coho	1,319	2,124	+61[b]
Steelhead	2,410	2,378	-1
Columbia River between Bonneville Dam and its confluence with Snake River			
Spring chinook	1,218	655	-46
Summer chinook	0	148	[c]
Fall chinook	70	201	[c]
Coho	231	344	[c]
Steelhead	1,834	1,479	-19
Columbia River above its Confluence with Snake River			
Spring chinook	1,801	758	-58
Summer chinook	909	286	-69
Fall chinook	485	115	-76
Coho	523	361	-31
Steelhead	1,485	938	-37
Snake River below Hells Canyon Dam			
Spring chinook	3,899	2,813	-28
Summer chinook	2,198	1,834	-17
Fall chinook	674	345	-49
Coho	481	379	-21
Steelhead	5,156	4,120	-20
Snake River above Hells Canyon Dam			
Spring chinook	1,865	0	-100
Summer chinook	1,865	0	-100
Fall chinook	371	0	-100
Coho	0	0	0
Steelhead	2,050	0	-100

[a]Habitat refers to natural spawning and rearing areas.
[b]Fishway at Willamette Falls constructed in 1971 increased habitat in the Willamette River Basin.
[c]Reason for increase in fish habitat not identified in original report.

Source: Northwest Power Planning Council 1986.

River Basin (eastern Oregon and Washington) and the Snake River drainage (southern Idaho), where well over 90% of the surface-water withdrawal in most areas is for irrigation (Jackson and Kimerling 1993). Overappropriation is common in western basins.

The history of irrigation in the Pacific Northwest dates back to early settle-

TABLE 3-4 Hypothetical Example of Potential Cumulative Mortality in Juvenile or Adult Salmon Migration in Relation to Number of Dams Requiring Passage[a]

Passage Mortality for Individual Dams (%)	Cumulative Mortality for Number of Dams Requiring Passage								
	1	2	3	4	5	6	7	8	9
5	5	10	14	19	23	26	30	34	37
10	10	19	27	34	41	47	52	57	61
15	15	28	39	48	56	62	68	73	77
20	20	36	49	59	67	74	79	83	86
25	25	44	58	68	76	82	86	89	92
30	30	51	66	76	83	88	92	94	96

[a]Mortality numbers for individual dams vary.

Source: Committee generated.

ment. In 1840, missions near Walla Walla, Washington, and Lewiston, Idaho, were the first sites to use irrigation for crop production. In 1859, the first irrigation project began in the Walla Walla River Valley; it was followed soon by projects in the John Day, Umatilla, and Hood River valleys of Oregon. Near the turn of the century, the Klamath irrigation project was begun in southern Oregon and northern California. An increasing demand for agricultural products in combination with the expansion of railroads throughout the region attracted commercial-scale farmers during the late 1800s and early 1900s (NPPC 1986). Federal legislation that encouraged the establishment of irrigation on newly acquired lands included the 1877 Desert Land Act and the 1894 Carey Act (Johansen and Gates 1967).

In many areas, irrigation techniques evolved from simple stream diversions to complex systems that used a variety of pumping and application mechanisms, such as sprinklers, storage reservoirs, groundwater pumps, and pressure-distribution devices. Technological advances in irrigation after World War II made it economically possible to cultivate lands that had previously been only marginally productive. The rapid increase in irrigated land during the 1900s (Figure 3-13) was due largely to an increase in federal multipurpose-reservoir projects as a result of the Reclamation Act of 1902 (NPPC 1986).

The areal extent of irrigated lands provides an indication of the volume of water withdrawn for irrigation, but it does not reflect changes in irrigation practices that result in more efficient use of water. Although annual water withdrawal has remained relatively constant over nearly 2 decades for Bureau of Reclamation projects (Figure 3-14), it is about 3-4 times that of the early 1950s. In 1990, total surface-water withdrawal for irrigation in the Pacific Northwest was about

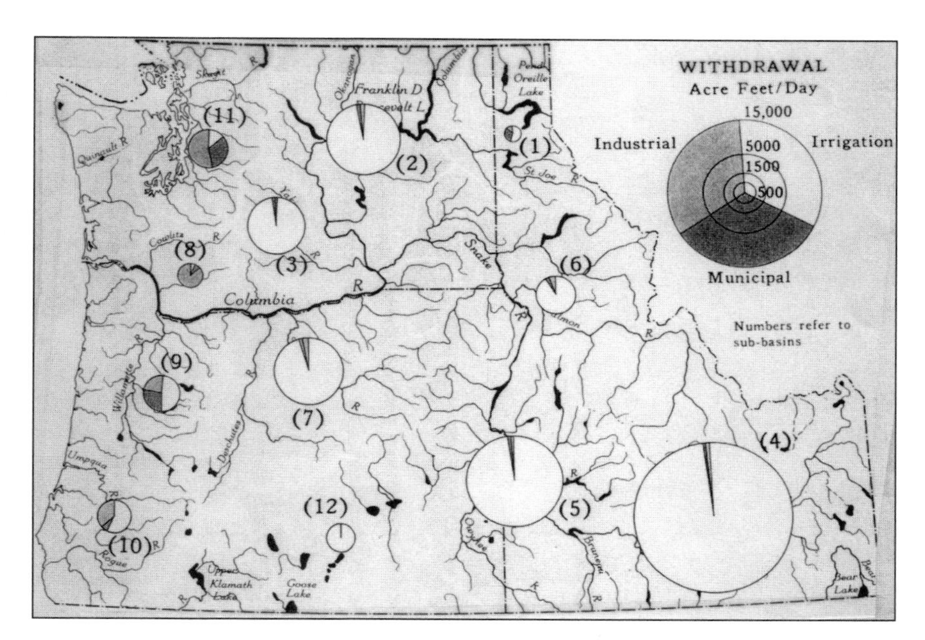

FIGURE 3-12 Surface-water withdrawal in subbasins of Pacific Northwest for irrigation, municipal, and industrial purposes. 1, Spokane; 2, Upper Columbia; 3, Yakama; 4, Upper Snake; 5, Central Snake; 6, Lower Snake; 7, Mid-Columbia; 8, Lower Columbia; 9, Willamette; 10, Coastal; 11, Puget Sound; 12, Oregon Closed. Source: Kimerling and Jackson 1985:77.

7% of the annual flow of the Columbia River at its mouth (Solley et al 1993). Pacific Northwest surface-water withdrawal of 27 million acre-feet in 1990 was the highest annual water use of any region in the United States—even higher than that in California.

Solley et al. (1993) estimated that in 1990 only one-third of the water withdrawn in the Pacific Northwest was returned to the streams and lakes. Water that returns to a stream from an irrigation project is often substantially altered and degraded (NRC 1989). Problems associated with return flows include increased water temperature, which can alter patterns of adult and smolt migration; increased salinity; increased pathogen populations; decreased dissolved oxygen concentration; increased toxicant concentrations associated with pesticides and fertilizers; and increased sedimentation (NPPC 1986). Water-level fluctuations and flow alterations due to water storage and withdrawal can affect substrate availability and quality, temperature, and other habitat requirements of salmon. Indirect effects include reduction of food sources; loss of spawning, rearing, and adult habitat; increased susceptibility of juveniles to predation; delay in adult spawning migration; increased egg and alevin mortalities; stranding of fry; and

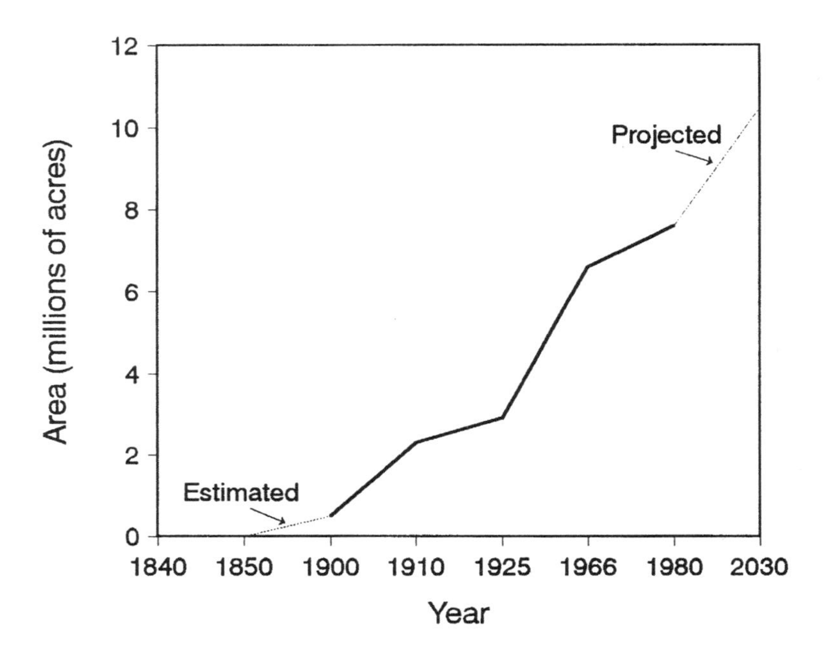

FIGURE 3-13 Area of irrigated lands in Columbia River Basin. Source: NPPC 1986.

delays in downstream migration of smolts (NPPC 1986). In some instances, irrigation withdrawal can result in the total dewatering of a stream and concurrent desiccation of aquatic habitats. In other situations, annually constructed instream diversion dams can block adults migrating upstream, prevent the redistribution of rearing juveniles within the stream system, or cause juveniles to enter the irrigation system.

The loss of juvenile salmon to irrigation intake systems has contributed to fish declines. Of over 55,000 water diversions in Oregon, fewer than 1,000 have protective fish screens; an additional 3,240 were recently identified as having a high priority for screening (Nichols 1990). In the summer of 1994, more than 80% of pumping sites taking water from the Columbia River on the Oregon shore failed to comply with requirements to protect migrating salmon (Roberta Ulrich, June 14, 1994, The Oregonian, Portland). However, the extent of fisheries losses resulting from the impingement of juvenile salmon on intake screens is essentially unknown.

ALTERING WETLANDS AND ESTUARIES

Since colonial times, wetlands in the United States have been considered a hindrance to productive land use. Swamps, bogs, marshes, and other wet areas

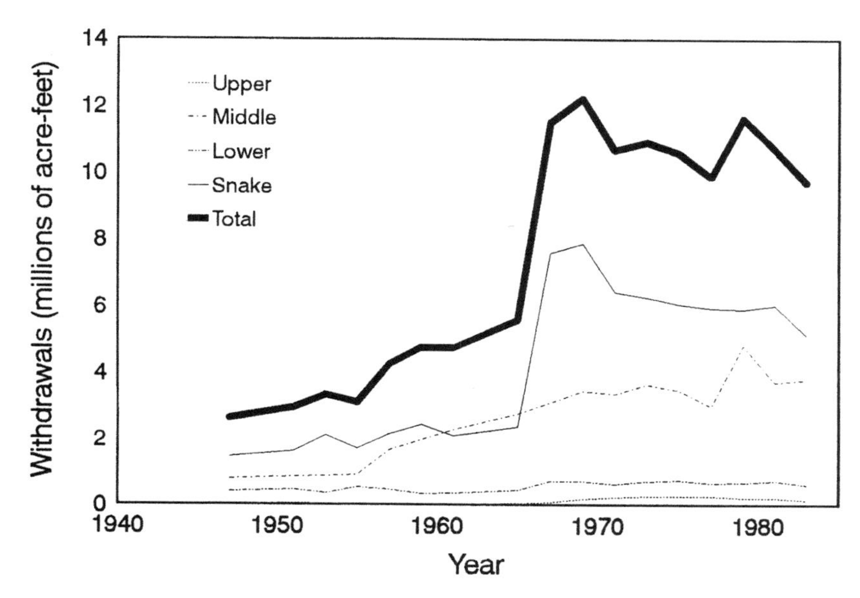

FIGURE 3-14 Surface-water withdrawal by Bureau of Reclamation for irrigation in Columbia River basin. Upper Columbia, above Chief Joseph Dam; Mid-Columbia, between confluence of Snake River and Chief Joseph Dam; Lower Columbia, below confluence with Snake River; Snake, Snake River basin. Source: NPPC 1986.

used to be considered wastelands to be drained, ditched, filled, or manipulated for other purposes (NRC 1995a). For the Pacific Northwest, wetland losses have been severe, with California experiencing the highest percentage loss of wetlands of any state. Of the estimated 5 million acres of wetlands that existed in California in 1780, only 454,000 acres remained in 1980, a loss of 91%. The amount of wetland area in Idaho declined from 877,000 to 385,700 acres, a reduction of 56% over the 200-year period from 1780 to 1980; in Oregon, from 2.26 to 1.39 million acres, a 38% loss; and in Washington, from 1.35 million to 938,000 acres, a 31% loss (Dahl 1990).

Agricultural land conversion and urban development—the same land uses responsible for most of the freshwater wetland losses—are the primary causes of estuarine wetland losses. Although most estuarine wetland losses result from conversions to agricultural land by ditching, draining, or diking, these wetlands are also experiencing increasing effects from industrial and urban causes. For example, historical changes in the lowlands of Humboldt Bay, California, indicate increasing encroachment of residential, commercial, and industrial land uses (Figure 3-15). Coastal salt marshes close to seaports and population centers in the Pacific Northwest have been especially vulnerable to conversion, with losses of 50-90% common for individual estuaries in Oregon and Washington. In the Puget Sound area, urbanization has caused even greater disruptions and conver-

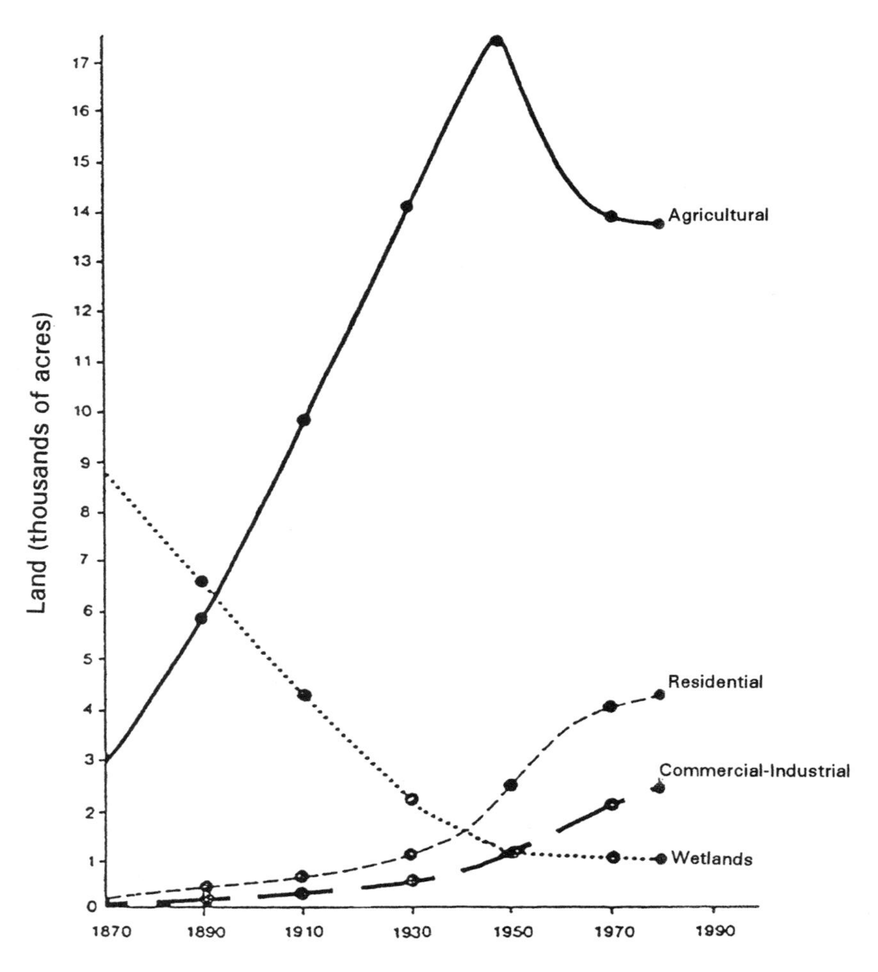

FIGURE 3-15 Historical changes in land use in lowlands surrounding Humboldt Bay, California. Source: Shapiro and Associates 1980, as reported in Boule and Bierly 1987.

sions of many estuarine wetlands. Although 24% of the Columbia River estuary had been converted from wetland habitat type between 1870 and 1983, tidal swamps and marshes together lost some 65% of their former area because of diking and filling (Thomas 1983). For the Salmon River estuary of Oregon, 75% of the wetlands had been isolated from the rest of the estuary by dikes (Frenkel and Morlan 1991). For salmon that prehistorically used freshwater and estuarine wetlands for rearing habitat, the conversions and losses of Pacific Northwest wetlands constitute a major impact.

SUMMARY AND CONCLUSIONS

The historical account of human development and natural resource use in the Pacific Northwest clearly illustrates formidable disruptions to the life cycles of anadromous salmon. The ecological fabric that once sustained enormous salmon populations has been dramatically modified through heavy human exploitation—trapping, fishing, grazing, logging, mining, damming of rivers, channelization of streams, ditching and draining of wetlands, withdrawals of water for irrigation, conversions of estuaries, modification of riparian systems and instream habitats, alterations to water quality and flow regimes, urbanization, and other effects. In many places throughout the landscape, human exploitation has lowered the productive capability of habitat, harvested animals and plants at an unsustainably intense level, or eliminated populations or habitat outright.

This characterization of the scope and magnitude of human impacts on the habitat and population of salmon in particular is robust even though the availability, completeness, and quality of historical records are neither consistent nor complete across the region. Only a few research efforts have attempted to understand some of the ecological consequences of past human activities for aquatic resources and fisheries; the studies that have been done are often of limited geographic coverage. Although various land uses and alterations were generally considered on an individual basis, their capability, in combination, to alter aquatic habitat or affect fisheries often involves complex interactions with other land uses and with natural disturbance regimes that operate across the region. Notwithstanding these limitations in our understanding and documentation, it is clear that human impacts on the land and waterways of the Pacific Northwest have basically and irreversibly altered the genetic and ecological constitution of anadromous salmon.

These human impacts have not only been widespread, but they have also been rapid by biological time scales. They should be expected to continue in the future, unless the momentum of human exploitation and transformation of the land and waters changes drastically.

The exponentially increasing human population and economic development of the region is likely to produce higher levels of resource impacts. In 1800, the collective population of a region bounded by the present-day states of Idaho, Oregon, and Washington stood at approximately 0.1 million, yet significant inroads into the habitats and populations of Pacific salmon had already begun. One hundred years later, the regional human population had climbed to 1 million with fishing, grazing, agriculture, and forest harvesting under way. At the end of the current century, the region's population will approach 10 million people (Figure 3-1).

From 1940 to 1990, census data indicate a regional (Idaho, Oregon, and Washington) population growth of 1.9% annually (Statistical Abstracts, USDC

Bureau of the Census). To the degree that the regional population growth continues beyond the year 2000, the expected population of 10 million people will continue to expand. If the population continued to grow at the rate it has in the past half-century, for example 1.9% per year, the population in 2100 would be over 65 million people.

The challenge is evident. Current and future efforts to save natural runs of salmon by reducing per capita impacts through conservation measures, improved land use practices, reduced hatchery competition, improved dam passage, better riparian protection, etc., could all be undermined by continued regional population and economic growth.

The salmon problem includes far more than simply numbers of people or their standard of living. From the perspective of a geologic clock that has registered natural disturbance patterns and transitions of ecosystems during the last 10,000 years, salmon thrived and sustained a wide distribution throughout the Pacific Northwest. However, the pace of time for salmon populations, as viewed through the kaleidoscope of ecological alterations and infractions, has been accelerating. Human disturbances of ecosystems that were once complex and productive may ultimately exceed the ability of salmon to adapt and maintain their populations. (In the Columbia River Basin alone, more than half of the basin that was originally accessible to salmon is no longer.) More important, the extent of environmental changes may exceed our understanding of what once existed, our ability to correct past mistakes, or our willingness to even try.

In sum, the salmon problem is about more than just a few species of fish. It is a question of cultural values, stewardship, and living with the land instead of off the land. In the 1970s, a Pacific Northwest River Basins Commission report (1972) expressed concern that

> projections of continued growth in population and economic activity in the Northwest would eventually lead to a major deterioration of the present high quality environment. . . . [L]and, energy, and air resource planning was lagging far behind water resource planning. [The committee presented] a view of the future of the Northwest based upon attainable balances between ecology and economics as an alternative to traditional projections of economic growth alone.

Thinking of the priorities of environmental protection and a market economy as competing with each other is counterproductive; they are probably in agreement more often than many people believe, especially over the longer term. In many respects, a sound economy depends on ecosystem functioning (Ashworth 1995, NRC 1995b).

If the Pacific Northwest is going to have more people, which may be inevitable, history has indicated that those increases will test our ability to sustain all that the fish need. Should we accept the challenge of trying to sustain native stocks of salmon across streams and ecoregions of the Pacific Northwest, that challenge is likely to test fundamentally our institutions, our views of resource use and economic development, and our social patterns and cultural beliefs.

4

Status of Salmon

The status of many specific salmon populations in the Pacific Northwest is uncertain, and exceptions exist to generalizations with regard to overall status. Nevertheless, a general examination of the evidence of population declines over broad areas is helpful for understanding the current status of species with different life cycle characteristics and geographical distributions.

Using a geographic information system based on state agencies' status reports from the American Fisheries Society and other sources, the Wilderness Society (1993) estimated current extinction risks of Pacific Northwest salmon throughout their historical ranges (Table 4-1). Classification of stocks followed Nehlsen et al (1991); "endangered" was equivalent to "high risk of extinction," "threatened" was equivalent to "moderate risk of extinction," and "of special concern" referred to populations not currently at risk, but not as secure as "healthy," for various known reasons, or for which there was incomplete information but the suggestion of depletion. Populations not known to be recently declining, often classified as "healthy" in state assessments (Nickelson et al. 1992, WDF et al. 1993), were also included. Percentages in Table 4-1 should be viewed as provisional, in that they were based on population assessments that were in many instances uncertain; however, they provide a picture of the relative status of different salmon species and runs. The following generalizations can be made from this information:

• *Pacific salmon have disappeared from about 40% of their historical breeding ranges in Washington, Oregon, Idaho, and California over the last century, and many remaining populations are severely depressed in areas where they were formerly abundant.* If the areas in which salmon are threatened or

TABLE 4-1 Current Status of Pacific Salmon over Their Known Historical Geographic Range in Washington, Oregon, Idaho, and California

	Status, % of Historical Range				
Species (Run)	Extinct	Endangered	Threatened	Special Concern	Not Known To Be Declining
Fall chinook	19	18	7	36	20
Spring/summer chinook	63	8	16	7	6
Coho	55	13	20	5	7
Chum	37	16	14	11	22
Sockeye	59	7	3	16	15
Pink	21	5	<1	0	73
Sea-run cutthroat	6	4	61	29	0
Winter steelhead	29	22	7	18	24
Summer steelhead	45	5	5	27	18
Overall	40	13	14	17	16

Source: Data from The Wilderness Society 1993.

endangered are added to the areas where they are now extinct, the total area with losses is two-thirds of their previous range in the four states. Although the overall situation is not as serious in southwestern British Columbia, some populations there also are in a state of decline, and all populations have been completely cut off from access to the upper Columbia River in eastern British Columbia. Even if the estimate of population losses of about 40% is only a rough approximation, the status of naturally spawning salmon populations gives cause for pessimism.

• *Coastal populations tend to be somewhat better off than populations inhabiting interior drainages.* Species with populations that occurred in inland subbasins of very large river systems (such as the Sacramento, Klamath, and Columbia rivers)—spring/summer chinook, summer steelhead, and sockeye—are extinct over a greater percentage of their range than species limited primarily to coastal rivers. Salmon whose populations are stable over the greatest percentage of their range (fall chinook, chum, pink, and winter steelhead) chiefly inhabit rivers and streams in coastal zones.

• *Populations near the southern boundary of species' ranges tend to be at greater risk than northern populations.* In general, proportionately fewer healthy populations exist in California and Oregon than in Washington and British Columbia. The reasons for this trend are complex and appear to be related to both ocean conditions and human activities.

• *Species with extended freshwater rearing (up to a year)—such as spring/summer chinook, coho, sockeye, sea-run cutthroat, and steelhead—are generally extinct, endangered, or threatened over a greater percentage of their ranges than species with abbreviated freshwater residence, such as fall chinook, chum, and pink salmon.*

• *In many cases, populations that have not declined are composed largely or entirely of hatchery fish.* An overall estimate of the proportion of hatchery fish is not available, but several regional estimates make clear that many runs depend mainly or entirely on hatcheries.

This committee assessed the current status of Pacific salmon from the Fraser River Basin in southern British Columbia to Monterey Bay in central California. This portion of the range includes nearly all salmon populations caught by fishers in Washington, Oregon, Idaho, and California. Historical records of population abundance vary greatly across this region; some river basins have extensive catch and escapement records extending back nearly a century; run sizes for other river systems are poorly known. Historical data on commercially caught species tend to be much more complete than those on species not heavily fished or species caught only by recreational angling (e.g., sea-run cutthroat trout).

The Endangered Species Committee of the American Fisheries Society (Nehlsen et al. 1991) identified 214 salmon stocks as being at risk of extinction and over 100 populations as being recently extinct in Washington, Oregon, Idaho, and California. The report was based on admittedly incomplete data, but it pointed out the widespread nature of salmon declines and the seriousness of the current problem. Although several populations already had been petitioned for protection under the Endangered Species Act (ESA) when the report was published, public interest in the "salmon problem" was heightened greatly by its appearance, and various state and federal agencies rapidly began to develop management plans to address the conservation needs of potentially listed populations. Nehlsen et al. (1991) also catalyzed efforts to assess further the current condition and causes of declines of salmon.

The remainder of this chapter discusses some of the difficulties in evaluating the status of wild populations and how these problems have been addressed in recently published status reports. It then summarizes regional trends in salmon populations in the Fraser River Basin, Puget Sound, the Columbia River Basin, and coastal river basins of Washington, Oregon, and California. Finally, general patterns in the overall condition of the species are presented on the basis of their geographical distribution and life-cycle requirements.

INTERPRETING HISTORICAL RECORDS

It is tempting to conclude that all salmon populations are depressed, but the picture is complicated. Salmon can be difficult to count; even if accurately counted, their populations are inherently variable owing to continual changes in freshwater and marine factors or to the cyclic nature of some species (e.g., pink and sockeye salmon). Long time-series records of catch or escapement, often spanning decades, are required for the statistical detection of large changes (50% or greater) in population abundance (Lichatowich and Cramer 1980). In addition to inherent variability, long-term changes in abundance might not be continuous or linear and so might not be clearly revealed with simple regression methods (Bledsoe et al. 1989). Short periods of record might suggest population increases or decreases when, in fact, long-term trends are in a different direction.

Population status can also be difficult to determine if catch statistics are the only source of information, in that these data can be misleading. One example is the often-displayed graph of salmon landings in the lower Columbia River (e.g., Kaczynski and Palmisano 1993:150). The graph ranges from nearly 50 million pounds before 1920 to less than 5 million in 1990. The authors noted that in-river fisheries were supplemented by offshore trolling in the middle to late part of this century, but casual readers might assume that the depiction of reduced catch reflects the magnitude of the decline of salmon in the Columbia River without recognizing that in-river catch has been largely supplanted by large ocean fisheries. Catch statistics tend to be biased in favor of large populations, either because fisheries tend to target them and ignore small populations with unusual run timing or because trends in catches of large populations mask simultaneous but sometimes opposite trends in smaller ones. An example is the trend in Puget Sound sockeye catches early in the twentieth century (Bledsoe et al. 1989:59). Dramatic declines in sockeye landings by Puget Sound fishers were apparent after 1913, but they were caused by rockslides on the Fraser River, in which most of the Puget Sound sockeye catch originated (Quinn 1994). Puget Sound fisheries on Fraser River sockeye masked trends of Puget Sound sockeye populations unless more refined catch data were examined.

THE STOCK CONCEPT

In the jargon of Pacific salmon fisheries, managers refer to stocks of salmon, i.e., populations or groups of salmon populations that are recognizable for management. The term *stock* has been used in various ways, but in some form the stock concept has been fundamental in understanding the population structure and management of Pacific salmon (Moulton 1939). A stock is considered the basic unit of salmon management (Moulton 1939) and has been defined as a recognizable or manageable group of animals (Larkin 1972, Waples 1991). The basic concept is that these local populations are largely reproductively isolated and over time become adapted to local conditions. In management practice, however, each of these local populations is not identifiable and stock usually refers to larger, recognizable groups of these basic population units. Salmon biologists have recognized this "fine-grained" structure of salmon populations. However, few management policies explicitly recognize the need to protect salmon at the level of local spawning populations, and most stock assessments are conducted at a much coarser geographic scale. The status reports in this chapter reflect this limitation and, in most cases, we do not know the status of the smaller local population units. In general, we use the term *stock* only when the report being cited specifically used that term.

Stocks have been identified in connection with geographical areas as small as individual tributaries (e.g., the winter steelhead of Oregon's Illinois River) and as large as entire ecoregions (e.g., sea-run cutthroat trout from Oregon coastal streams). For example, Nehlsen et al. (1991) recognized two stocks of sea-run cutthroat trout from all of Washington, excluding the Columbia River: one stock consisted of Hood Canal and Grays Harbor fish, and the other included all other populations. One reason for the inconsistency is that the available information differs greatly among species: commercial species tend to be better known than species fished only for sport, populations in rivers with dams and counting facilities tend to be better known than populations in streams without such facilities, populations close to urban centers tend to be better known than those in remote locations, and so on. Where detailed information is available on run sizes, salmon are often split into many stocks; but where information is lacking populations are often lumped together into only a few stocks. As a result, practical application of the stock concept has included several levels of genetic organization (Chapter 6 discusses this in more detail).

An additional difficulty in applying the stock concept to status assessment has arisen from the fact that many Pacific salmon demes are small. Although some demes number in the hundreds of thousands or even millions of fish in exceptional cases such as the Adams River, B.C., sockeye, others may consist of only a few adults that spawn in geographically limited areas, such as small headwater streams and portions of lake shorelines. These very small demes might be substantially isolated from other such demes and even possess local adaptations that distinguish them from other members of the species, but they are rarely considered stocks. Some small populations (e.g., Redfish Lake sockeye and Sacramento River winter chinook) have been treated as separate stocks for ESA purposes, but they are the remnants of much larger runs. In theory, the stock concept makes no allowance for deme size or metapopulation structure, in which populations consist of locally reproducing groups connected by some gene flow within a larger area. The management of stocks by state fisheries agencies has generally not recognized the geographical structure of salmon populations at such fine scales. Few management policies explicitly recognize the need to protect salmon at the level of individual demes, so most stock assessments are carried out on a much coarser geographic scale. In most cases, we do not know the status of salmon on the scale of local demes.

RISK ASSESSMENT

Beginning with Nehlsen et al. (1991), recent status reports have assigned salmon stocks to various risk categories on the basis of population trends and potential for loss of genetic integrity from introduction of maladapted genes from nonnative or artificially propagated populations. Graduated categories of risk have been used in which stocks in the highest-risk categories were considered at greatest peril of extinction. Different reports sometimes assessed the condition of the same stocks according to common data sets. However, different criteria of risk categories were often used, and that makes direct comparisons among reports difficult. The risk of extinction of stocks classified as of "special concern" in one report, for example, did not always correspond to the extinction risk of stocks with the same classification in another report. A comparison of the criteria used by Nehlsen et al. (1991), Higgins et al. (1992), Nickelson et al. (1992), WDF et al. (1993), and The Wilderness Society (1993) is given in Table 4-2.

Compilations of stock status must be read and compared carefully for several reasons (Quinn 1994). Compilations differ in subtle but important aspects of their definitions, and conclusions of the assessments can be distinctly different in areas of geographical overlap. For example, in the extreme case of stock extinctions, WDF et al. (1993) listed only one stock as recently extinct in Washington state whereas Nehlsen et al. (1991) listed 42 recent stock extinctions in Washington. Interestingly, the early chum salmon run to Chambers Creek in Puget Sound was said by Nehlsen et al. (1991) to be "at low levels . . . but appears to be rebuilding" but was classified as extinct by WDF et al. (1993). There are also major differences in how regional groups of stocks are judged. For example, WDF et al. (1993) listed 17 stocks of chum salmon in southern Puget Sound and classified 15 as healthy, one as extinct, and one of unknown status. In contrast, The Wilderness Society (1993) stated that "chum salmon are depleted or extinct in the rivers of southern Puget Sound . . .", and a petition to invoke ESA protection for chum salmon in some portions of southern Puget Sound was recently filed with the National Marine Fisheries Service.

Differences in judgment about the status of stocks are sometimes confounded by disagreement regarding the definition of *stock* or the number of populations being considered as a single unit. For example, Nehlsen et al. (1991) concluded that "Lake Washington" sockeye did not meet the criteria of risk of extinction, but WDF et al. (1993) recognized three stocks in the watershed (Cedar River spawners, Lake Washington and Lake Samammish tributary spawners, and Lake Washington beach spawners) and classified all of them as depressed.

Putting aside the differences and discrepancies among the reports, it is clear that a substantial number of wild salmon populations are in some jeopardy and that the status of many others is poorly known. We must first ask whether the loss of populations should concern us. In addition to the ESA's provision for distinct population segments to be protected, Alkire (1993) described the substantial economic value of salmon resources for both commercial (native and nonnative) and recreational fishers. They are also of great symbolic importance to native peoples and more recent settlers as well, representing clean water, forests, and the wonders of animal migration. And small populations might

TABLE 4-2 Comparison of Criteria Used to Assign Pacific Salmon Stocks to Different Status Categories

Nehlsen et al. (1991), Higgins et al. (1992), The Wilderness Society (1993):

High Risk of Extinction

- Populations' spawning escapements are declining; less than one adult fish returns from each parent spawner.
- Populations' recent escapements (within last 1-5 years) are under 200 in the absence of evidence that they were historically small.
- Population is likely candidate for listing as endangered under ESA.

Moderate Risk of Extinction

- Populations' spawning escapements appear to be stable after previously declining more than natural variation would account for and are generally in range of 200-1,000 spawning adults.
- Population is likely candidate for listing as threatened under ESA.

Special Concern

- Populations are believed to be vulnerable to minor disturbances, especially if a specific threat is known.
- Insufficient information on population trend exists, but available evidence suggests depletion.
- Continuing releases of nonnative fish are relatively large and potential for interbreeding with native population exists.
- Population is not at risk but requires attention because of unique characteristic.

Nickelson et al. (1992):

Special Concern

- Population is probably composed of less than 300 spawners, or
- Substantial risk exists for interbreeding between the population and stray hatchery fish in excess of standards established by Wild Fish Management Policy (Oregon).

Depressed[a]

- Available spawning habitat has generally not been fully seeded, or
- Abundance trends have declined over the last 20 years, or
- Abundance trends in recent years have been generally below 20-year averages.

Healthy

- Available spawning habitat has generally been fully seeded, and
- Abundance trends have remained stable or increased over last 20 years.

Unknown

- Insufficient data are available to judge population status.

WDF et al. (1993):

Critical

- Stock has declined to point where it is in jeopardy of significant loss of within-stock diversity or, in worst case, extinction.

Depressed

- Stock's production is below levels expected on basis of available habitat and natural variations in survival rates but above where permanent damage to stock is likely.

Healthy

- Stock is experiencing stable escapement, survival, and production trends and not displaying pattern of chronically low abundance.
- Stock is experiencing production levels consistent with its available habitat and within natural variations in survival for stock.
- Stocks have a wide range of conditions, from "robust to those without surplus production for harvest."

Unknown

- Information is insufficient information to determine status.
- Category includes "historically small populations [that] could be especially vulnerable to any negative impacts."

Extinct

- Stocks are known to have become extinct "during recent times."
- Only one stock (Chambers Creek summer chum) classified as recently extinct, but a number of stocks were not called extinct because of lack of agreement on whether they existed.

---

[a]This category supersedes the "special concern classification"; i.e., a population classified as depressed might also fit one of the criteria applied to the special concern category.

contain valuable genetic traits that could not be restored if the populations were lost and that could have great value to the aquaculture industry (Scudder 1989, Riddell 1993b).

Status reports have been published by agencies (Konkel and McIntyre 1987, Nickelson et al. 1992, WDF et al. 1993), scientific societies (Nehlsen et al. 1991, Higgins et al. 1992), industry organizations (Kaczynski and Palmisano 1993, Palmisano et al. 1993), and an environmental group (The Wilderness Society 1993). Although all the reports were written by fishery scientists, the committee notes that political pressures to classify salmon stocks as at risk or healthy have raised doubts among scientists as to the accuracy of the risk assessments. Stocks classified as healthy might not put fish and wildlife agencies in an unfavorable light or limit future management options; stocks classified as at risk could justify criticism of present policies or support ESA petitions. There are no simple means of verifying the accuracy of status reports. We conclude that the most prudent approach to risk assessment is to examine broad regional trends in populations and manage accordingly (and conservatively), rather than to rely on incomplete information where interpretation is open to question. Taken in total, the status reports are useful for identifying broad trends but should be viewed with caution at the level of individual populations.

FRASER RIVER BASIN

The Fraser River produces more salmon than any other river in the world (Northcote and Larkin 1989:172-204), including the seven species of anadromous salmon on which this report focuses. The Fraser River is smaller than the

Columbia but shares many biogeoclimatic features with it. Like the Columbia, the Fraser River Basin consists of wet coastal lowlands, canyon regions through the coastal mountains, and a dry, high interior plateau. The rivers have differed substantially in their development, most notably in the absence of dams on the mainstem Fraser River. Historical catches of Pacific salmon in or near the Fraser River tend, however, to show an abundance pattern similar to that observed historically in the Columbia. Because of these shared features between the Fraser and the Columbia and because so many Fraser River salmon are caught in U.S. waters, we describe it here.

Historical Fraser River catches by species are summarized in Figure 4-1 (Argue et al. 1986). In general, the early development of the fishery is evident: historically large catches during the early 1900s were followed by substantial declines in the 1920s and 1930s. That trend must be interpreted cautiously, however, in that it represents only the catch in the terminal area. For example, catches of chinook salmon in ocean troll fisheries would certainly reduce the rate of decline evident in the figure. However, it is generally true that salmon production in the Fraser River was declining throughout the basin until the middle 1970s.

The Fraser River has escaped the development of major dams on the mainstem, but it has not escaped impacts of human development. Some of the major point-source impacts occurred early in the development of British Columbia. Roos (1991) identified major impacts early in this century from gold-mining in the Quesnel drainage, extinction of the upper Adams River sockeye (Williams 1987) due to logging dams, siltation from early logging, and railway development through the Fraser canyon (Hell's Gate slide, Feb. 1914). The Hell's Gate rockslide accounts for the major loss of sockeye catch that started in 1916, and the loss of pink salmon populations in the Thompson River (Ricker 1989). Habitat impacts have, of course, accumulated with later development. For example, about 80% of the Fraser River delta wetlands have been lost to agriculture, urbanization, and flood control (Environment Canada 1986); and although the Fraser has avoided mainstem dams, more than 800 dams are licensed for agricultural water use in the upper Fraser drainage. It is difficult to assess the impact of those habitat changes on salmon production, particularly because they were simultaneous with the overfishing of the salmon populations.

During the 1980s, however, salmon production has been rebuilding for most species. Information on steelhead production is sparse, but steelhead are considered to be depressed. Recent returns of five species are summarized in Figures 4-2a to 4-2e. The sockeye, pink, and chum salmon figures are estimates of the total return of these Fraser resources (catch plus spawning escapements), but the chinook and coho figures account only for the terminal catch plus an index of spawning escapements. The indices are based on visual counts of spawners and undoubtedly underestimate the actual number of spawners. The figures for each species also include production from enhancement programs, but these are rela-

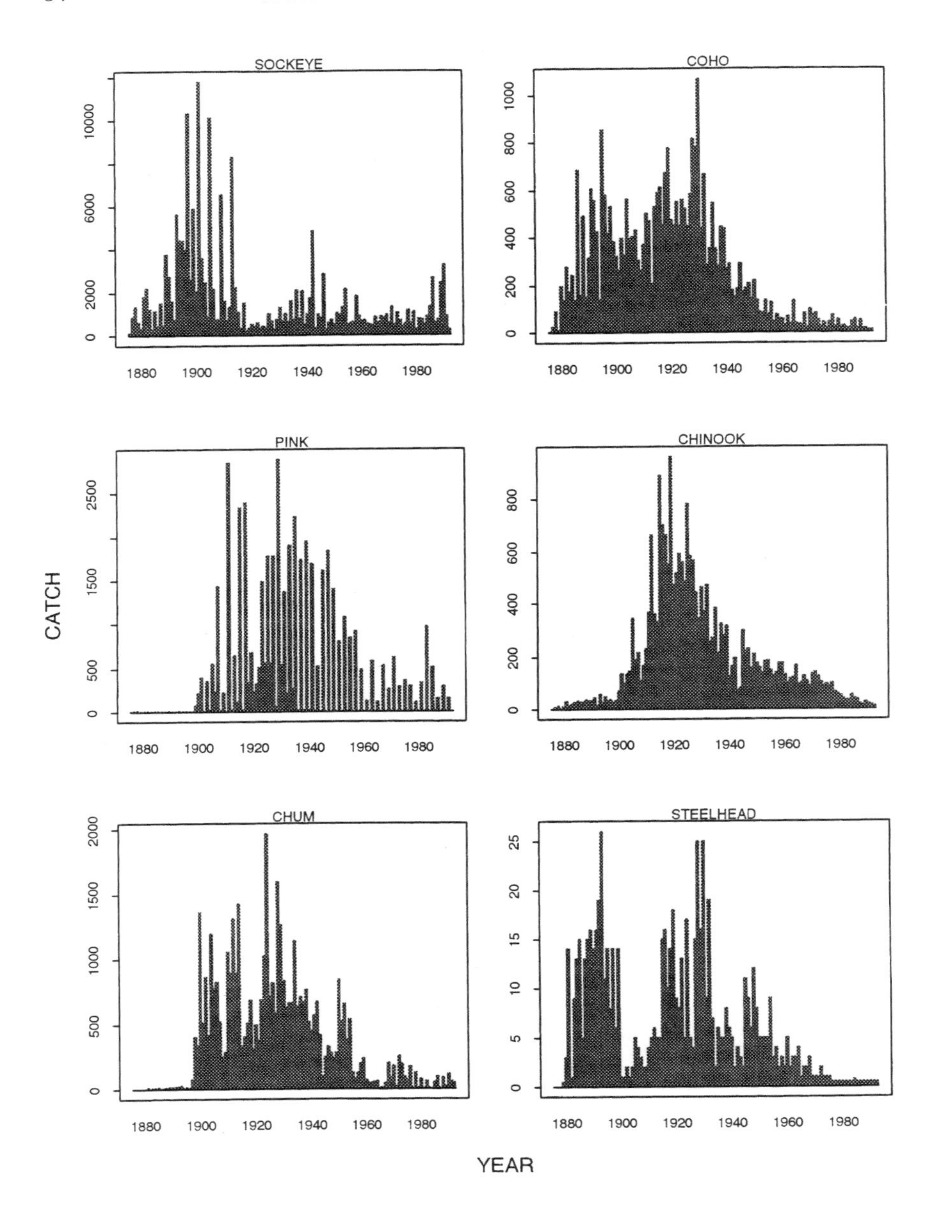

FIGURE 4-1 Historical commercial catch of Pacific salmon (numbers landed) in the Fraser River area (Statistical areas 28 and 29) between 1876 and 1985. Before 1951, the numbers were estimated from records of canned pack and product statistics; after 1951, the numbers are from records of fish sales (Canadian Department of Fisheries and Oceans annual catch statistics). Source: Argue et al. 1986.

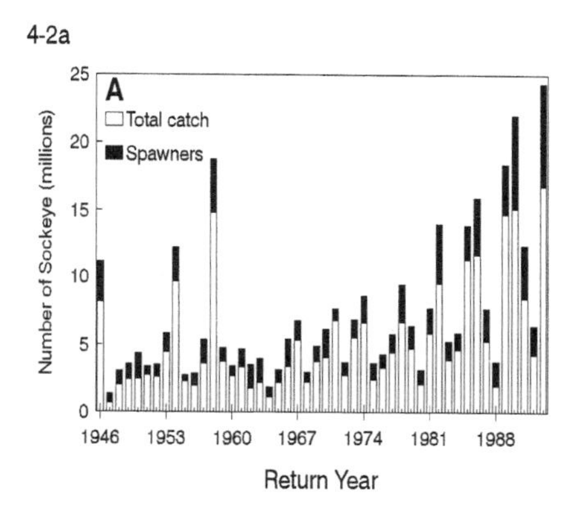

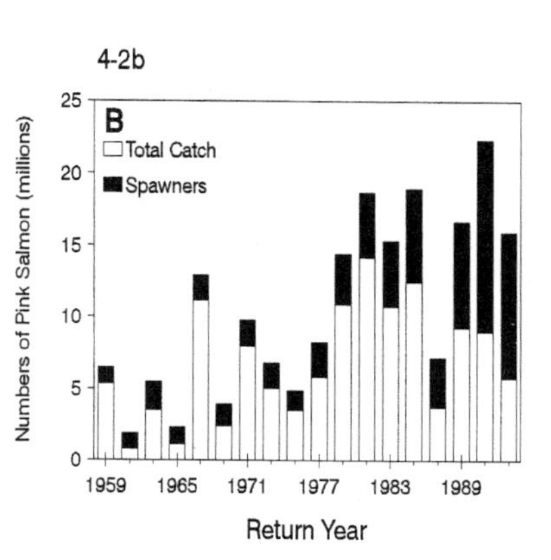

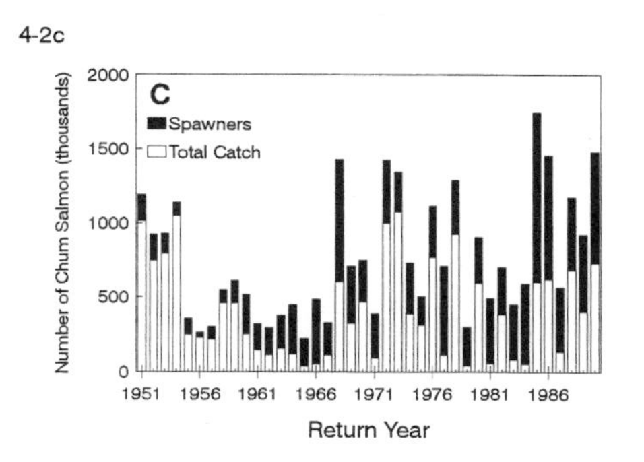

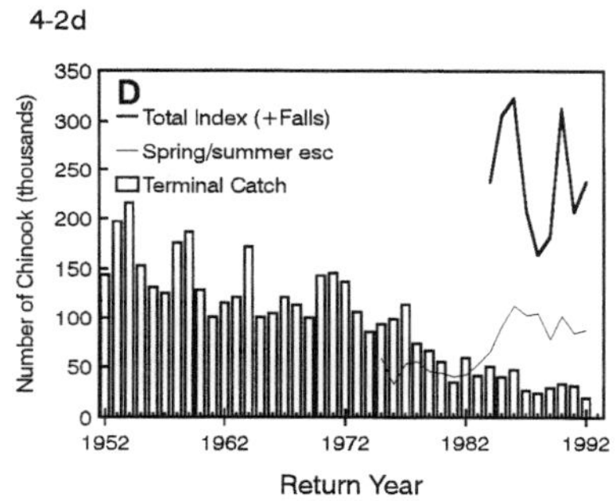

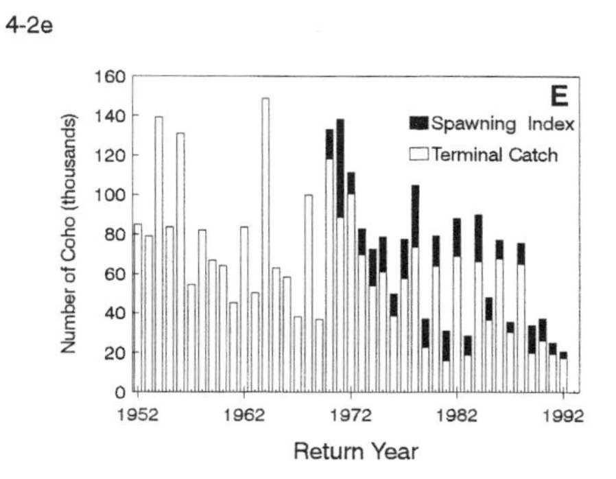

FIGURE 4-2a to 4-2e Long-term changes in catch and escapement from Fraser River, B.C. Source: (a), modified from Bisson et al. (1992), (b-e), committee generated from data provided by Canadian Department of Fisheries and Oceans.

tively small portions of the species production except for recent chum returns and a few years of coho.[1] Increases in production of sockeye salmon reflect reduced catch rates in fisheries to increase spawning-population sizes, good marine survival over almost 20 years, and the protection of the water courses. Similarly, the increased pink and chum production reflects management actions to increase spawning-population sizes. Terminal runs of chinook increased substantially after the Pacific Salmon Treaty (1985), but conservation programs had been implemented in the terminal area since 1980. The best representation of the terminal run of chinook is the solid line in Figure 4-2d; this line represents the terminal catch plus spawning index for spring and summer chinook and the estimated spawning population for the Harrison River fall chinook population. The escapement of chinook to the Harrison River has been estimated quantitatively since 1984 and constitutes one of the largest chinook populations in North America. The remaining concern for Fraser salmon is probably the returns of coho. As evident in Figure 4-2e, the return is highly variable and associated with the strict 2-year cycle of Fraser pink salmon production. Fraser pinks return in the odd years, and fisheries on Fraser pinks incidentally capture returns of Fraser coho. The Department of Fisheries and Oceans in Canada is developing a coho-conservation program for the Strait of Georgia and the Fraser River.

The production of salmon from the Fraser River is still less than occurred historically, and economic development and overfishing have resulted in some extinctions. However, the resource base (fish populations and habitat) is apparently adequate to support an increase in salmon production.

PUGET SOUND

The Puget Sound Basin includes river systems in Puget Sound, Hood Canal, and the Strait of Juan de Fuca. Summaries of the status of salmon in Puget Sound (WDF et al. 1993, Quinn 1994) suggest a wide range of population conditions, ranging from healthy to critical. Overall, 93 of 209 salmon stocks (44.5%) identified by WDF et al. (1993) were classified as healthy, 44 (21.2%) were depressed, 5.3% were in critical condition, and the status of 60 (28.8%) was unknown (Table 4-3). Only one stock, early chum salmon in Chambers Creek, was considered by WDF et al. (1993) to be recently extinct, although there have been many stock extinctions over the last 150 years (Nehlsen et al. 1991). Chum salmon populations were most often classified as healthy, followed by pink salmon. However, other species had about half or more of their stocks classified as depressed or critical (Table 4-3). Puget Sound was considered to have fewer depressed stocks than the Columbia River basin but more than the Washington

[1]Data on the total runs of Fraser sockeye and pink salmon were provided by the Pacific Salmon Commission.

TABLE 4-3 Status of Puget Sound Salmon, as Summarized by WDF et al. (1993)[a]

Status	Chinook	Chum	Coho	Pink	Sockeye	Steelhead	Total
Healthy	10	38	20	9	0	16	93
Depressed	8	1	16	2	3	14	44
Critical	4	2	1	2	1	1	11
Unknown	7	13	9	2	0	29	60

[a]Stocks listed in this table do not include 40 salmon stocks from Washington believed by Nehlsen et al. (1991) to have been extirpated in last 150 years. Also, one stock (of chum salmon) classified as extinct and five stocks of chinook in disputed status were not included.

TABLE 4-4 Overall Status of Washington Anadromous Salmon, as Summarized by WDF et al. (1993)[a]

	Puget Sound		Coast		Columbia River		All Washington	
Status	No.	%	No.	%	No.	%	No.	%
Healthy	93	44.7	65	56.5	29	26.1	187	43.1
Depressed	44	21.2	8	7.0	70	63.1	122	28.1
Critical	11	5.3	0	0	1	0.9	12	2.8
Unknown	60	28.8	42	36.5	11	9.9	113	26.0
Total	208	100.0	115	100.0	111	100.0	434	100.0

[a]Stocks listed in this table do not include 40 salmon stocks from Washington believed by Nehlsen et al. (1991) to have been extirpated in last 150 years. Also, one stock (of chum) classified as extinct and five stocks of chinook in disputed status not included.

coastal region (Table 4-4). However, almost all the state's critical stocks were believed by WDF et al. (1993) to be from Puget Sound. The Puget Sound region was also intermediate between coastal and Columbia River basins in the proportion of stocks whose status was unknown.

Because there is such a great diversity of conditions in Puget Sound, it is difficult to draw generalizations about the overall status of salmon populations. Some river systems have productive habitat and large, relatively stable salmon populations; other rivers have been heavily altered and have few or no healthy populations. For example, the river basins of northern Puget Sound had the fewest depressed salmon runs of any region in Washington, Oregon, Idaho, and California, according to The Wilderness Society (1993). However, the Elwha River on the north coast of the Olympic Peninsula has more extirpated salmon populations than any other river in Washington, except the Columbia River (Nehlsen et al. 1991), partly as a result of two mainstem dams without fish-passage facilities constructed a short distance above the river mouth early in the 1900s.

Slightly more than half the Puget Sound chinook stocks whose status was known were believed to be in a depressed or critical condition (Table 4-3). In addition, five stocks of spring chinook (Stillaguamish, Snohomish, Green, Skokomish, and Elwha rivers) were in "disputed" status because state agency biologists felt that they were extinct but tribal biologists felt that they still existed. The spring chinook from the Elwha River were particularly notable because this population is reputed to have included exceptionally large fish—up to 45 kg—and the Elwha River has been the focus of considerable attention regarding the possible removal of two mainstem dams. In general, spring chinook were more likely to be at risk than fall chinook. Spring chinook stocks classified as critical occurred in the North Fork Nooksack River, the South Fork Nooksack River, the White River (a Puyallup River tributary), and the Dungeness River. Those stocks were included in a recent ESA petition to list nine of the 12 critical stocks identified by WDF et al. (1993). Habitat loss and overfishing were cited in the petition as primary contributors to stock declines. About half the chinook populations in Puget Sound, including many of the larger populations, were supported (at least in part) by hatchery propagation.

Sixteen of the 37 coho salmon stocks from Puget Sound whose status was known were listed as depressed and one stock (Discovery Bay coho) was classified as critical (Table 4-3). According to WDF et al. (1993), only two U.S. coho stocks in the Puget Sound region—Skagit River and Deer Creek, a tributary of the Stillaguamish River—were purely wild; all other stocks have been supplemented with hatchery fish of nonnative origin at some time. About half the total number of coho stocks were being supported by hatchery production, although most of the coho in southern Puget Sound were of hatchery origin and most of the coho from Hood Canal and the Strait of Juan de Fuca were wild. Virtually all the status reports that discuss causes of declines (Nehlsen et al. 1991, Palmisano et al. 1993, The Wilderness Society 1993) cite habitat loss, overfishing, and negative influences from hatchery fish as contributors to the dwindling of wild coho populations. Protection of depleted populations of wild coho in Hood Canal tributaries has been an important reason for extensive recent cutbacks in salmon fisheries along the Strait of Juan de Fuca, through which Hood Canal coho must pass when returning to spawn.

Almost all the chum salmon stocks in Puget Sound whose status was known were healthy, according to WDF et al. (1993). The only three stocks classified as critical or depressed (Hood Canal, Discovery Bay, and Sequim Bay) were early-returning populations termed "summer" chum. Depression of early-returning chum populations is most likely related to overfishing (including incidental catch in fisheries that target other species) and to poor spawning conditions. About one-third of the chum populations in the Puget Sound region are supported by hatchery production.

There are fewer than 15 pink salmon populations in Puget Sound because this area is at the southern edge of the distribution of the species in North America, except for highly disjunct (and now extinct) populations in northern California. They have 2-year life cycles and most return in odd-numbered years to spawn (the single exception is a small, even-year spawning population from the Snohomish River). Most of the pink salmon stocks in Puget Sound have been classified as healthy (Table 4-3), but two stocks in Hood Canal and the Strait of Juan de Fuca were listed as depressed (Dosewallips River and upper Dungeness River), and two stocks were considered critical (lower Dungeness River and Elwha River). Virtually all the pink salmon populations are of wild origin, and none is currently supported by hatchery production. The largest salmon populations in Washington are pinks from northern and central Puget Sound rivers, where in odd-numbered years spawners can run in the hundreds of thousands (WDF et al. 1993). Like other salmon, pinks are vulnerable to disturbance of spawning areas and overfishing of depleted populations.

Four sockeye salmon stocks were identified by WDF et al. (1993) in Puget Sound—three from the Lake Washington system. Nehlsen et al. (1991) referred to an extinct sockeye population from Mason Lake in southern Puget Sound, but this population is not mentioned by WDF et al. (1993). All three Lake Washington stocks, which were partially of nonnative origin, were classified as depressed (Table 4-3), and the sockeye stock from the Baker River, a Skagit River tributary, was listed as critical. The Baker River sockeye is a native population now blocked from historical spawning areas by two dams. Adults are now captured at the lower dam and trucked to artificial spawning beaches in Baker Lake. According to WDF et al. (1993), as few as 92 sockeye have returned to the adult collection facility in recent years.

More stocks of steelhead (60) than of any other anadromous salmon are recognized in Puget Sound, but the status of half of them is unknown (Table 4-3). Half the stocks whose status is known were listed as depressed or critical. Surprisingly, most of the steelhead stocks from heavily urbanized portions of Puget Sound were classified as healthy or unknown by WDF et al. (1993); there were a few notable exceptions, including Lake Washington winter steelhead, a depressed population that is subject to heavy predation by sea lions at the Ballard Locks (Box 2-1), and Deer Creek summer steelhead, a critical stock that was recently petitioned for ESA protection and was made famous by the writings of Zane Grey. Few hatcheries propagate steelhead in Puget Sound, but large numbers of hatchery juveniles have been released throughout the basin. Most depressed steelhead populations occur in rivers of northern Hood Canal and along the Strait of Juan de Fuca; the origin of these stocks was unresolved between state and tribal biologists (WDF et al. 1993), but most are maintained solely by natural production.

Relatively little is known about the current status of sea-run cutthroat trout in Puget Sound. The species was not included in the WDF et al. (1993) stock inventory, and population trends are monitored at few locations. At least three separate sea-run cutthroat stocks in Puget Sound have been identified on the basis of electrophoretic comparisons of specimens from different areas: a northern Puget Sound stock, a northern Hood Canal stock, and a southern Hood Canal stock (Compton and Utter 1987). The species was considered at moderate risk of extinction by Nehlsen et al. (1991). In addition to the negative influences of habitat loss and overfishing, sea-run cutthroat appear to be highly vulnerable to competitive displacement by other salmon and might be affected by widespread coho and steelhead supplementation programs (Trotter et al. 1993).

COLUMBIA RIVER BASIN

Historical Trends

The Columbia River provides a case history for some of the most severe declines in salmon populations along the Pacific Coast. There the late spring and summer runs of chinook populations were historically the most abundant and heavily fished salmon in the basin (Thompson 1951, Van Hyning 1968, Chapman 1986). Chapman (1986) estimated peak runs in 1881-1885 at about 2.7 million spring-summer chinook. For all Columbia River salmon species, the Pacific Fishery Management Council (PFMC 1979) estimated predevelopment runs at 6.2 million (including all species except steelhead and sea-run cutthroat) on the basis of habitat availability. That number approximates predevelopment salmon (chinook, coho, sockeye, chum, and steelhead) estimates of 7.5 million by Chapman (1986) on the basis of peak catches and probable exploitation rates. The Northwest Power Planning Council (NPPC 1986) adopted a predevelopment-run estimate of 10-16 million fish—about double the numbers estimated by PFMC (1979) and Chapman (1986). Overfishing in the lower Columbia River rapidly depressed sockeye and chinook populations and then steelhead, coho, and chum. Habitat degradation and loss accelerated the decline. The number of naturally produced salmon in the Columbia River declined by the late 1900s to about one-eighth their predevelopment abundance, while total runs along the Pacific Northwest coast decreased by about two-thirds.

Before 1933, only catch data were available to indicate run sizes. Beginning with the completion of Rock Island Dam on the Columbia River in 1933, dam counts provided an additional data source. Completion of Bonneville Dam in 1938 provided a central location for estimating returns to the middle and upper Columbia River (WDFW and ODFW 1994). Coupled with in-river catch records, dam counts permitted calculations of approximate total run size. However, those estimates were better for some populations than for others. They permitted reasonable assessments of total adult populations of sockeye and steelhead, neither of which were caught in great numbers by ocean fisheries. For coho and chinook salmon, counts at dams and records of in-river catches yielded only partial adult estimates because information on ocean fisheries was lacking. Much later, in the 1970s, coded-wire tag information permitted managers to estimate ocean catch for some populations, so that total run size could be reconstructed. But those reconstructions were accurate, or even possible, only for particular hatchery populations. Ocean fishing of wild salmon could not be calculated, and distribution of hatchery fish could not properly be assumed to apply to wild fish.

Snake River

Sockeye salmon of the Snake River are listed as endangered under the provisions of the ESA. Sockeye passage at Ice Harbor Dam has varied widely but declined sharply after 1976 (Figure 4-3). The only production area now used by Snake River sockeye is Redfish Lake in the headwaters of the Stanley basin. Only four adults returned to the outlet trap there in 1991, one in 1992, and five in 1993; the Idaho Department of Fish and Game (IDFG) incorporated these adults in captive broodstock programs rather than permitting them to spawn naturally. Sockeye survivors of downstream migrants from Redfish Lake that were captured at the lake outlet and cultured in hatchery facilities to adulthood were released to spawn in Redfish Lake in 1993. Those downstream migrants might have consisted in part or wholly of the progeny of a group of residual sockeye or kokanee found in lake-shoal spawning areas. Those fish have an October maturation similar to that reported by Bjornn et al. (1968) for sockeye in the 1960s.

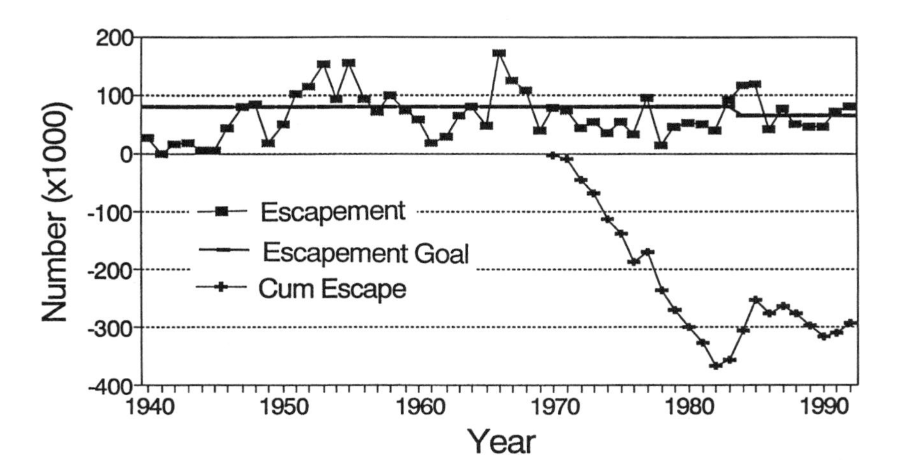

FIGURE 4-3 Counts of adult sockeye at Ice Harbor Dam (■ symbols), escapement goal (horizontal line), and the cumulative deviation of actual escapement from the escapement goal after 1970 (+ symbols).

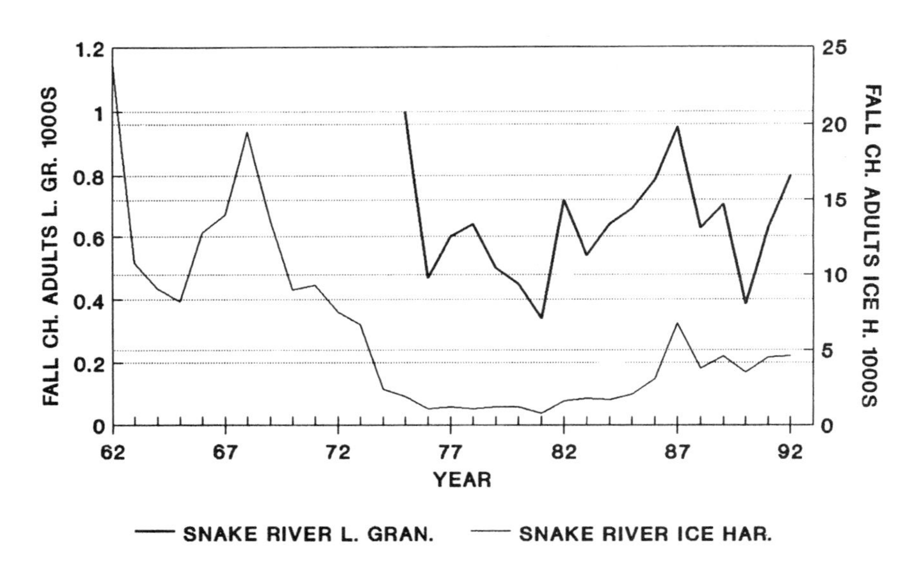

FIGURE 4-4 Fishway counts of fall chinook at Lower Granite and Ice Harbor dams.

The number of fall chinook returning to the Snake River, as counted at Ice Harbor Dam, declined from 1968 to 1976, remained low but stable through 1985, and then increased slightly (Figure 4-4), probably in response to production from Lyons Ferry Hatchery and straying from the Umatilla and Columbia rivers (Chapman et al. 1991). Adult counts at Lower Granite Dam, available since 1975, have ranged from about 350 to 1,000 and for the most recent 5 years averaged about 600 (Figure 4-4). High catch rates of fall chinook reduce the numbers of adults passing Lower Granite Dam. In the late 1980s, ocean fisheries took about 35% of fish, and river catch took 44-63% of the in-river run (i.e., the survivors of the ocean fishery), so fisheries took nearly 75% of adult recruits (Chapman et al. 1991). Catch rates have declined in recent years: perhaps 60% of fall chinook are caught. Fall chinook is the only ESA-listed Snake River salmon with high fishing rates.

Redd (spawning-bed or nest) counts for wild spring and summer chinook indicate declining numbers of spawners over the last 3 decades (Figure 4-5a and b). Redd counting in reasonably consistent index areas extends to 1957, although methods have not always remained the same (Chapman et al. 1991). The few counts before that date (see Welsh et al. 1965) cannot be compared directly with post-1957 counts (P. Hassemer, IDFG, personal communication).

The spring chinook run of 1994 appears severely depressed and is predicted to total about 20,000 fish at Bonneville Dam, whereas the escapement goal is about 80,000 fish. Counts of spring chinook jacks in 1994 were very low,

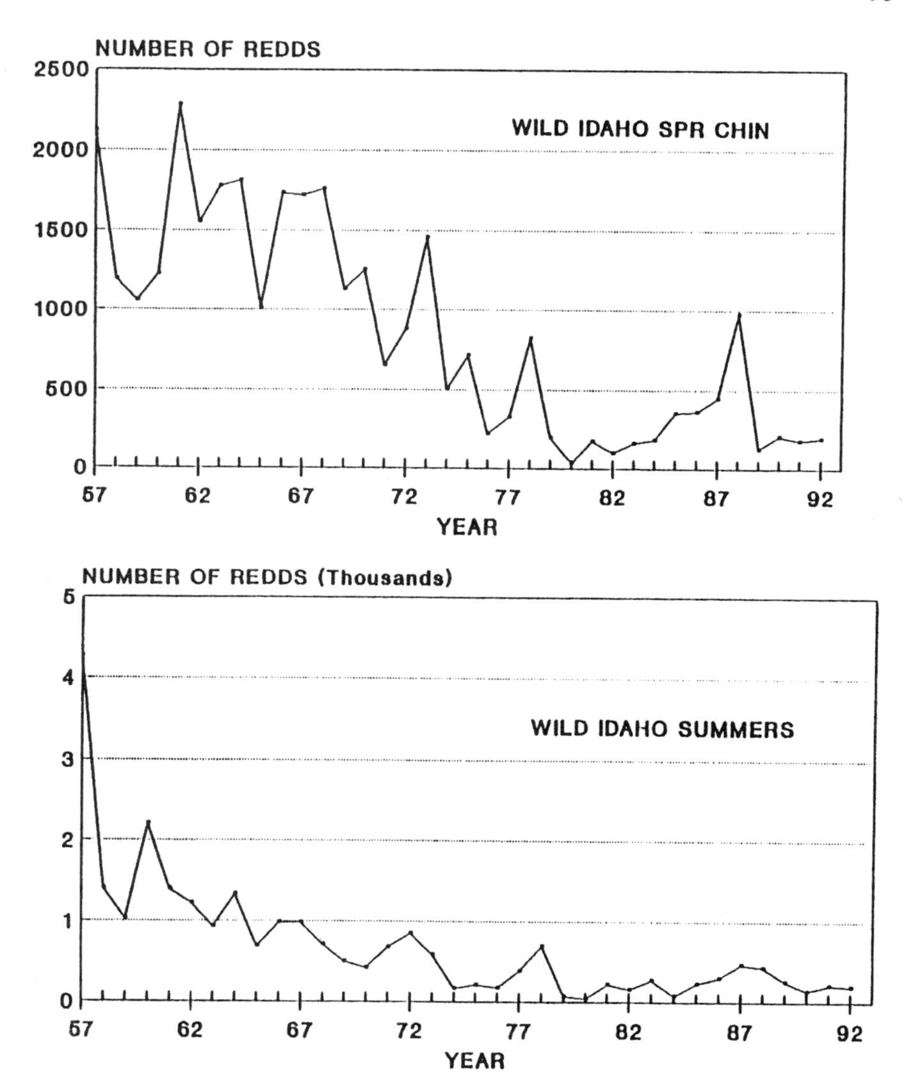

FIGURE 4-5 (a) Index area redd counts for wild spring chinook in Idaho. (b) Index area redd counts for wild summer chinook salmon in Idaho.

foreshadowing a greatly diminished adult run in 1995[2]. Most of the adults returning in 1995 will have migrated downstream as smolts in 1993, when fewer smolts were artificially transported downstream, because so much water was spilled over the dams in that year.

[2]The adult return for 1995 appears to be about half the 1994 run at press time.

For summer chinook, the highest redd counts in several Snake River index spawning areas occurred in 1957, the year in which completion of The Dalles Dam flooded Celilo Falls. Tribal fisheries at Celilo tended to take more summer and fall chinook than the earlier-migrating spring chinook because fishing access and efficiency increased as flows dropped. In other index areas, high counts occurred during 1957-1961, before other fisheries intensified (see WDF and ODFW 1992, Table 29). Summer chinook redd counts in the Snake River system declined in the 1970s and then rebounded from 1980 to 1988 until the effects of the extended drought that began in the winter of 1987-1988 again depressed the runs.

The IDFG counts of parr in late summer in index areas of streams that produce spring-summer chinook reflect escapement declines (Figure 4-5a and b). They also indicate habitat degradation in some streams, e.g., Bear Valley Creek (Figure 4-5a; Rich et al. 1992). Chamberlain Creek, a wilderness stream with no commodity-production uses, held more chinook parr than other Salmon River tributaries (Figure 4-5b). Declines in parr abundance reflect the combined effects of reduced spawner densities, drought in rearing areas, and conditions in the migration corridor during smolt emigration.

Richards and Olsen (1993) compared trends in Snake River spring chinook with trends in ocean production, emphasizing the 1980s. They concluded that trends were similar and that ocean ecological or interrelated inland climatic conditions affected both groups. Their analysis covered, at most, the years 1971-1991. That tends to mask the effects of reductions from the late 1950s to 1971 caused by intensive dam construction on the main Columbia and Snake rivers. Only Lower Granite and Little Goose dams remained to be constructed on the Snake River after 1971.

Two groups of steelhead use the Snake River. The first, called the A group, includes fish that pass Bonneville Dam before the end of August. They tend to spend 1-2 years at sea and be shorter than 81 cm (32 in). B-group steelhead pass Bonneville Dam mostly in September and are generally longer than 81 cm, having spent 2-3 years at sea. Some of the largest steelhead ever caught have belonged to the B group. Total numbers of wild and hatchery steelhead passing Ice Harbor Dam reached a low point in the middle 1970s but since have trended upward (Figure 4-6). Wild A- and B-group steelhead have been distinguished from hatchery fish at Bonneville Dam and in Zone 6 (upstream from Bonneville Dam) fishery only since 1984, when adipose fin removal permitted identification of hatchery fish. A-group steelhead use both the mid-Columbia and Snake rivers. Most B-group fish enter the Snake River. Numbers of B-group steelhead escaping the Zone 6 fishery have trended upward very slightly since 1986 (Figure 4-6). Beginning in 1991, efforts to protect wild fall chinook that originate in the Snake River have increased the B-group steelhead.

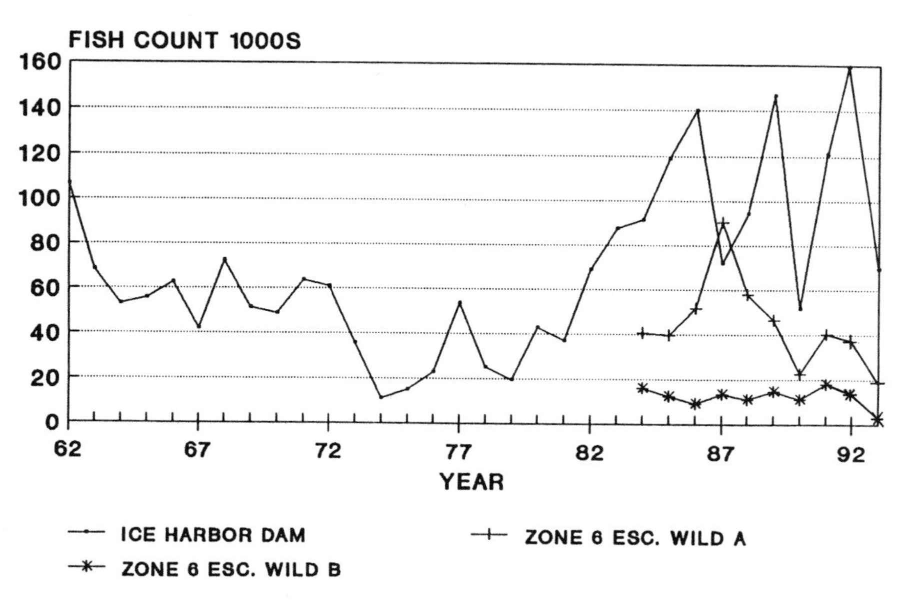

FIGURE 4-6 Steelhead escapements, upriver summer steelhead. Dam count June 1 - October 1. Source: WDF/ODFW 1992.

Middle and Upper Columbia River

Sockeye salmon in the Yakima River system were extirpated by construction of irrigation storage reservoirs without provision for fish passage. More recently, the National Marine Fisheries Service and the Bonneville Power Authority have investigated re-establishment of sockeye in some Yakima River reservoirs. In general, however, although Snake River sockeye declined sharply over the last 3 decades, Columbia River sockeye have sustained themselves better (Figure 4-7). Snake River sockeye pass eight dams, Wenatchee River sockeye pass seven, and Okanogan sockeye pass nine to reach spawning areas in the Okanogan River upstream from Wells Dam (Figure 4-8). It is not completely clear why abundance trends for Snake River and Columbia River sockeye differ, but several hypotheses exist. The Columbia River carries about one-fourth to one-third as many yearling or older hatchery salmon and steelhead smolts as does the Snake River. Fishery agencies release ten times more steelhead from Snake River hatcheries than from middle Columbia River hatcheries upstream from Priest Rapids Dam. Total releases of chinook salmon and steelhead from Snake River hatcheries average near 20 million fish. In the early 1960s, Raymond (1979) estimated that fewer than 6 million wild fish were produced from the Snake River drainage. It is possible that large populations of hatchery-produced fish may interact in negative ways with Snake River sockeye.

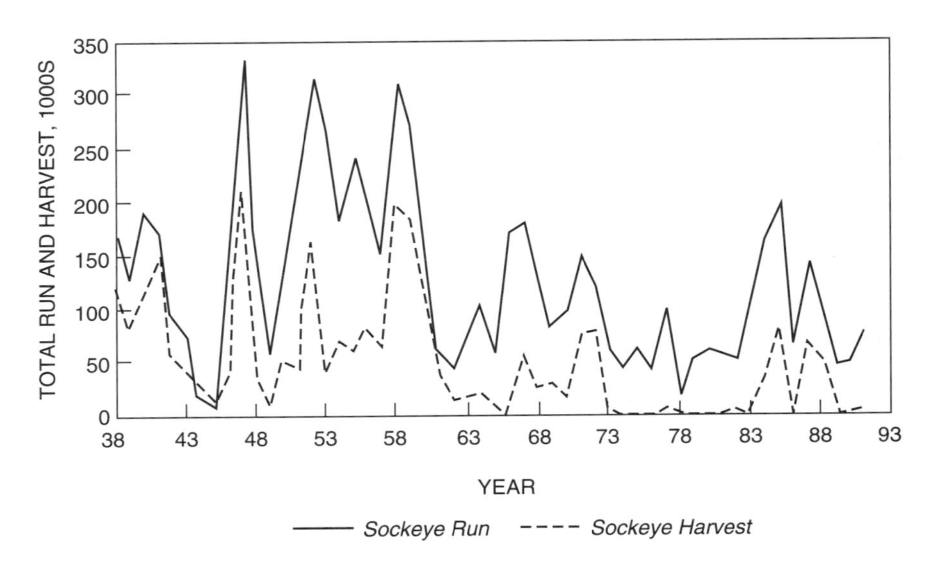

FIGURE 4-7 Estimated numbers of sockeye entering Columbia River and total sockeye catch.

The flow of the Columbia River during spring substantially exceeds that of the Snake River. Fish-deflection screens, vertical barrier screens, transportation, and conventional bypasses are absent from Columbia River dams upstream of the mouth of the Snake River but were present in the two uppermost Snake River dams (Little Goose and Lower Granite) during the rapid decline of the Snake River sockeye. Sockeye appear to suffer more damage in bypass systems than do other spring migrants (Johnsen et al. 1990).

Sockeye of the upper Snake River might have originated as residual sockeye in Stanley basin lakes after the partial removal of Sunbeam Dam in the 1920s, whereas Columbia River sockeye have had continuous access to the sea and to their natal areas. Residual sockeye might be less fit for the rigors of anadromy. All or some of the factors just discussed or variants of them might explain the demise of Snake River sockeye.

Although summer and fall chinook were treated separately by WDF et al. (1993), Utter et al. (1993) considered summer and fall chinook of the main Columbia River to consist of one population unit from the Hanford Reach through upriver areas. Thus, the summer-fall unit would include fish that managers have termed "upriver bright fall chinook" (URB) and "summer chinook" that spawn as far upstream as the middle reaches of the Wenatchee River, Methow River, and lower Similkameen River. A few summer and fall chinook also spawn in the

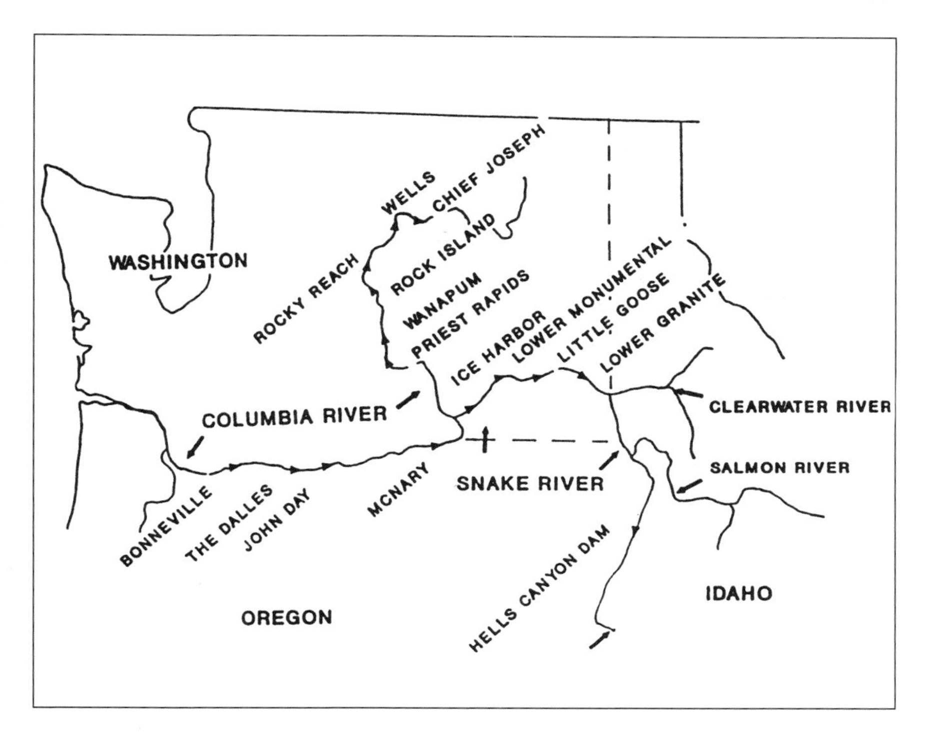

FIGURE 4-8 Location of mainstem hydroelectric dams on Columbia and Snake rivers in areas still accessible to salmon.

main Okanogan River. Summer-fall chinook populations in the middle Columbia River are more abundant now than they were in the 1930s, when counts became available at Rock Island Dam. However, hatcheries now contribute substantially to some of the runs. Five of the 14 summer and fall chinook stocks identified by WDF et al. (1993) from the middle and upper Columbia River were listed as depressed; the other nine were considered healthy.

There has not been an in-river fishery specifically for summer chinook in the Columbia River since 1964. Before then, the total catch of these populations probably amounted to over 70%, although the fish had to pass four to six dams that also killed some juveniles and adults. Ocean catch of these fish continues at about 35%. The runs are now not as resilient as they were in the 1930s, in that losses at hydroelectric dams cause the death of fish that formerly would have been available for fisheries or escapement. The strongest population is the naturally spawning fall run that uses the Hanford Reach between McNary Pool and Priest Rapids Dam, the last free-flowing reach in the middle Columbia River. From 1983 to 1991, numbers of spawners in the Hanford Reach ranged from about 50,000 to 165,000 (WDF et al. 1993). The reach is the largest natural production area for chinook salmon in the entire Columbia River Basin.

Summer-chinook runs at Rock Island Dam have been variable, averaging about 16,000 adults and jacks from 1933 to 1992 (Grant, Douglas, and Chelan Public Utility districts, personal communication; Figure 4-9). This population has been proposed for ESA listing. Over the same period, combined summer and fall chinook averaged 22,722 adults and jacks. The 1993 summer-chinook count at Rock Island was 13,401 adults and jacks (Fish Passage Center, September 17, 1993). Mullan (1990) stated that the average summer-fall chinook adult return in 1967-1987 to middle Columbia River tributaries was 15,497, consisting of the Wenatchee (12,012), Entiat (100), and Methow (3,385) rivers. In the same period, the total number of summer-fall chinook passing Rock Island Dam was about 16,500 adults. Thus, the hatchery supplement to the summer and fall run at Rock Island Dam was about 1,000 adults, or about 6% of the total run.

Numbers of middle Columbia River summer-fall chinook between McNary Dam and Priest Rapids Dam were fairly stable from 1962 to 1982 but then climbed rapidly and peaked in 1987 (Figure 4-10), when escapement reached 96,400 fish (PFMC 1993). The high returns in the 1980s apparently benefited from reduction in ocean fisheries, a result of the U.S.-Canada treaty, and perhaps from relatively high ocean survival. In-river fishing was also cut back to protect the summer component of the summer-fall chinook run and to reduce the catch of steelhead and fall chinook destined for the upper Snake River.

Spring chinook populations in the middle Columbia River were low in the 1930s and 1940s (Figure 4-11), when commercial fisheries took up to 86% of the

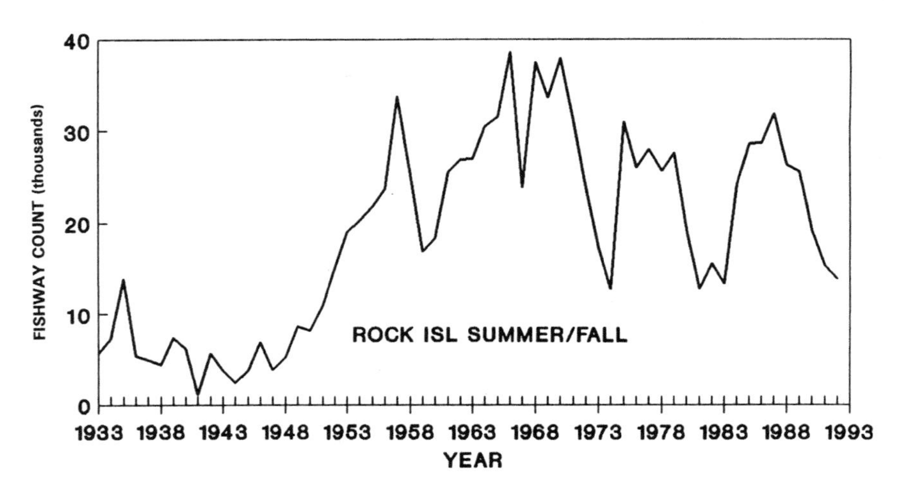

FIGURE 4-9 Fishway counts of adult summer and fall chinook at Rock Island Dam, 1933-1992.

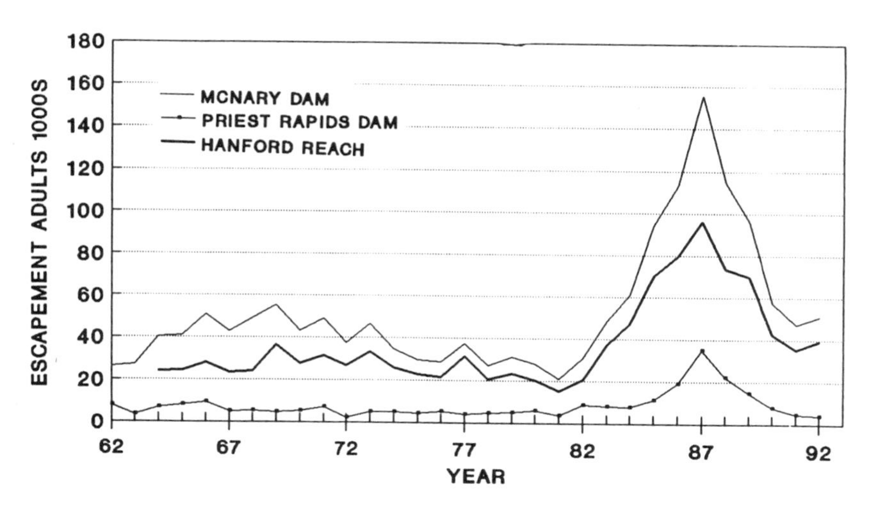

FIGURE 4-10 Estimated escapement of adult summer and fall chinook at McNary and Priest Rapids dams. Numbers of fish spawning in Hanford Reach was difference between counts at two dams.

runs. Populations grew with the Grand Coulee Fish Maintenance Project and with reduction in catches during the 1940s. Some of the increase resulted from absence of directed fisheries in the Columbia River and cessation of sport fishing in tributaries. Dam-related mortality of smolts and losses of adults between dams substituted for fishing mortality. Despite the upward trend in numbers of spring chinook passing Rock Island Dam during the 1980s, the overall number of upriver spring chinook entering the Columbia River has declined (Figure 4-11).

Of the 16 native stocks of spring chinook from the middle and upper Columbia River identified by WDF et al. (1993), 15 were classified as depressed and one (Asotin Creek) as critical. Many of the depressed stocks contained fewer than 500 spawning adults. Some of the stocks remained stable through the middle 1980s but have experienced severe short-term declines.

No naturally spawning coho populations remain in the Columbia River basin upstream of The Dalles Dam, although they once inhabited many subbasins in the middle Columbia and Snake River systems (The Wilderness Society 1993). They were extirpated by a combination of habitat loss, overfishing, and agency policy that relied on hatcheries. Likewise, chum populations have disappeared from the middle Columbia and Snake rivers. Nehlsen et al. (1991) listed chum extinctions in the Umatilla and Walla Walla rivers but did not note when the extinctions occurred or their causes. Pink salmon are not known from the Columbia River in historical times.

In the middle Columbia region, summer steelhead are now managed for both

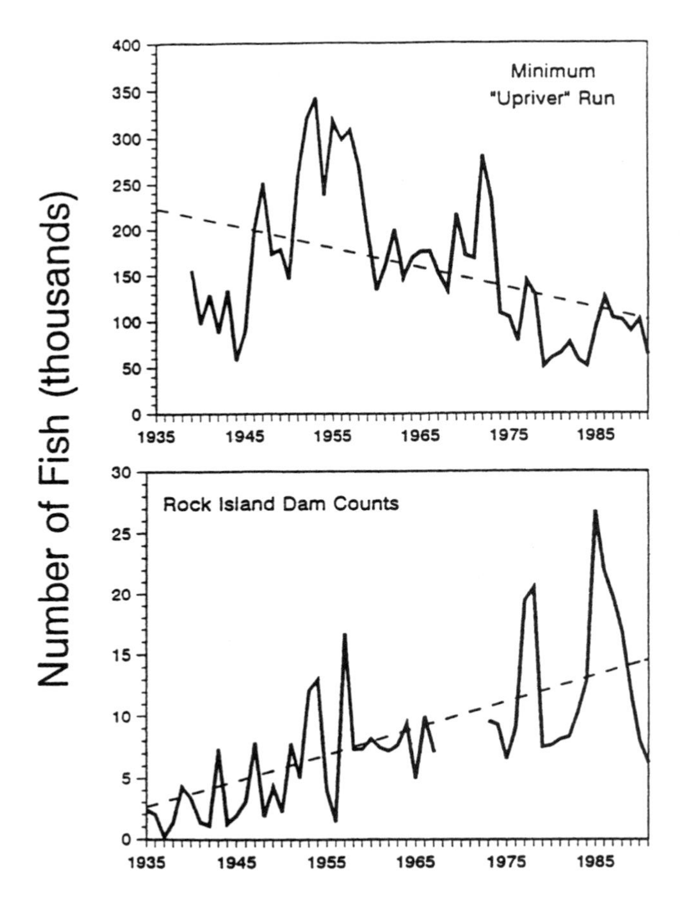

FIGURE 4-11 Estimated minimum number of upriver spring chinook entering Columbia River and ascending Rock Island Dam (dam counts not available for 1968-1972). From Oregon Department of Fish and Wildlife and Washington Department of Fisheries 1991c and unpublished data, Chelan County Public Utility District.

natural and hatchery production, a change from 10-15 years ago, when hatchery production dominated the management strategy of the Washington Department of Game (now the Washington Department of Fish and Wildlife). Adult summer steelhead for hatchery broodstock are trapped at Wells Dam, and progeny are distributed to the Wenatchee, Methow, and Okanogan rivers. Natural spawning

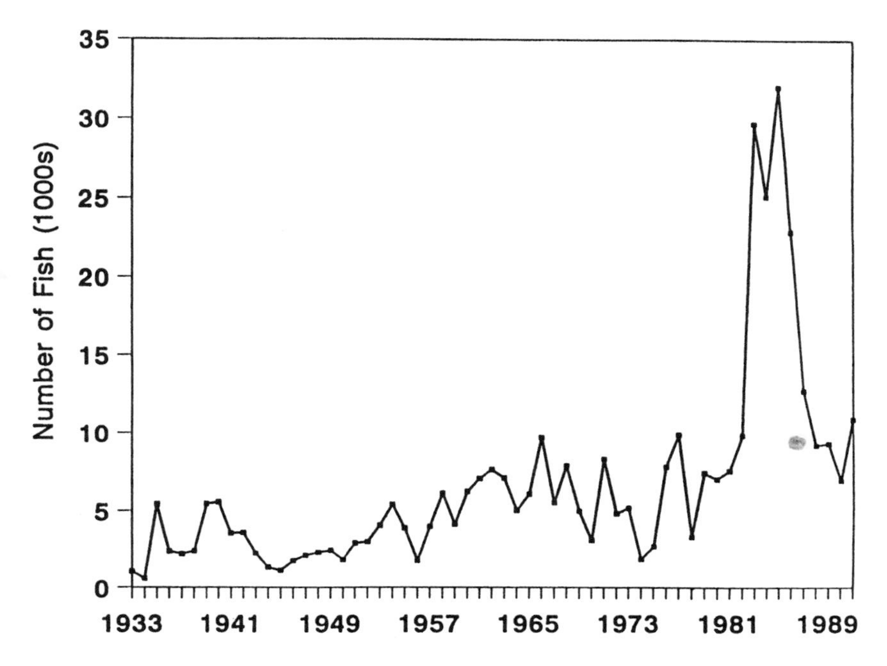

FIGURE 4-12 Number of steelhead ascending Rock Island Dam, 1933-1991. Unpublished data from Chelan County Public Utility District.

occurs, but only marked hatchery fish can be retained by sport anglers. Figure 4-12 shows trends in counts of steelhead passing Rock Island Dam. Steelhead in the middle Columbia are more abundant now than they were in the 1930s as a consequence of extensive hatchery production. However, Nehlsen et al. (1991) listed 11 extinct stocks of summer steelhead from the middle and upper Columbia basin and from the Snake River system, and 13 of 15 wild stocks identified by WDF et al. (1993) were considered depressed. The mid-Columbia steelhead are currently under consideration for ESA listing as threatened or endangered. Winter steelhead now occur only as far inland as Fifteenmile Creek near The Dalles, Oregon. Sea-run cutthroat trout apparently are not found above Bonneville Dam and are now extinct in the Wind and Klickitat rivers (Nehlsen et al. 1991).

Lower Columbia River

Many of the salmon populations inhabiting the lower Columbia River (all tributaries below Bonneville Dam) have been altered by hatchery and dam practices, and the total number of salmon and steelhead hatcheries operating in the lower Columbia River system is probably the greatest of any area throughout the range of Pacific salmon.

The National Marine Fisheries Service (Johnson et al. 1991) believed that native coho salmon in lower Columbia River tributaries had become extinct, partly because of widespread population transfers and broodstock selection practices (e.g., selection of early returning adults) at hatcheries. Although of mixed hatchery and wild origin, 17 naturally spawning stocks of coho in Washington tributaries of the lower Columbia River were recognized by WDF et al. (1993). All were considered depressed, but spawning counts were available for only three of these stocks.

In contrast, only two of the 17 stocks of chinook salmon from Washington tributaries of the lower Columbia were considered depressed by WDF et al. (1993); all others were classified as healthy on the basis of stable or increasing population figures. The two depressed stocks (South Fork Toutle and Green River fall stocks) were both part of the Toutle River system that drains the northern and western slopes of Mount St. Helens and were both affected by the 1980 volcanic eruption. Most of the other stocks were heavily supported by hatchery production.

Chum salmon populations are seriously depressed in the lower Columbia. Nehlsen et al. (1991) stated that only about 2,000 chum were spawning in lower Columbia River tributaries, a figure that represents 0.5% of their historical abundance. WDF et al. (1993) listed three Washington chum stocks from the lower Columbia and considered two of them depressed. Forestry and agricultural practices, urbanization, pollution, and overfishing in mainstem fisheries directed at coho and fall chinook were believed by Nehlsen et al. (1991) to have contributed to chum declines.

Five summer and 18 winter steelhead stocks were identified from Washington tributaries of the lower Columbia River (WDF et al. 1993). Of the summer steelhead stocks, two were considered depressed, and the status of the other three was unknown. Of the winter steelhead stocks, 12 were listed as depressed, two as healthy (the only two for which escapement goals had been established), and four as of unknown status. Nehlsen et al. (1991) observed that hatchery programs dominated steelhead in the lower Columbia. Although they did not list the native stocks in the Washougal, North Fork Lewis, Cowlitz, and White Salmon rivers as extinct, Nehlsen et al. (1991) felt that few native steelhead existed in these drainages. The overall depressed condition of steelhead in the lower Columbia River differs somewhat from the condition of coastal populations in Washington and Oregon, where greater percentages of populations are not declining.

Little is known about the status of sea-run cutthroat populations in the lower Columbia River. Sea-run cutthroat are known to spawn in a number of Washington tributaries, but long-term data are lacking. The species spends an extended period in freshwater and estuarine environments before migrating seaward and is sensitive to habitat degradation and estuary loss (Trotter 1989). It might also be vulnerable to negative interactions with hatchery fish and to overfishing (Trotter et al. 1993). Nehlsen et al. (1991) considered sea-run cutthroat to be depressed throughout the region.

COASTAL WASHINGTON, OREGON, AND CALIFORNIA

WDF et al. (1993) inventoried 32 chinook salmon stocks from the Washington coast: 22 fall and 10 spring-summer populations. Overall, about two-thirds of the stocks were classified as healthy according to the criteria in Table 4-2. However, a greater percentage of fall chinook (72%) than spring-summer chinook (40%) were considered healthy. Only a single fall chinook stock was listed as depressed, whereas 40% of the spring-summer chinook stocks were depressed. According to WDF et al. (1993), one-fourth of the coastal chinook stocks in Washington are supported by mixed wild and hatchery production.

Nickelson et al. (1992) recognized 55 wild populations of coastal chinook in Oregon, ranging from the Nehalem River on the north coast south to the Winchuck River near the California border. Life-history characteristics of coastal Oregon chinook vary greatly. Timing of spawning runs ranged from April to December, and age at maturity varied from 2 to 6 years. Juvenile chinook in Oregon coastal rivers usually spend less than a year in freshwater, and extended estuary residence is well documented. Populations from the Elk River northward usually rear in nearshore waters off British Columbia and southeastern Alaska; those from the Rogue River southward tend to rear off southern Oregon and northern California (Nickelson et al. 1992).

Of the 55 known naturally spawning chinook populations, 30 were considered healthy, eight were of special concern, eight were depressed, and the status of nine populations was unknown. Although the overall status of northward-migrating chinook seemed to be favorable along the Oregon coast (Lichatowich 1989), southward-migrating populations were in decline. According to Nickelson et al. (1992), 70% of the southward-migrating chinook populations—i.e., most populations from the Rogue River southward—were judged to be of special concern or depressed. In addition, runs of southward-migrating chinook have experienced wide variation in abundance over the last decade. For example, spring and fall chinook salmon entering the Rogue River increased from 30,000 in 1983 to about 200,000 in 1988 and then declined to about 30,000 in 1992.

Although many northward-migrating chinook populations from coastal Oregon were classified as healthy by Nickelson et al. (1992), there is evidence that populations are less abundant now than at the time of colonization. Lichatowich (1989) suggested that apparent increases in northward-migrating chinook could be an artifact of the period of record; if production at the time of the first stock inventories was already severely depressed by habitat destruction, comparisons between historical and present-day populations would imply that runs are stable or increasing. The Wilderness Society report (1993: Table A-2, A-4) compared estimates of natural runs of coastal chinook and coho in about 1900 (virtually all of which were wild fish) with estimates of current runs (which include both wild and hatchery fish). Of the 10 Oregon river systems examined, chinook in eight are less abundant now than in 1900, and seven of the eight populations were estimated to have declined to about half of their former abundance or less.

The Wilderness Society (1993) comparison of current and historical chinook abundances in selected Oregon rivers, as well as a similar comparison for coho, must be viewed somewhat cautiously because estimates of abundances now and in 1900 usually rely on comparisons of periods for which continuous records of abundance are lacking. Turn-of-the-century population estimates were based on reported catch weights, not numbers of fish caught; this introduced two potential sources of error: inaccuracies in reported total catch weights and assumptions concerning conversion of weight to numbers of fish. Nevertheless, decreases of 50% or more from estimated turn-of-the-century abundances to current abundances should probably be considered indicative of substantial overall declines in naturally spawning populations, especially because current populations include many hatchery fish. The committee recognizes, however, that 1900 may have been a time of relatively high ocean productivity for salmon in the Pacific Northwest (Smith and Moser 1988, CalCOFI Rep. 29, Baumgartner et al. 1992, CalCOFI Rep. 33), and that some of the differences in chinook and coho abundance along the Oregon coast between 1900 and the present might reflect differences in ocean conditions.

Moyle (1976) noted that chinook salmon once spawned as far south as the Ventura River in southern California but now occur only in the Sacramento-San Joaquin and some other river systems of northern California. The number of extant wild chinook populations in California has apparently not been established, but virtually all known populations have been reported to be at some risk of extinction (Higgins et al. 1992, Frissell 1993, The Wilderness Society 1993). Five chinook stocks from the Sacramento-San Joaquin River system and four stocks from the upper Klamath River system (actually spawning in southern Oregon) were declared by Nehlsen et al. (1991) to have become extinct within the last 150 years. According to The Wilderness Society (1993), all remaining spring and summer chinook populations in northern California are at high risk of extinction. Fall chinook, although not as imperiled in most areas, nevertheless are seriously depressed.

Coho salmon in Washington's coastal streams were considered by WDF et al. (1993) to be in generally good condition. Of the 26 known stocks, 17 were classified as healthy and nine as of unknown status. However, about 70% of the coho populations contained both wild and hatchery-spawned fish; hence there was a heavy reliance on hatchery production of coho on the Washington coast.

Coastal Oregon coho populations from the Necanicum River in the north to the Winchuck River in the south have been termed Oregon Coastal Natural (OCN) coho. Although 94 spawning populations are known (Nickelson et al. 1992), the most abundant OCN coho occur between the Nehalem and Coquille Rivers. Overall, OCN coho constitute the largest aggregate of coho populations in the United States outside Alaska. About half the 94 recognized populations occur in small streams that drain directly to the ocean. Some of these small streams were grouped by the Oregon Department of Fish and Wildlife according to geographic region and yielded a total of 55 coastal "populations." On the basis of spawner surveys over the last 20 years (Nickelson et al. 1992), six were classified as healthy, two were of special concern, 41 were depressed, and six were of unknown status. Since the middle 1970s, escapement of OCN coho to Oregon's coastal rivers and lake systems has usually been below the 200,000-adult target escapement established by the Oregon Department of Fish and Wildlife. The commercial and sport catch of these fish has also dropped markedly over the last 2 decades (Figure 4-13). Estimated declines of OCN coho since the turn of the century have been even larger. The Wilderness Society (1993) stated that runs in 10 coastal rivers declined from about 50,000-440,000 in 1900 to 5,000-55,000 fish currently, an average reduction of more than 80% in spite of extensive recent hatchery support for OCN coho. Although Nehlsen et al. (1991) noted only one coastal Oregon coho extinction, The Wilderness Society report (1993:25) stated that "in the past few years, detailed stream-specific surveys in Oregon and California have documented widespread extinctions of coho salmon in streams known to have supported spawning populations as recently as the 1960s and 1970s." Specific locations of the extinctions were not given, and it is not clear whether the statement referred to population units in small tributaries or to entire spawning runs.

Coho salmon in California apparently do not undertake extensive oceanic migrations but remain within a few hundred kilometers of their natal streams while at sea. They occur primarily in small to mid-size coastal rivers and creeks as far south as Monterey Bay. Attempts to increase runs of coho salmon in the Sacramento River by planting hatchery-produced fry in the 1950s were largely unsuccessful (Moyle 1976). Nehlsen et al. (1991) treated coastal coho populations from San Francisco to Oregon as a single stock with a moderate risk of extinction. Higgins et al. (1992) identified 20 separate coastal populations, about one-third of which were felt to be at high risk of extinction. The Wilderness Society (1993) considered native Sacramento River coho to be extinct and most of the coastal populations to be either threatened or endangered. Brown et al. (1994) reviewed the status of coho populations in California and concluded that coho had undergone dramatic declines from historic levels statewide. Coho salmon no longer pass Benbow Dam on the Eel River, although that river once supported runs of 5,000-25,000 fish. Brown et al. (1994) noted that the decline occurred despite substantial hatchery programs to raise coho at five hatcheries in California. Overall, although historical coho runs in California once totaled in the hundreds of thousands, estimated numbers of naturally spawning adults are now fewer than 5,000 (The Wilderness Society 1993).

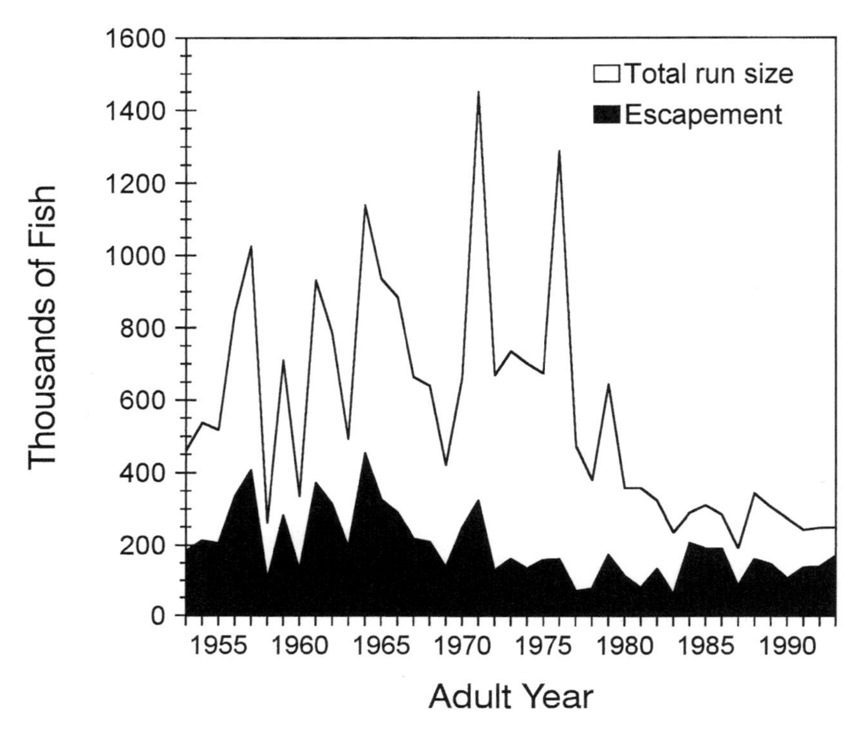

FIGURE 4-13 Estimated numbers of Oregon Coastal Natural (OCN) coho, 1953-1993. Area under upper line represents total number of adult or maturing adult fish; shaded area under lower line is number of naturally spawning salmon in Oregon's coastal streams. Difference between upper and lower lines represents commercial and sport catch, including in-river sport catch. From 1953 to 1959, escapement results were unavailable for Oregon coastal lake systems; therefore, abundance of lake-derived coho for each year was assumed to be 7.5% of total and 7.1% of escapement—averages of lake contributions to total run size and escapement during 1960-1970. Data courtesy of Oregon Department of Fish and Wildlife.

Chum salmon have been recorded along the West Coast south to the San Lorenzo River in Monterey Bay (Moyle 1976), but the current southernmost population of chum salmon is in the lower Smith River of northern California (Higgins et al. 1992). According to The Wilderness Society (1993), the Smith River contains the sole remaining self-reproducing California population, but adults occasionally stray into other rivers. There are 26 known wild chum populations on the Oregon coast, most of which vary greatly in abundance from year to year (Nickelson et al. 1992). Overall abundance is considerably less than in the early twentieth century. The chum run to Tillamook Bay in 1928 was estimated to be about 650,000 adults, but since the 1960s the maximum estimated run has been only 26,000 fish. Over the last 20 years, some runs have been relatively stable, although at reduced levels. Partly because there were no clear declines over this period in some Oregon rivers, Nickelson et al. (1992) declared 10 of the 26 chum salmon populations to be healthy, but 12 were listed as vulnerable because of their small run size. Of the 14 chum salmon populations known from the Washington coast, about two-thirds were classified as healthy by WDF et al. (1993), and the others were unknown.

Pink salmon are absent from the Washington and Oregon coast, and it is not clear whether they were found in abundance in any coastal streams within the last 2 centuries. Pink salmon do not now exist in California, although Nehlsen et al. (1991) reported that extinctions of Klamath and Sacramento pink salmon runs were recent. As late as the 1970s, a few stray pink salmon were observed in some northern California rivers, but the only reproducing population occurred in the Sacramento River system (Moyle 1976). The Sacramento River population was surely a zoogeographic enigma, being over 1,000 km south of the southern limit of pink salmon distribution in Puget Sound.

Sockeye salmon depend on lakes for much of their freshwater rearing. A few lakes are present in some coastal Oregon and California river systems, but no sockeye populations occur there now. Historically, small sockeye runs might have existed in the Sacramento River and in the upper Klamath River (Moyle 1976), but Nehlsen et al. (1991) did not list them among recent sockeye extinctions. Three sockeye stocks from the Washington coast are known; one is considered healthy, one is listed as depressed, and the status of the third is unknown (WDF et al. 1993).

The status of fully half the steelhead stocks in Washington coastal streams was unknown, according to WDF et al. (1993). Summer steelhead were the least understood; eight of the nine stocks were listed as of unknown status. About half the winter steelhead stocks were classified as healthy. Only two of the 31 coastal stocks were considered depressed. Only about 10% of the coastal steelhead populations are partially supported by hatchery production.

Winter steelhead occur in nearly all coastal Oregon streams. Summer steelhead are native only to the Siletz, Umpqua, and Rogue rivers. Some Rogue River steelhead exhibit an unusual life-history variation termed "half-pounders"; these spend only 3-4 months at sea before returning to the river immature. The only other populations of steelhead in which this life-history variation occurs are in the Klamath and Eel rivers of California. Otherwise, summer and winter steelhead exhibit a typically variable life cycle of 1-4 years of freshwater residence and 1-4 years at sea, although 2 years in freshwater and 1-3 years at sea is the norm. Steelhead can reproduce more than once in a lifetime (a phenomenon called iteroparity); however, the incidence of repeat spawning is low, ranging from 3-20% in coastal streams (Nickelson et al. 1992). Juvenile steelhead entering the ocean do not stay in coastal waters but move offshore, where they migrate to the Gulf of Alaska and the western North Pacific Ocean.

The Oregon Department of Fish and Wildlife identified a number of wild Oregon coastal steelhead populations, some of which were very small and poorly known. However, its 20-year analysis of current status generally included only streams in which the annual sport catch exceeded about 200-300 fish (Nickelson et al. 1992). Of the 24 wild winter steelhead populations selected for analysis, 19 were considered depressed and five healthy. Two-thirds of the summer steelhead populations were also considered depressed. A strong correlation was noted among winter steelhead population trends from 1980 to 1990 in most streams. After higher than average runs in the mid-1980s, in the 3-year period from 1988-1990, 20 of the 24 winter steelhead populations were below the 20-year average in abundance for all 3 years, and four of five summer steelhead populations (two populations were established from nonnative populations) were below average for all 3 years.

Historically, steelhead occurred in coastal rivers throughout the entire length of California, ranging as far south as the Tijuana River in northern Mexico. Summer steelhead are now limited to the Eel River and streams northward; winter steelhead are now found as far south as the Ventura River in southern California. Nehlsen et al. (1991) recognized five summer steelhead stocks at moderate to high risk of extinction in northern California. Higgins et al. (1992) identified 11 summer steelhead stocks at risk, eight of which were classified as endangered. Winter steelhead populations in northern California are generally in much better condition than summer steelhead. Most are thought to have been fairly stable in recent decades, but nearly all populations have declined somewhat over the last 5 years (Wilderness Society 1993). In central and southern California, winter steelhead are at low levels, although the populations might always have been relatively small in this warm, dry region.

Light (1987) assessed coastwide abundance of steelhead and estimated an average annual abundance of about 1.6 million steelhead adults, of which half were wild and half were of hatchery origin (Figure 4-14). However, the proportion of wild steelhead was greatest in areas with lower human populations (Alaska and British Columbia) and in California, where huge areas of steelhead habitat have been lost in the Sacramento and San Joaquin rivers. Essentially, California now relies on coastal rivers and streams for most steelhead production. Light (1987) noted that abundance in the 1980s was about the same as it had been in the 1970s but that the fraction of hatchery fish was greater in the later period. Richards and Olsen (1993) found similar production trends in steelhead, citing the Washington Department of Wildlife (1992). They concluded that ocean conditions were primarily responsible for recent declines in steelhead abundance.

Sea-run cutthroat trout are distributed along Pacific Northwest coastal streams south to the Eel River. Like steelhead, sea-run cutthroat rear in fresh water for up to several years, but unlike steelhead, they sometimes spend extended periods in estuaries and most do not travel far from natal streams while at sea (Trotter 1989). Repeat spawning occurs in sea-run cutthroat trout, perhaps to

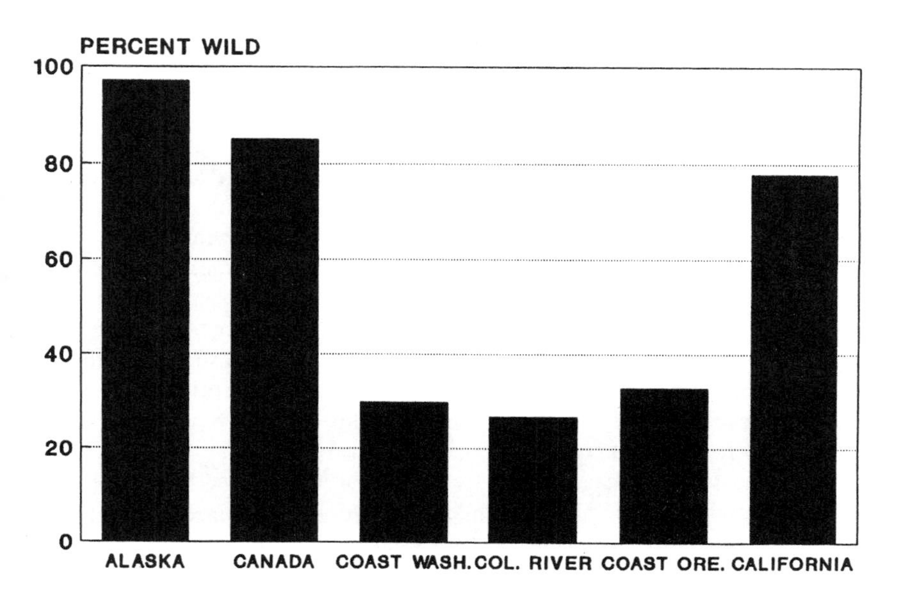

FIGURE 4-14 Percentages of wild steelhead in Alaska, British Columbia, coastal Washington and Puget Sound, Columbia River Basin, coastal Oregon, and California. Source: Light 1987.

a greater extent than in steelhead. Anadromous adults sometimes mate with nonmigratory stream-dwelling cutthroat. Of all species of Pacific salmon, sea-run cutthroat trout are perhaps most poorly known, because they are not commercially fished and recreational anglers are not required to maintain punchcard records. No recent extinctions of sea-run cutthroat trout in Washington, Oregon, or California were noted by Nehlsen et al. (1991), but these authors believed the species to be in decline.

Ninety two sea-run cutthroat populations from the Oregon coast have been identified. Nickelson et al. (1992) believed that many of these populations had declined before 1980, in that population surveys of central Oregon streams between 1980 and 1990 showed no substantial trends of decline. Sea-run cutthroat from the North Umpqua River have recently been petitioned for status review under the ESA. Returning adults at Winchester Dam declined from about 1,000 in 1946-1956 to fewer than 100 wild fish by 1960 and have remained at extremely low levels. Higgins et al. (1992) identified four stocks from northern California that were felt to be declining, and The Wilderness Society (1993) considered all coastal sea-run cutthroat to be threatened. The status of sea-run cutthroat was not assessed by WDF et al. (1993) for Washington's coastal streams, but Trotter et al. (1993) believed the species to be in general decline in Washington, in part because of supplementation of streams with hatchery-produced coho salmon.

WILLAPA BAY—A CASE STUDY

Although no single river basin is representative of the Pacific Northwest coast as a whole, the Willapa Bay Basin is an instructive example of the interactions among habitat alteration, fisheries, and hatchery management in a major drainage system without dams. The Willapa Bay Basin is a large, highly productive estuary and watershed on the southwestern Washington coast. At mean high water, the bay covers about 120 mi^2 (311 km^2), and the surrounding watershed encompasses 1,060 mi^2 (2,745 km^2). There are 745 rivers and streams with over 1,470 linear miles (2,366 km) in the basin (Phinney and Bucknell 1975, Suzumoto 1992). There are at least 15 known stocks of salmon in the basin (WDF et al. 1993), mostly fall chinook, chum, coho, and winter steelhead.

Land- and water-management activities in the basin have been typical of coastal areas with neither dense human populations nor large hydroelectric projects. Much of the watershed is commercially forested and has been logged since before the turn of the century. Lowlands have been farmed and diked for pasture. Habitat in many streams in the basin has been altered because of those management actions (Pyle 1986), but to date there are no comprehensive surveys of stream conditions in Willapa Bay tributaries. A fishing fleet has operated in the basin since early in the century, catching salmon originally by trapping and netting and later primarily by gill netting. There are about 19,000 inhabitants of the basin, most of whom depend on the natural resources of this productive area. In spite of the typical panoply of human impacts, the Willapa Bay ecosystem remains in remarkably good condition. The bay is one of the cleanest estuaries in the contiguous United States and commercial shellfish production on the extensive mudflats is among the highest in the world.

Of the 15 salmon stocks identified by WDF et al. (1993), nine were considered healthy, five of unknown status, and only one (North River chinook) depressed. By most measures, then, Willapa Bay is believed by many to be a healthy, productive system for salmon and other natural resources.

However, wild salmon inhabiting the Willapa Bay Basin are in serious trouble. Nehlsen et al. (1991) listed native Willapa Bay coho at high risk of extinction from high catch rates in mixed-stock fisheries and negative interactions with hatchery fish. The primary fisheries-management objective in the Willapa Bay Basin is the production and capture of hatchery fish, and hatcheries have been present there for almost a century. Of the six salmon hatcheries constructed since 1895, three are still operational. Early in the 1900s, combined catches of chinook, coho, and chum salmon in Willapa Bay occasionally exceeded 250,000 returning adults; by the 1960s, the terminal fishery captured only 30,000-40,000 fish (Suzumoto 1992). An aggressive program to increase hatchery production was undertaken, and by the 1980s, total catches were 100,000-240,000 salmon, with the greatest numerical increases occurring in coho and chum. Total numbers of eggs taken at the hatcheries from returning adults increased from fewer than 5 million in the 1950s to more than 40 million eggs in the 1980s, and the number of hatchery smolts went from fewer than 5 million to more than 20 million in the same period. Willapa Bay Basin hatcheries recently produced 11% of the fall chinook and 7% of the coho released by all state-operated hatcheries in Washington (Suzumoto 1992). Many of the hatchery fish are released directly from rearing facilities, but chinook, coho, and chum fry have all been planted in the basin's streams to increase returns to hatchery trapping locations and to bolster declining wild runs. Weirs at the hatcheries prevent many fish from spawning naturally. In some years, fish are allowed to pass upstream, but only after the annual egg-take quota is met.

Some of the hatchery releases were offspring of fish from outside the basin. At least six nonnative coho populations and eight nonnative chinook populations are known to have been cultured and released from the Willapa, Nemah, and Naselle hatcheries from 1952 to 1990 (Suzumoto 1992). Chum salmon at the hatcheries have been native Willapa Bay Basin populations, although hatchery chum propagation was discontinued in the 1980s. Streamside egg boxes were used in the 1970s and 1980s to augment chum and coho in a number of basin streams; more recent enhancement projects have used in situ egg-incubation chambers (plastic buckets) that hold up to 10,000 eggs.

Catch rates of Willapa Bay Basin chinook and coho have been estimated to be in excess of 70% (Table 4-5), a high fishing rate that has occurred among salmon populations elsewhere along the Pacific Northwest coast in rivers with large hatchery outputs (Nickelson et al. 1992). Although most of the catch has

TABLE 4-5 Average Catch and Catch Rates of Coho and Chinook in the Willapa Bay Basin for the Years 1971-1991[a]

	Coho	Chinook
Escapement	31,600	16,500
Catch in Willapa Bay	44,500	14,200
Washington mixed-stock catch	42,000	17,000
Alaska and B.C. catch	24,000	19,000
Total run size	142,100	66,700
Total catch	111,000	50,000
Catch rate	78%	75%

[a]Catches in Willapa Bay and escapement from Suzumoto (1992). Interception by fisheries in British Columbia and Alaska based on assumed interception rates of 40% for fall chinook and 20% for coho.

TABLE 4-6 Comparison of Historical Numbers (Based on Early 20th-Century Estimates) of Spawning Salmon with Recent Escapement Goals and Hatchery Use of Returning Adults for the Willapa Bay Basin[a]

	Historical Escapement	Recent Escapement Goal	Recent Goal (% of Historical Run Size)	Hatchery Take (% of Run)	Number Naturally Spawning Adults	Number Spawning per Kilometer
Chinook	72,000-122,000	15,000	12-21%	80%	3,000	2.4
Coho	64,000-108,000	8,000	7-12%	80%	1,600	0.6
Chum	215,000-366,000	35,400	10-16%	0%	35,400	42.0

[a]About 2,450 km of streams was assumed to be available to coho, 1,170 km to chinook, and 830 km to chum salmon.

Source: Suzumoto 1992.

been of hatchery origin, exploitation rates of wild fish have also been high. Compared with historical estimates of salmon returns, which occasionally exceeded 500,000 fish, recent escapements are only a small fraction of former runs (Table 4-6). Coho appear to be most severely depressed, but both chinook and chum salmon are also far below estimated escapements of early the twentieth century. Large numbers of coho and chinook are removed from streams for artificial propagation, so the actual number of naturally spawning salmon in the basin is less than 10% of the historical runs.

Naturally spawning coho have not been counted annually routinely, because the basin has been managed for hatchery production of this species; but spawning counts of chinook and chum are available. From 1968 to 1991, chum spawning counts for the Willapa Bay Basin averaged about 28,000 which agrees reasonably well with the escapement goal of 35,000 (Table 4-6). However, average annual chinook spawning counts from 1987 to 1991 averaged about 15,000 (Suzumoto 1992), a figure considerably greater than the estimate of 3,000 naturally spawning fish in Table 4-6. Suzumoto (1992) pointed out that most of the chinook and coho observed in recent spawning surveys were strays from hatcheries. Nevertheless, the estimate of naturally spawning chinook in Table 4-6 might not accurately reflect the most recent survey information.

Even allowing for error in the estimates of naturally spawning chinook and coho, the number of salmon using the basin's streams is still far below what the ecosystem is capable of supporting. For example, the Oregon Department of Fish and Wildlife recommends spawning densities of 24 adult coho per kilometer (about 40 adults per mile) for adequate seeding of Oregon's coastal streams.

Recent coho escapement to the Willapa Bay tributaries might be only about one-tenth that target (Table 4-6). Management of Willapa Bay salmon for hatchery production has resulted in an underuse of the natural rearing capacity of the drainage system. If most of the naturally spawning coho and chinook in streams with artificial production facilities are hatchery strays, the number of wild fish and the genetic integrity of native populations must indeed be low.

In addition to causing substantial declines of wild populations of salmon in the Willapa basin, the combination of habitat alteration, high catch rates, and removal of fish at hatcheries might be depriving the aquatic ecosystem of an important seasonal source of nutrients. To judge by reductions in naturally spawning fish over the last century, the basin has lost more than several thousand metric tons of salmon tissue each year (Table 4-7). Present loadings of salmon carcasses and their nutrients are now generally less than 10% of historical levels (Tables 4-7 and 4-8). It is likely that this absence of nutrient capital has further reduced the capacity of the Willapa basin to produce fish, shellfish, and other important aquatic resources and has led to a long-term decline in ecosystem productivity.

The current condition of wild salmon in the Willapa Bay Basin illustrates a systemic problem along the Pacific Northwest coast: habitat, hatchery, and fishery management decisions have failed to protect the natural capacity of these areas to produce salmon. Large annual investments in artificial-production facilities coupled with various degrees of habitat losses and high exploitation rates have driven wild populations down while increasing the importance of hatchery runs. So dependent is the basin on hatchery salmon production that if the flow of hatchery smolts were stopped, the runs would probably experience major declines.

TABLE 4-7 Comparison of Historical and Recent Return of Salmon-Carcass Biomass to Willapa Bay Basin Streams[a]

	Mean Body Weight (kg)	Historical Escapement to Streams	Historical Biomass Returned to Streams (metric tons)	Recent Escapement to Streams	Recent Biomass Returned to Streams (metric tons)
Chinook	8.9	72,000-122,000	641-1,086	3,000	27
Coho	3.7	64,000-108,000	237-400	1,600	6
Chum	5.0	215,000-366,000	1,075-1,830	34,500	172
Total		351,000-596,000	1,953-3,316	39,100	205

[a]Historic run-size information from Suzumoto (1992). Current salmon run sizes based on fishery escapement goals and assumation that 80% of chinook and coho are taken for hatchery use and removed from ecosystem.

TABLE 4-8 Historical and Recent Annual Loading of Salmon-Carcass Phosphorus, Nitrogen, and Total Biomass to Willapa Bay and Its Tributaries[a]

	Delivery to Streams (kg/km of stream length)		Delivery to Willapa Bay (kg/ha of surface area)	
	Historical	Recent	Historical	Recent
Phosphorus	3.0-5.0	0.30	0.23-0.38	0.02
Nitrogen	82-140	9.0	6.3-10.70	0.69
Total biomass	823-1,400	86.0	63-107	6.90

[a]Carcass biomass assumed to be 0.364% phosphorus and 10.0% nitrogen by wet weight.

5

Values and Institutions

INTRODUCTION

The "salmon problem" comes at a time of intensive questioning of values and ethics concerning human relationships with nature. American conservation policy generally is grounded in moral values (Callicott 1991). To a large extent, the values are and have been anthropocentric. "Preservationists," in the moral tradition linking Thoreau and Emerson with John Muir, emphasize the aesthetic, spiritual, and psychological values of preserving wilderness. Around the beginning of this century, a more democratic and practical moral position was found among "conservationists," such as Gifford Pinchot, who established the U.S. Forest Service. In accordance with the "gospel of efficiency" (Hays 1959) and the philosophy of utilitarianism, *conservation* became defined as the "greatest good of the greatest number for the longest time" (Callicott 1991:45). Nature became "natural resources"; and science, with its own world view, became the basis for determining the "best" uses of those resources. This moral position predominates and is exemplified in the law, policies, and practices of state and federal resource-management agencies and in the training of most natural-resource scientists. However, its anthropocentrism has long been challenged, as in Aldo Leopold's formulation of a more ecocentric "land ethic" (Leopold 1944), in a sustained, highly diverse moral and political position that we now think of as "environmentalism." Environmental issues have been addressed through choices made within economic, political, and individual ethical frameworks; an important component of political choice has been planning, a concept that has substantially influenced federal actions in the Northwest.

The values and the ethical positions held today by people involved in and affected by the salmon problem of the Pacific Northwest encompass all that history and more; these values encompass what Castle (1993) has called a "pluralistic, pragmatic and evolutionary approach to natural resource management." A first step is recognizing and articulating that pluralism, because problems in managing and protecting fish populations are due in part to the failure to articulate divergent interests, goals, and values and to address them explicitly (e.g., Larkin 1977, Rothschild 1983, Barber and Taylor 1990). This chapter describes how the widely varied ways that humans intervene in salmon populations are linked to socially validated values.

From a policy perspective, the salmon problem is one of long-standing and serious conflict in fact, interest, and values. Disputes over fact reflect scientific uncertainty and difficulty in developing consensus among scientists over interpretation of facts (Ludwig et al. 1993). Interest conflict is related to the clustering of preferences and values around particular uses of salmon and their habitats, such as flood control, timber harvest, economic return from hydroelectric power, and agricultural, industrial, or municipal use. Examples of interest conflict might be a position favoring dam drawdown over flow augmentation to protect irrigators or a position favoring harvest restrictions rather than a change in river flow to protect the operations of aluminum companies. People often invoke widely held values to protect particular interests, but values are genuine sources of conflict in themselves. Some people hold the value that it is unnatural to transport salmon in barges, and thus oppose that option. Those with a more utilitarian value system argue that it is the cheapest and most effective way to get salmon through the system of Snake River dams. Value conflict stems from different assessments of the desirable goals of public action. For example, there is a value conflict between the goal of relatively inexpensive electricity in the region and the goal of protecting endangered populations of salmon.

INDIVIDUAL PREFERENCES AND PUBLIC VALUES

Different people want different things from the oceans and rivers. In the aggregate, and through various social processes, those different desires and goals are expressed as public values. For example, members of the public differ in how they say they weigh economic against environmental concerns or in how much they think government should be involved in protecting the creatures and habitats of nature. These and other values are revealed in surveys, political choices (voting and letter-writing), and consumer choices.

Two recent surveys are pertinent to attempts to measure how people value salmon and measures to protect or rehabilitate them. The first, a forestry-related survey, was completed in the fall of 1991 by faculty at Oregon State University (Steel et al. 1992, 1993). It polled a representative sample of national and Oregon publics about natural-resource issues, including fisheries. The second is the

Western Rangelands Survey, conducted in the spring of 1993 (Steel and Brunson 1993), a national and regional telephone survey. For each survey, about two-thirds of sampled households responded. More than 2,000 responded to the forestry survey, and more than 1,300 to the rangelands survey.

Data presented in Table 5-1, from the 1991 forestry and 1993 rangelands surveys, indicate values toward salmon, salmon-habitat protection, and protection of endangered species shared widely in both national and Oregon samples. The data indicate regional support and even stronger national support for protection of salmon and critical habitat associated with salmon. Less than a majority supported setting aside endangered-species laws.

In the 1991 OSU survey of natural-resource and forestry issues, respondents were asked to select their own positions on a scale regarding the importance of managing for environmental and economic considerations (Table 5-2). Nearly half who responded favored an equal balance between environmental and economic components in natural-resource management. However, 42% of the national sample and 37% of the Oregon sample expressed a preference for an environmental perspective.

The data from the two surveys suggest public values that support protection for wildlife, salmon, and other fish on public lands. The groups that most strongly favor protection include urban residents, younger people, women, environmental-group members, and respondents who report that they go fishing often (Table 5-3). Among Oregon residents, the majority surveyed support greater efforts to protect wildlife habitat and salmon, but at proportions slightly lower than respondents of the nation as a whole. There are a few notable exceptions, however: the oldest groups, men, and people who depend on the ranching and timber industry for their livelihoods tend to favor extractive uses of natural resources.

The differences by age could be explained by either a generational shift or the tendency of people to become less environmentally conscious as they age and assume larger financial obligations. Longitudinal studies, however, suggest that the age differences reflect a societal shift in the direction of stronger support for environmental values.

HOW SALMON ARE VALUED

Surveys estimate the range and depth of public opinion, which might or might not be well informed by scientific appraisals of the value of natural resources. From a scientific perspective, wild salmon populations are an example of an ecosystem's natural capital. Like any capital asset, natural capital embodies value in its two components: the stock and the flow of benefits provided by that stock. The stock of wild salmon provides a flow of benefits in the form of reproductive services, genetic diversity, habitat nourishment, and food for wildlife and humans. Our greatest success has been in designing incentives to capture the flow of human-food benefits from wild salmon populations. Our correspond-

TABLE 5-1 National and Oregon Concern for Wildlife, Salmon, and Their Habitat[a]

	National (%)			Oregon (%)		
	Agree	Neutral	Disagree	Agree	Neutral	Disagree
1991 OSU Forestry Survey	78	10	12	55	21	25
A. Greater protection should be given to fish and wildlife habitats on federal forest lands.						
B. Endangered species laws should be set aside to preserve timber jobs.	17	18	65	37	15	48
1993 WSU/USU Rangelands Survey						
C. Greater protection should be given to fish such as salmon in rangelands.	76	10	14	59	21	20
D. Endangered species laws should be set aside to preserve ranching jobs.	20	17	65	39	14	47
E. The loss of streamside vegetation is a serious problem on federal rangelands.	83	10	7	78	16	5

[a] Scales have been collapsed for presentation purposes. With the exception of E, all differences between national and Oregon responses are significant at $p < 0.001$.

Source: Smith and Steel 1995.

TABLE 5-2 Economic Versus Environmental Tradeoffs: National and Oregon Samples Compared

Many natural resource management issues involve trade-offs between natural environment conditions (wildlife, old growth forests) and economic considerations (employment, tax revenues).
"Where would you locate yourself on the following scale concerning these issues?"

	National (%)	Oregon (%)
"The highest priority should be given to maintaining natural environmental conditions even if there are negative economic consequences."	42	37
"Both environmental and economic factors should be given equal priority in forest management policy."	47	44
"The highest priority should be given to economic considerations even if there are negative environmental consequences."	11	19

Source: Steel et al. 1992.

ing failure has been in protecting the flow of indirect and nonhuman benefits. One way to present the salmon problem is to say that the value of the Pacific Northwest's salmon-capital asset has depreciated over time as its productivity has declined (Alkire 1993). A major problem is that the market, which we rely on for signals about the status of resources, does not account for the full range of costs and benefits of environmental resources, such as salmon. That is called a market distortion. When such market distortions exist, some resources are underpriced and overused, and others overpriced and underused.

Salmon's total value has three parts: direct value (e.g., catching wild salmon for food), indirect value (e.g., the contribution of genetic diversity to reproduction), and option value (e.g., the future contribution of wild salmon to fishing, future genetic diversity, or having wild populations in the future, or combinations of future alternatives). Wild salmon's existence value is the intrinsic value that people place on simply knowing that wild populations exist or on supporting stewardship of wild populations as a bequest to future generations (Pearce and Turner 1990, Pearce 1993).

Direct values are easiest to calculate because they are amenable to quantification through expression in money exchange. They include the subsistence value of salmon for food and the commodity value of salmon in market exchange. The latter is the most usual measure of salmon value because monetary measures are at hand. Salmon has commodity value to fishers, processors, and distributors; restaurants, suppliers, and boatbuilders; and tour operators, fishing guides, and charter boat operators, and others who market either salmon or the privilege of trying to capture or view salmon. Further direct value is added through the

TABLE 5-3 National and Oregon Support for Protection of Salmon and Wildlife on Federal Timber and Rangelands[a]

	% Agree 1991: "Greater protection should be given to fish and wildlife habitats on federal forest lands."[b]		% Agree 1993: "Greater protection should be given to fish such as salmon in rangelands."[c]	
	National	Oregon	National	Oregon
Gender				
Women	86	64	84	65*
Men	71	44	69	53*
Age Cohort				
18-29 years	88	55	87	60
30-45 years	83	61	85	59
46-60 years	68	47	72	57
61 plus years	64	47	63	47
Residence				
Urban	79**	63	78*	63**
Rural	73**	52	69*	51**
Interest				
Environmental group member	84	76	84	77
Timber/ranching dependent	60	48	65	43
Outdoor Recreation				
Fishing	80	57	80	84
No fishing	76	50	74	51

[a]Differences noted by * are significant at $p < 0.01$ or better based on χ^2. Those marked with ** are significant at $p < 0.05$ or better based on χ^2. All other differences are significant at $p < 0.001$ or better based on χ^2.

[b]Source: Steel et al. 1992.

[c]Source: Steel et al. 1993.

impact of salmon-based income spent in an economy. There are estimated to be over 8,000 fulltime work years involved in the west coast salmon industry (NMFS 1995). Fishing is an important component of Oregon's coastal-community economies, second only to the wood-products industry in its income impact (Radtke 1992). However, the economic contribution of salmon is declining with reductions in allowable catch levels and in market prices. The direct economic value of 1993 commercial salmon fisheries is only 10% of the 1976-1993 average value. The direct economic value of 1993 recreational fisheries is 30% of the 1976-1993 average (PFMC 1994).

Indirect and option values, expressed in wild salmon's contribution to future harvest or to genetic diversity, often compete with direct values in resource-management policy. Existence values also must compete with direct values of salmon. Existence values are very important in the Pacific Northwest, probably increasingly so as salmon becomes more scarce. Many people, including some involved in commercial fishing, value salmon and the activity of fishing for salmon for nonmonetary reasons (Smith 1981).[1] The ideas of salmon fishing as a lifestyle for commercial fishers and of salmon fishing as recreation for anglers express existence values. Existence values also include the beauty of salmon runs and the appreciation of salmon as wildlife—values that are shared within a broader public. Option and existence values are notoriously difficult to assess in monetary terms because, for value to be assigned, people must be asked what they would be willing to pay to maintain salmon's amenity and existence values or how much they would have to be compensated to give up the amenity and existence values of salmon. Contingent valuation—the technique most commonly used to assess option, amenity, and existence values—is difficult to administer without bias and is sometimes used to disputed effect (Pearce and Turner 1990, Pearce 1993).

Direct Regional Economic Value

The "exvessel" value—the value at the first sale of the vessel's catch—of West Coast landings of commercial salmon varies with quantity caught and price. The exvessel value of ocean-caught salmon, which was about $98 million in 1979, fell to about $12 million in 1983 with the El Niño depression of landings, rose to about $80 million in 1988, and fell to about $6.6 million in 1994 (PFMC 1995:IV-6).

Exvessel prices have also declined; reduced landings since the late 1980s have not been accompanied by increases in price. In California, prices per pound of both chinook and coho declined overall from 1988 to 1993; and in Oregon and Washington, prices have declined since 1990 (PFMC 1994). Paradoxically, the price decline, which has exacerbated the revenue effects of low landings, is caused by an oversupply of salmon on the world market. West Coast salmon prices are not determined by catches in the Pacific Northwest but by quantities landed in Alaska and British Columbia or farmed in Norway and Chile. Commercial ocean landings in both Alaska and British Columbia are at historically high levels, and production of farmed salmon in Norway, Chile, and elsewhere reached 1 billion pounds in 1994, exceeding U.S. commercial landings for the first time (Johnson 1995). In world markets, more salmon is available now than at any time in the last 50 years. Large quantities of salmon depressed prices on world markets while continuing recessions in Japan, Europe, and the United

[1]An indirect measure of the importance of nonmonetary benefits is the extent to which people engage in fishing even though it is not profitable (Stevens 1972, Liao and Stevens 1975, F. Smith 1973, 1979, Radtke 1992).

States lowered demand for salmon. The near-term outlook is for continued large supplies of both captured and farmed salmon with correspondingly low salmon prices worldwide (Pacific Fishing 1994).

Continued declines of commercial ocean landings and recreational ocean trips in the Pacific Northwest have caused a decline in salmon-related income. Total state-level income impacts associated with ocean commercial and recreational fisheries were 70% lower in 1993 (PFMC 1994) than in 1979-1988, when expectations for salmon's economic contribution to coastal economies were high. Total income generated by salmon fishing includes income to businesses that rely directly on salmon fishing for income (commercial fishers, and charter boats), marine businesses that depend indirectly on salmon fishing for income (marine suppliers), and community businesses that benefit from the infusion of salmon-generated money into the economy. Along the coast, there is a wide variation in the economic dependence on salmon and a corresponding variation in the extent to which the economic impact of the decline has been felt.

The ocean's natural variability combines with human influences on biological productivity to create a high level of uncertainty about salmon yields, which generates economic uncertainty about salmon as a source of income. Fishers have adapted to the uncertainty by diversifying their fishing over species, times, and areas. Trollers who once specialized in salmon fishing now troll for albacore and might jig or set longlines for rockfish. Gillnetters fish in Alaska, as well as on the Columbia River. Some trollers and gillnetters capture a greater share of the landed value of their catch through such activities as direct marketing, custom canning, and smoking. Salmon trollers and gillnetters have also diversified to nonfishing activities. Salmon fishing combines with a number of nonfishing occupations because it is so seasonal.

The recreational fishery also has diversified in response to uncertainty. Charter boats and individual anglers have branched out into nonsalmon species, primarily rockfish. Management actions have supported that change. In California, Oregon, and Washington, commercial fishing for some rockfish species has been curtailed to provide exclusive areas for recreational fishing and relieve pressures, particularly that caused by charter boats, on salmon.

Diversification strategies have been hindered by regulations and license restrictions. Shortened seasons give advantages to highly mobile fishers, but catch quotas distributed throughout the year tend to benefit local communities of fishers, who can subsist on small catches close to their home port, and license limitations restrict the movement of fishers between fisheries.

Indirect and Option Values

Symbolic values are poorly represented in most analyses of values and processes of valuation. Salmon have both material and symbolic or cultural importance to society; they affect families, larger communities, the region, and the nation. In this sense, the commodity values noted above are social values, and so are noncommodity values. Included in social values are the values of salmon for subsistence and nutritional health, their values for recreation and for tourism, and their spiritual values in American Indian life and ceremony. Moreover, there are important symbolic links between ethnic, community, and regional identities and salmon. Salmon have provided social continuity and heritage for many Americans—the American Indian tribes and non-Indian fishing communities that depend on salmon fishing, the generations of sports anglers proud of their pursuits of steelhead and other salmon, the general public of the Northwest who have adopted salmon as a regional symbol, the airport shops that sell smoked salmon and salmon artifacts to tourists wanting souvenirs, and so on. Salmon are featured in art and song in the Pacific Northwest to an extent shared by few other fishes anywhere (Holm 1965, Stewart 1979, Williams 1980, Jay and Matsen 1994).

Salmon have symbolic value to those outside the region, as well. The heroic feats of salmon, in their long migrations from headwaters stream to ocean, led back again by long-remembered odors of their natal stream, vaulting waterfalls and dodging bears to mate and die (Hasler 1966), tattered relics of their former beauty—this is one of the great sagas of animal behavior. The social values of salmon include their roles as symbols and indicators of environmental quality, such as the role of the return of salmon runs as a sign of improved water quality on the Willamette River of Oregon (Gleeson 1972, Starbird and Georgia 1972, F. J. Smith 1973). They are also both symbols and substantive foci in relationship to broader issues of habitat conservation, biodiversity, and the fate of threatened and endangered species.

The biological and ecological values of salmon fit into categories of indirect and option values that are underpriced in the marketplace. Salmon are essential parts of the aquatic and riparian ecosystems that they inhabit. Young salmon are food for fish and birds. Adults bring carbon and calories from the ocean to nutrient-poor lakes and streams. The spent carcasses of adults that have spawned provide food for bears, foxes, wolves, ravens, and eagles. In other words, salmon have value beyond the human sphere and in ways only recently recognized by human institutions of North America.

Biological diversity is a nonrenewable resource (Lichatowich 1992, Wilson 1992, Riddell 1993b). Its principal biological values include adaptedness in existing populations and the potential for further evolution, the maintenance of the spatial and temporal bases of production, the knowledge gained from studying diverse organisms, and indeed the services that organisms provide to other inhabitants of the earth, including the ecological services that support human activities.

Another way to express the different types of values embodied in salmon is to consider them in the context of ecology and economics. Lichatowich (1992) distinguishes between the economies of human systems ("industrial economies")

TABLE 5-4 Characteristics of Industrial and Natural Economies

Industrial Economy	Natural Economy
Fossil fuel as primary source of energy	Solar radiation as primary source of energy
Large, centralized production facilities *(economies and monocultures)*	Dispersed production *(stability through diversity of scale, the principle of spreading risk)*
Emphasis on production *(linear extraction of resource, continuous growth)*	Emphasis on reproduction *(renewable cycles, limits to growth)*
External improvements and changes *(humans dominate, shape course of diversity, determine scope for change)*	Internal improvements and changes *(natural processes and genetic evolution)*
Independent spheres of economic activity *(individualism)*	Interdependent parts *(contextual)*
Global imbalance *(greenhouse effect, holes in ozone)*	Global gas exchange *(GAIA effect)*

Source: Lichatowich 1992.

and natural systems ("natural economies") to illustrate that the two systems have different dynamics that can work to the detriment of salmon. The characteristics of industrial and natural economies are summarized in Table 5-4. The attributes of human economic systems that have encouraged growth and expansion—individualism, extraction, scale economies, biological simplification—have eroded the health of natural economies that rest on interdependence, renewal, dispersed production, and diversification.

Resource Values and Public Choice

Markets reflect some of salmon's values but not all. Many nonmarket values of salmon are not easy to measure and compare, and so management decisions concerning salmon often do not adequately reflect the importance of salmon to society; nonmarket values especially are underrepresented. As a result, decisions about resource use may not achieve societal goals. To correct the discrepancy between social values and resource use, attempts can be made to design policies that reflect the full range of resource values. International environmental agreements concerned with accounting for the full costs of pollution call this the "polluter pays" principle. A corresponding application of the principle to Pacific

Northwest salmon might be called the "full value" principle, incorporating the full range of wild salmon's values into decisions about its protection and use.

Full value is a public, not a private, question. Private choices are made with individual preferences about products whose attributes and prices are known; however, they require well-defined property rights, which do not exist in the United States for public resources, such as salmon and navigable waterways, and are imperfect for other resources, such as groundwater and grazing lands. Consequently, public choices are central to the salmon problem. Public choices have to take into account many owners with multiple preferences, attributes that are not fully observable and sometimes unknown, and prices that reflect only part of the resources' full value to society. As we have seen, wild salmon have not only commodity values (represented in market prices) but many other values that remain unpriced. Unpriced resource values are easy to ignore and are often unrepresented in public decisions about environmental resources, but they are nevertheless real. The concept of full value points to the problem of "externalities"—the problem that some costs and benefits are beyond the accounting of the decision-making unit.

Value Over Time and Generations

As a practical matter, society's willingness to pay for resource values is an expression of living generations. But environmental resources have stock and flow benefits that extend indefinitely over time. The long time horizon (involving several salmon lifetimes) raises the issue of distributing resource benefits between current and future generations. Although it is natural for members of the living generations to place greater value on present benefits and discount the value of expected future benefits, desire for resource sustainability and concern for future generations demand that expected future benefits be given greater weight than they now receive. Such discounting makes benefit-cost analysis difficult, even when the full range of values is included (Alkire 1993, Pearce and Turner 1990). One source of difficulty is that environmental variability affects the economic and scientific spheres in different ways. Environmental variability creates economic uncertainty, which causes people to discount the future more heavily, and this leads to pressures to increase rates of immediate, direct use. And environmental variability creates scientific uncertainty about biological processes, which can be perceived to call for a cautious approach and lead to pressures to lower rates of immediate, direct use. The resulting tension between economic and scientific responses to uncertainty adds complexity to decisions about appropriate rates of resource use. That tension is widespread in decisions concerning wild salmon in the Pacific Northwest.

INSTITUTIONS AND VALUES

There is a large gap between what people in the Northwest and throughout the nation seem to want (to judge from attitude surveys and political behavior), which includes clean and aesthetically pleasing environments and a lot of salmon, and what they get, which has included the demise or near demise of many salmon populations. One reason for the gap is that people also want other things, some of which might be contradictory, such as increased disposable income and continued population and economic growth, which are clearly problematic for the health of salmon and their habitats. Moreover, conflicts are often founded on differing goals and values, and conflicts make it difficult to attain even widely shared objectives.

A second major reason for the gap is a long, contentious history of political, legal, and cultural institutions that perpetuate the kinds of behavior that continue to endanger wild things and their habitats (Wilkinson 1992). Accordingly, institutional change must be at the core of effective response to natural-resource problems.

By institutions, we refer to cultural rules, conventions, norms, laws, and practices that define and play roles in regulating and giving meaning to individual and group behavior (Bromley 1989). Although courts, Congress, state legislatures, federal and state fish and wildlife agencies, the Bonneville Power Administration, the Pacific Fisheries Management Council, and the Columbia River Inter-Tribal Fish Commission might be thought of as particularly relevant "institutions" in the popular sense, it is useful to think about them instead as management organizations that derive their existence and forms from institutions.

At one level, the institutions include procedural rules, including the rule of openness to public review and response, which is a fairly recent development in federal water management (see Blumm 1981). They also include the "working rules" and "rules of the game" of organizations. At another level, institutions include legal relations of rights, duties, privileges, immunities, and responsibilities (Bromley 1989)—expectations about one's behavior and the behavior of others. Property rights are particularly important, in that they create a complex checkerboard of private and public, exclusive and inclusive, accepted and contested rights across the migratory paths of salmon.

"LORDS OF YESTERDAY"

The institutions of the West have played a major role in shaping the wants of individuals, creating and taking away rights, guiding allocation decisions, and solving and creating problems. In *Crossing the Next Meridian*, Wilkinson (1992) created the metaphor "lords of yesterday" for institutions, rooted in the past, that play a major role in the problems facing the West today, including declining fish runs. Wilkinson (1992) identified several lords of yesterday—institutions developed for understandable reasons in the nineteenth century that were co-opted, distorted, and outdated but are still solidly entrenched through the power of those whose interests are being served. They are institutions designed for frontier conditions and intended to promote settlement and private industry; they are arguably out of step with conditions today but resistant to change.

One such institution is the 1872 General Mining Law, which created unique property interests in public lands with little provision for environmental protection and hardly any public return for private gain. Another is the dedication of vast areas of federal public land in the West to cattle ranching with a fee structure that subsidizes private ranchers and has encouraged grazing, which has had negative effects on riparian stream habitats for fish. A third is a system of forest management guided by the principle of "wise use." A fourth is the peculiarly western system of water rights: water users in much of the West gained a vested property right to water and combined into various organizations (e.g., irrigation districts) to control the use of public water. Subsidized by federal reclamation projects, western water-rights holders (as varied as agriculturalists and cities) have had few incentives to worry about water quality or the flows needed by salmon. Their special, private interests are not always the same as the larger public interest (Johnson 1989).

Lords of yesterday that most directly affect salmon and their rivers are manifested in a long-held attitude that dams and water diversions are more important than salmon or the people who depend on them, including the American Indian tribes, and in an almost blind faith in technological solutions like hatcheries and fishways. The history of that attitude has included inattention to the cumulative consequences of particular events and decisions and delay in the development of cooperative and comprehensive management; the inattention and delay are major contributors to the current salmon problem (Wilkinson 1992). A brief history of attempts to arrest or make up for declines in salmon runs shows the institutions of the past and their consequences for salmon; Chapter 3 provides more detail.

As Euro-Americans settled in the Pacific Northwest and the resulting economic activities began to affect fisheries in the late nineteenth century, those interested in and concerned about salmon often provided alternative perspectives and proposed interventions. Those attempts at stewardship focused initially on the direct causes of declining salmon abundance but by the 1930s took a more comprehensive approach as the intertwining of the fate of the salmon and a landscape modified by human enterprise became apparent. The efforts were given their contemporary shape in the 1970s as the challenges of both fishery management and freshwater-habitat governance were taken up by the federal government, the region's principal property-owner and conservator of endangered species.

In the nineteenth century, canners in Astoria lamented the fact that there was much more gear than needed to land the annual catch; many more fish were being caught than could be packed. After peaking in 1883 and 1884, catches began to drop. The 1893-1894 report of the Oregon Fish and Game Protector stated that

> it does not require a study of statistics to convince one that the salmon industry has suffered a great decline during the past decade, and that it is only a matter of a few years under present conditions when the Chinook salmon of the Columbia will be as scarce as the beaver that was once so plentiful in our streams (Fish and Game Protector 1894).

Evidence of the decline included smaller catches, the use of smaller fish in processing, and the substitution of fish that would not have been used earlier because they were not believed to be of adequate quality. The Fish and Game Protector's report for 1893-1894 included a number of recommendations, some of which retain their currency (Fish and Game Protector 1894:27):

- Repeal all laws for the protection of salmon and enact a new law comprehending the entire subject.
- Provide that all dams in streams ascended by salmon should be provided with fishways.
- Provide for the protection of salmon and other fish by requiring all persons or corporations operating or having in charge any mill race, irrigating or mining ditch connecting with any river, lake, or other waters in the state, to put in and maintain . . . a wire screen . . . as will prevent any fish from entering.

A century later, the laws are complex, and jurisdictions overlap. Many dams thwart fish passage, and numerous irrigation ditches remain unscreened throughout the Columbia River Basin.

In 1903, Miller Freeman started the *Pacific Fisherman*, a trade magazine for all Pacific fisheries. Freeman was an ardent advocate for the salmon industry. He voiced strong support for conservation, saying that "there can be no compromise with conservation." He wrote further that "to have a fishing industry, you must have fish; and if you use your fish wisely you can have them forever" (50th Anniversary Number, Pacific Fisherman, August 1952, p. 4). He constantly tried to keep people focused on the future of the Pacific Coast's fishing industries. Yet despite his being an articulate spokesperson for conservation, Freeman's goal was not realized. "There will never be an end to the fisheries' problems," Freeman forecast. "As solutions are found, new posers arise" (50th Anniversary Number, Pacific Fisherman, August 1952, p. 4).

Many of the causes of decline have been known for more than a century, but getting action in an industrial economy has been difficult. Fishery regulations have always lagged behind the actions of fishers and the needs of the fish. The first steps to control fishers were taken in 1866 and 1871 when Washington State wrote rules to limit gear that blocked fish passage. In 1877 and 1878, Washington and Oregon introduced minimum mesh sizes and closed periods.

Another part of the salmon problem has been that the ecosystems needed by the fish were also required to feed the growth of the economy. River-basin resources—such as water, riparian lands, and timber—were developed for purposes that did not take account of the needs of salmon or of people staking their livelihoods on the sustained abundance of salmon. Fishery management fell behind what was needed to protect the fishery; it is still behind.

The lords of yesterday are sustained by the economic and political interests of those who benefit from their allocations of property rights. Two important values of American society provide the basis for those allocations: technological optimism and political pluralism. Each affects how science is used in the processes of public policy.

Technological Optimism

American confidence in human ability to invent and implement effective solutions to a wide span of problems is of historic proportions (Winner 1977, Norgaard 1994). The conversion of salmon habitat in the Northwest to human purposes has been monumental, as documented later in Chapter 7. Governments and residents of the region have thought that they could have both economic development and salmon because they believed that artificial production of juvenile salmon in hatcheries could compensate for the loss of spawning and rearing habitat and that modifications of dams and their operations could maintain the migration routes left open after the mainstem dams had been built. Federal support of Northwest hatcheries has a long history, dating back to the 1890s (Cobb 1930, Mitchell Act of 1950) and continuing with the Lower Snake River Compensation Plan (1976, see Platt and Dompier 1990) and the Columbia River Basin Fish and Wildlife Program (NPPC 1982, 1991). State hatchery programs (Brown 1982), many supported by payments from nonfederal hydropower projects, have substantially contributed to the belief that human intervention could compensate for lost natural production as well. Along the mainstem Columbia and Snake rivers, the U.S. Army Corps of Engineers invested hundreds of millions of dollars in an attempt to retrofit the dams with fishways, bypass technology, and juvenile-salmon transportation programs that would reduce dam-related deaths of migrating juvenile and adult fish.

The pattern of technological attempts to offset other human impacts is not limited to the Columbia River; it is widespread throughout the Northwest, from California to British Columbia and Alaska (e.g., Black 1993, Hilborn 1992, Meffe 1992). A consistent theme of this technological optimism has been neglect of scientific rigor (Ludwig et al. 1993, Winton and Hilborn 1994). Hatcheries and other means intended to benefit fish and wildlife were rarely monitored or evaluated. Management objectives or other ways of stating hypotheses about effectiveness were not formulated. Undocumented judgments of agency personnel, often made without supporting evidence, were accepted as expert opinion. Historical experience that would have prevented the re-enactment of errors was not taken into account. All that seemed unimportant at first: adult fish appeared to be abundant in the oceans and the river reaches below the dams, and objections to the shifting of fish production from inland streams to the lower Columbia were politically unimportant (Cohen 1986, Lothrop 1986). As salmon abundances have declined and American Indian treaty rights have gained legal standing, however, the inadequacy of the scientific record has become glaring and finally crippling (Salmon Technical Team 1993).

Political Pluralism

The lords of yesterday have also been defended within a political system organized around interests (Truman 1971). That approach has given the United States a resilient and long-lived government, and its competitive, relatively open structure has permitted the articulation of a wide range of concerns by many segments of a diverse society (Crenson 1971, Wildavsky 1979). Yet it is important to note that the implicit rules of competition in such a pluralist system stress advocacy rather than scientific debate.

Advocacy might be well suited to debates over equity and fairness, but it can be debilitating when science is at stake. In some arenas, political debates have fostered intellectual corruption: a finger-pointing style emphasizing partial disclosures of scientific knowledge to advance positions and to deflect responsibility from one's own agency or interest group; interference with testing of hypotheses, which leaves inadequate knowledge as the basis of decisions; and a range of pressures on technically trained people in and out of official positions. The patterns are common in policy arenas where scientific and technical knowledge are intertwined with political disputes (cf. Miles 1987). The undermining of scientific integrity that results is often accepted as a necessary cost of settling social disputes within an established political culture. But the cost inherent in this social commitment to pluralist competition is substantial where salmon are concerned.

"PRINCES OF TODAY"?

Current institutions—"princes of today"—also have profound effects on natural resources and how they are used and managed. Examples include the public trust doctrine and American Indian rights as shaped by court cases; the concept of cooperative planning and management; the very notion of resource planning; the strong presence of the federal government in resource development; regionalization of management systems and the advent of system planning and ecosystem approaches; international regimes for salmon management; and categorical imperatives about species protection embodied in national legislation.

Public Trust and American Indian Rights

Laws and court rulings have upheld as well as challenged water rights and coastal-management institutions, raising the question about whether the concept of *public trust* can be applied beyond its original scope, rights of navigation and fishing in intertidal foreshores and navigable or tidal waters, to environmental protection, freshwater rivers, and surrounding habitats (e.g., Johnson 1989, Johnson et al. 1992).

Since the early 1970s, federal courts and Congress have also crafted legal institutional frameworks related specifically to fish management in the Northwest. Some institutions are about allocation principles. A key example is the legal rulings concerning American Indian treaty rights to salmon. The historic fishing case *United States v. Oregon* (302 F.Supp. 899) was initiated by tribes of the Columbia River; Federal Judge Robert C. Belloni's ruling (1969) was the first to provide a practical interpretation of the legal concept of "usual and accustomed." The ruling was the crucial building block for other federal-court decisions important to the region's tribes. In 1974 in *United States v. Washington* (384 F.Supp. 312), which was initiated by Washington tribes, Federal Judge George Boldt decided that Belloni's notion of "fair and equitable share of the resource" meant 50% of all the catchable fish destined for the tribes' traditional fishing places. The next year, Belloni applied the 50/50 standard in *United States v. Oregon*. *United States v. Oregon* remains under the federal court's continuing jurisdiction. Defendants are the states of Oregon, Washington, and Idaho; the plaintiffs are the United States, the four Columbia River treaty tribes, and the Shoshone-Bannock tribe of southeastern Idaho, whose status in the case is different. *United States v. Washington* has finally moved to a cooperative-management agreement between the treaty tribes and the state of Washington (Clark 1985, Cohen 1986, Blumm 1990), which, apart from its shaky start and challenging coordination tasks, has the potential to be the basis of and model for further cooperation in salmon management. Thus, the Northwest Indian Fisheries Commission and the Washington Department of Fisheries and Wildlife jointly carried out the first general inventory of wild salmon and steelhead stocks in 1992 and are similarly comanaging a second-phase study of interactions between hatchery and wild fish (WDF et al. 1993). Accordingly, both treaty rights and cooperative management have emerged as central institutional principles.

Resource Planning

Declining fish populations are now treated as part of a comprehensive public responsibility for the environment. The institutional and conceptual framework for that approach arose, however, from planning for efficient economic exploitation—a set of ideas that put domestication at the center of stewardship. In the 1930s, the idea of public planning gained prominence in the United States. As governor of New York, Franklin Roosevelt said, "In the long run, state and national planning is essential to the future prosperity, happiness and existence of the American people" (Bessey 1963:16). The assumption made by advocates of planning is that human societies can anticipate problems and, through planning, avert the undesired consequences. A continuing theme is that if we could only achieve the planning ideal, we could foresee and solve the problems.

Natural-resource planning grew out of the concept of conservation, an approach to economic exploitation of the landscape stressing scientific management and multiple use. Applied to public policy by President Theodore Roosevelt, forester Gifford Pinchot, and geologist-hydrologist W.J. McGee in first decade of the twentieth century, conservation promised to allow growth while maintaining environmental quality.

Federal planning efforts in the region focused on water-resources development, but fishers and fish habitat received low priority. A 1925 act of Congress directed the U.S. Army Corps of Engineers and the Federal Power Commission to prepare general development plans called "308 reports." The 1932 308 report for the Columbia basin recommended a 10-dam system. For the decade 1933-1943, regional planning was conducted under the auspices of the National Planning Board (NPB). In 1934, NPB encouraged formation of the Pacific Northwest Regional Planning Commission. In 1943-1953, the strong centralized planning of the commission declined in favor of voluntary cooperation, and the commission was disbanded. Regional planning lived on in the Columbia Basin Inter-Agency Committee (CBIAC), organized in 1946. Interest in planning declined further during 1953-1963, but the CBIAC "continued . . . as the only intergovernmental regional agency active in the resources planning and development field in the Pacific Northwest" (Bessey 1963:109). The Pacific Northwest River Basin Commission supplanted the CBIAC as the regional planning unit in the 1960s and was superseded by the Northwest Power Planning Council in 1981.

Until the 1980s, planning in the Pacific Northwest emphasized domesticating the region's water resources for hydroelectric generation, irrigation, navigation, and flood control. The health and sustainability of fish stocks had a lower priority. Instead, economic development and expanded settlement were encouraged in the Pacific Northwest, an area noted for its natural beauty and relatively low population density. In 1930, the Pacific Northwest, with about 9% of the nation's land area, had only about 2.5% of its people (Bessey 1963). By 1990, the region's population had grown more rapidly than the nation's and represented 3.5% of the nation's population. The Columbia River Basin east of the Cascades remains less densely populated than the United States as a whole, although the Puget Sound lowlands and the Willamette Valley have urban densities comparable with those of other urbanized areas of the country.

Implied in all the planning discussions was the assumption that humans have the ability to foresee problems and to correct problems of their making. However, 60 years of planning shows persistent failure to get ahead of the impacts.

Planning, like management, has turned out to be more reactive than anticipatory (Bella 1991).

By some metrics, the region's planning efforts have been a success. For example, the aggressive water-development policy that led to the damming of the Columbia River and its various tributaries (and the widespread construction of hatcheries as mitigation for blocked salmon runs) was largely accomplished within a half-century. By any standard, such an effort was monumental, requiring the sustained and concerted efforts of politicians and government (Reisner 1986). But planning efforts that were designed to achieve rapid economic development were also major contributors to the demise of the region's salmon populations.

That failure of foresight reflects a deeper bias in how economies grow. Human technologies allow economic actors in the Pacific Northwest to ignore and partially circumvent natural limits on energy materials and services that can be harnessed from the region's natural resources. First, regional limits on inputs (e.g., on steel needed for building hydropower dams) can be circumvented by importing materials and services. However, the environmental costs of using these external inputs accumulate within the region. For example, dams enable diversion of water away from fish to support agriculture. Second, the capability to bring in inputs from external sources encourages development of production (e.g., fruits and vegetables) for external sale. Consequently, the region's economy depends increasingly on maintaining external inputs to support an economy that relies partly on export of goods. From the perspective of biota and natural resources, the drawback of this approach is that the economic incentive to increase the output of exportable goods is accompanied by a social disincentive to account fully for the environmental costs of the growing economic activity. One overlooked cost is the depletion of natural resources—especially the destruction of their ability to regenerate naturally.

Today, population growth and economic growth continue to be dominant regional economic goals, despite the increasingly open acknowledgment that "our prosperity has had a price: dramatically reduced salmon runs" (NPPC 1992a:5).

The Federal Presence in Regional Stewardship: Changing Resource Values

The prosperity of the Columbia River Basin has been tied to its dams. Like the salmon that were the mainstay of its American Indian culture, the annual harvest of water from snow and rain, collected and harnessed by the dams, became the lifeblood of the basin's industrial economy. Until 1980, it seemed that the fate of the river would remain tied to that of the power system. Then came the energy crisis in the Pacific Northwest. What happened then illustrates how ideas about natural resources can emerge in crisis.

The economy of the Pacific Northwest is greatly influenced by the federal government, whose policies have been a shaping force in regional affairs since the Great Depression. With Grand Coulee and Bonneville dams, two major projects of the 1930s, the federal government transformed the regional economy. By the time water-resource development in the Columbia basin ended in the 1970s, three-fourths of the power in the Northwest came from falling water, and electricity rates were the lowest in the nation.

The New Deal in the Northwest was anchored in the Bonneville Power Administration (BPA), a federal agency created in 1941 to market the enormous output of the national government's dams on the Columbia. But BPA soon became a regional economic force whose influence was rooted in the Federal Columbia River Power System, which generated nearly half the electricity produced in Washington, Oregon, Montana, and Idaho. Although the dams themselves are operated by the U.S. Bureau of Reclamation and the Army Corps of Engineers, BPA power sales have usually governed river operations. BPA is largely invisible to the retail consumer, marketing more than half its power to utilities, which supply businesses and households. Two-fifths of BPA's sales are made directly to Northwest industrial customers, mostly large aluminum refiners, which account for 40% of the nation's aluminum production capacity.

From the turn of the century, when steam-driven turbines came into service, until 1970, electric power was a story of increasing scale and declining cost. Growth in demand brought decreasing cost per unit sold, and growth benefited producer and consumer alike. In the 1970s, however, the benefits of growth vanished as costs of construction, borrowed money, and environmental damage all rose ominously. In the Pacific Northwest, industry leaders and their critics turned to the federal government for solutions. The product of their political labors was the Northwest Power Act of 1980.

The act was designed to solve a set of social problems by technological means. As demand for power grew during the 1970s, more power plants seemed necessary to utilities. The utilities proposed federal legislation to enable them to build more plants in 1977. Yet citizen activists, whose voices were growing rapidly in power and influence, argued that energy conservation could meet the demand for power at lower environmental and economic cost. The search for compromise took more than two years. Toward the end of the search, American Indian tribes and commercial and recreational fishers demanded that the damage to the Columbia's fish runs be repaired. Rather than choosing among partially conflicting claims, Congress sought to accommodate them all. The result was a complex piece of legislation, whose execution has taken turns unanticipated by those who fashioned the act's key compromises. The Columbia River Basin has been called the largest experiment in ecosystem rehabilitation in the world (Lee 1993a). Underlying this venture is the value underwritten by the Pacific Northwest Electric Power and Planning Conservation Act of 1980 (Northwest Power Act), which sets forth the principle that hydropower users are responsible for protecting and restoring fish and wildlife resources that are adversely affected by the federal Columbia River power system. The Northwest Power Act tried to reorganize the indirect and direct values of salmon by making the cost of protecting and restoring the salmon fishery a cost of doing business to be shared by all electric-power consumers in the region and by according equitable treatment to anadromous fish when compared with power production as a goal (Wilkinson and Conner 1983, Lee 1993a).

By 1982, when the Northwest Power Act was in the early stages of implementation, the expected power shortage that had motivated its enactment had evaporated. Demand was far below expectations because the economy was in recession. Rates were rising rapidly, however, to pay for the costs of power plants already being built. Energy conservation gained plausibility. Rising rates and falling demand meant that instead of a deficit, there was a surplus of power through the 1980s. Instead of a financing mechanism to build new power plants, the Northwest Power Act became the blueprint for a laboratory of energy and environmental conservation.

The regional stewardship framed by the act revolves about the Northwest Power Planning Council and the two plans it has promulgated. Chartered by the four Pacific Northwest states, the council is composed of two members from each state, appointed by the governors under procedures established by state law. Under some circumstances, the council has the unusual authority to restrict or redirect the actions of federal agencies. The council is in effect an interstate compact, a form of government organization that shares both state and federal authority. Council policies are embodied in two regional plans, both of which are products of extensive public discussion.[2]

The council's first task is to formulate a plan to guide electric-power development, including energy conservation. Three versions of the plan have been issued, the most recent one in 1991. Their central premise is regional cost-effectiveness planning that minimizes costs across the Pacific Northwest's many utilities, a premise that the region's fragmented industry would not naturally follow. The regional approach emphasizes a portfolio perspective: the objective is not a single power plant or conservation program, but a set of available alternatives. The least costly options should be exercised, and costs should be reckoned on a regional, life-cycle basis.

Second, the council sought to bring back the salmon of the Columbia River

[2]Although public participation declined as electric-power rates stabilized in the middle 1980s, many divergent interests are at stake in the energy-dependent sectors of the economy and in the environmental management of the Columbia River Basin, so a broad spectrum of observers continues to pay attention. Thus, the pluralist approximation to democracy—characterized by one political scientist as majority rule by the minority that cares—arguably remains valid. Nearly all observers agree that the council's plans articulate a public interest that goes beyond the claims of utilities or other interest groups.

Basin. The Northwest Power Act articulated a policy that is easily stated but difficult to define and pursue: electric-power consumers are obliged to fund, through the BPA, a program "to protect, mitigate [sic], and enhance fish and wildlife to the extent affected by the development and operation of any hydroelectric project of the Columbia River and its tributaries." That mandate produced a wide-ranging effort that was specified in the program created by the Northwest Power Planning Council. The program has been a systematic but not coordinated attack. The design calls for action at all the principal points at which human actions affect the fate of salmon: during fishing, at hatcheries and spawning grounds, and along migration routes in the mainstem river. But the limited authority conferred by Congress on the council has made coordination of the rehabilitation effort indirect at best.

The Northwest Power Act limits the contribution of electric-power ratepayers to damages attributable to hydroelectric power generation. In 1986, the council drew on anthropological and historical studies of the wilderness Columbia and legal analyses of the contemporary river and set the responsibility of present-day ratepayers at 8-11 million adult fish per year. The loss of so many fish, beyond the remaining 2.5 million returning to the river, could be ascribed to hydroelectric power generation. The practical meaning of the responsibility is unclear, however, in that the biological capability of the remaining habitat and technically feasible hatchery sites might fall well below 8 million fish.

More specific guidance was formulated in the 1987 version of the Northwest Power Planning Council's Columbia River Basin program, which set out an ecosystem-scale approach. The invoking of the Endangered Species Act (ESA), beginning in 1990, forced a rethinking of ecosystem-scale plans to rescue stocks nearing extinction. In light of the continuing difficulties of responding to the requirements of the ESA in the Pacific Northwest, however, an understanding of the earlier ecosystem approach remains relevant. The most tangible objective of that plan was an "interim" goal of doubling salmon populations over an unspecified time. That goal was "a signal that the program is a long-term, serious effort to solve complex problems not amenable to quick-fix remedies." Implicit in those words were times of biological significance—several generations of salmon. The council program emphasized learning and sustainability to limit cost and risk. Rapid learning—including effective evaluation—would lower costs. Aiming at sustainable increases in fish populations implied practices that lowered risks to salmon gene pools. And avoiding an explicit deadline to achieve doubling permitted time for learning how to rebuild fish populations in a biologically sound and economically sensible way.

The heart of the ecosystem approach lay in a 4-year process known as system planning. System planning involved trying to think about the interactions among the hundreds of activities that affect the abundance and health of the basin's fish and wildlife, including changes outside the scope of the program, such as land use and shifts in water rights. System planning emphasized the conservation of the genetic heritage of the salmon population but also envisioned rebuilding populations rapidly by artificial means to enable resumption or expansion of fishing. The principal hope lay in supplementation, a technique of releasing hatchery-bred juveniles into underpopulated streams before the fish migrate to sea.[3] Supplementation was thought to combine some of the advantages of hatchery culture, especially the high survival of young fish before they are released, with some of the strengths of natural production because the selective forces that determine which fish survive after release are largely natural. Supplementation thus promised effective use of existing and new hatchery capacity and the hope of rebuilding wild stocks in their native streams and at population levels that would permit fishing.

The keystone of the Columbia River Basin program is an augmentation of the flow of the river called the water budget. Before the dams were built, flow was heavily concentrated in the spring, when mountain snow rapidly melts. Spring floods carried juvenile salmon to the ocean. The trip is made much longer now by the slower flow of water in reservoirs than in the unimpounded river. The slowdown exposes juvenile fish to predators for a longer time. In addition, salmon must make a physiological changeover in going from fresh to salt water; if arrival in salt water is delayed, the fish might revert to a freshwater constitution, stop growing, and never reach legal-catch size. For years, the fish and wildlife managers and American Indian tribes had requested higher flows in the springtime migration season. But the requests carried no authority, and the dams, controlled by utilities and the U.S. Army Corps of Engineers, were usually run to optimize power revenues.

The water budget is a more generous compromise for the Columbia River than for the Snake River because the upper Columbia discharges more water, and a much smaller fraction of its water is diverted for irrigation. In practice, even the water dedicated to the water budget in the Snake River drainage has often been

[3]The term *supplementation* has been used in various ways. Some view it as an effort to augment natural production by complementing seeding of natural spawners with fish bred and cultured in hatcheries to a stage before the smolt stage, thus filling vacancies in natural habitat. This differs from more usual hatchery practice in that the natural habitat, with all its natural-selective forces, provides for at least part of the growth of the young fish to smolt stage, rather than a hatchery. The NPPC's system planning approach seems to have leaned toward that meaning. Others view supplementation as any augmentation of populations in natural production areas—with adults, eggs, fry, or juveniles, including smolt releases. Little information exists on supplementation under either meaning (Chapter 12, Bowles 1995) and its use has been condemned (White et al. 1995). Unknowns include long-term effects on fitness of wild demes (chapters 6 and 12) and various ecological, behavioral, and management consequences of supplementation and outplanting procedures on naturally produced cohorts (chapters 11 and 12). Supplementation efforts, such as those in the Yakima basin (Fast et al. 1991), should be undertaken on modest scales with adaptive management, as described in Chapter 12 and by Bowles (1995) and White et al. (1995), until their effects and effectiveness are better understood.

unavailable. The biological benefits of the water budget are hard to see, in part because it is small in comparison with natural fluctuations.

The Power Planning Council program identified measures to be taken by the American Indian tribes and government agencies that exercise management responsibility for fish and wildlife, by hydropower project operators and regulators, and by those charged with land and water management. Most of the program measures have been funded by BPA, at a total annual cost in 1994 of about $350 million (see Box 9-1)—about 1% of the agency's annual revenues. Although modest from that perspective, the program's funding is unprecedented for fish and wildlife, as is its spatial scope—overlapping four states and covering a land area the size of France. About $54 million is spent directly by BPA through contracts with the American Indian tribes and fish and wildlife agencies. The remainder of the cost is incurred as lost revenues—earnings foregone because water is released to benefit fish rather than power-users.

In 1990, the federal presence changed course sharply, as petitions were filed to list five populations of Columbia River salmon under the terms of the ESA. These salmon petitions raised the specter of severe economic dislocations because the federal government's duty to protect an endangered species supersedes its role as operator of economically important facilities, such as irrigation systems and hydroelectric-power dams. By 1994, it seemed clear that the battle over the salmon of the Columbia will lead to substantial modifications in how the river is operated.

Whether those modifications succeed in resuscitating the endangered stocks, the federal presence has fundamentally altered the treatment of fish and wildlife in the Pacific Northwest. American Indian treaty rights, ignored for more than a century, are widely accepted. More than 40,000 stream-miles in the Columbia River Basin have been put off limits to hydropower development. Fishery-resource management agencies in tribal, state, and federal governments exercise substantial influence in the policies and budgetary choices of BPA. There have been important changes in power-system operation and planning in response to the Columbia River Basin program. The impact of those measures can be gauged by different yardsticks: in lost power revenues of over $150 million per year in 1993 or in power resources displaced or subordinated to fishery protection (the equivalent of a medium-size coal plant).

An effort to achieve balance among competing values and interests is embodied in institutions that provide regional frameworks for resource management given the large spatial scale of the life cycles of salmon and the habitats that they need. The cost is about $100 per adult salmon in the Columbia drainage. "Even though economists assessing the costs and benefits of comparable programs have arrived at the same approach as the one set forth in the Northwest Power Act, the sheer magnitude of the dollar figure fuels debate" (Lee 1993a:50). Moreover, the effects have so far been disappointing, particularly with respect to the listing of Snake River runs as endangered species:

> As in the case of energy conservation, salmon rehabilitation has broken new ground. As also with the efforts in electric power, it is not clear that what is being done will be enough to achieve a workable, sustainable balance. The river is in an unnatural, partly managed condition. Increased flows in the spring—the remedy recommended by the fisheries advocates—may or may not suffice to rescue the salmon. The mounting disillusion with hatcheries demonstrates the fragility of our understanding and the unexpected, even perverse effects of earlier attempts to mitigate damage (Lee 1993a:50).

FISHERIES MANAGEMENT INSTITUTIONS

As noted above, there is a long history of institutions and organizations devoted to management of salmon catches at the state level. The fragmentation that ensued contributed to the difficulties in managing migratory species throughout their ranges. The Magnuson Fishery Conservation and Management Act (MFCMA), together with the Pacific Salmon Treaty of 1985 and the joint management of fishing by states and American Indian tribes, has created a fishing-management regime that embraces all the fisheries that target salmon throughout their migratory range. These institutions set up new arenas for conflict and for cooperation. This section discusses the regional fishery management system set up by the 1976 MFCMA and the international regime set up by the Pacific Salmon Treaty. Chapters 10 and 11 discuss fishing management in greater detail.

The MFCMA is one piece of enabling legislation for regional response in marine and anadromous fisheries management. It expanded federal jurisdiction over fisheries to the zone 3-200 mi from shore. The Pacific Fishery Management Council (PFMC), one of eight regional councils created by the MFCMA, is designed to be a venture in participatory, decentralized, cooperative-management planning based on scientific and user-group advice and dependent on federal-agency approval and implementation. PFMC has 16 voting members, including state and federal fishery agencies in the region and gubernatorial appointees who are supposed to represent the balance of state and fishery interests and to be knowledgeable about the fisheries. PFMC also coordinates state fish- and wildlife-agency policy concerning anadromous fishes; California, Oregon, Idaho, and Washington are the states involved. The regional management council system established by the MFCMA might have been a model for the Northwest Power Planning Council (Wilkinson and Conner 1983:56, footnote 212).

In recent years, the entire regional fishery management-council system has come under attack because of alleged "capture" by special interests in the fisheries (e.g., Safina 1994, see also discussion in NRC 1994). However, the councils provide the basis not only of regional coordination of fishery-management efforts but also of effective user participation in fishery management. PFMC has been unusually successful in the use of participation by members of the fishing industry in the planning process in ways that enhanced both that process and the effectiveness of management, especially compliance (Hanna, in press). PFMC has, however, had a history of failing to meet escapement goals, which suggests that it, like other fishery-management agencies, tends toward "risk-prone" decision-making in the context of scientific uncertainty and political pressures (cf. Ludwig et al. 1993, Rosenberg et al. 1993).

A fourth important value arena concerns international salmon management, namely, the 1985 Pacific Salmon Treaty between the governments of the United States and Canada (Agreement January 1985, Annex IV amended May 1991), the culmination of two decades of negotiations and fish wars. The treaty is the mechanism for balancing catch allocations between the salmon-producing nation and the catching nation. Northward-migrating chinook from British Columbia streams are caught in Alaskan waters; Columbia River chinook are caught off British Columbia; and Fraser River sockeye are caught by U.S. fishers in U.S. waters or beyond the 200-mi limit.

The only prior cooperative-management agreement was for sockeye and pink salmon from the Fraser River and nearby areas—the "Fraser Panel area"—under the "Sockeye Commission" (International Pacific Salmon Fisheries Commission). (Until 1977, the United States was party to the International North Pacific Fisheries Convention of 1953, for high-seas protection of salmon; it withdrew when the 200-mi zone of extended jurisdiction was created.) The 1985 treaty set up a new Pacific Salmon Commission. A scale of management that reflects the full range of salmon movement calls for that kind of international institution (Morgan 1987), but the regime is not guaranteed to work. The principles of the 1985 treaty are to optimize production of salmon and to achieve equity in the number of fish intercepted by the two countries. The main stumbling block is a dispute over how to ensure that each country gets either fish or compensation equivalent to its salmon production. Chapter 10 discusses the difficulties in getting agreement between fishery interests in Alaska and the lower 48 states on how to negotiate with Canada. Aboriginal fishing rights further complicate the situation.

Biodiversity and Endangered Species

Recognition of the value of biodiversity (or the cost of extinction of species) led to the Endangered Species Act (ESA) of 1973, which, like the Marine Mammal Protection Act of 1976, made statutory policy relevant to the salmon problem in the Pacific Northwest. Therein is the beginning of the current crisis. On April 2 1990, the Shoshone-Bannock tribe of Idaho petitioned NMFS to list Snake River sockeye as an endangered species; it was closely followed by environmental groups petitioning to list the spring, summer, and fall runs of chinook salmon in the Snake River and coho salmon in the lower Columbia River. In stating that economics may not play a role in the decision to list a species as endangered or threatened, the law reflects a social decision that preservation of species is a categorical imperative.[4] Thus, by congressional act, a major change in perception occurred: flora and fauna deemed to be endangered or threatened were lifted from the utilitarian values realm of decision in terms of tradeoffs or costs and benefits and into the extrinsic "rights-based" realm of categorical need for protection (Thompson et al. 1994). Not only do different human groups have rights in relation to the claims of others, but nonhuman biological organisms have rights. Those rights might conflict with human claims, even as the merits that we impute to the nonhuman world shift our notions of what is valuable.

Compensation, Liability, and the Law

The question of who should pay for salmon rehabilitation (Berry and Rettig 1994) is another complex value issue that affects the nature of our stewardship over salmon in the Pacific Northwest. Law interacts with politics to result in decisions about rights and liabilityand about claims and compensation. Numerous important legal questions are involved; we offer only a few illustrative examples here. One is the balance of federal and state powers over river-flow management, a question that affects dam management in the Northwest. A May 1994 U.S. Supreme Court decision (*PUD No. 1 of Jefferson County et al. v. Washington Department of Ecology et al.*) allowed the state of Washington to set minimal water- flow levels for salmon protection. (An embedded legal question was the relationship between water quality and water quantity.) Another is the extent to which the public-trust doctrine, which applies to all navigable or tidal waterways in the United States, can be used to protect water flows and quality required for salmon (Wilkinson 1980, Johnson 1989, Johnson et al. 1992). The issue of tribal treaty rights and their relationships to endangered species issues might be noted as well.

Who pays for mitigation and for compensation, and who should be compensated?[5] In the case of salmon, mitigation involves project design and "retrofitting" to minimize harmful effects of projects on fish; compensation has come to mean artificial propagation (with its own unintended and harmful consequences) to make up for losses due to dams, irrigation, and other practices. Increasingly, however, policies will require mitigation and compensation for social and economic losses (Berry and Rettig 1994). For example, what about irrigators in the upper Snake River Basin whose irrigation water is lost through flow augmentation? What about salmon fishers in the lower Columbia River whose boats, gill

---

[4]Economics can, however, be considered in designating critical habitats (Hyman and Wernstedt 1991). Current debates over the ESA indicate that this categorical imperative is not universally supported.

[5]"Mitigation refers to measures taken during planning, construction, operation, or implementation of a policy in order to avoid or reduce adverse effects. In contrast, compensation refers to payments made to offset losses that occur despite mitigation efforts" (Berry and Rettig 1994:4).

nets, and gill-net rights are suddenly worthless? What about power consumers asked to pay higher rates for electricity?

One newer aspect will be the question of whether private-property rights have been taken away—the "takings" issue. That is a subject of intense activity, with changing definitions and understandings that make "equity" an issue that can no longer be ignored. Salmon rehabilitation also poses new questions and levels of uncertainty about the future of investments (Berry and Rettig 1994:11). It is likely to result in more litigation and increased pressure on governments for compensation.

VALUES AND ANALYSIS

The values in policy conflicts are couched in arguments about fact. Consequently, scientists, purveyors of fact and scientific interpretation, are actors in salmon-policy debate. Although the methods of science are objective, scientists do have values that play important roles in the policy process (NRC 1992b). Among the strengths, and potential weaknesses, of science and engineering is the widely held value placed on rational, deductive thinking and discovery of truth by the testing and observation of highly simplified versions of the natural world. In the extreme, this can lead to mechanistic and deterministic views of nature that obscure flux and indeterminacy (Botkin 1990, Norgaard 1994). It also can be used to support an unwarranted dichotomy between humans and nature (with humans being factored out of the equation as a source of complicating noise). The problems inherent in applying reductionist science to problems that are multidimensional, changing, and seen as different by many different groups (Rittel and Webber 1973, Wilson and Morren 1990) are central to the salmon situation.

Economic analyses can be biased by a narrow focus on monetary benefits and costs. Economic efficiency as a policy objective might not capture the full range of extrinsic nonmonetary benefits and costs and is only one of many possible objectives. Although properly done benefit-cost analyses should reflect, to the greatest extent possible, all the known benefits and costs of a particular action, nonmonetary as well as monetary, in practice this is not reliably the case.

Even within scientific disciplines, one can find differences. In ecology, for example, there are biotechnologists versus bioconservatives (Sagoff 1992). The biotechnologist values tend to accept control of nature for the benefit of human society and view the proper role of the science of ecology as prediction and control. The bioconservative values are more concerned with protecting nature in its "uncontrolled state" and might believe that "nature knows best." Those differing but equally science-based approaches can lead to profound differences in how people weigh interventions—e.g., in the salmon context, barge transport versus dam removal or hatcheries versus reducing mortality and restoring habitat.

Any understanding of goals and values requires an interdisciplinary approach to analysis and policy evaluation because of the wide range of values and perspectives inherent in the problem and in planning for rehabilitation or recovery (Hyman et al. 1993:1). For several reasons, attempted solutions often resort to narrow definitions of the problem. The result is usually more conflict. Trying to force all goals and values into some sort of quantifiable form within the same decision framework is often the basis of "scientific management," as FEMAT (1993:VII-18) described for responses to the environmental movement on the part of the U.S.D.A. Forest Service. The basis of the Forest Land Planning and Management Act, like many other legislative attempts to deal with problems, is the notion that rationality is the product of comprehensive assessments and planning whose goal is to increase the accountability of management decisions to public values, science, and ecological reality. Indeed, many approaches to environmental management, including federal legislation, have this rational viewpoint. But conflicts over natural resources continue and perhaps are increasing despite and sometimes, perversely, because of the quantifying, rational approach. As noted in the FEMAT socioeconomic environmental impact statement, adding more rows to the linear program models did not lead to politically responsive decisions (FEMAT 1993:VII-20).

Sometimes the problem stems from the persistence of narrow definitions of value of natural resources (such as timber and salmon) as commercial commodities. Often one finds isolated rural communities pitted against an urban leisure class or rare and endangered wild species despite facts that might point to common interests or the possibility of common solutions. Organizations can also have institutional commitments to programs that make it difficult to take a flexible approach to problem-solving. Moreover, a problem can stem from institutional inertia in dealing with the fact that nature is often indeterminate and prone more to surprises and chaotic lurches than to calibrated and predictable responses to human attempts at "sustained yield" management.

Problems like these emphasize the need to develop more appropriate interdisciplinary approaches. The standard effort usually brings natural scientists and economists together (sometimes not for long, splitting them into independent working groups). Its history is one of disappointment and failure and includes recent attempts to use interdisciplinary approaches to the Pacific Northwest salmon problem (Hyman et al. 1993). A more interactive interdisciplinary approach that incorporates the expertise required to deal with nonbiological and nonmonetary issues is required.

The idea of rebuilding the salmon runs of an industrialized ecosystem is heroically optimistic—a hope that might not have occurred to anyone except those who had rehabilitated the Willamette River Basin in Oregon or Lake Washington near Seattle. Those environmental successes came through the disciplined execution of the planning paradigm that has been fitfully applied to the much larger Columbia River Basin. The extension of those experiences to the multijurisdic-tional, multifunctional situations of the Pacific Northwest would require coordinated action and learning on a new, larger scale—a scale on which planning and action have been tried but have not been successful. A more explicit appreciation of the values, interests, and institutions involved in this undertaking is required. Chapter 13 explores this further and urges constructive change in institutions that include cooperative management, bioregional governance, and adaptive management.

6

Genetics and Conservation

Managing salmon requires an understanding of the biological dynamics of the populations in which they occur and reproduce. In particular, knowledge of the structure of the genetic variation in salmon is needed to make decisions about how to identify and protect the local reproductive units, which are the fundamental biological units. That knowledge is needed to guide conservation strategies. Without it, the safest conservation strategy requires conserving virtually everything.

Even the first people who thought about conservation of western North American salmon were aware of differences between local populations and of the importance of the differences in the success of the fish. R.D. Hume packed canned salmon with a booklet that he wrote on conserving the valuable salmon resource (Hume 1893):

> I firmly believe that like conditions must be had in order to bring about like results, and that to transplant salmon successfully they must be placed in rivers where the natural conditions are similar to that from which they have been taken.

Moulton (1939) and Thompson (1959, 1965) provided the first modern description of the importance of local populations and genetic diversity for the management of Pacific salmon. They pointed out that each local population was genetically adapted to its own environment ("home-stream colony") and had a characteristic level of abundance around which it varied. In any river system, some colonies are capable of supporting exploitation, and others are not; commercial exploitation is likely to lead to the disappearance of some colonies (i.e., the extinction of local populations).

This chapter considers the fundamental role of genetics in the long-term sustainability of salmon. It examines the organization of genetic diversity within species and stresses the importance of recognizing the local reproductive group as the primary demographic and genetic unit.

The objective of increasing run sizes (NPPC 1987) and the biological need for preserving existing genetic diversity have led to the characterization of a choice of "meat versus museums" or productivity versus genes (Backman and Berg 1992). That is, we are faced with a choice between increasing production and protecting wild and natural runs. But that simple view is incorrect and dangerous. Genetic diversity is part of the fabric of a biological resource. The resource (production) cannot be separated from its genetic basis. Sustained salmon and steelhead productivity can be maintained only if the genetic resources that are the basis of such productivity are maintained. Those interested in protecting genetic diversity are not interested in maintaining genes (i.e., DNA sequences) for their own sake; rather, they are interested in protecting genetic diversity because it is necessary for the long-term persistence of the fish themselves.

Examination of the 1987 Columbia River Basin Fish and Wildlife Program reveals that genetics has been part of the development of the program (NPPC 1987). However, a closer reading indicates that genetics has not been incorporated into planned activities. There is little sign of recognition that all management activities (e.g., activities related to fishing, passage, habitat, and production) affect genetic resources and that genetic diversity is part of the fabric of the resource. It is important that potential genetic effects of all management activities be considered in making decisions.

STRUCTURE OF GENETIC VARIATION

Genetic diversity within a species occurs at two primary levels: genetic differences between individuals within local breeding populations called *demes* in population genetics, and genetic differences between breeding populations. The total genetic diversity within a species is often viewed as a hierarchy of levels (see Figure 6-1 for a simplified structure). The two levels of genetic diversity listed above are of primary importance in the genetics and evolution of a species. Genetic differences between individuals within a local breeding population are the basis for natural selection and adaptive evolution, and genetic differences between breeding populations reflect local adaptation to past environments and random events. The genetic differences between breeding populations represent the broadest pool of genetic variation and are a valuable component of diversity. Adaptation to future environments is dependent on variation within the local breeding population as well as variations between populations.

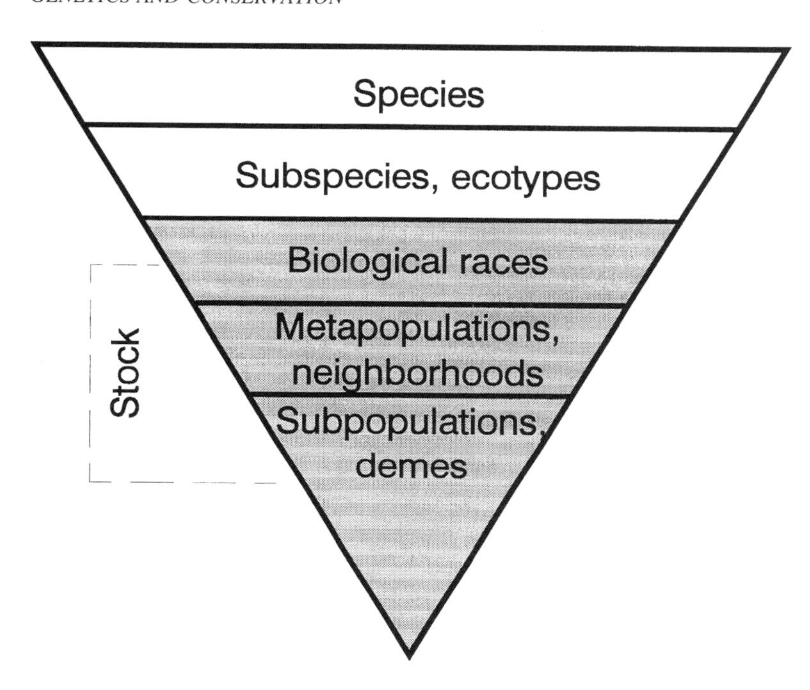

FIGURE 6-1 Schematic representation of structure of genetic diversity in Pacific salmon. Inverted triangle emphasizes that locally adapted and largely reproductively isolated, local breeding populations are basic unit of diversity. Varied operational definitions of stock would place this term in the range identified by dashed lines. Source: modified from Riddell 1993b.

Animal and plant breeders over the centuries have used artificial selection among individuals within breeding populations to improve their stocks. One simply breeds individuals that have some favorable variations in the characters wanted. These variations are due almost universally to both environmental and genetic influences. A breeder looks at successive generations to see whether the selection process has worked. Even though the scientific basis of selective advance was not understood until this century, artificial selection has produced extraordinary results. Humans have profoundly changed the inherited characteristics of dogs, cattle, grains, ornamental plants, and many other species.

Darwin was the first to emphasize that natural ecological processes select for hereditary changes in a manner very similar to that of human selection. In nature, it is the harshness and challenge of a continually changing environment that determines which individuals successfully reproduce under natural conditions. Natural selection results in improved adaptation to environmental conditions.

Simply put, both artificial and natural selection result in descent with genetic change; this is biological evolution.

For the evolution and continued existence of species, genetic differences between populations are as important as genetic differences between individuals within a population. Consider the extreme where no differences exist between local populations. In that case, a species consists of many copies of the same genetic population and is extremely vulnerable to environmental change. For example, a new disease might be introduced to which most individuals are genetically susceptible; the disease would jeopardize all populations and therefore the entire species. However, in the usual case, where genetic differences do exist between local populations, it is likely that some populations would have a higher frequency of genetically resistant individuals and thus would be relatively unaffected by the disease.

LOCAL REPRODUCTIVE UNITS

Because of homing, the fundamental unit of replacement or recruitment for anadromous salmon is the local population (Rich 1939, Ricker 1972). That is, an adequate number of individuals for *each* local reproductive population is needed to ensure persistence of the many reproductive units that make up a fished stock of salmon. The homing of salmon to their natal streams produces a branching system of local reproductive populations that are largely demographically and genetically isolated. The demographic dynamics of a fish population are determined by the balance between reproductive potential (i.e., biological and physical limits to production) and losses due to natural death and fishing. "Population persistence requires replacement in numbers by the recruitment process" (Sissenwine 1984), so fishery scientists have focused on setting fishing intensity so that adequate numbers of individuals "escape" fishing to provide sufficient recruitment to replace losses.

The distinction between a local breeding population and a fished stock is critical (Beverton et al. 1984). Whereas a local breeding population has a specific meaning—a local population in which mating occurs—*stock* is essentially arbitrary and can refer to any recognizable group of population units that are fished (Larkin 1972, Chapter 4). The literature has often been unclear on this distinction (e.g., Helle (1981) defined a stock as a local breeding population, citing Ricker (1972) as his authority). In practice, it is extremely difficult to regulate losses to fishing on the basis of individual local breeding populations. Thousands of local breeding populations make up the West Coast salmon fishery, and many of these are likely to be intermingled in any particular catch. Nevertheless, the result of regulating fishing on a stock basis and ignoring the reproductive units that together constitute a stock is the disappearance or extirpation of some of the local breeding populations (Clark 1984).

The homing behavior of anadromous salmon results in a complex pattern of genetic differentiation among local breeding populations because individual fish that home to different streams cannot breed with each other. The importance of the demic structure of a group of populations (or *metapopulation* as described below) for managing a fishery is related to how often fish stray or spawn in a nonnatal stream, although it is often not clear how much small differences in where or when salmon home are due to environmentally caused differences, random events, or genetics (Quinn 1993). It is difficult to measure empirically the amount of straying among natural local breeding populations of salmon (Quinn 1990). However, analysis of genetic variation in a network of local breeding populations can provide insight into the pattern and amount of straying among local breeding populations.

The chinook of the Klamath River drainage show a complex pattern of genetic differentiation at 36 variable protein loci detected electrophoretically; Bartley et al. (1992a) described genetic variation in chinook from a larger geographical area, but we have analyzed their data only for the Klamath River. Samples were collected from 10 locations throughout the drainage (Figure 6-2). To simplify the presentation, we analyzed the eight loci in which the mean frequency of the most common allele (or a form of gene) was less than 0.95 (Table 6-1). We pooled all other alleles at each locus to standardize the statistical analysis among loci. There are highly significant differences in allele frequencies among the 10 samples at all eight loci individually ($p < 0.001$; contingency chi-square test, and—except for populations 8, 9, and 10, which show no evidence of genetic differentiation between each other—the distributions of allele frequencies are significantly different between all other pairs of populations (contingency chi-square test summed over all eight loci).

To display the pattern of genetic differentiation between the 10 populations, we used principal-component analysis of the covariance matrix of allele frequencies at the eight loci. About 75% of the total variation is explained by the first two principal components (Figure 6-3). In general, the samples tend to cluster according to geography (Figure 6-3); the importance of geographical distance is especially apparent for the first principal component, which explains about 61% of the total variation. Thus, straying appears to be more common among populations that are closer to each another, consistent with patterns documented by Quinn et al. (1991) and Pascual and Quinn (1994).

Those data indicate substantial genetic differentiation among chinook populations from the Klamath River (except for population samples 8, 9, and 10, from the extreme upstream portion of the river basin). The genetic differentiation results from the homing tendency of chinook and the resulting reproductive isolation among the breeding populations. Bartley et al. (1992a) estimated that the amount of genetic differentiation in the 10 samples is consistent with an average of about four migrants per generation among local populations. The average age of sexual maturity (i.e., generation time) in chinook from the Klamath River is

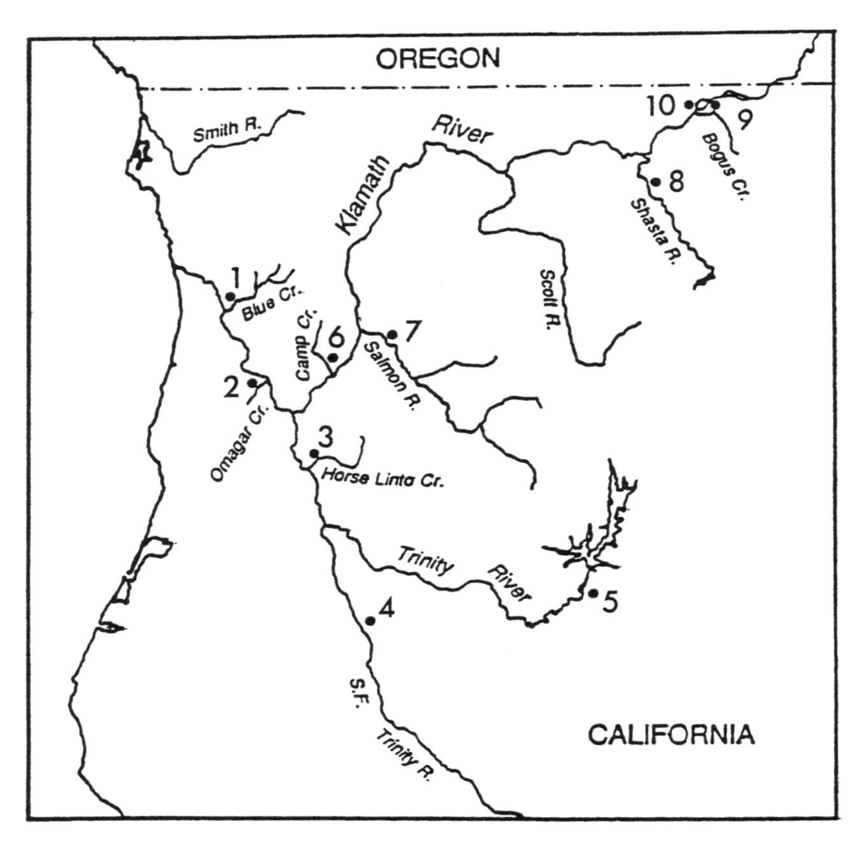

FIGURE 6-2 Collection sites of 10 samples of chinook in Klamath River drainage. Source: Bartley et al. 1992a.

about 4 years (Healey 1991). Therefore, the estimated rate of straying among these local breeding populations is about one stray per year within each reproductive population.

LOCAL ADAPTATION

One important reason to protect local populations is that they are *locally adapted* to the streams that support them. In other words, evolution has made a local breeding population better able to survive and reproduce in its home stream than in other streams. These adaptations are of great importance. Re-establishing new populations through introductions once the local populations have been lost has proved to be extremely difficult. And even if a newly introduced popu-

TABLE 6-1 Loci in Which Mean Frequency of Most Common Allele Was Less Than 0.95 in 10 Samples of Chinook Collected From Klamath River Drainage[a]

Sample	Locus								
	mAH-4	*GPI-2*	*MDHP1*	*MDHP2*	*PGK-2*	*PGM-4*	*SOD-1*	*DPEP1*	*N*
1	0.985	0.765	0.315	0.859	0.400	0.490	0.755	0.895	100
2	0.775	0.615	0.390	1.000	0.380	0.565	0.815	0.770	100
3	0.905	0.915	0.465	0.827	0.380	0.665	0.904	0.940	100
4	0.980	0.885	0.330	0.859	0.320	0.505	0.845	0.930	100
5	0.942	0.929	0.150	0.726	0.292	0.363	0.917	1.000	120
6	0.943	0.769	0.333	0.564	0.186	0.528	0.852	0.824	106
7	0.929	0.888	0.245	0.622	0.189	0.495	0.968	0.964	98
8	0.955	0.945	0.212	0.562	0.155	0.592	1.000	1.000	100
9	0.938	0.945	0.228	0.558	0.185	0.667	1.000	1.000	128
10	0.899	0.949	0.247	0.598	0.146	0.586	0.990	0.990	99

[a]Sample locations are shown in Figure 6-2. *N* = number of fish analyzed.

Source: Bartley et al. 1992a.

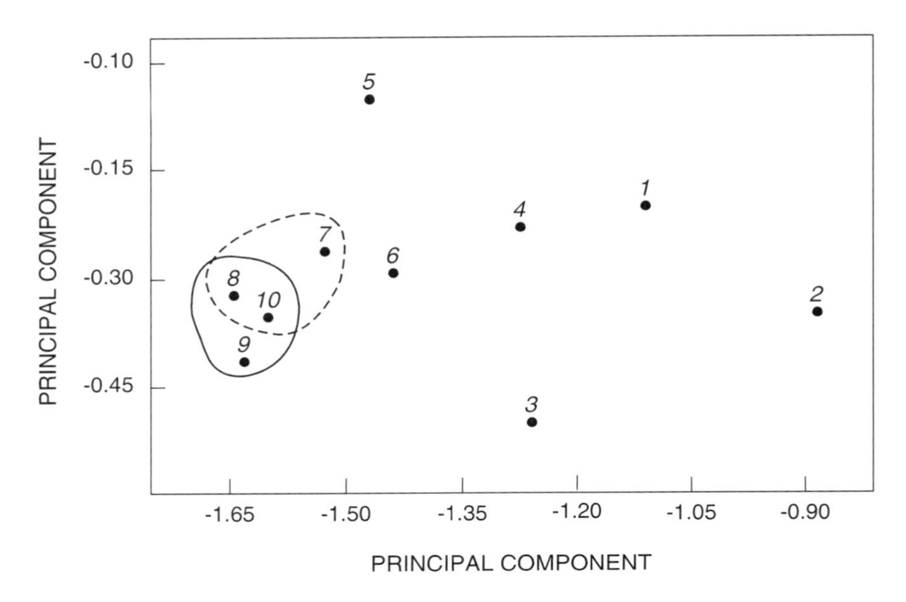

FIGURE 6-3 Plot of first two principal-component scores derived from allele frequencies at eight loci in Table 6-1. Samples enclosed by solid line are not statistically different. Samples enclosed by dashed line have statistically different allele frequencies ($0.05 > p > 0.001$); all other pairs of samples have highly significant differences in allele frequencies ($p < 0.001$).

lation is initially successful, it might not be adapted to the range environmental conditions that have happened in the past and can be expected to occur again in the future. That is, important local adaptations might be of importance only in extreme environments that are unlikely to occur within the time frame that we are concerned with but that are extremely likely to occur on an evolutionary time scale.

Adaptation to local conditions is demonstrated by the timing of spawning among populations of sockeye in the Fraser River system (Brannon 1987). Although the timing of spawning varies little from year to year in a given spawning location, there are great and consistent differences among spawning areas because of adaptations to the most favorable local conditions for incubation, timing of emergence, and juvenile feeding (Burgner 1991). The timing of spawning for local populations in the Fraser River is influenced primarily by the temperature regime of the spawning site (Figure 6-4). Spawning is later in the warmer incubation environments. That pattern of differences among populations results in similar emergence timing of the progeny during the next spring because of the greater rate of development at higher temperatures. Smoker et al. (in press)

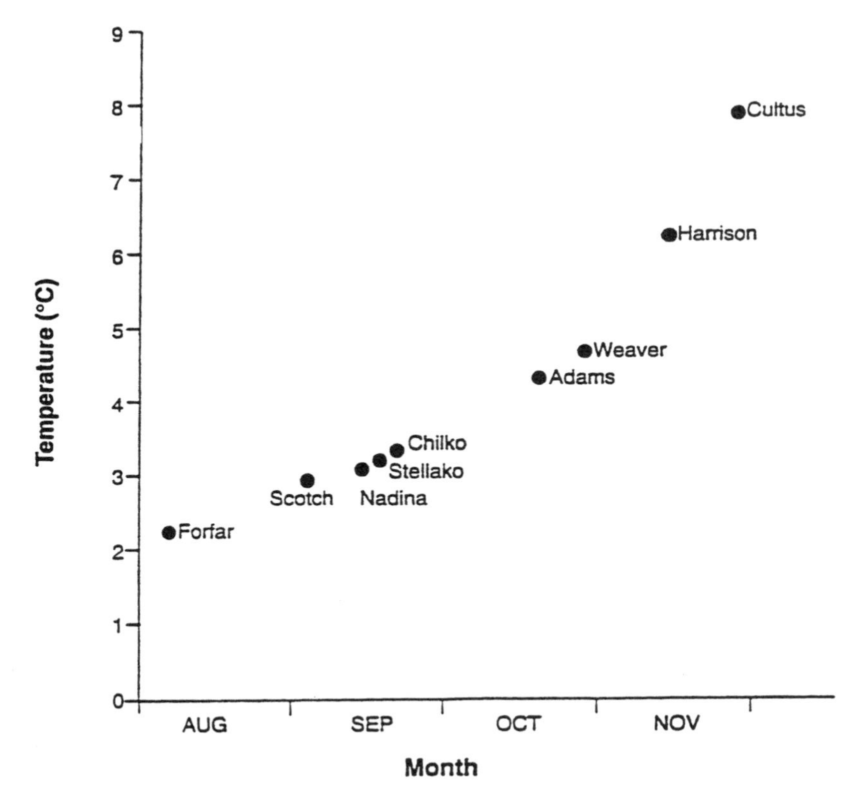

FIGURE 6-4 Spawning times and mean incubation temperatures of nine local breeding populations of sockeye from the Fraser River. Source: Brannon 1987.

showed similar genetic variation in the timing of migration in pink salmon in Alaska.

Stearns (1976) has defined a life-history "strategy" as a set of coadapted reproductive traits resulting from selection in a particular environment. Some anadromous salmon have spectacularly complex life histories. For example, Snake River sockeye emerge in freshwater at an elevation of 2,000 meters, migrate to a nursery lake, and generally spend 2 years growing in the lake. They then smoltify and migrate some 1,500 km downstream to the ocean, where they spend 2 or 3 years. In the ocean, they undergo long feeding migrations. On approaching sexual maturity, they return to the mouth of the Columbia River and then retrace their journey of 1,500 km upstream to their nursery stream, where they spawn.

The life history of those fish includes three distinct habitats: the nursery lake, the ocean, and the spawning stream. The fish undergo complex behavioral, physiological, and morphological transformations in transition from one habitat to another. In addition, they must undergo four major migrations: from stream to lake, from lake to ocean, a feeding migration in the ocean, and from ocean to natal stream. The timing of each of those events must be precise. Migration up and down the freshwater system to the ocean must be timed to correspond with appropriate water flows and the availability of food. In some cases, arrival in the freshwater habitat can precede actual spawning by many months. A number of studies have demonstrated that nearly all those aspects of life history are influenced to some degree by genetic differences among individuals and populations (reviewed in Taylor 1991 and Levings 1993).

A well-studied aspect of local adaptation in salmon is the migratory behavior of newly emerged fry. Studies with sockeye, rainbow trout, cutthroat trout, and arctic grayling (a freshwater member of the salmon family, *Thymallus arcticus*) have demonstrated innate differences in migratory behavior that correspond to specializations in movement from the spawning and incubation habitat in streams to lakes favorable for feeding and growth (Raleigh 1971, Brannon 1972, Kelso et al. 1981, Kaya 1991). Fry emerging from lake-outlet streams typically migrate upstream on emergence, and fry from inlet streams typically migrate downstream. Quinn (1985) showed that differences in compass-orientation behavior of newly emerged sockeye correspond to movements in feeding areas (Figure 6-5).

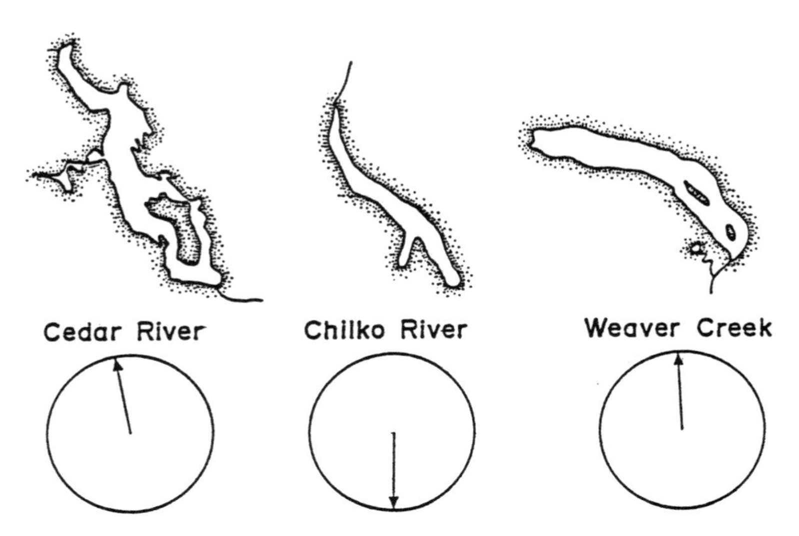

FIGURE 6-5 Mean compass orientations of newly emerged sockeye fry from three populations: Cedar River in Washington and Chilko River and Weaver Creek in Fraser River system in British Columbia. Compass orientation of experimental fry corresponds to the direction that will facilitate migration in the nursery lakes. Source: Quinn 1985.

Strong evidence exists that spawning populations of anadromous salmon exhibit highly specific local adaptations for a number of traits such as the complicated homing behavior, temperature adjustments, unique local mating behavior, and adjustment of smolts to local feeding conditions. These adaptations are most likely to be quantitative characters that are dependent on the effects of many genes, each of which has only a small effect individually (polygenes). These principles are basic to genetics of populations but are known mostly from the study of species better suited than salmon to this kind of genetic analysis.

For the above reasons, we would expect it to be difficult to "replace" a local population with transplants from nonlocal populations. The more complex the life cycle, the more difficult it would be to replace a local population. Complexity would be influenced by the number of migrations and major habitat shifts. A break in continuity between any one of these stages would cause failure of reproductive success. For a newly introduced population to be successful, at least a few individuals must be able to complete the entire life cycle and reproduce.

The experience with attempted introductions of salmon populations supports these conclusions. Attempts to establish self-sustaining anadromous populations generally have failed (Steward and Bjornn 1990, Burgner 1991, Heard 1991, Ridell 1993b). For example, a logging dam was built in 1908 on the Adams River, a tributary of the Fraser River in British Columbia (Riddell 1993b). The dam blocked access of sockeye to the upper Adams River from 1908 to 1921; this run had been among the largest sockeye runs in the Fraser River system. The upper Adams River area has 1.2 km^2 of spawning area, which should be sufficient to support 6 million adult sockeye per year on the basis of the productivity of other sockeye populations in the region. Sixteen attempts between 1949 and 1975 to reintroduce sockeye to these spawning areas from other areas have not been successful in reestablishing this run. Today, very few fish return to spawn in the upper Adams River.

In contrast with the lack of success with anadromous sockeye, introductions of kokanee (nonanadromous sockeye) have been successful. The greater success of introduced kokanee is probably related to their having a simpler life history than anadromous sockeye. Similarly, introductions of trout into lakes and streams throughout the intermountain west of North America have been generally successful (Allendorf and Leary 1988).

METAPOPULATION STRUCTURE

The individual local breeding populations within a drainage basin or other geographical area are usually connected in a higher level of organization by exchange of individuals through "straying." The set of local breeding populations connected by exchange of individuals is a metapopulation or a "population of populations" (Hanski and Gilpin 1991); an example of a metapopulation is the fall-run chinook of the lower Columbia River (Pascual and Quinn 1994). The local breeding population is the primary demographic and genetic unit, as discussed in the previous sections. However, the metapopulation organization becomes very important when time scales longer than several generations are considered.

From a demographic perspective, metapopulation dynamics is the balance between extinction and recolonization of local breeding populations. That is, individual local breeding populations are expected to have a limited time of persistence on an evolutionary time scale. New populations are expected to be established by strays from other local breeding populations within the metapopulation. The metapopulation concept is thus closely linked with the process of local extinctions and the establishment of new local breeding populations. The geographical arrangement of salmon into discrete spawning populations where environmental conditions are appropriate for successful reproduction makes the metapopulation model appropriate for salmon. Local breeding populations of salmon are small enough and exist in such variable environments that they are likely to have relatively short persistence times.

From a genetic perspective, the concept of the metapopulation is similar to Wright's view of species subdivided into many small local breeding populations that breed largely within themselves but are connected with other local breeding populations by migration and genetic exchange (Wright 1951, 1969). Wright (1940) also considered the potential evolutionary importance of local extinction and recolonization of vacant habitat patches in his analysis of genetic population structure. Partial isolation of local breeding populations allows the evolution of adaptations to local environmental conditions, and the infrequent exchange of individuals ensures that all the alleles within a metapopulation will be present in each local breeding population and can be acted on by natural selection (Allendorf 1983).

Local genetic adaptations play an important role in the demographic dynamics of metapopulations. As discussed previously, efforts to re-establish local breeding populations by transferring fish from other locales has generally not been successful. That lack of success has at least a partially genetic basis in that the transferred fish do not have the appropriate genetic adaptations to complete their life cycle in the habitat to which they are transferred. Thus, straying itself might not be adequate for recolonization; the straying individuals must be from similar-enough environmental conditions for their progeny to be able to complete their life cycle and successfully reproduce in the new environment.

The total genetic diversity in a species is the sum of the variability over many hierarchical levels from the smallest local breeding population to the metapopulation to larger geographical areas that contain many metapopulations. It is impossible to draw sharp boundaries between different levels of spatially structured populations, but it is useful to identify the mechanisms that operate in the population dynamics on different spatial scales (Hanski and Gilpin 1991). For example, Figure 6-6 shows the distribution of chinook on different geographical

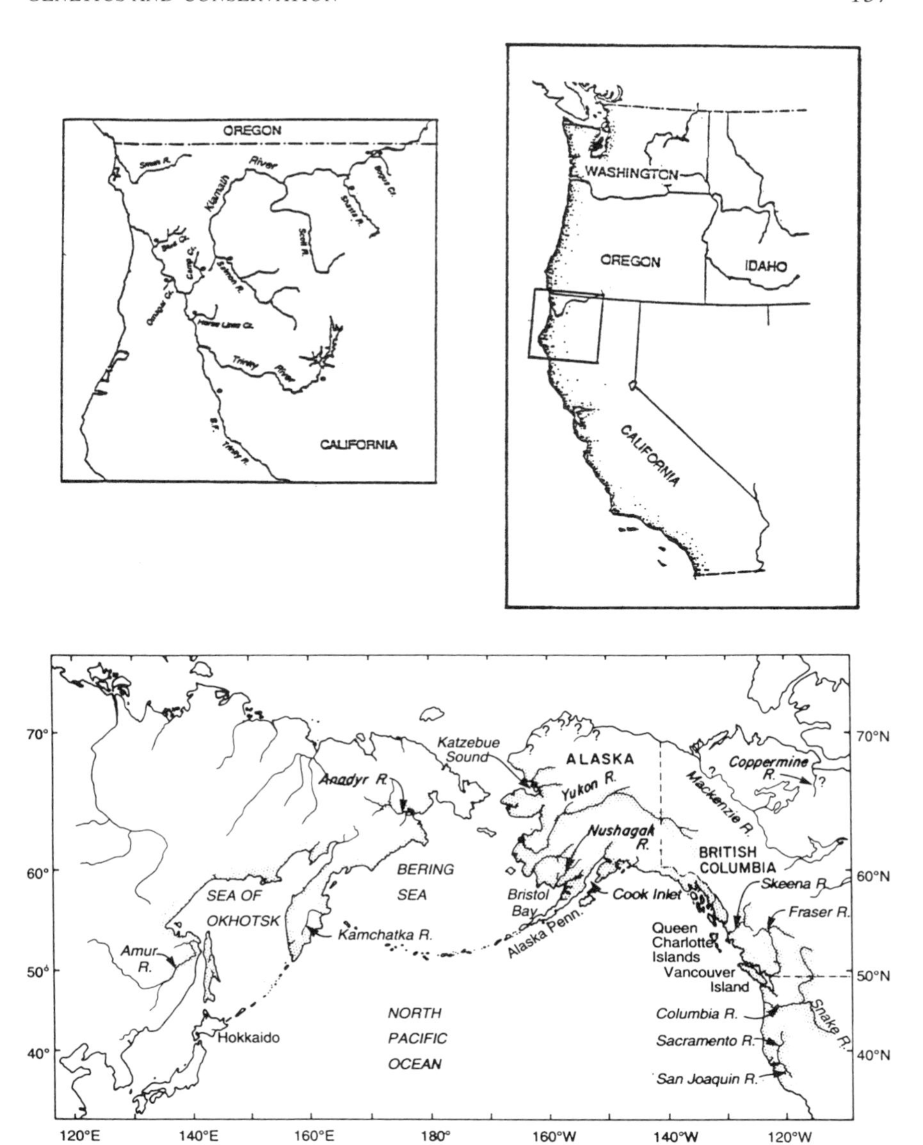

FIGURE 6-6 Three scales of geographical population structure of chinook. A, metapopulation of chinook in Klamath River drainage (Source: Utter et al. 1992). B, U.S. West Coast (Source: Utter et al. 1992). C, North Pacific Ocean and Bering Sea, showing distribution of chinook salmon through their natural range (Source: Healey 1991).

scales beginning with the metapopulation of chinook in the Klamath River drainage (A). The chinook in western North America (B) consist of many such clusters of local breeding populations or metapopulations. Finally, the entire range of chinook (C) contains all the genetic diversity present in the species.

The different geographical subdivisions of chinook have adaptive differences that are partially genetically based. For example, chinook returning to a coastal river in California are adapted to survive and reproduce in an environment very different from a river of comparable size in Alaska. On a smaller scale, it is extremely likely that the enormous chinook that used to inhabit the Elwha River on Washington's Olympic Peninsula were genetically different from the smaller chinook of other rivers in the region. The importance of geography in the distribution of genetic variation in chinook was examined by Utter et al. (1992), who found that 43% of the variability in allele frequencies at 24 polymorphic loci in nine local breeding populations of chinook from the West Coast of the United States is explained by the correlation between geographic distance and genetic similarity (Figure 6-7).

Studies of biochemical variation do not permit any interpretation of the adaptive significance of genetic differences among local breeding populations, because we do not completely understand the relationship between biochemically detected genetic variation and phenotypic variation. It is likely that the pattern of

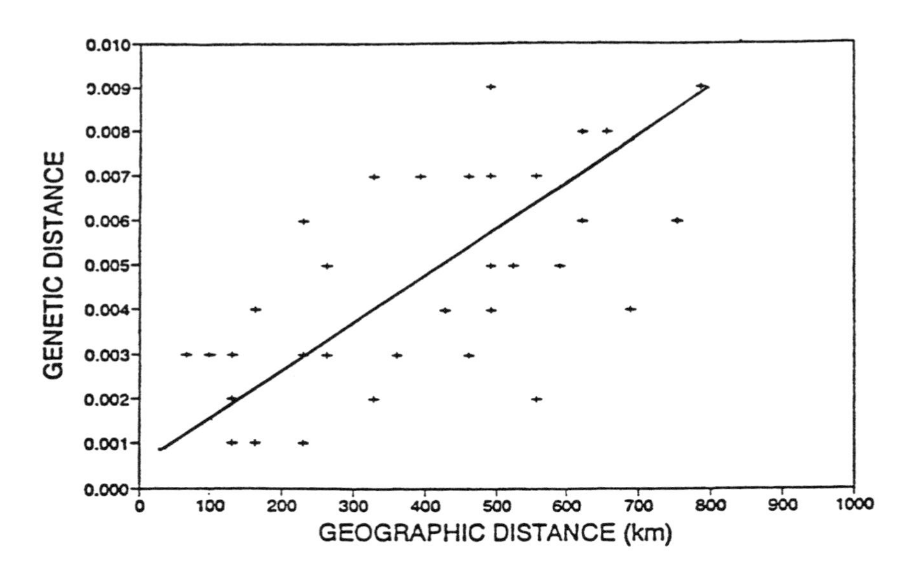

FIGURE 6-7 Correlation between geographic and genetic distance for nine local breeding populations of chinook from coastal drainages of California, Oregon, and Washington ($r^2 = 0.43$; $P < 0.001$). Line is principal axis of correlation. Source: Utter et al. 1993.

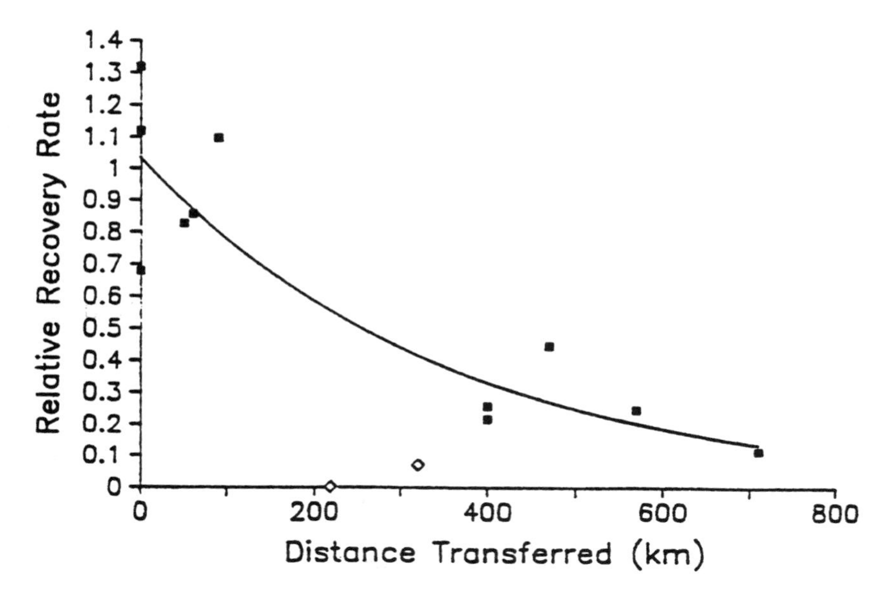

FIGURE 6-8 Relative recovery rate (in comparison to local fish) for transferred hatchery coho versus distance transferred. Source: Reisenbichler 1988.

genetic divergence at protein loci among chinook populations largely reflects the operation of genetic exchange and genetic drift on selectively neutral, or nearly neutral, genetic variation. However, as gene flow decreases and genetic divergence increases, the probability that populations will acquire adaptations to their local environment increases. For example, Reisenbichler (1988) showed that the distance from the home stream is inversely related to the success of hatchery-reared coho (Figure 6-8).

LEVEL OF GENETIC ORGANIZATION TO BE CONSERVED

The study of salmon genetics does not by itself lead to an answer to the question, What should we conserve? It does make clear that genetic variation—evolutionary potential—is present at all population levels, and it does make clear how important natural population-genetic structure and dynamics are. The Endangered Species Act (ESA) leads to attempts to decide what level of genetic organization should be conserved by specifying that it applies to "distinct population segments" of vertebrate species. For that reason, the concept of "evolutionarily significant unit," or ESU, has been introduced (see, e.g., Waples 1991, Waples in press). This concept has formed the basis for the policy on distinct population segments for salmon adopted by the National Marine Fisheries Service. The policy has been used in ESA listing determinations for Pacific salmon (summarized by Waples, in press).

The ESU concept provides a consistent, scientifically based framework for interpreting the meaning of distinct populations of salmon under the ESA. The unifying theme of the ESU concept is conservation of the evolutionary legacy and potential of the biological species—that is, the genetic variability that is a product of past evolutionary events and that represents the reservoir on which future evolutionary potential depends. The goal is thus to ensure viability of the biological species by conserving enough of its basic components to allow the dynamic processes of evolution to proceed.

Waples (1991) advocated a holistic approach to identifying ESUs and provided specific guidance on how to integrate genetic, phenotypic, life-history, ecological, and geographic information. The ESU concept is inherently hierarchical and applies equally to local breeding populations and to groups of populations (metapopulations). The focus is on units that are largely independent over evolutionarily important periods, and the "bottom-line" test for an ESU is whether its loss would represent a significant loss of ecological and genetic diversity to the species as a whole. It is important to recognize (as did Waples, in press) that decisions about what constitutes "significance" and about the resource tradeoffs implicit in recovery plans are largely societal decisions that cannot be based on scientific grounds alone.

A recent National Research Council report (NRC 1995b) developed the concept of the evolutionary unit (EU). The EU concept is similar, but not identical, to the ESU; differences between them are largely a matter of emphasis. The EU concept does not stress reproductive isolation as a criterion because reproductive isolation is often more difficult to assess directly than it is in anadromous salmon. The EU was meant to be applicable to all animals and plants, while the ESU was developed for anadromous salmon. The NRC report concluded that the application of either the EU or ESU concept would lead to similar results most of the time and also recommended that identifying an EU be separate from deciding whether it is in need of protection (NRC 1995b).

One aspect of the problem was discussed by Allendorf and Waples (in press) in analyzing the importance of local adaptation to conservation programs. They pointed out a paradox: small, locally adapted populations (demes) are clearly essential units for conservation efforts, but the kinds of local adaptations that they contain have evolved many times in salmon in a relatively short time and hence are likely to evolve again. Such recurrent adaptations may actually be different genetic solutions to an environmental challenge and different demes adapted to the same challenges may have different genetic complements. For some future challenge, an appropriate solution may require components of several of them, even if they appear to be adapted to the same environment. If that is true, why should we expend a lot of time and money on conserving wild runs?

One answer to the question concerns the time scale of concern. Although salmon have adapted relatively quickly and frequently in evolutionary time—perhaps over periods of hundreds to a few thousand years and even faster where they have been introduced into new environments, such as New Zealand (Quinn and Unwin 1993)—the depletion of salmon and their habitats in the Pacific Northwest is happening over periods of decades. It is not appropriate to imagine that a process that takes hundreds or thousands of years can compensate for one occurring at least 10 times as fast. In addition, human impacts on the environment have probably affected the types of environmental variations and their variability, and that change is also likely to decrease the ability of populations to adapt to local conditions.

Another answer concerns the number and sizes of natural populations. The local adaptations that produced the diversity of salmon life histories in the Pacific Northwest took place in a large and undepleted metapopulation structure. Today, parts of the metapopulation structure are missing, other parts are reduced in size, some local breeding populations have been extirpated, and many areas are populated largely or only by hatchery fish. It is therefore likely that even given hundreds or a few thousand years, local adaptation would not occur as quickly as it did in the past. So, although the evolutionary plasticity of salmon gives us hope that rehabilitation is possible, it is not a reason to diminish efforts to conserve diverse wild runs if long-term sustainability of salmon in the Pacific Northwest is a goal.

Finally, it is likely that some of the responses to local conditions is phenotypic, as well as genetic, at least in its early stages. If the genetic variability in salmon populations is reduced by local extinctions, increases in the proportions of hatchery fish, and reductions in population sizes, it is likely that phenotypic (or environmentally caused) plasticity will be reduced as well. Indeed, the relationship between phenotypic plasticity and genotypic variability in fishes has been a research topic for many years (e.g., Alm 1959, Policansky 1983, Stearns 1989). A better understanding of the relationship in salmon would clearly be helpful in any rehabilitation and conservation program.

EFFECTS OF HUMAN ACTIVITIES ON GENETIC DIVERSITY

The genetic structure of salmon populations in the Pacific Northwest has been modified drastically over the last 100 years. Many locally adapted populations have been lost because of dams and loss of habitat. The remaining wild spawning populations have been modified to an unknown degree by fishing, interbreeding with fish released from hatcheries, and habitat modifications.

Relatively few of the local breeding populations of salmon that existed 150 years ago in the Pacific Northwest exist today. It is difficult to estimate accurately what has been lost. Pacific salmon have disappeared from about 40% of their historical range in Washington, Oregon, Idaho, and California (Chapter 4). Many local breeding populations in the remaining 60% of the historical range have been extirpated. Williams et al. (1992) listed 106 stocks of salmon and steelhead that have been extirpated in this region. Most of these stocks probably comprised two or more local breeding populations (Williams et al. 1992). In addition, many local breeding populations have been swamped by introgression from hatchery salmon, the original local breeding population having been replaced by genetic material from straying hatchery fish. For example, coho still spawn and reproduce in the lower Columbia River. However, it appears that the native local breeding populations of these fish have been replaced by feral hatchery fish (Johnson et al. 1991).

In addition, a substantial proportion of the remaining demes of salmon are threatened by imminent extirpation (Williams et al. 1992). Nehlsen et al. (1991) clearly summarized the current status of salmon and steelhead in their review:

> With the loss of so many populations prior to our knowledge of stock structure, the historic richness of the salmon and steelhead resource of the West Coast will never be known. However, it is clear that what has survived is a small proportion of what once existed, and what remains is substantially at risk.

The continual erosion of the locally adapted groups that are the basis of salmon reproduction constitutes the pivotal threat to salmon conservation today.

Extant salmon populations have also been genetically modified by human actions over the last 100 years. Most management practices affect some aspects of the genetic makeup of salmon resources. Perhaps the largest effect has been produced by hatcheries, described briefly by Allendorf and Ryman (1987) and discussed in detail in Chapter 12. But other management practices also have effects. The potential and actual effects of fishing have been reviewed recently by Nelson and Soulé (1987), Thorpe (1993), and Policansky (1993). For pink salmon in particular, fishing appears to have changed the genetic makeup of many stocks and resulted in smaller adults. It has likely affected other species as well and might have changed the frequency of early-maturing small males (jacks) of several salmon species.

Changing the habitat of salmon also has the potential for indirectly causing genetic changes in their populations, although this has received little attention. There is evidence of selective effects of hatchery environments (see Chapter 12), and there is every reason to believe that changes in water flows, temperatures, and chemistry and other changes in the physical and biological environments of salmon in streams caused by human activities have resulted in genetic changes. In summary, a large array of human activities have affected salmon genetics in such a way as to reduce genetic diversity at all levels of population structure.

CONCLUSIONS

Sustained productivity of anadromous salmon in the Pacific Northwest is possible only if the genetic resources that are the basis of such productivity are maintained. We have already lost a substantial portion of the genetic diversity that existed in these salmon species 150 years ago. The possible genetic effects of any actions must be considered when any management decisions are made. The local reproductive population, or deme, is the fundamental biological unit of salmon demography and genetics. An adequate number of returning adults for every local breeding population is needed to ensure persistence of all the reproductive units. The result of regulating fishing on a metapopulation basis and ignoring the reproductive units that make up a metapopulation is the disappearance or extirpation of some of the local breeding populations and the eventual collapse of the metapopulation's production.

The metapopulation model of geographical structure is important for salmon because of the geographical arrangement of salmon into discrete spawning populations adapted to the environmental conditions in which they reproduce. Local demes of salmon are small enough and exist in variable-enough environments for it to be likely that they will have relatively short persistence times on an evolutionary scale. Although the deme is the functional unit of salmon genetics and demography, the cluster of local populations (the metapopulation) connected by genetic exchange via natural straying is the fundamental unit on an evolutionary time scale. *This conclusion is crucial because it leads to many other conclusions and recommendations about salmon management.* For example, most of this report's conclusions and recommendations about hatcheries, fishing, and habitat rehabilitation are founded on the importance of maintaining appropriate diversity in salmon gene pools and population structure, which has not been adequately recognized.

7

Habitat Loss

DIMENSIONS OF THE PROBLEM

Salmon habitat in freshwater is defined by physical and chemical characteristics of the environment during the portion of the life cycle spent in streams, lakes, or estuaries. It is generally taken to include

- Water quality: temperature, dissolved oxygen, turbidity, nutrients, and environmental contaminants.
- Properties of flow: velocity, turbulence, and discharge.
- Geological and topographic features of the stream and its valley: width and depth, streambed roughness, particle size composition, riffle and pool frequency, and floodplain characteristics.
- Cover: shading, interstitial hiding spaces, undercut banks and ledges, woody debris and aquatic vegetation.

For many of those features, streamside vegetation plays an important direct and indirect role in affecting local habitat characteristics. Some biotic components of the environment are influenced by physical habitat conditions, including prey, predators, competitors, and pathogens.

In this report, *altered habitat* is habitat that has been changed by human activity but is still accessible to salmon; lost habitat is habitat that used to be accessible but is no longer. Habitat alteration and loss that lead to reduced salmon production can occur when either of two conditions exists: anthropogenic perturbations transform freshwater spawning or rearing habitat to an unnatural or otherwise unproductive state or human intervention prevents natural disturbances from creating or maintaining habitat that is important for salmon production. Although most anthropogenic changes in habitat result in impairment of the productivity of aquatic ecosystems (Reice et al. 1990), some lead to increased production by improving survival or growth of one or more life-history phases. And some types of alterations do not directly affect salmon habitat but cause changes in the species composition of the aquatic community that might or might not be favorable to salmon (Reeves et al. 1987) or shift conditions from those favorable to one salmon species to those favorable to another (Lichatowich 1989). The important point is that habitat can be altered by the direct effects of human perturbations and by human prevention of natural disturbances (Sousa 1984, Wissmar and Swanson 1990). Either can impair salmon production, especially when its spatial or temporal scale differs fundamentally from that of the natural disturbance regime of an area.

Habitat alterations can have positive or negative outcomes that are often difficult to predict. For example, removal of streamside vegetation, a frequent consequence of human alteration of riparian zones, results in increases in solar radiation and water temperature. Higher light levels and warmer water can promote algal growth (Gregory et al. 1987), which leads to increased invertebrate production and more food for rearing salmon (Hawkins et al. 1983). Higher light levels also tend to enhance foraging efficiency of stream-dwelling salmon (Wilzbach 1985). The resulting increased growth rates might confer improved overwinter survival and increased smolt size (Holtby and Scrivener 1989). Large smolts, in turn, might be better able to escape predation in nearshore environments and have higher survival rates at sea (Pearcy 1992). All those processes potentially improve productivity. However, increased temperatures have also been shown to reduce growth efficiency when food is scarce (Brett et al. 1969) and to favor competitive dominance of cyprinid fishes, such as redside shiners (*Richardsonius balteatus*), over salmon (Reeves et al. 1987). And outcomes can be complicated by the presence of exotic species or pathogens, which also tend to be favored by higher temperatures (Li et al. 1987). Those processes potentially limit salmon production.

Because so many physical and biological factors, of which temperature is only one, are influenced by the removal of streamside vegetation and because interactions between these factors are still poorly understood, predictions of the specific consequences for salmon of altering stream temperatures to salmon are often prone to error. Models of the impact of habitat change on salmon generally suffer from an inability to predict the consequences of interacting ecological processes (Mathur et al. 1985, Fausch et al. 1988); this is especially true when models are extrapolated to geographical regions beyond those in which their quantitative relationships were developed (Shirvell 1989).

Because habitat loss is widely acknowledged to have contributed to the decline of virtually every species of Pacific salmon in western North America

(Nehlsen et al. 1991), the lack of precise knowledge of relationships between various types of habitat change and salmon populations need not be a barrier to improved environmental management. Different land uses (e.g., forestry, agriculture, grazing, mining, and urban and industrial development) are practiced at different locations in a river basin, but they share some effects with respect to habitat alteration and loss. This chapter identifies some important types of habitat alteration and loss, discusses how these changes influence the functioning of aquatic ecosystems, and identifies specific consequences for salmon.

NATURAL VERSUS ANTHROPOGENIC DISTURBANCES AND WATERSHED PRODUCTIVITY

Natural disturbances play a crucial role in the various life-history phases of salmon. Pacific salmon evolved in freshwater environments that included a variety of natural disturbances, including seasonal high flows and floods, glaciers, droughts, wildfires, volcanism, landslides and debris flows, and seasonally extreme temperatures. Their adaptations to life in frequently disturbed freshwater ecosystems reflect, in part, a need to cope with unusual events. Such adaptations include relatively high fecundity and large eggs, which permit extended intragravel residence by alevins during periods of unfavorable stream conditions; excellent swimming abilities of both juveniles and adults; occasional straying from natal streams by adults; and differentiation into locally adapted populations. Salmon with prolonged freshwater life cycles appear to be somewhat more flexible in their habitat requirements than those with abbreviated or lacustrine freshwater life cycles (Miller and Brannon 1982). For example, Reimers (1973) identified five distinct life-history strategies involving different periods of riverine and estuarine residence in a single population of fall chinook salmon in southern Oregon (Table 7-1). Multiple life-history strategies within populations might be an effective means of hedging against unusual events.

Not all disturbances result in diminished salmon production. Some cause short-term population declines but ultimately lead to increased productivity concurrently with habitat and trophic recovery (Gregory et al. 1987, Schlosser 1991). Natural disturbances can alter habitat in ways that stimulate salmon production and maintain environmental heterogeneity (Naiman et al. 1992). Wildfires and some types of soil disturbances increase nutrient availability and so enhance primary production (Walstad et al. 1990). Floods entrain particulate organic matter and both large and small woody debris from riparian zones. High flows cleanse spawning gravel of fine sediment and scour new pools. Wildfires open forest canopies, provide large woody debris, and create opportunities for early successional plant communities in riparian zones (Agee 1993). Windstorms and windthrow provide recruitment of large woody debris and increase the complexity of local habitats. Disturbances of many types increase the transport of nutrients, organic matter, and large woody debris to estuaries (Sibert 1979, Simenstad

TABLE 7-1 Major Variations in Fall Chinook Salmon Life Histories in Sixes River[a]

Life-History Variation	Description
1	Emerge from gravel, move directly downstream through main river and estuary and into ocean within few weeks.
2	Emerge from gravel, move into main river (or possibly stay in tributaries) for rearing until early summer, move into estuary for short period, and finally move into ocean before period of high productivity in estuary during late summer and autumn.
3	Emerge from gravel, move into main river (or possibly stay in tributaries) for rearing until early summer, move into estuary for extended rearing, and finally enter ocean after experiencing improved growth in estuary during late summer and autumn.
4	Emerge from gravel, stay in tributary streams (or, rarely, in main river) until autumn rains, and then move directly to ocean.
5	Emerge from gravel; stay in tributary streams (or, rarely, in main river) through summer, autumn, and winter, and then enter ocean during next spring as yearlings.

[a]A coastal Oregon stream.

Source: Reimers 1973.

et al. 1982). All of those processes are important to maintaining fish production in aquatic ecosystems (Gregory et al. 1991) and are necessary for normal ecosystem functions and diverse aquatic communities (Poff and Ward 1990). Although a large natural disturbance can cause a temporary decline in salmon populations, productivity might rebound to exceed predisturbance levels for extended periods (Bisson et al. 1988). Some types of natural disturbances that have beneficial long-term effects on salmon habitat, such as floods and wildfires, damage property and threaten lives and so are aggressively controlled and suppressed. It is therefore important to view human activity not only as a cause of habitat change, but also as potentially hindering natural disturbance patterns and recovery processes from creating and maintaining productive and diverse habitat.

Productivity declines when habitat alteration and loss impair the successful completion of life-history stages in the context of a watershed's landscape, its natural disturbance regime, and its anthropogenic changes. If a salmon population exists close to the environmental tolerance limits of its species—for example, at the edge of its range either geographically or with respect to riverine environmental gradients—relatively minor changes in key habitat characteristics resulting from natural climatic events or from human activity can influence population viability or expression of full evolutionary potential (Table 7-2 and 7-3). Many of the known extinctions of salmon populations over the last century have occurred near the edges of geographical ranges (Nehlsen et al. 1991, The Wilderness Society 1993), and many of these have apparently been caused or accentuated by human-related habitat losses. Understanding the effects of habitat alteration must include considering changes in the context of an area's natural disturbance regime (Table 7-2).

Habitat disturbances can be "cumulative" in the sense that different factors acting sequentially or concurrently can limit population size or growth during different phases of freshwater and estuarine rearing periods (Elliott 1985). To some extent, populations can adjust to alteration in or loss of habitat in a compensatory fashion; after a period of decreased survival, reduction in competition can lead to increased survival or growth (Chapman 1966). However, not all factors can be compensated in this manner and interactions between different types of habitat change may exacerbate the damage each would do independently (Niemi et al. 1990, Hicks et al. 1991). Furthermore, anthropogenic changes to habitat may occur so fast that natural selection processes are unable to adjust and compensate.

Human activities change the frequency or magnitude of disturbances (Table 7-2), and result not only in loss of or alteration in habitat but in substantial changes in natural recovery processes (see Figure 7-1a). Natural disturbances large enough to have an important impact require recovery intervals that might include periods of high production followed by re-establishment of density-dependent regulating mechanisms and biological controls that cause a return to predisturbance levels. There have been relatively few long-term studies of stream-dwelling salmon after large natural disturbances (Hanson and Waters 1974, Waters 1983, Elliott 1985, Bisson et al. 1988), but available evidence suggests that 10 years or more might pass after a large disturbance before salmon populations return to the normal range of predisturbance abundance.

Frequent anthropogenic perturbations of various intensity superimposed on natural regimes of less-frequent disturbances (Table 7-2) can hinder recovery processes and prevent populations from returning to their former abundance (Figure 7-1). Such perturbations gradually "ratchet" populations downward, a pattern typical of salmon populations in areas of progressive encroachment on riparian zones or areas with chronic input of sediment (Cederholm et al. 1981). Frequent, relatively small perturbations tend to increase the year-to-year variability of salmon populations. Hartman and Scrivener (1990) concluded that population instability was one of the most serious long-term consequences of logging for coho and chum salmon in Carnation Creek, British Columbia. Characteristic declines occur because populations do not have time to recover fully before the next large disturbance.

Very large anthropogenic impacts can cause so much damage to salmon populations or their habitat that abundance declines precipitously and does not recover (Figure 7-1). Such changes are characteristic of extensive habitat losses that might occur, for example, if a large portion of a river system were blocked. Sockeye salmon populations in the Fraser River underwent a major crash in 1913-1914 when rockslides caused by railroad construction in the canyon at Hell's Gate blocked much of the upper river, including most of the spawning grounds. Sockeye and other salmon that used the upper Fraser River remained at critically low densities until construction of fishways in the 1930s (Ricker 1987). Damage to or loss of habitat was so great that natural recovery was precluded until the fishways were completed. Many of the stock extinctions noted by Nehlsen et al. (1991) resulted from similar very large anthropogenic perturbations.

The spatial and temporal scales of anthropogenic habitat alterations that are imposed on salmon populations often differ in both frequency and magnitude from natural disturbance regimes. It is the natural disturbance regimes to which local populations are adapted and that have historically powered the creation of new, productive habitat: These characteristics must be retained or replicated if freshwater salmon habitat is to be sustained (Hill et al. 1991).

SEDIMENTATION

Sediment can enter watercourses by various mechanisms, and inputs can be chronic or episodic. Mobilization of soil particles through surface and gully erosion delivers small particles (fine sediment) to the stream network. Surface erosion is normally associated with precipitation but can occur chronically if human activities generate continuous runoff of sediment-rich water to streams. The erosion of large volumes of hillslope material, a process termed mass erosion, occurs when large upper soil movements (often rapid), such as landslides, and deeply seated slope failures, such as earth slumps, deliver coarse and fine sediment, large woody debris, and fine organic matter to streams.

Both surface erosion and mass erosion are normal processes (Leopold et al. 1964); their frequency depends mostly on the geology and erosiveness of soils and underlying rock and on the intensity and duration of rainfall and snow melt (Swanson et al. 1987). Some areas have naturally high erosion rates; examples include sandstone-dominated coastal river basins in northern California and western Oregon, granitic sediments in northern and central Idaho, and glacial-lacustrine deposits in northwestern Washington. That kind of area is often among the most sensitive to erosion caused by anthropogenic perturbations, such as logging and road building (Figure 7-2).

Improvements in road-construction and logging methods can reduce erosion rates. Rice (1992) documented an 80% reduction in mass erosion from forest roads and about a 40% reduction in mass erosion from logged areas in northern California due to improvements in forest practices beginning in the middle 1970s. However, the potential for continued alteration in and loss of salmon habitat

TABLE 7-2 Approximate Occurrence Rates of Different Types of Natural and

Type of Disturbance	Approximate Recurrence Interval (years)	
	Natural	Anthropogenic
Daily to weekly precipitation and discharge patterns	0.01 - 0.1	0.001 - 0.1
Seasonal precipitation and discharge; moderate storms; ice formation	0.1 - 1.0	0.01 - 1.0
Major floods; storms; rain-on-snow events	10 - 100	1 - 50
Debris avalanches and debris torrents	100 - 1000	20 - 200

Anthropogenic Disturbances

Physical and Chemical Factors Influenced by the Disturbance	Habitat Effects
Stream discharge; channel width and depth; storage and transport of fine particulate organic matter; fine sediment transport and deposition; nutrient concentrations; water current velocity	Minor alteration of particle sizes in spawning gravels; minor variations in rearing habitat; minor temperature change; altered turbidity; altered primary productivity
Bank-full flows; moderate channel erosion; high base-flow erosion; increased mobility of sediment and woody debris; local damming and flooding; sediment transport by anchor ice; gouging of stream bed by ice movement; reduced winter flows with extensive freezing; seasonal nutrient concentrations	Changes in frequencies of riffles and pools; changes in particle sizes in spawning gravels; increased channel width; flooding of side channels; removal (or sometimes addition of) cover; relocation of holding areas. In areas affected by ice: decreased water temperatures; lower primary and secondary productivity; egg dewatering or scour during anchor ice formation and breakup
Inputs of sediment, organic matter and woody debris from hillslopes, riparian zones and streambanks; localized scour and fill of streambeds; lateral channel movement; streambed mobilization resulting in redistribution of coarse sediment and flushing of fine sediment; redistribution of large woody debris; inundation of floodplains; transport of organic matter and large woody debris to estuaries	Changes in the frequencies of riffles and pools; formation of large log jams; burial of some spawning sites but creation of new areas suitable for spawning; increased amounts of fine particular organic matter for processing by the benthic community, resulting in increased secondary production; destruction or creation of side-channels along the floodplain; increased secondary production and cover habitat in estuaries
Large, short-term increases in sediment and large woody debris inputs; extensive channel scour; large-scale movement and redistribution of substrate, fine particulate organic matter and large woody debris; damming and obstruction of channels at the terminus of the torrent track; accelerated streambank erosion, resulting in channel widening; destruction of riparian vegetation; very large short-term increase in suspended sediment; subsequent summer temperature increases from vegetative canopy removal	Extensive loss of pool habitat in the torrent track; loss of spawning gravels; loss of habitat complexity along edge of stream; destruction of side-channels and other overwintering areas; creation of new cover in the terminal debris dam; creation of new spawning areas in the sediment terrace upstream from the debris dam; short-term loss of aquatic invertebrates; possible damage to gills from heavy suspended sediment load; increased primary production

TABLE 7-2 Continued

Type of Disturbance	Approximate Recurrence Interval (years)	
	Natural	Anthropogenic
Beaver activity	5 - 100	0 (removal of beavers)
Major disturbances to vegetation *Windthrow*	100 - 500	50 - 150 (buffer strip blow-down)
Wildfire	100 - 750	50 - 150 (timber harvest rotation)
Insects and disease	100 - 500	50 - 150 (timber harvest rotation)
Slumps and earthflows	100 - 1,000	50 - 200

Physical and Chemical Factors Influenced by the Disturbance	Habitat effects
Channel damming; obstruction and redirection of channel flow; flooding of streambanks and side-channels; entrainment of trees from riparian zone; creation of large depositional areas for fine sediment; conditions that promote anaerobic decomposition and denitrification, resulting in nutrient enrichment downstream from the pond	Enhanced rearing and overwintering habitat; increased water volumes during low flows; refugia during floods; possible blockage to upstream migration by adults and juveniles; elevated summer temperatures and lower winter temperatures; local reductions in dissolved oxygen, including areas under ice in winter; increased production of lentic invertebrates in pond; increased primary and secondary production downstream from pond
Increased sediment delivery to channels; decreased litterfall; increased inputs of large woody debris; decreased riparian canopy; increased retention of sediment and fine organic matter; reduced litterfall	Increased pool habitat; localized sedimentation; increased in-channel cover; increased summer temperatures and decreased winter temperatures; creation of eddies and alcoves along channel margins; increased secondary production
Increased sediment delivery to channels; inputs of large woody debris; loss of riparian canopy and vegetative cover; short-term increase in fine particulate organic matter and nutrients; decreased litterfall; increased peak discharge; short-term increase in summer flows from reduced evapotranspiration; short-term increase in biochemical oxygen demand in stream substrate	Increased sedimentation of spawning and rearing habitat; increased pool habitat and in-channel cover; increased water volume in summer; increased summer temperatures and decreased winter temperatures; increased secondary production; reduced dissolved oxygen in spawning gravels; scour of eggs and alevins in spawning gravels
Inputs of large woody debris; loss of riparian canopy and vegetative cover; decreased litterfall; short-term increase in summer flows from reduced evapo-transpiration	Increased pool habitat and in-channel cover; increased summer temperatures and decreased winter temperatures; increased water volume in summer; increased primary and secondary production
Low-level, long-term contributions of sediment and large woody debris to streams; partial blockage of channel; local baselevel constriction below point of entry; shifts in channel configuration; long-term source of nutrients	Sedimentation of spawning gravels; scour of channels below point of entry; accumulation of gravels behind obstructions; possible blockage of fish migrations; increased pool habitat in coarse sediment and large woody debris depositional areas; destruction of side channels in some areas, creation of new side channels in others; long-term maintenance of aquatic productivity

TABLE 7-2 Continued

Type of Disturbance	Approximate Recurrence Interval (years)	
	Natural	Anthropogenic
Volcanism	100 - 1,000	
Climate change	1,000 - 100,000	10 - 100 (thermal discharges, riparian canopy removal, channelization)

Source: Modified from Swanston 1991.

resulting from forestry activities continues. The FEMAT report (1993) noted that federally owned forest lands within the range of the northern spotted owl contain about 180,000 km of roads, a substantial portion of which constitutes potential threats to riparian and aquatic habitats, mostly through sedimentation. An estimated 250,000 stream crossings are associated with the road system, and most of the crossing structures might be unable to withstand storms with a recurrence interval of less than 25 years (FEMAT 1993). Road failures often result in debris torrents in small streams and can be particularly damaging to coho, steelhead, and sea-run cutthroat habitat.

Increased erosion from land use is not limited to the relatively steep forested terrain of the Pacific Northwest. For example, glacial sediment deposits in eastern Oregon and Washington are widely farmed for the production of dryland crops. However, because extensive areas are left fallow (i.e., barren of vegetative cover) each winter, winter rainfall, particularly on frozen soils, causes much surface erosion and sediment movement to streams. Similarly, urbanization, mining, excessive grazing, and other land uses can increase sediment production well beyond background levels.

Increased erosion can impair the reproductive success of salmon in several

Physical and Chemical Factors Influenced by the Disturbance	Habitat effects
Increased delivery of fine sediment and organic matter; scour of channels from mudflows; formation of mudflow terraces along rivers; destruction of riparian vegetation; damming of streams with creation of new lakes; increased nutrients	Sedimentation of spawning gravels; loss of pool habitat from mudflows, but creation of pool habitat in areas with tree blowdown; creation of new overwintering habitat and side channels along mudflow terraces; short-term potentially lethal sediment and temperature levels during eruptions; long-term increases in primary and secondary production; formation of migration blockages; long-term benefits to lake-dwelling species
Major changes in channel direction, gradient and configuration; stream capture; long-term changes in temperature and precipitation regimes	Major changes in frequencies of dominant habitat types; shifts in species composition related to preferences for temperatures, substrates, and streamflows; faunal transfers from stream capture; reproductive isolation may lead to stock differentiation; founder effects

ways. Spawning areas can be buried by large quantities of coarse and fine sediment, and spawning migrations can be blocked by landslides (Swanson et al. 1987). Intrusion of fine sediment into the egg pocket of redds can cause smothering of eggs and entrapment of alevins (Chapman 1988). Likewise, the rearing capacity of streams is often damaged by increased erosion (Hicks et al. 1991). Fine sediment can fill interstitial spaces and prevent their use by juvenile salmon for cover (Platts and Megahan 1975). Turbidity impairs foraging efficiency (Noggle 1978), causes emigration (Bisson and Bilby 1982), and disrupts social behavior (Berg and Northcote 1985). Long-term increases in sediment in conjunction with other impacts on riparian systems can result in pool-filling. Sedell and Everest (1991, see also FEMAT 1993) surveyed streams in national forests in Washington and Oregon that had been originally surveyed from 1935 to 1945 (Figure 7-3). They found that there had been about a 60% reduction in the frequency of large, deep pools (greater than 2 m deep and 43 m^2 in surface area), a loss attributed to filling with sediments and decrease in pool-forming processes. Nickelson and Hafele (1978) found a statistically significant correlation between densities of juvenile coho and the volume of pools in western Oregon streams. Bjornn et al. (1977) added sand to a natural pool in an Idaho stream, experimen-

TABLE 7-3 Spatial Scales, Recovery Times, and Some Biological Recovery Mechanisms of Stream Organisms Following Natural and Anthropogenic Disturbances in the Pacific Northwest

Nature of Disturbance	Spatial Scale	Examples
Natural		
	Small	Flood of 1- to 3-year recurrence interval; local windstorm; minor landslide
	Large	Major wildfire; dam-break flood (small streams); flood of 50- to 100- year recurrence interval (large streams)
Anthropogenic		
Acute sublethal	Small	Minor landslide or streambank erosion; short-lived toxicant (local use); temporary water withdrawal
	Large	Short-lived toxicant (widespread use); various flood-control practices
Acute lethal	Small	Short-lived toxicant (e.g., spill); major debris torrent
	Large	Introduction of pathogen to drainage system; channelization
Chronic sublethal	Small	Point-source sediment inputs; local thermal change; migration blockage in tributary
	Large	Increased erosion at the watershed level; widespread loss of riparian vegetation; habitat simplification; multiple water withdrawals; dams
Chronic lethal	Small	Frequent discharges of long- or short-lived toxicants; chronic anoxia; temperature or flows beyond tolerance limits
	Large	Frequent introductions of pathogens; frequent discharges of long- or short-lived toxicants; chronic anoxia; temperature or flows beyond tolerance limits

Source: Modified from Poff and Ward 1990.

tally reducing pool volume by half and water deeper than 0.3 m by two-thirds, and found that fish numbers declined by about two-thirds.

However, stream habitat can be improved if mass erosion delivers coarse sediment to streams that have little structural roughness, thereby creating new cover and increasing channel complexity (Everest et al. 1987, Swanson et al.

Relative Recovery Time	Biological Recovery Mechanisms
Short	Behavioral avoidance and refuge-seeking; increased growth among survivors; rapid recolonization of disturbed area
Moderate to long	Adjustment of populations and community structure to new habitat conditions; species migrations and new population establishment
Fast	Behavioral and physiological avoidance; refuge-seeking; rapid recolonization
Long	Physiological acclimation; selection for tolerant species; behavioral avoidance and refuge-seeking; shifts in community organization
Moderate to long	Behavioral avoidance; recolonization; new species establishment
Long	Population and community adjustments; selection for tolerant species
Moderate to long	Local population and community adjustments; colonization by tolerant species; behavioral and physiological acclimation
Long	Population and community adjustments throughout system; selection for tolerant species; species migrations
Very long	Selection for tolerant species
Very long	Colonization by rare, resistant species

1987). Thus, the specific effects of accelerated erosion depend on the nature and timing of sediment delivery (i.e., magnitude and frequency), the size of the particles eroded into the stream, and the prior condition of the stream itself.

Some watershed studies have clearly shown that sedimentation lowers survival of salmon eggs and alevins in the gravel (Hartman and Scrivener 1990).

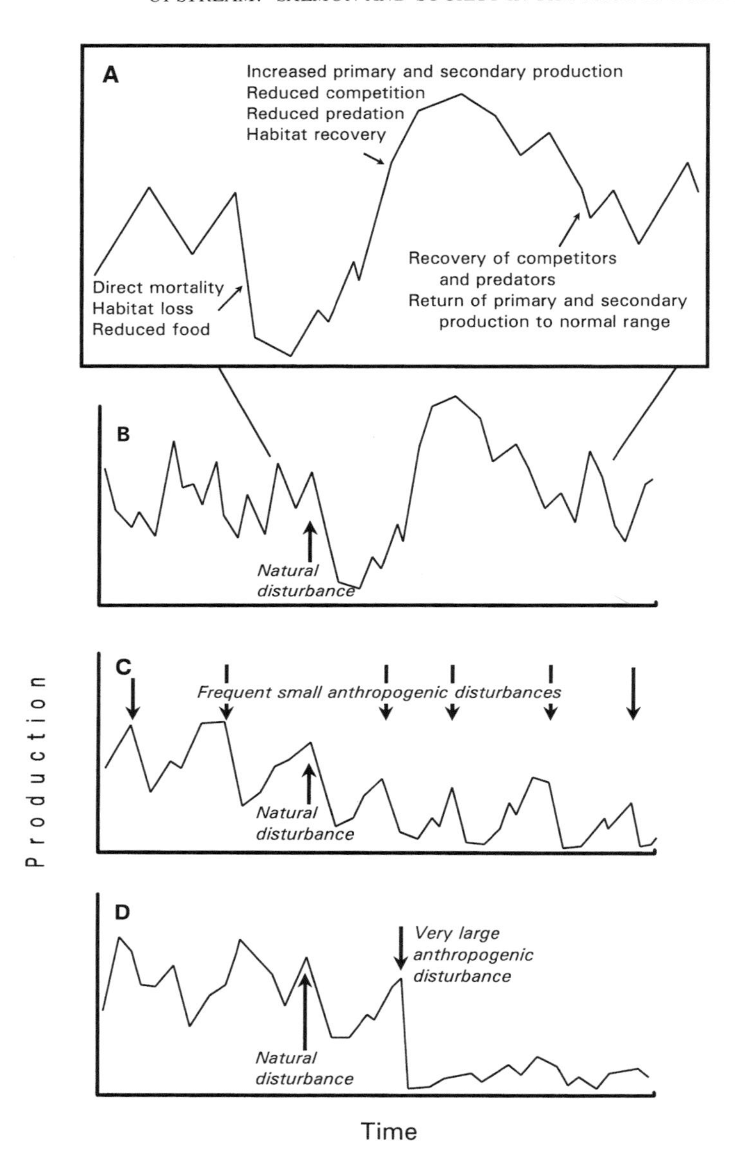

FIGURE 7-1 Hypothetical response of fish populations to natural disturbances of moderate to large intensity (b), frequent small anthropogenic impacts superimposed on natural disturbance regime (c), and single very large anthropogenic impact superimposed on natural disturbance regime (d). Box (a) above graphs shows some natural biophysical processes that operate during initial depression and later recovery stages of disturbance response.

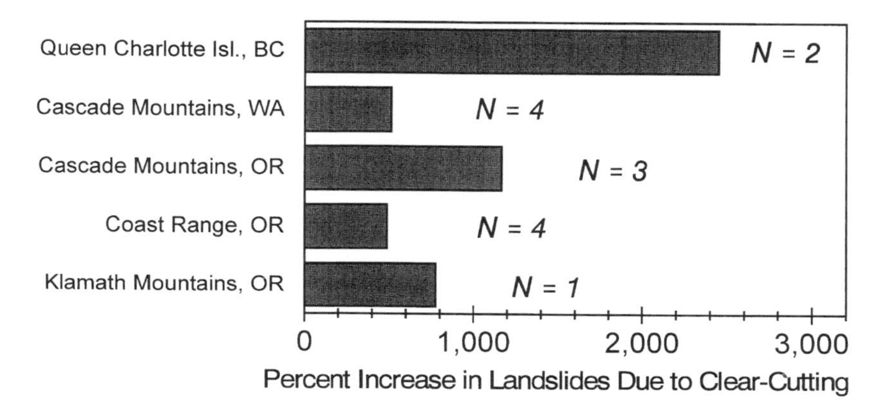

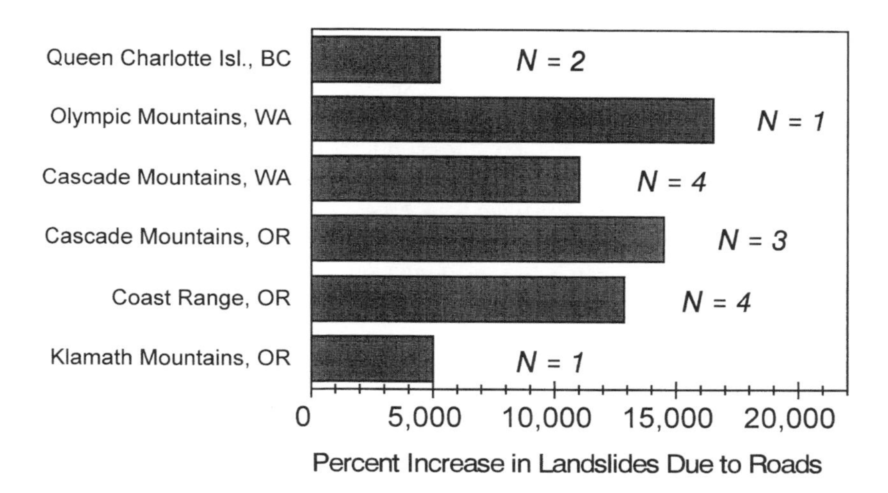

FIGURE 7-2 Increased frequencies of landslides caused by clear-cutting and roads in Pacific Northwest. Each bar is mean of results of existing studies of erosion rates on forested watersheds. Source: variety of sources summarized by Pentec Environmental, Inc. 1991, for Washington Department of Natural Resources.

Bjornn and Reiser (1991) summarized studies of spawning requirements of Pacific salmon and noted that the area used by each spawning pair ranged from less than 1 m^2 for pink salmon to 20 m^2 for chinook salmon. They concluded that substrate-particle size composition does not constitute a comprehensive measure of the suitability of stream gravel for spawning; water depth, velocity, and proximity to cover were also important in determining suitability. Citing unpublished data of R. A. House (Bureau of Land Management), Bjornn and Reiser (1991) reported that although gravel made up about 25% of the total substrate of an Oregon stream, only about 30% of the gravel—i.e., 8% of the total area—was suitable for spawning by coho salmon. It appears unnecessary for all stream gravels to be relatively free of fine sediment for successful spawning to occur. Furthermore, salmon have the ability to purge spawning areas of fine sediments during the spawning act itself (Everest et al. 1987). Thus, the proportions of fine sediment needed to decrease salmon production in streams are not known with certainty. Most research has been limited to the effects of fine sediment on reproductive success, with the result that the effects of sedimentation on other aspects of stream habitat and the food web of streams are poorly known.

STREAMBANK EROSION

Erosion of streambanks can affect fish habitat by introducing new sediment into the stream, by eliminating undercut streambanks and other types of cover along the margin of the channel, and by altering the width and depth of the channel. Localized streambank erosion is a natural consequence of channel meandering for most streams and rivers (Dunne and Leopold 1978), but damage caused by excessive erosion of streambanks and riparian zones can affect virtually all physical and biological processes in streams (Platts 1991).

The frequency and extent of streambank erosion in a watershed can be exacerbated by several types of anthropogenic perturbation. Among the most common are disturbances to streambanks resulting from livestock grazing and watering. Livestock grazing in semiarid regions of eastern Oregon, Washington, and Idaho continues to be very destructive to riparian vegetation in streams that support salmon populations (Beschta et al. 1991, Kauffman et al. 1993). The area affected by grazing can be considerable. Federal and nonfederal range land accounts for about one-third of the total land area of the Pacific Northwest (Table 7-4). A 1984 U.S. Department of Agriculture report of stream and riparian conditions in southeastern Washington (cited by Palmisano et al. 1993) indicated that about 1,100 km of the 10,600 km of streams on nonforested range land was severely eroded because of grazing. Platts (1991) compared the results of 21 studies of the effects of livestock grazing on riparian zones and fish populations in western North America and found that 20 documented substantial damage to riparian vegetation and 18 found substantial decreases in fish populations. Resistance of streams to damage from large storms was also seriously impaired where grazing had reduced or eliminated vegetation along streambanks. Although damage to riparian zones and streambanks has diminished with improvements in grazing practices in recent years (Platts 1991), many streams in range land still contain severely degraded habitat.

In forested areas, damage to streambanks can be caused by the removal of streamside vegetation. The root systems of forest vegetation are important for providing long-term bank stability and resisting the erosive effects of high flows

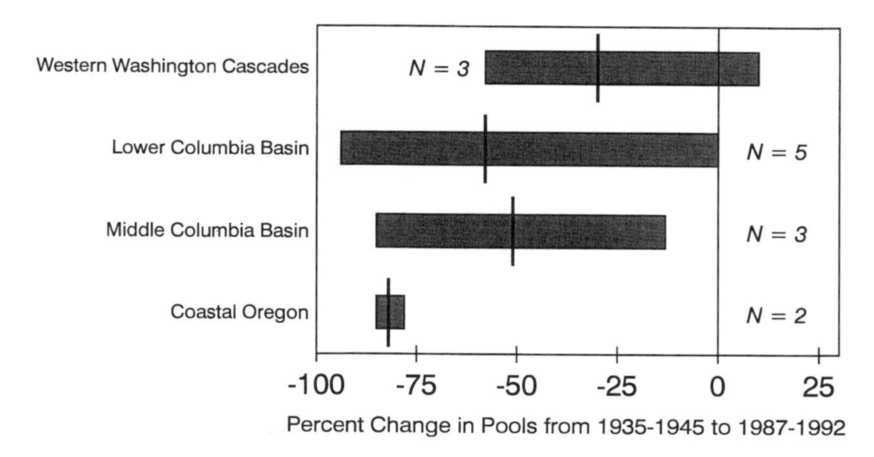

FIGURE 7-3 Changes in occurrence of large, deep pools in 3rd- to 5th-order streams in selected river basins in Pacific Northwest. Sample size refers to the number of river basins surveyed. Source: Data of Sedell and Everest 1991, and FEMAT 1993.

(Sedell and Beschta 1991). Hence, the logging of riparian trees—particularly those close enough to the channel, where their roots help to reinforce and bind streambank sediments—can initiate widening of stream channels and concurrent loss of undercut banks.

Damage to streambanks can also result from urban development and recreational activities adjacent to streams. Clark and Gibbons (1991) listed types of recreational activities that potentially affect riparian vegetation and streambanks. They included hiking, horseback riding, use of on- and off-road vehicles, camping, fishing, boating, water sports (swimming and use of temporary dams), removal of debris, bathing, and dish-washing. In heavily used recreational areas, damage to streambanks and riparian zones can be severe.

STREAMBANK ARMORING AND CHANNELIZATION

Just as streambank erosion can alter fish habitat, prevention of all lateral channel movement can change the pattern of responses to natural disturbances in streams. To prevent bank erosion and lateral channel adjustments, streambanks are often armored with erosion-resistant materials—such as rocks, concrete, or logs—to protect property or bridges from damage during high flows. Streams and rivers with armored banks commonly appear as straightened channels with few breaks in the current, particularly along the margins. The energy of the flow in channelized streams is focused along the thalweg (the deepest part of a channel), where high rates of downcutting can occur. Channelized streams typically have little physical complexity, because large roughness elements (e.g., woody

TABLE 7-4 Area and Percentages of Oregon, Washington, and Idaho in Different Land Uses

	Oregon		Washington		Idaho		Total	
	km^2	%	km^2	%	km^2	%	km^2	%
Federal lands								
Commercial forests	559	22.3	211	12.0	387	17.4	1,157	17.8
Other forests[a]	198	7.9	172	9.8	291	13.1	661	10.2
Range land	532	21.2	68	3.9	635	28.5	1,234	19.0
National parks	7	0.3	73	4.1	4	0.2	84	1.3
Non-federal lands								
Crop land	176	7.0	314	17.8	264	11.8	754	11.6
Irrigated land	71	2.8	66	3.7	142	6.4	278	4.3
Commercial forests	480	19.1	511	29.0	165	7.4	1,156	17.8
Pasture	77	3.1	58	3.3	55	2.5	190	2.9
Range land	370	14.8	226	12.8	267	12.0	863	13.3
Urban and industrial	38	1.5	63	3.6	19	0.9	121	1.9
Total	2,508	100.0	1,762	100.0	2,229	100.0	6,498	100.0

[a]Areas assigned to this category were estimated before designation of additional federal forest lands in late successional and old-growth reserves, spotted owl and murrelet conservation reserves, and riparian reserves in April 1994 Record of Decision for Amendments to Forest Service and Bureau of Land Management Planning Documents Within the Range of the Northern Spotted Owl; effect of new policy is to increase area of "Other forest" and decrease area of "Commercial forests" on federal lands in Oregon and Washington.

Source: Modified from Pease 1993.

debris) that obstruct the flow are typically transported downstream during periods of high discharge.

Some types of armoring materials can provide habitat for juvenile salmon. For example, rock riprap consisting of large boulders with many interstices is used for cover in both summer and winter by salmon and steelhead in middle Columbia River tributaries (Mullan et al. 1992). Where streambanks are armored with smooth, continuous structures, there are no hiding places for young fish, no eddies that serve as holding areas in swift currents, and few suitable rearing sites.

INSTREAM MINING

In the late 1800s, mining of gold and silver was a major industry in the Pacific Northwest. Placer deposits were heavily worked, thus stream and riparian alterations were not only locally devastating, but resulted in vast quantities of sediment being washed downstream. Johnson (1984) indicated that late 1800s gold mining activities in the Rogue River basin destroyed large areas of coho habitat. Similarly, hydraulic mining along the Sacramento River had so damaged fish runs by 1866 that the region's first cannery, which had opened only two years previously, was shut down and moved to the Columbia River (see Chapter 3). Even today, evidence of channel alteration from early mining operation persists throughout streams and rivers of the Pacific Northwest.

While instream mining operations generally cause significant modification to channel morphology, streambanks, bed material composition, water quality, and other habitat features, few studies have addressed their direct effects upon salmon. Although present-day mining impacts on fisheries are largely not known, they are probably relatively minor compared to historical turn-of-the-century impacts (Kaczynski and Palmisano 1993, Palmisano et al. 1993).

Another form of instream mining that has become relatively common over the last half century is aggregate extraction for sand, gravel, and cobbles. Palmisano et al. (1993) indicated that mining of sand and gravel occurs along or in nearly every major salmon stream in Washington. The extraction of aggregate material from streams and rivers has local effects, such as altering bed elevations and channel morphology, that in turn affect other characteristics (e.g., spawning habitats, streambank morphology, channel patterns, riparian vegetation) and processes (e.g., bedload transport rates, connectivity of hyporheic and subsurface water zones). The persistent removal of aggregate material, particularly when rates of removal exceed rates of replenishment, can cumulatively result in major changes to a stream or river system (Collins and Dunne 1987).

Stream protection regulations associated with the removal of aggregate material from channels of the Pacific Northwest were generally absent until recent decades, with the exception of U.S. Army Corps of Engineers Section 10 permits for work in navigable waters. In Oregon, there were no state stream protection regulations for gravel removal operations before 1965 (Kaczynski and Palmisano 1993). While Oregon has issued over 4,000 permits for gravel removal since 1967, with a total permitted volume of over 800 million cubic yards, there has been essentially no environmental monitoring and assessment by state agencies of resulting environmental impacts (WRRI 1995 a, b); most regulatory efforts by the state have simply focused on managing on-site impacts. Furthermore, little research has been undertaken to evaluate the short- or long-term effects of aggregate removal upon either channel characteristics (Collins and Dunne 1987) or anadromous salmon (Kaczynski and Palmisano 1993, Palmisano et al. 1993).

DIKING, DRAINING, AND FILLING

Diking, draining, and filling—primarily for urban and industrial development, agriculture, and creation of pasture land—are most common in estuaries and tidal sloughs but also occur in wetlands and floodplains. Loss of estuarine and riverine wetland habitat can potentially affect all salmon. Those most likely to be affected are coho, which can use riverine wetlands and estuaries for overwintering (Tschaplinski and Hartman 1983), and chum, chinook, and sea-run cutthroat trout, which rear in estuaries for extended periods (Simenstad et al. 1982, Trotter 1989, Healey 1991, Salo 1991).

Loss of estuarine habitat from draining and filling can result not only in loss of rearing area but in substantial alteration of the food base of estuarine communities. Sherwood et al. (1990) estimated that 77% of the 10,500 ha of tidal swamps, 63% of the 6,500 ha of tidal marshes, and 7% of the 17,000 ha of tidal flats of the Columbia River estuary were diked or filled between 1870 and 1970. They suggested that those activities reduced production of emergent vegetation by about 80% and benthic algal production by about 15%. Placement of structures (rock jetties and pile dikes), in addition to diking or filling, to improve navigation had reduced the tidal prism (salt wedge) by about 15%, simplified the complex network of tidal channels, and focused the flow into navigation channels. Such changes, with alterations in sediment and water transport characteristics resulting from upstream impoundments and water uses, have had a profound effect on benthic invertebrates that contribute substantially to the biological productivity of the Columbia River estuary for salmon (Simenstad et al. 1992).

Loss of estuarine habitat through filling has been proportionately greater in some rivers than in the Columbia River. Although highly variable (Table 7-5), losses of estuarine habitat have exceeded 90% of the historical area of some Puget Sound river systems (Simenstad et al. 1982). And reductions in estuarine tidal marshes by diking and filling have exceeded 50% of the historical area in seven of the 15 Oregon coastal estuaries examined by Boule and Bierly (1987). The greatest losses have occurred in areas that are heavily industrialized and urbanized, particularly in rivers with shipping ports at their mouths. Reductions have probably been less in coastal rivers without large cities and shipping ports (Palmisano et al. 1993).

Small streams in urban areas can also be altered by land-filling. Lucchetti and Fuerstenberg (1993) analyzed the drainage network of Thornton Creek, a third-order tributary to Seattle's Lake Washington, between 1893 and 1977. They found that Thornton Creek had lost all major wetlands and 60% of the open channel network, including all first-order tributaries, to urban development. The remaining stream system was constrained by loss of riparian vegetation, streambank armoring, and an extensive series of culverts and underground pipes. Although cutthroat trout continued to survive in Thornton Creek, the last adult coho salmon was reported in 1979, and none has appeared since then despite repeated fry plantings by school and volunteer groups. Conditions in this stream might be typical of small streams in heavily urbanized areas, where habitat loss has been extensive and permanent (Booth 1991).

Conversion of riverine wetlands to agricultural fields and livestock pasture and navigation improvements along rivers have transformed river valleys from marshy, densely vegetated areas with highly complex river channels to simplified drainage systems most of whose flow is confined to the mainstem (Sedell and

Luchessa 1982). Overall wetland losses in some areas have been great; for example, only about 9% of the wetlands present before Euro-American colonization remain intact in California (Dahl 1990). Beechie et al. (1994) estimated that 54% of the riverine slough and wetland habitat available to coho salmon in Washington's Skagit River was lost from the floodplain because of diking, draining, and filling for agriculture and creation of pasture. The total area of lost slough habitat was about twice the combined losses of tributary habitat due to water withdrawals, impassable culverts, and inundation by a major reservoir. Estimated annual losses of juvenile coho salmon caused by elimination of rearing habitat in Skagit River valley sloughs ranged from 220,000 to 560,000.

FLOOD CONTROL

Prevention of damage by flooding is usually achieved through the use of flood-control dams, dredging to increase channel capacity, and dikes and levees to prevent rivers from overtopping their banks and spilling out onto the floodplain. Specific effects of dams are discussed in Chapter 9, but flood-control measures can have a serious effect on salmon habitat. Dikes along river channels impair connections between rivers and floodplains that supply large woody de-

TABLE 7-5 Changes in Areas of Selected Puget Sound Estuaries (Including Hood Canal) and Grays Harbor (a Coastal Washington Estuary) from 1800s to 1970s

	Area (ha)		
Estuary	Predevelopment	Amount in 1970s	Change (%)
Puget Sound			
Nooksack	445	460	+3
Lummi[a]	580	30	–95
Samish[a]	190	40	–79
Skagit	1,600	1,200	–25
Stillaguamish	300	360	+20
Snohomish[a]	3,900	1,00	–74
Duwamish[a]	260	4	–98
Puyallup[a]	1,000	50	–95
Nisqually	570	410	–28
Skokomish	210	140	–33
Dungeness	50	50	0
Washington Coast			
Grays Harbor[a]	19,500	13,600	–30

[a]Estuaries with substantial industrial development.

Source: Simenstad et al. 1982 as cited in Palmisano et al. 1993.

bris, fine organic matter, and dissolved nutrients to the drainage network (Pinay et al. 1990); these materials are important habitat elements, providing both refuge and major food sources for riverine and estuarine communities (Vannote et al. 1980).

Juvenile salmon can spend large portions of their freshwater residence in floodplain environments (Levy and Northcote 1982; Brown and Hartman 1988), especially in winter. Survival and growth are often better in floodplain channels, oxbow lakes, and other river-adjacent waters than in mainstem systems (Peterson 1982). Isolation of rivers from floodplains can in some instances be extensive (Sedell and Froggatt 1984). The Pacific Northwest River Basins Council (1972; as quoted in Kaczynski and Palmisano 1993) estimated that some 3,000 km of dikes, 5,000 km of bank protection, and 5,800 km of channel modifications (dredging) for flood control had been completed by the early 1970s, and the numbers are likely greater today. Sedell and Luchessa (1982) used maps and river-clearance records from the 1800s to document that coastal river valleys then were filled with mazes of floodplain sloughs, beaver ponds, marshes, and wetlands. Flow was channeled into river mainstems away from floodplains to improve river navigation and for flood control.

ALTERED STREAMFLOW

Human activities can increase or decrease flows, cause streamflows to become more or less variable than the natural discharge regime, and alter the timing of seasonal runoff patterns. Changes in flow result from water impoundments, water withdrawals, enlargement or shrinkage of the effective drainage network (an example might be a watershed in which many roads effectively function as first-order channels), increasing the imperviousness of the soil surface, altering the rapidity of runoff, altering groundwater quantities or movements, altering the depth of coarse and fine sediment in the stream (which can affect the amount of surface discharge relative to subsurface flow), altering streamside and hillslope vegetation via forest-management practices (which can affect evapotranspiration, snow interception, and fog drip), and the conversion of forest land to other uses (such as agriculture and urbanization). Salmon are very sensitive to changes in streamflow and time their life-cycle movements according to local discharge regimes (Northcote 1978, Groot and Margolis 1991).

Increases in peak discharges are often a concern when sensitive eggs and alevins are still in stream gravels because bedload movement can cause mortality (Everest et al. 1987, Chamberlin et al. 1991). Juvenile salmon might seek cover in interstices of stream gravels to avoid predators or when conditions for feeding become unfavorable in winter (Chapman and Bjornn 1969, Hillman et al. 1987). Recently emerged fry can occupy the interstices of 2- to 5-cm gravel, and yearling and older salmon require particles larger than 7.5 cm (Bjornn and Reiser

1991). Mobilization of gravel and cobble substrates during peak discharges can kill salmon directly if they happen to be present.

In small- to medium-sized channels, high flows can alter pool and riffle habitat and can flush sediment and organic matter from streams (Sullivan et al. 1987). Some flushing is beneficial in that it removes fine sediment from the streambed, but excessive flushing can remove much of a stream's organic matter and adversely affect secondary productivity (Sedell et al. 1978, Murphy and Meehan 1991). Frequent movement of large woody debris during periods of peak discharge can deposit logs on upper streambanks, where habitat benefits are minimized (Bisson et al. 1987).

Many human activities contribute to increasing peak flows. For example, the seasonal removal of vegetation and litter cover from agricultural lands usually leaves them prone to surface runoff, which is generally more rapid than subsurface flow. Similarly, soil compaction can increase the amounts of precipitation that enter streams through surface runoff. Although soil compaction can occur with almost any type of land use, effects are often most pronounced in urban and industrial settings, where extensive roads and paving can effectively double the frequency of hydrologic events that are capable of mobilizing stream substrates (Booth 1991).

Logging affects surface and groundwater hydrology in complex ways (Chamberlin et al. 1991), and studies have indicated that the frequency and magnitude of stream discharge peaks are sometimes increased after harvesting (Beschta et al. 1995). Forestry activities—including road construction, timber falling and yarding, slash burning, and mechanical scarification—can all cause water to reach streams more rapidly (Harr et al. 1975, Harr and McCorison 1979). Internal changes in soil structure occur after logging as water is piped along channels of decaying tree roots and moves faster downslope (Hetherington 1988). Logging roads and landings form relatively impermeable surfaces. Roadside ditches collect water from roads, and subsurface flow is intercepted by road cuts (Harr et al. 1975). Forest canopies intercept snowfall, shade the snowpack on the ground, and decrease ground-level wind velocities. Clearcutting contributes to increased snow accumulation in logged areas because snow is no longer retained in forest canopy. In areas with snow-dominated hydrology (e.g., forested areas of Idaho, eastern Oregon, and Washington), the timing and rate of snowmelt are generally advanced in clearcuts, and this effect persists for several decades. Increased snowmelt peaks during warm rains are especially pronounced in the transient snow zone—the zone of alternating rain and snowfall at intermediate elevations in coastal and western Cascade Range watersheds (Harr 1986). Rain-on-snow runoff exacerbated by logging can generate extremely high streamflow peaks and can contribute to extensive streambed scour and habitat alteration (Harr et al. 1989).

At the other extreme, changes in discharge regimes from a wide variety of land uses can result in unnaturally low streamflows. Water withdrawals for irrigation, hydroelectric production, urban and industrial consumption, and other uses can shrink stream channels and reduce or eliminate fish spawning and rearing habitat. Logging can have a variable effect on summertime flows (Hicks et al. 1991).

Redd dewatering or reduced intragravel flows can also be caused by reductions in discharge, which can lead to egg and alevin desiccation or to reductions in water exchange rates in the environment of the embryos. Salmon eggs can survive exposure to low flows, and even some periods of no surface-water flow, but survival rates depend on intragravel permeability, humidity, temperature, and dissolved oxygen (Bjornn and Reiser 1991). Redd dewatering can contribute to freezing conditions in the egg pocket during cold periods—a further cause of death. Salmon often select spawning sites in areas of subgravel water upwelling, which helps to buffer the risk of dewatering or freezing and ensure adequate water movement past embryos. However, if no upwelling areas exist naturally or if human activities have eliminated them, flow reductions can reduce reproductive success.

Effects of reduced discharge on streams' capacity for rearing juvenile salmon are complex. Reduced flows can lower stream volume and lead to crowding, which might cause increased aggression, competition emigration, and predation (Chapman 1962, 1966). Although the relationship between stream volume and salmon carrying capacity is often unclear, there is some evidence that densities of salmon can be influenced by the amount of water present in streams. Kraft (1972) diverted water from a Montana stream during summer to cause a 90% reduction in discharge. Many brook trout moved from runs to pools; average numbers of trout in partially dewatered runs decreased by about 60%. Smoker (1955) found a positive correlation between the 1935-1954 commercial catch of coho salmon and summer streamflows two years previously (the time when juveniles resided in streams). However, a significant positive correlation between summer flows and coastal Oregon coho was not detected by Scarnecchia (1981). In most instances, the potential effects of flow reductions on rearing salmon will be mediated by the amount of pool habitat and by biological factors, such as food availability, population levels, competitors, and predators.

Altering the flow regime of a stream can interfere with the upstream migration of adult salmon. Salmon runs are often timed to take advantage of particular streamflows in smoothing fish passage and avoiding predators (Northcote 1978). Upstream migration can occur on both rising and falling portions of storm hydrographs (Shapovalov and Taft 1954), or it can be timed to coincide with low or high flows to surmount potential passage barriers. Flow alteration can interfere with environmental cues that trigger run timing for species that spawn in both large rivers, such as chinook (see Healey 1991), and small streams, such as coho (see Sandercock 1991). Salmon often congregate near stream mouths when spawning is near and are stimulated to migrate upstream by their reproductive condition, changes in water discharge, and appropriate stream temperatures. In species that enter freshwater shortly before spawning, such as chum salmon (Salo 1991), long periods of unfavorable flow can reduce reproductive success. Murphy (1985) and Heard (1991) documented large mortality in pink and chum salmon caused by a combination of anoxia and low flow in intertidal reaches of southeastern Alaska streams. It is not known whether these die-offs were related to flow changes caused by human disturbance.

ALTERED GROUNDWATER

The role of groundwater in maintaining the productivity of stream ecosystems has only recently become appreciated; for a particularly thoughtful review, see Stanford and Ward (1992). Interactions between terrestrial and aquatic ecosystems are often strongly influenced by subsurface water movement and biogenic nutrient transformations (Gilbert et al. 1990). Rivers and their valleys are linked by exchanges of water between phreatic (true groundwater) and hyporheic (river-influenced groundwater) zones, and these linkages influence nearly all aspects of the physical and chemical habitat of aquatic organisms (Ward and Stanford 1989). Particularly important are nutrient transformations that occur in hyporheic and phreatic zones; these transformations ultimately regulate nutrient availability to primary producers in streams and rivers. Interchanges between surface water and groundwaters are critical to maintaining productivity in both large and small lotic systems and can be strongly disrupted by human activities that remove groundwater or inhibit the movement of water into or out of rivers and floodplains.

The specific effects on salmon of groundwater alteration by anthropogenic perturbations are poorly known. Sites of groundwater upwelling in streams are preferred spawning areas for chum salmon (Lister et al. 1980) and can become less suitable if upwelling characteristics are changed. In the Olympic Peninsula, spring-fed streams originating as groundwater seeps at the base of hillslopes serve as important overwintering areas for coho salmon (Peterson and Reid 1984, Cederholm et al. 1988). Roads crossing hillsides often cut into soils and intercept subsurface water, transporting it quickly downslope in ditches (Ziemer 1981), and might prevent it from draining to springs and streams along the valley floor. Warming of groundwater after logging in the Carnation Creek watershed raised late-winter stream temperatures (Hetherington 1988) and accelerated egg development; this caused early emergence of coho salmon fry and losses due to excessive downstream movement in years with frequent early-spring freshets (Holtby 1988). Loss of riparian vegetation and associated channel changes in arid regions through livestock grazing lower the zone of water-saturated soil and reduce subsurface storage of water; these effects can allow small streams to run dry in summer (Elmore 1992). Other than those few case-study reports, little is known about how altered groundwater affects salmon. It is likely, however, that some of the greatest impacts of alterations in groundwater on salmon habitat result from changes in nutrient dynamics and consequent loss of aquatic productivity (Triska et al. 1989, Stanford and Ward 1992) and are expressed as reductions in growth rates and species diversity in streams and rivers.

ALTERED RIPARIAN VEGETATION

Removal of vegetation along streambanks and floodplains can be caused by a variety of human activities. Like several other types of habitat alteration, loss of streamside vegetation can affect salmon in a variety of ways. Riparian vegetation mediates key interactions between aquatic and terrestrial ecosystems and in many respects controls the productivity of streams by influencing water, sediment, and nutrient dynamics; shading; inputs of fine particulate organic matter and woody debris; and the stability of streambanks and floodplain terraces (Beschta 1991, Gregory et al. 1991).

The direct influence of riparian vegetation on streams declines with increasing distance from the channel and with the height of the dominant tree species. Ecological functions provided by riparian vegetation are achieved at different distances, depending on the type of function and the width of riparian vegetation needed for the function to take place naturally. The height of dominant trees influences the potential distance over which riparian vegetation directly affects stream channels, e.g., tall trees potentially contribute shade, particulate organic matter, and large woody debris at greater distances from streams than do short trees. Areas capable of producing large, tall trees thus possess wider functional riparian zones than areas in which trees do not grow as large. For this reason, FEMAT (1993) used the height of dominant late-successional tree species that would naturally grow in a particular riparian zone as the basis for recommending streamside buffers needed to safeguard ecological functions instead of suggesting a fixed linear distance from the streambank to the outer margin of the buffer strip that would not allow for differences in potential tree growth between regions. According to FEMAT (1993) estimates, the width of vegetated buffers needed to achieve greater than 90% effectiveness for different ecological functions ranges from as low as 0.5 tree heights for protection of streambank root strength to about three tree heights for some microclimate characteristics (Table 7-6). In coastal regions of the Pacific Northwest, late-successional trees in riparian zones typically reach 70 m or more in height; therefore, protection of the full range of ecological functions and characteristics provided by riparian vegetation might require intact buffers of natural vegetation in excess of 70 m.

The buffer widths summarized in Table 7-6 are applicable to protecting ecological functions whether streams contain fish or not. However, many current land-use regulations maintain a higher level of riparian protection based on the presence of fish at a particular site or on whether there are important fish resources or public water supplies downstream. In areas next to streams, termed streamside management zones, some level of activity, such as logging or grazing, is permitted as long as minimum requirements for environmental protection are met. These requirements are often related to shading and streambank integrity (Bisson et al. 1992) but do relatively little to safeguard other processes. The width of streamside management zones required by state land-use laws is much less than the width of natural vegetation needed to provide full ecological protection. For example, the forest-practice regulations of California, Oregon, and Washington (Table 7-7) require narrower buffers next to streams than would be needed to fulfill the buffer requirements of Table 7-6 for aquatic and microclimate functions. Although streams on federally owned lands are protected by federal standards and guides for forestry practices, the level of protection given to non-fish-bearing permanent and intermittently flowing headwater streams is usually less than that afforded fish-bearing streams. Protection of forested riparian zones on federal lands is generally greater than on state-owned and privately owned lands (Robinson 1987, FEMAT 1993). Thus, there are two sets of double standards for protection of riparian forests: one set for fish-bearing and non-fish-bearing streams and the other set for federal lands and state or private lands.There is a third set of double standards: Riparian protection requirements on federally owned, state-owned, and privately owned commercial forests, taken in the aggregate, are far more restrictive than requirements along streams where agricultural and urban or industrial land uses are dominant. In California, Oregon, and Washington, state environmental regulations concerning forestry practices have been in place since the early 1970s, but until recently none of those states has enacted an agricultural-practices act explicitly protecting riparian vegetation. Rather, agricultural activities are limited by the water-quality requirements of the Clean Water Act, state water-quality standards, and voluntary compliance with best management practices. Although forestry-practice regulations deal, however incompletely, with riparian-zone protection, agricultural-practice regulations tend to emphasize meeting water-quality thresholds for drinking water and protection of aquatic biota but do not address riparian protection. Urban and industrial wastewater discharges are also regulated according to water-quality requirements, and protection of riparian zones in urban areas is often left up to local ordinances. Much of the historically most productive salmon habitat exists in lower river valleys and coastal lowlands where riparian zones are given the least protection (Sedell et al. 1990).

ALTERED THERMAL REGIME

The amount of thermally altered habitat in Pacific Northwest streams and rivers is not known, but it is probably proportional to the extent of water impoundments, riparian canopy alteration, heated discharges, and channel changes, such as widening and shallowing. In small streams, temperatures potentially lethal to salmon were measured after logging and burning in an Oregon coastal watershed during the 1960s (Hall and Lantz 1969). In the Salmon Creek drainage

TABLE 7-6 Estimated Widths of Riparian Forests Needed To Achieve > 90% of Full Function Provided by Streamside Vegetation to Stream Ecosystems in Forested Watersheds of Pacific Northwest.

Ecological Characteristic	Distance from Channel (in late-successional dominant tree heights[a]) Needed to Achieve > 90% Effectiveness
Functions directly affecting stream ecosystems	
Contribution of leaf litter and other material	0.7
Shading	1.0
Contribution of large woody debris	1.0
Intact root systems along streambanks	0.5
Protection of microclimate	
Soil moisture	0.5
Radiation	1.0
Soil temperature	1.5
Air temperature	2.0
Wind speed	3.0
Relative humidity	3.0

[a] About 70 m for late-successional coniferous forests in coastal regions.

Source: FEMAT 1993.

of the Oregon Cascades, summertime stream temperatures at the mouth of the basin underwent a long-term increase during a period of increased harvest activity (Beschta and Taylor 1988). The FEMAT report (1993) presented current and historical summer temperatures in 46 forested watersheds in western Oregon and Washington. Maximum temperatures in recent years had risen more than 2°C above historical maximums in 85% of the watersheds, and modal summer temperatures were beyond the range of historical conditions in 54%. Maximum temperatures exceeded 20°C (potentially stressful for salmon) in 70% of the streams and 25°C (potentially lethal) in 20%. It is likely that river basins in the Pacific Northwest are now thermally altered on a scale that under natural conditions would exist only in the presence of an exceptionally large climate change.

Effects of temperature changes on salmon and their habitat are summarized in several reviews (Beschta et al. 1987, Bjornn and Reiser 1991, Hicks et al. 1991). Thermal alterations potentially affect the survival and growth of virtually every stage of the freshwater life cycle. Clear patterns have yet to be established, but increased temperatures at northern latitudes along the West Coast have tended to benefit salmon production in small, naturally cold streams by increasing available food resources (e.g., Holtby 1988), whereas salmon productivity at southern latitudes usually declines when temperatures extend into thermally stressful

TABLE 7-7 Current Requirements for Riparian Protection on State-Owned and Privately Owned Forests in California, Oregon, and Washington

Stream Classification	Minimum Riparian Protection Zone Width Each Side of Stream (m)
California[a]	
Class I (fish-bearing)	23-46[b]
Class II (non-fish-bearing)	15-30[b]
Classes III & IV (no aquatic life)	None
Oregon[c]	
Type F (fish use or fish and domestic use together	15-30[d]
Type D (domestic use only)	6-21[d]
Type N (no fish or domestic use)	0-21[e]
Washington[f]	
Types I-III (fish-bearing)	7-30[g]
Type IV (non-fish-bearing)	None[h]
Type V (intermittent or ephemeral)	None

[a]Up to 50% of overstory and 50% of understory may be removed; exceptions for greater removal are given.
[b]Determined by slope steepness.
[c]The Oregon Rules permit timber harvest within riparian management zones as long as conifer basal requirements are met.
[d]Different riparian management area widths are specified for small, medium, and large Type F and Type D streams.
[e]For some small Type N streams, understory vegetation and nonmerchantable conifers must be left within 3 m of the stream.
[f]Widths of riparian zones along Washington's streams are determined by process called watershed analysis; some timber harvest is allowed within riparian management areas.
[g]Designed to recruit, on average, 70% of historical levels of large woody debris.
[h]Vegetative buffers may be required along lower reaches of Type IV waters for temperature protection and to buffer streams from applications of forest fertilizers.

Source: FEMAT 1993.

ranges (e.g., Burns 1971). Unless food is extremely plentiful (a situation that rarely occurs in nutrient-poor coastal river systems), higher temperatures decrease lower salmon growth efficiencies (Brett et al. 1969, Bisson and Davis 1976, Wurtsbaugh and Davis 1977), which must be offset by increased food production for temperature increases to be beneficial.

Increased temperatures can influence the migratory behavior of spawning adults. Bjornn and Reiser (1991) cited several studies that show delays in the upstream movement of sockeye, chinook, and steelhead adults because natal streams became too warm. Berman and Quinn (1991) found that adult chinook salmon behaviorally regulated their body temperature by pausing in areas of cool water during upstream migrations; they concluded that reduced availability of cool areas could potentially reduce spawning success.

Altered thermal regimes might have an even greater effect on the production of salmon through effects on interspecific interactions between salmon and non-salmon fishes, principally competition and predation. Reeves et al. (1987) demonstrated that juvenile steelhead in laboratory streams were aggressively displaced from preferred foraging sites by juvenile redside shiners, a species of minnow, when experimentally increased temperatures became physiologically less favorable for steelhead. In a forested river system in southern Oregon, they found that the outcome of steelhead-redside shiner competition followed a thermal gradient in which stream warming resulted in the retreat of steelhead to higher-elevation, cooler tributaries.

Many river systems in the Pacific Northwest contain a variety of introduced species that originated in warmer waters in eastern North America (Li et al. 1987), which were planted to provide a greater diversity of recreational angling opportunities. Their effects on native salmon populations are often poorly understood, but many species are known to prey on young salmon or compete with them for food or rearing sites at some stage in their life history. Temperature increases—whether caused by riparian canopy removal, water impoundment, agricultural and urban runoff, or heated industrial discharges—create conditions favorable to many warm-water game species and might enable them to gain a competitive advantage or facilitate their predation on juvenile salmon.

Altered thermal regimes can change other characteristics of habitat in streams, rivers, and estuaries by altering the structure of plant and invertebrate communities (Bisson and Davis 1976). Aquatic plants and macroinvertebrates have specific temperature preferences. Changes in thermal regime can alter species composition in ways that might or might not be favorable to salmon production.

DECREASED LARGE WOODY DEBRIS

Perhaps no other structural component of the environment is as important to salmon habitat as is large woody debris, particularly in coastal watersheds. Numerous reviews of the biological role of large woody debris in streams in the Pacific Northwest (e.g., Harmon et al. 1986, Bisson et al. 1987, Gregory et al. 1991) have concluded that woody debris plays a key role in physical habitat formation, in sediment and organic-matter storage, and in maintaining a high degree of spatial heterogeneity ("habitat complexity") in stream channels. Loss of large woody debris from streams usually diminishes habitat quality and reduces carrying capacity for rearing salmon during all or part of the year (Hicks et al. 1991). As with temperature, the exact manifestation of the effects of woody-debris loss on salmon is often difficult to predict.

Two general trends with respect to loss of woody debris are clear. First, the distribution and abundance of large woody debris have been extensively altered in most river systems. Headwater streams have lost woody debris through several processes, most related to logging activity. Years of splash damming, a method of floating logs from upland logging sites to downstream mills, have scoured channels and removed much of the instream woody debris that was present. Sedell and Luchessa (1982) found that some coastal river systems in Oregon and Washington contained hundreds of active splash dams from the early to middle 1900s. In some areas without splash dams, increased frequency of landslides caused large debris torrents (Figure 7-2), scoured stream channels, and created conditions similar to those resulting from splash dams and log drives. When it became apparent that accelerated hillslope erosion had caused numerous debris torrents culminating in large, impassable logjams, fishery management agencies in the middle 1900s undertook aggressive programs of debris removal to facilitate adult salmon spawning migrations. In addition to removing the logjams, stream-cleaning crews often removed large woody debris after logging even when there was little evidence that the debris actually constituted a migration barrier (Narver 1971).

Harvest of timber from riparian zones in coastal and western Cascade watersheds created ideal conditions for early-successional tree species, such as red alder, which replaced late-successional conifers as the dominant form of riparian vegetation over large areas (Kauffman 1988). Recruitment of new debris to streams from alder-dominated riparian zones was more rapid than from conifer-dominated stands, but the hardwood debris was smaller, was more prone to breakage, and decomposed faster than conifer debris (Bilby 1988), so streams in second-growth forests became progressively debris-impoverished after removal of the old-growth stand (Grette 1985, Veldhuisen 1990). Rotational harvest ages of forests on many industrial forest lands (40-60 years) have been short enough to preclude re-establishment of dominant conifers in riparian zones (Andrus et al. 1988). The combined effects of those anthropogenic perturbations have led to a large-scale reduction in the quantity and quality of large woody debris in many forested headwater tributaries and to a substantial decline in woody debris floated downstream to lowland streams (Sedell et al. 1988).

Mainstem rivers and estuaries historically also contained great amounts of large woody debris (Gonor et al. 1988). Some of the debris was produced in uplands and fluvially transported to depositional sites along rivers and their floodplains, but woody debris was also recruited from riparian zones adjacent to rivers and estuaries and entered channels through natural processes of floods, windstorms, fires, and beaver activity. Often, the lower reaches of rivers contained massive accumulations of debris that formed huge drift dams. Much of the wood in Pacific Northwest river systems was removed for navigational improvements and flood control in the late 1800s and early 1900s; for example, 368 km of the Sacramento River, 88 km of the Willamette River, and 24 km of the Chehalis River (Washington) were cleaned for river navigation from 1867 to 1912 (Sedell et al., 1990). Rivers and estuaries containing large volumes of woody debris were characterized by spatially complex and diverse channel systems and highly productive salmon habitat, but many of these areas were lost in the early twentieth century.

The second trend related to alteration of large woody debris abundance has been simplification of stream channels and loss of pool habitat (Figure 7-3). Simplification occurs when loss of small-scale spatial heterogeneity leads to channel conditions characterized by uniform substrate, depth, and velocity; loss of sediment and organic-matter retention capacity; elimination of backwaters, eddies, and side channels; and loss of instream cover. In other areas of North America where habitat simplification has taken place, streams support fewer species and are less resistant to community disruption from natural disturbances (Karr et al. 1985, Schlosser 1991). Pacific Northwest streams have fewer species than most other regions (Moyle and Herbold 1987) but have a great diversity of locally adapted populations. Streams with simplified channels usually contain fewer species than streams with structurally complex channels (Bisson et al. 1992), or they might be suitable for one age group but not multiple age groups—an important factor for salmon rearing two years or more in streams. Reeves et al. (1993) demonstrated that Oregon coastal watersheds with histories of forest management had lower salmon diversity than unmanaged watersheds. Typically, simplified channels with scarce woody debris support abundant populations of underyearling salmon but contain few yearling and older fish (Bisson and Sedell 1984, Hartman and Scrivener 1990, Reeves et al. 1993). Reduction in numbers of older salmon in simplified streams is often related to loss of pool habitat and winter cover resulting from elimination of large woody debris (Bisson et al. 1987). Of the 43 intermediate-size tributaries with histories of forest management examined in the FEMAT report (1993) for western Oregon and Washington, two-thirds had modal pool frequencies below the range of frequencies believed to have existed historically (generally 25-60%). In about half the watersheds, pools comprised less than 20% of channel areas.

MIGRATION BARRIERS

Dams are an important class of migration barriers and are discussed in Chapter 9. Many smaller barriers to migration are probably unknown. Nehlsen et al. (1991) noted that a substantial fraction of 106 stock extinctions might have resulted from migration blockages. They quoted from a story told by a Twana Indian who was born about 1865 but referred to the extinction of a sockeye run in southern Puget Sound in 1852:

> There were some sockeye in Mason Lake, south of Hood Canal Puget Sound area. These ran up Sherwood Creek from Allyn on Case Inlet. They'd hang around the lake till ripe, then run up the creeks from there. The Squaxon got them with a weir in Sherwood Creek. Finally a pioneer named Sherwood built

a little dam in the creek and stopped the fish, and they named the creek after him.

Many small populations might have been extirpated by similar activities in the 1800s and early 1900s. Although historical records are infrequent and usually rely on anecdotal information from people with little ability to identify salmon species, small dams probably contributed to the extinction of many local breeding populations, as in the case of Atlantic salmon in eastern North America. Likewise, landslides that blocked migratory routes eliminated some runs before the middle 1900s, but records were often poorly kept or lost. In many instances, nonnative stocks of salmon have been introduced into formerly blocked river systems (Johnson et al. 1991).

Rearing habitat of juvenile salmon can also be lost to blockages. Some of the most productive rearing sites in streams are located in backwaters along the edge of the channel and in side-channel areas (Sedell et al. 1984, Sedell and Beschta 1991). Highways built next to streams and rivers often disrupt access to these off-channel sites by physically isolating them from the main channel or by including culverts that are impassable for juvenile salmon. Culverts that are designed to pass adult salmon might create speeds that exceed the sustained swimming abilities of juveniles. Furniss et al. (1991) give the sustained swimming speeds of juveniles of several salmon species; they range from about 20 cm/s for juvenile coho 5 cm long to 70 cm/s for juvenile sockeye 13 cm long. Water speeds in many culverts are too great to allow juvenile passage at any time except during periods of low streamflow; in others the outfall of the culvert might be suspended too high above the water for juveniles to enter.

Unscreened water diversions constitute a potential migration blockage if downstream-migrating juvenile salmon are entrained in diverted water. Nichols (1990) identified over 3,000 unscreened water diversions in Oregon, including 1,300 on coastal rivers, that potentially affected salmon-rearing streams. In addition to blocking migrations, water withdrawals potentially influence available rearing habitat. Kaczynski and Palmisano (1993) reported that about 60% of the water diversions in Oregon were for irrigation and 20% for urban uses. Palmisano et al. (1993), citing several studies by the National Marine Fisheries Service, stated that about 70% of Washington's water diversions lacked proper screening in the late 1970s and that 30% continued to be improperly screened or designed even after efforts to improve screening.

WATER POLLUTION

Before enactment of the federal Water Pollution Control Amendments to the Clean Water Act in the 1970s, fish kills in the United States occurred with some regularity. Dissolved-oxygen concentrations in Oregon's Willamette River in the 1940s and 1950s often dropped to anaerobic levels because of sewage and industrial discharges, creating an uninhabitable zone along a substantial reach of the river until nutrient discharges were controlled (Warren 1971). Anaerobic conditions often occurred in upper Grays Harbor, the estuary of Washington's Chehalis River system, during the 1920s and 1930s in response to effluents from two sulfite pulp mills, three municipal sewage-treatment plants, and agricultural runoff (Eriksen and Townsend 1940). One pulp mill, built in 1928 near the mouth of the Hoquiam River in Grays Harbor, exerted a biochemical oxygen demand of 115,000 kg/d, a load equivalent to the raw sewage produced by 1.4 million people (Seiler 1989). Water quality was degraded during low river discharges from May to October in Grays Harbor and was severely damaging to chinook, coho, and steelhead; but it apparently did not substantially affect chum salmon, which emigrated earlier than the other species and did not rear in the upper estuary. Pollution-abatement efforts have reduced sewage and industrial discharges over the last two decades and the upper estuary is no longer anaerobic in summer, but experimental releases of smolts from hatcheries upstream have shown that a pollution block still exists in Grays Harbor and that exposure of smolts to water of poor quality has reduced seawater adaptation, increased infestation by a trematode parasite, lowered disease immunity, and possibly increased vulnerability to predation by birds and squawfish. Smolts in the Chehalis River system survive at roughly half the rate of smolts from a nearby, relatively unpolluted river (Seiler 1989). The case study of Grays Harbor has been well documented and might be representative of the effects of water-quality degradation on salmon in lower rivers and estuaries with heavy urban and industrial development. Although the concentration of pollutants in wastewaters is now regulated more strictly than before, the volume of pollutants in water could be equal to or greater than volumes existing before water-quality laws were enacted. Servisi (1989) estimated that the volume of wastewater discharged into the Fraser River had tripled since 1965.

Mining is another source of water pollution in Pacific Northwest rivers. Nelson et al. (1991), citing a study by the Environmental Protection Agency, reported that in 1961-1975 at least 10 million fish were killed nationally by mining-related water pollution, although the number of salmon included was not given. Nelson et al. (1991) provided a thorough discussion of the different types of water pollution resulting from mining activity: in western North America, metals and radionuclides from mining wastes can be highly toxic to salmon; some highly toxic metals, such as copper and zinc, are also highly synergistic—their combined effects greatly increase lethality. Furthermore, metals can "bioaccumulate" in fish tissues, causing long-term stress and posing potential health threats to people consuming the fish. Sediment can be an important byproduct of mining activity. In an Idaho tributary of the Salmon River, Konopacky et al. (1985) found that dredging for rare earths generated 500,000 m^3 of sediment, which smothered important downstream spawning and rearing areas of chinook salmon and steelhead. Spaulding and Ogden (1968) estimated that hydraulic mining for gold in the Boise River, Idaho, generated 116.5 million kilograms of sediment in 18 months.

LOSS OF REFUGES

Because natural disturbances were an important part of the freshwater environment of Pacific salmon, many populations, particularly those at the edges of the range, underwent periodic expansion and contraction in response to local extinctions and periods of recolonization. That process fostered genetic diversity and versatility and enabled salmon to be resilient and locally variable (Scudder 1989). Over the last century, many small populations have become extinct as a result of human activity (Nehlsen et al. 1991, Frissell 1993), and the geographical distribution of some species has become highly fragmented (The Wilderness Society 1993). Within the confines of a river basin, the ability of salmon to recolonize areas of local population loss, such as a third- to fourth-order tributary system, might depend on the presence of refugia (survival areas) containing high quality habitat and relatively stable populations (Sedell et al. 1990). At present, watersheds without substantial anthropogenic perturbations are limited almost solely to national parks and designated wilderness areas, and in the Pacific Northwest states these are usually at elevations above the occurrence of anadromous salmon. High-quality habitat might exist in small patches within river basins that are subject to land and water management, but often habitat refugia widely distributed throughout the system there are insufficient for colonization of nearby disturbed sites (Sedell et al. 1990).

Concern for the continued viability of Pacific salmon on federally owned forest lands has led to the establishment of "key watersheds" in which high priority is given to protecting stream habitat (Figure 7-4) (Reeves and Sedell 1992, FEMAT 1993). Protection of habitat watersheds will be achieved by controlling erosion and by leaving large riparian buffers adjacent to both fish-bearing and non-fish-bearing streams. However, the distribution of key watershed reserves is limited primarily to headwater drainages where national forests are located (FEMAT 1993); few are in lower river valleys and coastal lowlands. In the Pacific Northwest landscape, the latter kinds of environment lack refugia with high-quality habitat for salmon (Frissell 1993), and there seems to be little hope of future establishment of such areas without considerable public resolve and financial commitment.

SUMMARY

Habitat for salmon in Pacific Northwest river basins has been lost or extensively altered over the past 150 years. As a result of alteration and degradation of biophysical conditions, many salmon populations have been extirpated or de-

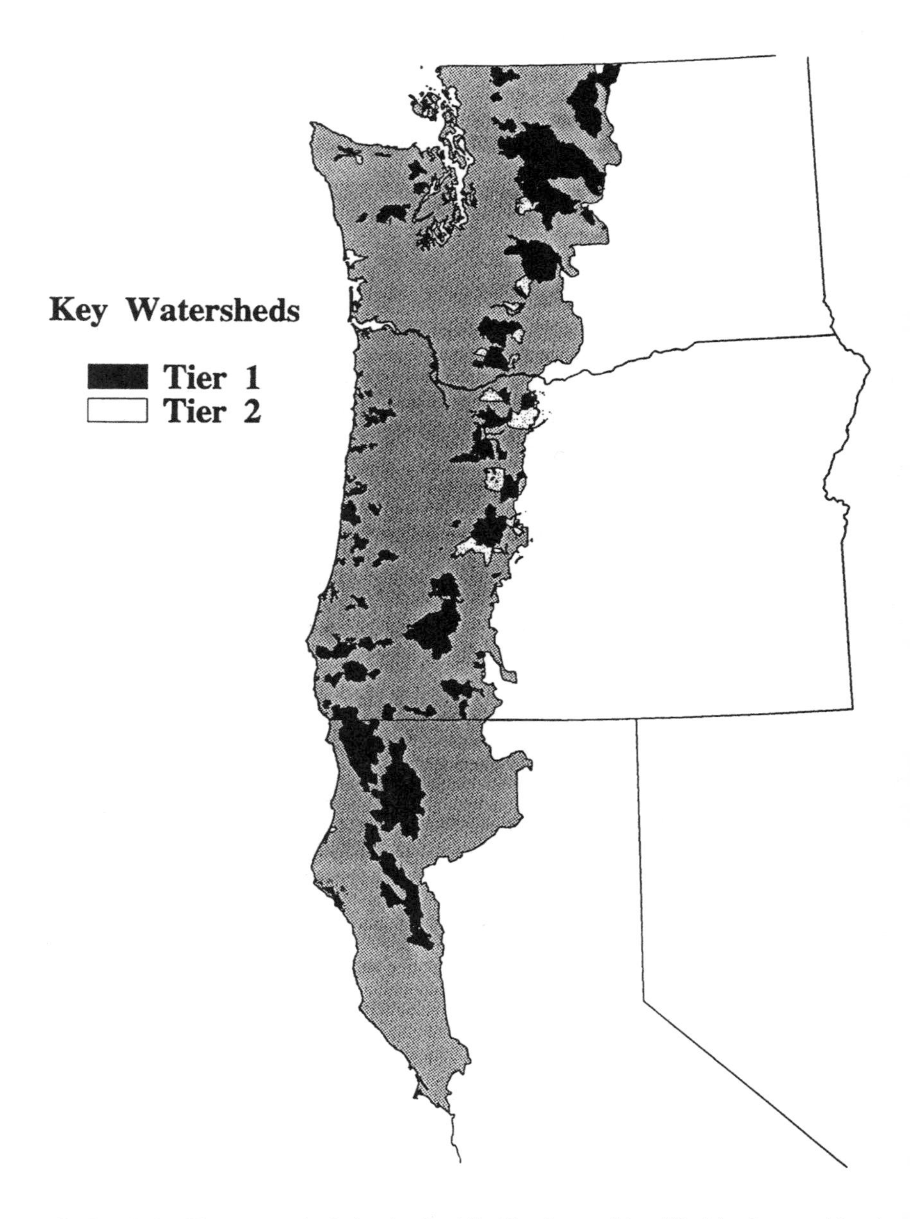

FIGURE 7-4 Key watersheds in the Pacific Northwest identified in the president's forest plan. The shaded area indicates the known range of the northern spotted owl (*Strix occidentalis caurina*). Source: FEMAT 1993.

pleted. Although the full extent of the modifications of salmon habitat will never be known, some generalizations are possible.

As a result of human development, the current condition of most river basins in California, Idaho, Oregon, and Washington is significantly different from the conditions in which salmon evolved. Stream-habitat alterations and losses associated with many types of land uses have included increased sediment loading, higher and more variable water temperatures, reduced amounts of large woody debris, reduced and simplified riparian plant communities, new barriers to migrations, lower streamflows during some periods and higher peak flows at other times, loss of stream-bank integrity, simplified channel structure, and reduced small-scale habitat heterogeneity. These changes have generally reduced the productivity of river basins for salmon, although some changes have occasionally increased production.

Anthropogenic habitat disturbances have often resulted in simultaneous changes in a wide variety of functions, processes, and habitat characteristics. It has usually been impossible to identify which habitat changes have had the greatest effect on salmon. Human-caused disturbances have interacted with natural ones: for example, effects of widespread removal of riparian vegetation, loss of ground cover, or decreased stability of hillslopes have sometimes not been manifested until after a heavy rainfall or snowfall. The simultaneous alteration of many factors had important implications for habitat rehabilitation. First, it is important to avoid attempting only to improve individual aspects of habitat (e.g., pools in streams) without addressing other aspects that might be equally degraded and are critical to salmon production. It is not enough to improve a pool in a stream, for example, if the water is too warm to support salmon. In many cases, these single-factor approaches are not effective (NRC 1992a). Second, rehabilitating watershed processes to the extent possible given human development, including the re-establishment of riparian functions—such as providing shading, organic matter, and large woody debris—is probably more effective in improving salmon habitat over the long-term than substituting artificial structures for ecological functions.

Habitat changes caused by human activities have occurred at far different spatial and temporal scales than natural disturbances in the Pacific Northwest. These differences between anthropogenic and natural disturbance patterns have interfered with the abilities of salmon to survive—and recover from—changes in their habitat. Anthropogenic perturbations have been significant causes of direct mortality for juvenile and adult salmon. Examples include excessive sediment inputs, scouring of redds, creation of migration blockages, high water temperatures, and toxic discharges. Catastrophic habitat loss can also occur as a result of natural disturbances, but the frequency and spatial scale of natural disturbances—unlike those of many human-caused disturbances—are such that salmon's behavioral and physiological characteristics allow their populations to be resilient.

Human activities have prevented natural disturbance regimes from creating or maintaining productive salmon habitat. In addition to imposing new disturbance regimes, the land- and water-management actions of an increasing human population have systematically prevented natural disturbances from providing crucial raw materials for productive habitat. Examples include flood control and wildfire suppression, both of which have interfered with processes that provide woody debris and nutrients to river systems. In addition, stream processes have been altered to the extent that they cannot respond to natural disturbances in a normal manner. Rehabilitating ecologically productive watersheds will require allowing natural disturbances to occur to the greatest extent possible.

Habitat alteration has changed the outcome of interactions between salmon and other species. Increased water temperatures, for example, have favored warm-water species at the expense of cold-water species such as salmon. In addition, many Pacific Northwest river basins contain nonnative fishes introduced from eastern North America and elsewhere, which are often better adapted to warm temperatures than salmon and which can prey on or outcompete young salmon. Some habitat alterations unfavorable for salmon have resulted from exotic animals or plants.

Habitat protection—especially with respect to riparian zones—has been very uneven across different types of land uses and ownerships. Overall, streams on public lands receive greater protection than those on private lands; those on forested lands often receive greater protection than those on agricultural and range lands; and streams and sloughs in urban and industrial areas have generally received the least protection. Water-quality requirements also differ according to the predominant use or according to various federal, state, and local discharge regulations. There often are large variations in the degree of protection afforded in different places within a river basin. Habitats on private and public lands are important to salmon, and rehabilitation programs that focus only on public lands will be less effective than those involving private lands as well (NRC 1995b). Therefore, cooperation between private and public landowners is important to protecting and rehabilitating habitat at a watershed scale. Development of cooperative habitat-conservation agreements between public and private landowners and resource managers will help identify critical habitat areas in need of rehabilitation while providing site-specific flexibility for landowners to provide different types of protection measures (see also Chapter 13 and NRC 1995b).

So much habitat has been lost or altered that relatively few areas of high-quality habitat remain, especially in large river valleys and in coastal lowlands—typically home to large numbers of people. Therefore, protecting and rehabilitating enough habitat to provide refuges for salmon and sources for recolonization of other areas that might be improved in the future will entail operating in a context of human development—in other words, ways must be developed for people and salmon to live together. Although many of the best remaining sites are in forested headwaters, nodes of good habitat are also needed in large river valleys and coastal lowlands. These sites can be identified and locally based protection measures can be developed. In some cases, landowner incentives can be used as a financial carrot in place of the more traditional regulatory stick. An example of such a program is King County's *Waterways 2000*, an aquatic conservation program funded from county real-estate taxes. It provides with tax reductions to landowners for dedicating part of their property to riparian protection, outright purchases of important greenways and conservation easements, and public education emphasizing good stewardship of the county's streams. The initial phase of the program has already helped to protect good habitats in six important salmon watersheds in this urban area, which includes Seattle (King County Surface Water Management Division 1995).

8

Habitat Management and Rehabilitation

From tributaries to mainstem rivers and from headwaters to estuaries throughout the Pacific Northwest, habitat changes have resulted from the cumulative effect of various land uses, cultural developments, and other factors operating on a wide range of temporal and spatial scales. Specific cause-effect relationships involved in habitat degradation are not easy to decipher. Restoration strategies for improving habitats are not always clear, and there is often disagreement among professional biologists, other technical specialists, land-managers, policy-makers, and the general public regarding how to proceed.

Programs to restore aquatic habitats have increased in recent years as the extent and magnitude of impacts of management activities and cultural practices on wetlands, streams, rivers, and estuaries have become more widely recognized. In a recent review of aquatic-ecosystem restoration, the National Research Council (1992a:17) defined restoration as the

> reestablishment of predisturbance aquatic functions and related physical, chemical, and biological characteristics. Restoration is different from habitat creation, reclamation, and rehabilitation—it is a holistic process not achieved through the isolated manipulation of individual elements.

The present committee concurs with that definition of restoration and emphasizes that in-channel restoration of aquatic habitats can seldom be accomplished without considering the accompanying riparian zone or other portions of a watershed that affect the aquatic system.

WATERSHED INFLUENCES

An essential concept in stream ecology is that terrestrial components of the environment have a profound influence on aquatic systems. Those components can include watershed characteristics, sediment production, and nutrient cycling. In the upper reaches of the Columbia River Basin, the hydrology of mountainous watersheds is dominated by snowmelt regimes that tend to produce distinct seasonal hydroperiods with predictable frequency and duration. In contrast, coastal watersheds along the Pacific tend to have a rain-dominated hydrology with sharply defined storms that are much less predictable in timing, duration, and magnitude. The hydrologic regimes for the two geographic areas are remarkably different, as they are for other areas of the region, but anadromous salmon have adapted their life strategies to all of them. In the Pacific Northwest, many dams and stream diversions and a wide variety of land uses have contributed to altering the flow regimes of streams and rivers. The natural hydrologic regime associated with a particular watershed has an important influence on instream functions and processes and habitat characteristics. Hence, where the natural disturbance pattern has been altered through dams, irrigation diversions, or other flow modifications, some degree of restoration of the natural hydrologic regime might be required before restoration of aquatic functions and habitat characteristics can occur (Hill et al. 1991). In estuarine and depressional wetlands, re-establishing hydroperiods within the range of natural conditions is critical to the restoration of such systems (Kusler and Kentula 1990).

Similarly, a variety of land-use practices (e.g., timber harvesting and roading in steep terrain, grazing, dryland farming and agriculture, and urban development and construction) can increase sediment production or alter the timing of its entry into stream systems (Everest et al. 1987, Swanson et al. 1987, McNabb and Swanson 1990). Reducing sedimentation is often another prerequisite for restoring aquatic habitats. In some instances, the transport of sediment to downstream reaches can be hindered or prevented. Dams of various sizes—from small stockwater ponds too numerous to inventory to mainstem Columbia River dams—all tend to impound and store sediment from upstream reaches, preventing the normal downstream transport and storage of fluvial sediment by a stream or river system. The larger of such structures also tend to flatten hydrograph peaks and so further alter instream sediment transport, particularly that of coarser bedload sediments. Where hydrologic and sedimentation regimes have been seriously altered, a fundamental objective of restoration would be to restore the dynamics of natural flow regimes enough to re-establish the processes and functions of both aquatic and riparian ecosystems.

The riparian environment associated with Pacific Northwest streams and rivers is of primary importance to the functioning of aquatic systems (Johnson et al. 1985, Mutz and Lee 1987, Salo and Cundy 1987, Abell 1989, Gresswell et al. 1989, Chaney et al. 1990). Riparian zones can have pronounced effects on the biological, chemical, and physical components of the aquatic system. From a biological perspective, the structure and species composition of streamside vegetation influence the local characteristics of both the channel and its aquatic habitat. Vegetation provides shade and moderates stream temperatures, provides a source of carbon for instream organisms through annual leaf-fall, and, at least for forested systems, provides the large woody debris that is often a key habitat feature (discussed in Chapter 7). However, riparian vegetation does much more. The underground root systems of streamside vegetation bind soil particles and provide stability to stream banks, thus influencing channel structure. Plant stems and near-surface roots provide flow resistance during periods of overbank flow and thus promote sediment deposition and floodplain development. Root systems, in combination with large woody debris, provide channel roughness elements that not only promote sediment storage but encourage the hydraulic exchange of streamflow and hyporheic (underground) flows. Chemical transformations of various kinds occur in the highly variable oxidation-reduction environment associated with riparian areas. Because streamside vegetation has such an important influence on the characteristics and productivity of aquatic habitats, restoration of these habitats requires a commitment to the restoration of riparian vegetation functions and processes where they have been substantially altered by human influences.

HABITAT-MANAGEMENT OPTIONS

The influx of Euro-Americans has altered the characteristics and functioning of stream systems in most freshwater salmon habitats in the Pacific Northwest. The last two centuries and particularly the last 50 years have seen rapid transition of watersheds. Although the degree of alteration varies widely throughout the region, habitat impacts and losses are common. The variability of impacts indicates that various options (Figure 8-1) are needed for improving conditions for sustaining anadromous fish populations.

Protection

Anadromous salmon have adapted over thousands of years to types of habitats that existed in the Pacific Northwest before Euro-American settlement. They thrived in naturally functioning aquatic-riparian ecosystems. Where wild populations continue to survive and maintain healthy populations, the *protection* of intact and functional aquatic-riparian habitats should have a high priority; where protection is the desired management option to sustain fish and other aquatic organisms, human influences need to be prohibited or minimized.

Restoration

Where aquatic-riparian habitats have been degraded by human activities but have the potential to recover the characteristics that make them functionally equivalent to a pristine system, *restoration* might be the management target; a

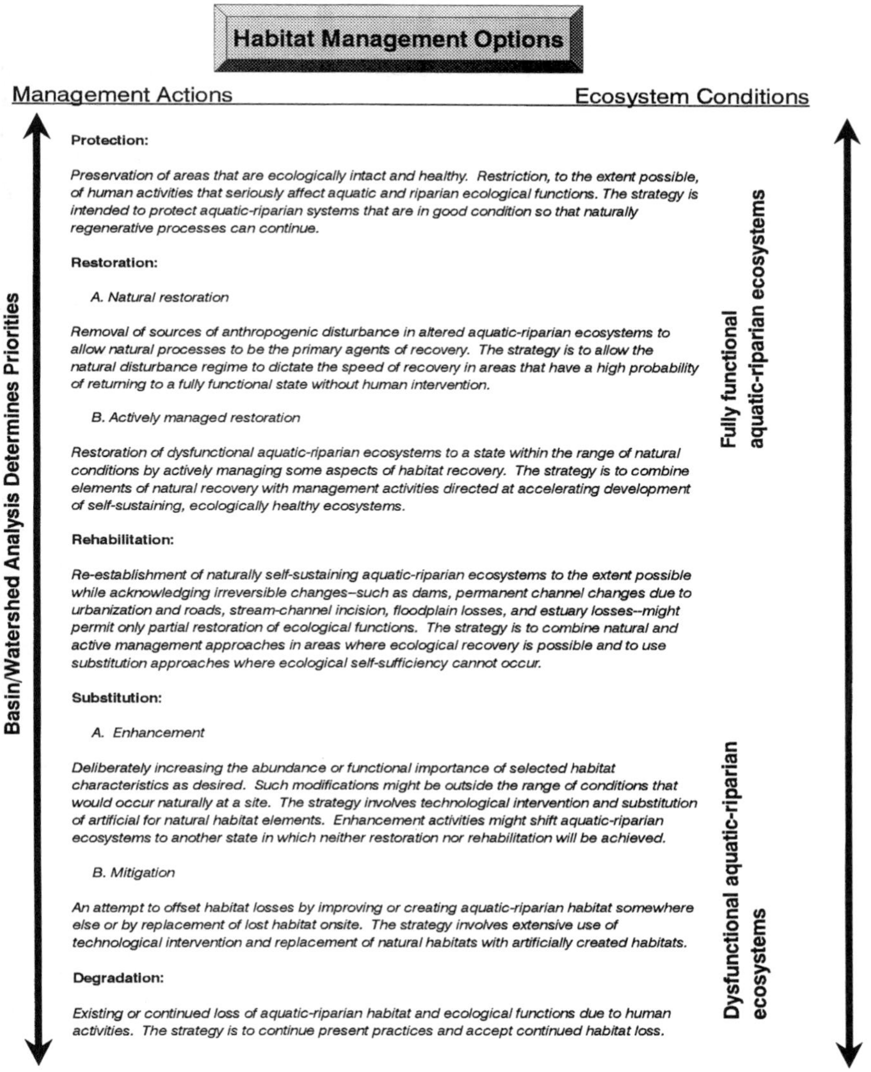

Habitat Management Options

Management Actions | Ecosystem Conditions

Basin/Watershed Analysis Determines Priorities

Protection:

Preservation of areas that are ecologically intact and healthy. Restriction, to the extent possible, of human activities that seriously affect aquatic and riparian ecological functions. The strategy is intended to protect aquatic-riparian systems that are in good condition so that naturally regenerative processes can continue.

Restoration:

A. Natural restoration

Removal of sources of anthropogenic disturbance in altered aquatic-riparian ecosystems to allow natural processes to be the primary agents of recovery. The strategy is to allow the natural disturbance regime to dictate the speed of recovery in areas that have a high probability of returning to a fully functional state without human intervention.

B. Actively managed restoration

Restoration of dysfunctional aquatic-riparian ecosystems to a state within the range of natural conditions by actively managing some aspects of habitat recovery. The strategy is to combine elements of natural recovery with management activities directed at accelerating development of self-sustaining, ecologically healthy ecosystems.

Rehabilitation:

Re-establishment of naturally self-sustaining aquatic-riparian ecosystems to the extent possible while acknowledging irreversible changes–such as dams, permanent channel changes due to urbanization and roads, stream-channel incision, floodplain losses, and estuary losses--might permit only partial restoration of ecological functions. The strategy is to combine natural and active management approaches in areas where ecological recovery is possible and to use substitution approaches where ecological self-sufficiency cannot occur.

Substitution:

A. Enhancement

Deliberately increasing the abundance or functional importance of selected habitat characteristics as desired. Such modifications might be outside the range of conditions that would occur naturally at a site. The strategy involves technological intervention and substitution of artificial for natural habitat elements. Enhancement activities might shift aquatic-riparian ecosystems to another state in which neither restoration nor rehabilitation will be achieved.

B. Mitigation

An attempt to offset habitat losses by improving or creating aquatic-riparian habitat somewhere else or by replacement of lost habitat onsite. The strategy involves extensive use of technological intervention and replacement of natural habitats with artificially created habitats.

Degradation:

Existing or continued loss of aquatic-riparian habitat and ecological functions due to human activities. The strategy is to continue present practices and accept continued habitat loss.

Fully functional aquatic-riparian ecosystems

Dysfunctional aquatic-riparian ecosystems

FIGURE 8-1 Alternative approaches to habitat management based on existing watershed conditions and desired level of improvement.

return to natural streamflow and sediment input might be possible. Generally, such areas have experienced adverse impacts of historical watershed and land-use practices; if the land use can be modified to reduce or eliminate potential impacts on the watershed's hydrology and sediment production, the prognosis for long-term re-establishment of a natural disturbance regime would be good. In other instances, natural hydrological and sediment-production processes have been not been impaired, but riparian vegetation and channel structure have been changed, and the recovery of riparian vegetation to a more natural condition might be the management goal (Table 8-1).

TABLE 8-1 Management Strategies Appropriate for Different Habitat Objectives

Management Area	Aquatic-Riparian Management Strategy
Class 1 waters	Protection
Bear the closest resemblance to waters unaltered by modern human activities, contain a complete set of native biota, and have a high degree of natural protection. Each contains a complete set of native fauna, a diversity of habitats, and enough area to maintain viable populations of the largest and most mobile species. Management goal: keep as pristine as possible, recognizing that some biotic change is inevitable or necessary.	1) Identify aquatic-riparian ecosystems in good ecological condition through some method of watershed analysis. 2) Implement measures designed to prevent adverse human impacts, including the proscription of all potentially damaging activities within the ecosystem (e.g., no new roads in roadless areas).
Class 2 waters	Restoration
Modified by human activity but contain mainly native organisms and have reasonable potential to be restored to Class 1. Management goal: maintenance of natural diversity and prevention of further degradation, but allow potentially compatible uses (e.g., low impact recreation, selective logging, nonriparian grazing).	1) Inventory riparian and aquatic habitat characteristics throughout watersheds. 2) Establish substantial undisturbed riparian zones adjacent to streams, lakes and wetlands where future high impact management activities will occur; allow natural recovery of ecosystem functions (e.g., federally proposed "PacFish" buffers). 3) Remove high impact anthropogenic disturbances that are now occurring (e.g., fencing riparian zones in grazing areas). 4) Identify opportunities for accelerating the development of desired ecological conditions by actively managing riparian zones (use caution, however) or reconnecting rivers with their floodplains.

TABLE 8-1 Continued

Management area	Aquatic-riparian management strategy
Class 3 waters	Rehabilitation/Enhancement
Appear natural, but their biotic communities have been significantly and probably irreversibly altered. Unlikely ever to be restored to Class 1 but can be refuges for native species or migration corridors for anadromous species. Vulnerable to change and cannot be relied upon for long-term preservation of species. Management goal: maintenance of supplemental populations and gene pools, sources of organisms to stock restored waters, and "wild" areas that can sustain fairly heavy public use.	1) Inventory riparian and aquatic habitat characteristics throughout watersheds. 2) Perform watershed-scale evaluation that considers (a) the geomorphic setting of the river and its valley, (b) the natural disturbance regime of the region, (c) the historical patterns of anthropogenic disturbances, and (d) condition of the watershed relative to some type of unmanaged reference site. 3) Prioritize and implement, at the watershed scale, plans to achieve habitat goals.
Class 4 waters	Enhancement/Mitigation
Artificial aquatic refuges created and/or managed for protecting species that otherwise would likely become extinct. Simulate original environments but require continuous management and monitoring; should be regarded as temporary solutions for saving species or for providing back-up populations for species with limited wild populations. Management goal: short-term back-up for Class 2 and 3 waters.	1) Inventory riparian and aquatic habitat characteristics throughout watersheds. 2) Perform watershed-scale evaluation that considers (a) the geomorphic setting of the river and its valley, (b) the natural disturbange regime of the region, (c) the historical patterns of anthropogenic disturbances, and (d) condition of the watershed relative to some type of unmanaged reference site. 3) Prioritize and implement, at the watershed scale, plans to achieve habitat goals.
Class 5 waters	Mitigation
Artificial refuges with no attempt to recreate natural conditions. Management goal: maintain small areas with some semblance of important habitat characteristics; often species-directed.	1) Inventory riparian and aquatic habitat characteristics throughout watersheds. 2) Perform watershed-scale evaluation that considers (a) the geomorphic setting of the river and its valley, (b) the natural disturbange regime of the region, (c) the historical patterns of anthropogenic disturbances, and (d) condition of the watershed relative to some type of unmanaged reference site. 3) Prioritize and implement, at the watershed scale, plans to achieve habitat goals.

Source: Adapted in part from the aquatic diversity management area concept of Moyle and Yoshiyama (1994).

Management policies directed at the restoration can potentially proceed along two major pathways: *natural restoration* (referred to as "passive restoration" by Kauffman et al. 1993) and *active restoration.* In natural restoration, removal of the sources of anthropogenic disturbances is all that is necessary for full restoration of the system. For example, where agricultural practices occur in riparian areas, the cessation of such practices might allow the long-term re-establishment of riparian vegetation and associated functions. Natural disturbances would combine with the establishment, growth, and succession of riparian plants to assist in restoring aquatic habitats. Similarly, the removal of grazing from streamside zones or a change to grazing policies that allow full recovery of riparian plants along streams might be all that is needed to restore aquatic-riparian functions for many rangeland streams. For forested riparian systems that have previously experienced great amounts of logging, the establishment of no-harvest buffers might provide for the restoration of aquatic-riparian functions and characteristics. The time required for restoration to occur will depend on the local conditions (e.g., species of riparian plants, climate, geomorphic characteristics of stream and valley, and hydrologic disturbance pattern). However, the intent of natural restoration is to use fully the natural abilities of physical processes (channel adjustments, bank building, scour and fill, etc.), chemical processes (nutrient transformations), and biological processes (establishment, growth, and succession) to restore the functioning of aquatic-riparian systems. Because of the multitude of variables, microclimates, and other conditions associated with recovering riparian systems, natural restoration allows those processes to occur at a level dictated by the local capability of the system.

The first step in active restoration also involves the removal or elimination of activities that are causing degradation. However, where monitoring or observation indicates that recovery will not be complete or might require much time, additional management practices can be considered. Active restoration incorporates practices designed to fill an ecological void or accelerate natural recovery. For example, large woody debris may have been removed from the channel when a forested riparian zone might have been harvested years earlier. Although the growth rates and species composition of the second-growth riparian forest could be providing the desired functions within their natural range, the scarcity of large wood in the stream might not be overcome for many decades. In this situation, addition of large woody debris in configurations normally expected for the stream might be undertaken. In other instances, if riparian areas have been used for crop production, eliminating agricultural practices might initiate natural recovery of riparian functions and aquatic habitat; but because native plant species that would be characteristic of the local riparian system are infrequent, natural restoration could take a very long time, so planting native species of riparian plants obtained from locally adapted genetic stock might accelerate recovery. In rangeland areas where prolonged grazing and other practices have caused the disappearance of willows, cottonwoods, or other key riparian plants along streams, active management might be needed to re-establish a native vegetation community.

The practices involved in active management can vary widely, but the intent is to assist or accelerate the restoration of aquatic-riparian functions and related physical, chemical, and biological characteristics that support natural communities and maintain aquatic productivity. Such practices are intended to aid in re-establishing a sustainable aquatic-riparian ecosystem and aquatic habitat that will be, for all practical purposes, functionally equivalent to pristine conditions.

Rehabilitation

In many wetlands, streams, rivers, and estuaries in the Pacific Northwest where habitat alteration and loss have been extensive, restoration itself is not feasible. Natural disturbance regimes might have been altered to such an extent that there is little opportunity for restoration. For example, hydrologic and sediment transport regimes could have been affected by dams, irrigation diversions, changes in fire frequency, conversion of lands to agricultural practices, etc.; introduced plants could have replaced native riparian species; channel incision could have lowered local groundwater tables and affected hyporheic interchanges with the stream; estuaries could have been filled; and road construction, agricultural practices, or urban development could have reconfigured channel sinuosity or shifted stream location. In many of those situations, self-sustaining aquatic-riparian ecosystems that can provide important habitat for anadromous salmon might still be possible, but increased levels of human effort (time, money, and management persistence) will probably be required because of the extent and magnitude of the changes. Habitat management can then be directed toward re-establishing self-sustaining conditions that are able to provide some of the ecological requirements of anadromous fishes—i.e., *rehabilitation*. Rehabilitation of habitat features can occur, but full restoration to predisturbance functions and characteristics is unlikely (Figure 8-1).

An important aspect of the rehabilitation option is that not all the predisturbance aquatic-riparian functions can be restored. For example, if a dike had been constructed along a stream's edge, moving the dike back from the channel would allow the return of streamside vegetation and some floodplain functions, such as temporary storage of floodwaters, sediment deposition on floodplain terraces, and improved interactions between stream and groundwater. Even though the channel might not be able to develop full predisturbance sinuosity, the prior density of side channels, or full floodplain functions, a major improvement in aquatic-riparian functions and characteristics would be achieved by the repositioning of the dike. Rehabilitation projects constitute important opportunities for developing improved and sustainable habitats for salmon and other aquatic-riparian biota.

Substitution

Enhancement

Substitution approaches to habitat management (Figure 8-1) are generally directed toward selectively altering or modifying habitat features to offset the effects of anthropogenic impacts. In the last decade, numerous instream habitat *enhancement* projects have been undertaken throughout the Pacific Northwest. Many involve the placement of gabions (wire baskets filled with rock), boulders, riprap, large wood, or other large structures in stream channels. The addition of such roughness elements might be outside the range of conditions–with respect to size, spatial distributions, orientation, etc.–that would be expected to occur naturally in the setting. For example, the placement of boulders or logs in meadow systems that historically did not have these large roughness elements is outside the range of expected natural conditions. Exotic materials, including gabions and geotextile fabrics, are commonly used in enhancement projects. For lakes, the addition of fertilizers to boost production would constitute an enhancement approach. Enhancement can provide important opportunities for improving fish habitat in some instances, but often it has not been very successful in improving conditions that sustain productivity. The National Research Council's 1992 report on the restoration of aquatic ecosystems includes a strongly worded caution about the practice of enhancement (pp. 222-223):

> Practitioners of species-centered stream management generally introduce artificial structures into stream and river environments to modify banks, channel, bed, or current in hopes of improving salmonid or other game fish productivity.
>
> When this work is done without a profound understanding of the interactions among stream hydrology, fluvial geomorphology, and fish, the least detrimental consequence may be that mechanical structures emplaced in the stream at considerable expense and trouble could be of limited durability and longevity.

The term *enhancement* implies improvement and betterment of a system, but it is important to be aware that if ecosystem needs are misinterpreted, a stream-habitat enhancement project can actually shift an aquatic ecosystem from one degraded state to another (Figure 8-1). For example, fully spanning logs with bank revetments might be placed in a stream that is deficient in pools to provide additional pool habitat. If the new pools are without cover and losses to predation are increased, if the logs create waterfalls that become barriers to juvenile or adult fish movements, or if the channel can no longer adjust to high flows and sediment transport by altering sinuosity or creating natural pools, the long-term consequences of the enhancement project might simply be another form of habitat degradation.

Mitigation

Where habitat losses are unavoidable, *mitigation* is a management option that attempts to minimize or offset the effects of loss by creating new habitat. Although the concept of mitigation is relatively simple, its application often is not. Losses of habitat at a particular site are seldom balanced by mitigation at another site; i.e., replacement of physical habitat rarely includes replacement of all relevant ecological interconnections. The ecological ledger is not linear and cannot be "balanced" on the basis of simple tabulations of side channels, lengths of streams, or areas of spawning gravels created. If a key component of a required habitat feature is severely altered or destroyed, there might be no suitable means of mitigating that impact.

Habitat-management projects tend to focus on the characteristics and needs of a specific stream reach, but the condition of individual habitats, stream reaches, and entire tributary systems within a watershed must be considered for restoration planning to be effective. It might do little good to invest time and money in restoring spawning gravel for salmon if rearing habitats are scarce or downstream barriers severely limit accessibility to them. It might do little good to emphasize riparian restoration when excessive sediment production from upstream land uses or streamwater withdrawals are adversely affecting instream habitats. Habitat managers need to be aware of how habitat alterations affect the life-history stages of salmon and other aquatic organisms on various spatial and temporal scales. They must also understand how conditions change as a result of a wide range of anthropogenic perturbations and natural disturbance patterns. The complexity of environmental factors and options available to habitat managers presents a major challenge to the reduction of anthropogenic impacts.

WATERSHED ANALYSIS

Fish habitat upstream in a river basin might be little affected by human activities, but habitat lower in a watershed could experience the cumulative effects of a wide range of human activities or other factors. Understanding factors that affect the availability and quality of aquatic habitat at lower sites is much more difficult. In addition, the importance of various contributing factors will probably change in emphasis or magnitude from watershed to watershed. Thus, there is an increasing need to understand cumulative effects not only on a site-specific basis, but also across entire watersheds. Only through a broad geographic perspective can the unique qualities of each watershed and their spatial and temporal effects on aquatic habitats be effectively understood.

A recent development in forest-management planning has been the procedure of watershed analysis to evaluate resources and the potential environmental impacts of land-management proposals. The general goal of watershed analysis is to combine habitat-inventory information with environmental-hazard assessments over a relatively large area, usually encompassing a fourth- to sixth-order stream network, so that land-use prescriptions can be based on stewardship objectives and opportunities for habitat restoration can be identified on somewhat larger geographical scales than are normally used. For forested basins, watershed analysis can lead to management prescriptions that provide greater environmental protection than standard forestry rules. The procedure was created in the Timber, Fish, and Wildlife Program in Washington State to address the cumulative effects of logging-related activities and has been incorporated into the state's forest-practices laws (Washington Forest Practices Board 1993). Idaho is developing a cumulative-effects analysis and control process designed to protect water quality from forested watersheds so that beneficial uses are supported (Idaho Department of Lands 1994). A comprehensive study is under way in Oregon to identify the cumulative effects of forest practices on air, soil, water, fish, and wildlife (Beschta et al. 1995), and recent revisions to Oregon's forest-practices rules include a watershed-analysis option under certain circumstances. At the federal level, management guidelines proposed by the Forest Ecosystem Management Assessment Team (FEMAT) for regulation of national forests in western Oregon, western Washington, and northern California include watershed analysis as an important underpinning for future resource prescriptions and management (FEMAT 1993).

Watershed analysis as now envisioned requires a landscape-scale perspective, although it is too early to determine whether its application will result in substantive improvements in habitat protection or in a more comprehensive approach to aquatic-habitat restoration. Its success will ultimately depend on how managers translate habitat-inventory and hazard-assessment information into prescriptions and on the strength of the commitment to effective monitoring (the Washington watershed-analysis procedure encourages monitoring but does not require it). Watershed analyses and assessments now being implemented (Washington Forest Practices Board 1993) and those suggested by FEMAT (1993) are designed to promote efficient regulation, continued use of forest resources for timber production, and the protection of forested ecosystems.

Considering entire watersheds in the regulation of forest practices has many benefits, but it also presents several practical and conceptual difficulties (Washington Forest Practices Board 1993:v):

> 1. Watershed ecosystems involve a complex dynamic between many watershed and biological processes operating at many spatial scales. Scientific understanding of these processes is limited, and comprehensive reliable techniques for evaluating watersheds are lacking.
>
> 2. The physical and biological characteristics of a watershed and sub-areas within it reflect the local geology, terrain, climate, vegetation and so on. Consequently, every watershed is unique, with its own distribution of these factors as well as effects due to the history of past disturbance including natural events or land use.

3. Because of these differences in landscape features, the sensitivity of watersheds and sub-areas within them to forest practices also varies from place to place. While one location may generate no likelihood of local or cumulative effects from an activity, the same activity conducted in the same way in another location with heightened sensitivity could have both local and cumulative impacts.

Although those difficulties appear to represent barriers or constraints to watershed analysis, they actually provide compelling arguments for doing it. Spatial and temporal variability; the dynamic interactions between physical and biological processes; the unique attributes of each watershed (geology, terrain, climate, vegetation, fish populations, history of land use, and natural disturbances); the local sensitivities of watersheds, sub-areas, and stream reaches to management practices; and other factors all suggest that some form of comprehensive analysis is necessary for ecologically sensitive management planning. A simple "stream-reach analysis" is obviously inadequate for understanding watershed-scale concerns, processes, and cumulative effects that are driven by land-use practices or by institutional and social programs.

The use of watershed analysis for improving habitat should be directed at providing the public and managers with information that will identify a range of issues and opportunities for Pacific Northwest streams. Because a wide range of spatial and temporal scales needs to be considered for a given watershed, such analysis is expected to yield both strategic and tactical approaches to improving habitat. Several types of information should be considered in the analysis:

- *Spatial context.* The highly varied nature of fish habitat requires that geomorphic characteristics of aquatic habitats be identified throughout a basin. The setting of various stream reaches (e.g., constrained vs. unconstrained, sinuous vs. braided, incised vs. unincised, and bedrock-controlled vs. alluvial floodplains) can constitute an important spatial context from which to understand both reach and watershed scale problems, issues, and habitat-management opportunities. Without a spatial perspective, inappropriate and counterproductive habitat manipulations or alterations of selected reaches might be selected. A geographic information system (GIS) might be useful for displaying and analyzing some types of spatial information, but the large variability in conditions associated with specific stream reaches indicates that on-the-ground assessments of habitat and related factors should have high priority in watershed analysis.
- *Temporal context and disturbance regimes.* Characterizing the temporal variability of flow patterns and sediment yields, long-term channel adjustments, climatic patterns, vegetation succession patterns, fire history, or other factors provides an important perspective on the dynamic features of a particular watershed and stream system. The peak-flow regime of a given watershed is especially important because the instream biota, riparian vegetation, and many channel-forming processes are fundamentally tied to the frequency and magnitude of

hydrologic events. In other instances, the occurrence of low flows can affect habitat quality and productivity, and managers need to recognize the implications of recurring drought and its effects on habitat dynamics and quality. For most watersheds, long-term data are generally lacking or are of varied quality; extrapolation of temporal information from sources outside a watershed is often necessary. Managers need to develop an understanding of the randomness of natural-disturbance regimes and incorporate that understanding into strategies for the maintenance or improvement of sustainable aquatic habitats.

- *Riparian vegetation and reference sites.* Because of the importance of riparian vegetation to many channel and aquatic-habitat characteristics, an understanding of riparian plant communities is fundamental. Furthermore, the mechanisms by which riparian vegetation interacts with natural-disturbance patterns and land-use practices need to be thoroughly understood on stream-reach and watershed scales. Because many riparian plant communities have been affected by historical land-use practices, reference sites that consist of ecologically intact and functional aquatic-riparian systems should be identified. It may not be possible to restore or rehabilitate all aquatic habitats in a watershed to the same functional level, but reference sites are necessary to increase understanding the complex interactions between streamside vegetation, channel characteristics, and aquatic habitats. They provide fundamental information on types of processes, functions, and desired future conditions of intact riparian systems.
- *History of impacts.* Institutional, scientific, or social records of human land use and changes in aquatic and riparian ecosystems in a given watershed are often incomplete. However, a thorough understanding of historical practices and their effects is very helpful. Important insights into current conditions are often gained when historical information is developed and used to understand the magnitude and extent of human perturbations. It is important for both society and fishery managers to understand the magnitude and extent of changes that have occurred over periods of decades or longer.

Watershed analysis requires gathering a large amount of inventory information to obtain an improved understanding of how spatial and temporal patterns, cause-effect relationships and other interactions, and cumulative effects occur in a particular watershed or stream reach and how they affect aquatic habitats. In watershed analysis, the best available technical understanding and scientific information can be focused on a particular watershed. The process can also highlight limitations of databases, lack of information, monitoring needs, and so on. Most important, watershed analysis can be adapted to the goals of managing a particular resource or group of resources. It can provide information about managing a specific reach of stream; in the case of anadromous salmon, whose range transcends institutional and land-ownership boundaries, results of a watershed analysis can be joined with other socioeconomic considerations to develop habitat-management priorities on a watershed scale.

Watershed analysis can provide a relatively clear understanding of existing aquatic resources and factors that affect them, but social preferences and institutional constraints might preclude implementation of particular solutions to habitat problems. For example, dewatering of a stream by irrigation diversions seemingly could be solved by shifting to irrigation techniques that are more efficient and less consumptive. Any water savings might be retained in the stream to maintain summer rearing habitat. However, improved water-use efficiencies might instead allow a landowner to increase the amount of area under irrigation with no net change in the amount diverted. If the streamwater were no longer used by the landowner, it might simply be diverted by another landowner immediately downstream with a junior water right. In essence, what could have seemed like a relatively simple approach to solving an instream problem begins to involve important institutional barriers. The institutional constraint of "use it or lose it," which is so deeply etched in western water law, precludes what in some instances might be a simple solution to an instream habitat problem. When the original water laws were promulgated, concerns for fish and aquatic habitat had a low priority. Similar institutional constraints occur with respect to other resources and land-management practices. Unless the role and importance of various social, economic, institutional, and population factors that affect aquatic habitats are considered, the potential for effective maintenance or improvement of habitat for anadromous salmon and other aquatic organisms will be greatly constrained.

Whereas watershed analysis might provide important resource perspectives previously unavailable to land managers, it is important to point out that watershed analysis is currently designed only for drainages with forestry operations (e.g., Washington Forest Practices Board 1993). There are no institutional or legal means of applying the watershed-analysis approach to the management of nonforest lands. And it is not known whether the methods used to assess the consequences of forest management and provide recommendations for habitat restoration are fully applicable to streams and lakes surrounded by land used in other ways. Thus, there is no institutionally sanctioned means of assessing habitat and identifying opportunities for restoration in the case of over multiple ownerships encompassing a variety of land uses. The freshwater-habitat needs of anadromous fish, however, do not stop at forest boundaries.

OPPORTUNITIES AND CHALLENGES

Numerous opportunities await those interested in improving and restoring aquatic habitats in the Pacific Northwest. In some instances, substantial recovery can be accomplished by the simple cessation of streamside management practices that cause local degradation of freshwater habitats; restoration might proceed rapidly and be easily observed, or useful recovery might take a long time. Most degraded aquatic habitats take years or even decades to have their natural production fully restored. Given current knowledge about habitat restoration and rehabilitation and the large amount of freshwater habitat in need of improvement, habitat managers should be more concerned about whether a particular degraded habitat is improving (moving in the right direction) and less concerned about whether it has attained a specific desired condition. Aquatic habitats are complex and dynamic; they vary over large temporal and spatial scales in response to local structure, vegetation, and unpredictable natural disturbances. Attempting to "micro-manage" individual stream reaches to meet a perceived potential is probably unrealistic and unneeded.

Some of the changes in terrestrial and aquatic ecosystems in the Pacific Northwest caused by Euro-American development are permanent, some are declining in importance with respect to their effects on aquatic habitats, and others are growing in importance. Nevertheless, degraded aquatic habitats can be improved in large portions of the Northwest's stream systems. Most of the length of a stream network comprises relatively small streams, which provide water, nutrients, organic matter, and sediment to downstream reaches. Minimizing adverse impacts on these small streams and restoring their ecological connections are not only technically feasible but of paramount importance if conditions in many larger streams and rivers are to be improved.

Habitat managers often view restoration practices from spatial and temporal perspectives that are too limited, do not match the life histories of the salmon species of concern, and fail to address important ecological processes that are responsible for maintaining natural productivity. The approach to habitat improvement has often involved introducing habitat elements to streams and lakes in an attempt to boost productivity, e.g., placing structures in streams, excavating spawning channels or riverine ponds, or adding fertilizer to rearing lakes. Such habitat enhancement is used when there is evidence that existing habitat is deficient in some way and when it is believed that creation of new conditions will eliminate a major bottleneck in salmon production. Enhancement, based on a "limiting-factor" analysis of the current situation (Reeves et al. 1989), assumes that information about the factors controlling the abundance of the populations in question is sufficient for managers to rehabilitate degraded habitat in a cost-effective manner. In many cases, however, habitat rehabilitation has been undertaken without either formal or informal analysis of limiting factors but instead has been based on the presumed extent of degradation and the efficacy of existing restoration methods. In other cases, habitat is modified with the intent of increasing productivity beyond what would be expected in normal conditions. The whole-lake fertilization project of British Columbia's Salmon Enhancement Program, which was intended to improve survival and growth of juvenile sockeye salmon in oligotrophic (nutrient poor, unproductive) lakes (Hyatt and Stockner 1985), is an example of an attempt to increase salmon production above natural levels.

Despite the large amounts of time and money that have been devoted to habitat restoration and enhancement by federal agencies, state agencies, and others, few projects have been shown unequivocally to increase salmon populations (e.g., Hilborn and Winton 1993). Some of the reasons for the failure to demonstrate success are the high inherent variability of freshwater salmon production, erroneous assumptions about production bottlenecks, including the ocean, and the time required to demonstrate the effect of restoration, and unwillingness to monitor biological responses to project implementation effectively (Lichatowich and Cramer 1980, Hall and Knight 1981, Sedell and Beschta 1991, Hilborn and Winton 1993).

A central difficulty for many habitat-alteration projects is a general failure to match the scale of the project to the scale of life histories of salmon in a river basin. In many cases, the scales do not match. For example, many habitat-restoration projects in Oregon and Washington are targeted at increasing the rearing capacity of streams for coho salmon or steelhead. The methods might involve constructing pools and placing large woody debris in the channel (House and Boehne 1986) or, when coho are the species of concern, creating riverine ponds for overwintering (Cederholm et al. 1988). It is easy to demonstrate that enhanced habitat is occupied by target species, but such projects are typically applied only to a small portion of the total drainage network inhabited by salmon because of their high cost and the boundaries imposed by land ownership. As a result, overall benefits (if any) of restoration and enhancement projects cannot provide measurable signals in terms of increased smolt production or run size.

The "limiting-factor" approach has been used by fishery biologists to assess habitat deficiencies and as a basis for proposing enhancement projects. It provides a means of identifying site-specific habitat defects, such as excessive water temperature, lack of pools, insufficiency of large woody debris, excess of fine sediment, or lack of flushing flows. Those might be important factors, but the approach has often resulted in additions of whatever was believed to be in short supply (Beschta et al. 1991, Beschta et al. 1993, Kauffman et al. 1993). The simple alteration of physical features in streams or lakes does not necessarily promote restoration of habitat when riparian and watershed management practices continue to exert their effects on the aquatic ecosystem. Attempts at improving habitats by adding in-channel roughness elements without eliminating management practices that are causing habitat degradation are likely to fail. For example, some short-term habitat benefits might be achieved by adding large woody debris to streams, but the benefit can be only temporary from an ecological perspective unless riparian management practices ensure the long-term recruitment of large woody debris from the riparian zone.

Limiting-factor analyses often ignore the hydrogeomorphic features of a particular stream. For example, boulders, logs, or other roughness elements might be used in inappropriate locations, such as wet meadows, where they were probably never present. Habitat treatments that have been developed in one ecoregion are freely transferred to another or transferred from one channel type to another with little regard to the geomorphic context of a stream reach or the character of the streamside vegetation (e.g., House and Boehne 1986, Seehorn 1992). The apparent inability of limiting-factor analysis to incorporate a more holistic perspective of instream habitat might underlie the current shift toward "ecosystem management" that is being proposed for federal forest ownership in the Pacific Northwest (FEMAT 1993).

The concept of "desired future conditions" (DFCs) has been suggested as a means of defining habitat goals, such as the number of pools per mile of stream or the abundance of large woody debris. Identifying DFCs requires that the characteristics of fully functional aquatic ecosystems be known for each site. DFCs might overemphasize the physical attributes of a stream reach unless sufficient recognition is given to other factors or temporal processes that resulted in the attributes (e.g., species composition, structure, and successional patterns of riparian vegetation, natural flooding regimes, and fire) or to how a particular reach of stream operates within a larger spatial context. Although DFCs on a stream-reach scale might remain a component of habitat-improvement goals, care must be exercised in their application across varied ecological and structural templates. Landscape-scale DFCs are needed to protect the integrity of local breeding populations where trends within river basins indicate repeated patterns of habitat alteration. Some examples of DFCs include

- Increased percentage of riparian zones with late-successional forest characteristics where early-successional forests are dominant.
- Increased percentage of reaches with free-flowing discharge regimes in river basins where flows are largely controlled.
- Increased connections of rivers to floodplains and side channels where riverbanks have been extensively contained within dikes and levees.
- Reduced watershed erosion where human activities have accelerated sediment inputs.
- Increased areas of high-quality habitat throughout river basins that provide protection to local demes.

In recent years, increasing attention has been directed at understanding and improving riparian habitat in Pacific Northwest streams (e.g., Elmore and Beschta 1987, Salo and Cundy 1987, Gresswell et al. 1989, Chaney et al. 1993, FEMAT 1993). Because naturally functioning riparian ecosystems are crucial to sustainable and productive aquatic habitats, riparian plant communities along stream channels are often protected by setting a specific distance from the channel within which anthropogenic disturbances are minimized or excluded. That can be useful for planning purposes, but the specific width of buffer strips, riparian reserves, or streamside management areas might change locally, depending on structural features, stream size, desired level of protection, and location within the drainage system. As one goes downstream, discharge increases and floodplains widen and riparian reserves or buffer strips tend to increase (FEMAT 1993). Both large and small streams need space to allow their channels to adjust to continually changing flow and sediment loads. Where artificial confinement of channels and degraded riparian areas have occurred, the ability of a channel to respond to natural events is severely constrained. Streams given room to interact with streamside vegetation can usually be recovered, but restoration of aquatic habitat where roads, buildings, riprap, berms, or other types of structures encroach on a stream is often unsuccessful.

The level of risk associated with a specific management practice or restoration effort should be considered and acknowledged. For example, in the design of a bridge or other structure of high value, a safety factor is often incorporated into the final design to account for unknown factors and the longevity of the structure. Similarly, the dimensions of riparian protection zones should also include safety factors to allow for natural disturbances, uncertainties about the ecosystem of interest, and changes in public values. If an additional margin for error is allowed, the probability of habitat improvement becomes greater and options for future management decisions are increased.

PROPERTY RIGHTS AND HABITAT PROTECTION ON PRIVATE LANDS

Land ownership carries the right to undertake a wide range of activities. But streams, fish, and other organisms that use aquatic environments are generally considered to be publicly owned resources. Where private lands abut streams, rivers, wetlands, and estuaries, the demarcation between private and public rights can become ambiguous. Ecosystem boundaries are spatially irregular and can change with time. Owing to their wide-ranging life-cycle requirements, salmon often cross many property lines during their migrations to and from spawning and rearing areas. Although much of their time in freshwater might be spent in streams, rivers, and lakes on publicly owned lands, salmon can reside in or pass through privately owned lands. Recent efforts to protect habitat in federal forests have led to formulation of an aquatic-conservation strategy that calls for a network of wide buffer strips adjacent to all streams and lakes (FEMAT 1993). Depending on the number of tributaries in a watershed, such a system of buffer strips can account for up to about half the total land area. The aquatic conservation strategy in FEMAT (1993) has been endorsed by the Snake River Salmon Recovery Team (1994) as an appropriate habitat-conservation measure for endangered salmon in the Snake River basin. If applied to privately owned forests inhabited by salmon, the system of large buffer strips recommended by FEMAT (1993) would undoubtedly impose severe financial hardships–reduced return on investment for shareholders in forest-products companies or reduced incomes for small woodlot owners. Applying similar restrictions to nonforest landowners in rangeland, agricultural, or urban areas could produce similar hardships.

Yet there is little doubt that over the last century land and water uses on many privately owned lands have continued to degrade aquatic habitat and resulted in loss of the natural production capacity of these waters (Lichatowich 1989, Thomas et al. 1993, Moyle and Yoshiyama 1994). Uniform and consistently applied habitat-conservation strategies are not practiced on the scale of river basins, the scale most relevant to the metapopulation structure of Pacific salmon. The dilemma is clear. How can private-property rights be respected while adequate habitat is provided for salmon across the landscape?

The committee believes that progress toward solving the dilemma is possible and recommends that attention be given to developing a *more equitable and more uniform system of habitat-protection requirements on private ownerships* across all land uses, establishing *joint planning groups* for entire river basins (or subbasins), where private landowners can participate in land-use policy decisions, investigating various *incentives* for landowners to practice improved environmental stewardship, and expanding programs that involve *the public* in monitoring and habitat-conservation projects. Those steps would benefit not only salmon but virtually all public values associated with aquatic-riparian ecosystems.

At present, different environmental laws apply to privately owned lands under different types of management. Forestry operations are regulated by state forest-practices acts. Range and agricultural lands are regulated by voluntary best-management-practices or provisions of water-quality laws. Urban-industrial lands are regulated by pollution-control and local greenway ordinances. The intent of such laws might or might not include safeguarding the ecosystem processes necessary to maintain aquatic productivity; for example, some laws require landowners to do only what is necessary to make water safe for human consumption. The committee suggests that habitat for salmon and other important aquatic resources will benefit greatly from a system of environmental management that transcends the type of use for which private lands are zoned and that acknowledges the need to protect interactions between aquatic and terrestrial ecosystems. A system need not require identical protection measures (e.g., buffer-strip width) in every situation. Local conditions and landscape-level considerations can provide landowners with some management flexibility. However, more emphasis must be given to protecting the quality of the land (and especially the riparian zone) that affects freshwater ecosystems if environmental conditions are to improve and habitat degradation is to be reversed. New management systems might take the form of an integrated land-use practices act that applies to all private-land uses and that includes provisions for protecting both riparian zones and water quality. The committee believes that providing greater consistency in protecting aquatic habitat on private lands should have high priority in state and local governments.

A second means of improving habitat protection on privately owned lands is to involve property-owners more fully in environmental-policy matters so that they have more ownership in habitat-conservation decisions. Property-owners are often concerned with environmental quality and participate in outdoor recreational activities, such as hunting and fishing. Landowners might resent being told what they can and cannot do with their land, but they usually understand the value of having abundant natural resources nearby. They might be reluctant to embrace federal and state regulations administered by bureaucrats who are not familiar with local conditions, but they often take pride in displaying improved environmental stewardship to their friends and neighbors. As opposed to being told "this is how you must do it," private landowners are usually more responsive to "this is what you should protect, but you can be creative in helping to design protection measures that fit your situation." Local planning organizations and conservation boards (e.g., soil-conservation districts) could be useful forums for stimulating private-landowner involvement. A few local property owners practicing improved habitat conservation can act as examples and catalysts for others.

On the watershed or river-basin scale, private landowners can participate in joint planning organizations that help to set environmental policy and promote environmental stewardship. Many such organizations have formed in Oregon, Washington, and British Columbia with the goal of improving fish and wildlife habitat through rehabilitation projects and increased riparian protection (Krueger 1994, C.L. Smith 1994). Membership often includes representatives of federal and state fish and wildlife agencies, conservation organizations, tribes, private landowners, and commercial fishers. In the face of such diverse interests, progress can be slow (Halbert and Lee 1990), but those involved value having a seat at the table and are more likely to accept responsibility for reaching consensus on private-property issues (Pinkerton 1993).

A third approach to improving habitat on privately owned land is an expanded system of conservation incentives. Incentives can include tax deductions for riparian protection, conservation easements, and cost-sharing programs for restoration projects. Some of the most successful examples have been associated with rangeland where cost-shared riparian fencing has lowered livestock damage to streamside areas (Elmore 1992). Conservation easements can also provide financial incentives for habitat protection. Some incentive programs have not succeeded, however. Oregon instituted a Riparian Tax Incentive Program (RIPTIP) in the 1970s to provide tax relief to private landowners for increasing riparian protection. The value of the tax incentive in many cases was insufficient to encourage property-owners to participate; less than 50 miles of streams had received additional protection when the program was terminated. Most property-owners felt that the program did not provide enough reward to offset the financial losses that resulted from forgone management opportunities. The failure of RIPTIP suggests that incentive programs should not be based strictly on tradeoff between environmental and economic interests, because the latter will most often prevail.

Finally, the public can be encouraged to participate more fully in monitoring programs and habitat-restoration projects. Dedication to habitat protection is increased when people take part in long-term habitat monitoring (Hellmund 1993). They begin to recognize the importance of less-visible but nonetheless critical linkages between aquatic and terrestrial ecosystems. Regulatory agencies often fail to enlist public involvement in monitoring programs, believing that the programs must remain solely in the hands of technical specialists. But simple habitat features can be monitored by lay persons, and public participation can be a powerful tool not only for expanding habitat databases but also for environmental education. A better-educated public is far more likely to recognize the long-term value of protecting aquatic resources through individual actions that promote environmental stewardship than a public that tends to lay the problem at someone else's doorstep.

Likewise, management organizations can enlist the aid of property-owners in carrying out habitat-restoration activities, such as replanting native vegetation in riparian zones. Public interest in enhancing salmon is generally high, as evidenced by the growing number of local fishery enhancement organizations. Many of the projects have emphasized small-scale artificial production techniques (e.g., egg boxes and small rearing ponds), but it should be possible to generate more interest in restoring degraded habitat, particularly in riparian zones. Such projects might not produce immediate benefits in returning adult salmon, but private landowners can take comfort in the knowledge that their efforts will help to maintain natural productivity so that their children will be able to enjoy wild fish.

BURDEN OF PROOF

Many of the improvements in watershed-management practices that have occurred in past decades have required that substantial impacts or harm be demonstrated before changes would occur. For instance, early stream-temperature studies showed that removal of riparian shade during logging could have important adverse affects on summertime stream temperatures (e.g., Brown and Krygier 1970), but only after the evidence was clear were shade requirements instituted. Other studies found an association between accelerated sediment production and particular types of roads and logging practices (for summaries of research, see Ice 1985, Swanson et al. 1987). Management practices associated with forested areas have continued to evolve when research results linked particular practices to potentially adverse changes in sediment production, channel stability, water quality, flow regimes, and fish habitat. Changes in forest practices came only after major impacts had been identified. Furthermore, the burden of proving damage was generally on state agencies that were claiming that particular practices had detrimental effects. Similar situations exist for rangelands and agricultural areas; important impacts of management practices must be well documented (e.g., continued violation of a water quality standard) before changes in the practices are considered.

A major change with regard to the burden of proof has apparently occurred on many federally owned forest lands (FEMAT 1993). Relatively large riparian buffer strips are intended to provide high levels of protection for fisheries and other riparian-dependent wildlife species. Although the dimensions and configurations of the riparian reserves can be changed, such change can occur only after a watershed analysis has been conducted (FEMAT 1993). Alterations in riparian buffers, particularly decreases in width, are allowed only if it can be demonstrated that alterations will not adversely affect water quality, wildlife, fisheries, or other aquatic organisms. This shifting in the burden of proof is a major change in how forest resources on federal lands are managed. The extent to which the shift will carry over onto state-owned or privately owned lands or other types of land use is not known.

HABITAT MANAGEMENT AND FISHERIES MANAGEMENT

There is critical interaction between habitat management and fisheries management. Many habitat alterations do not directly extirpate populations, but instead reduce survival rates so as to reduce sustainable exploitation rates. Determinations of sustainable exploitation rates are based on analyses of long-term population performance; therefore, there is increasing risk of catch rates remaining too high as habitat and survival deteriorate more rapidly than productivity assessments can be updated. It is important that habitat degradation and reduced ocean productivity not be used as excuses to continue overfishing, especially where information gathering and adaptive responses are delayed even if management agencies are determined to respond as wisely as possible. Greater attention should be paid by fisheries managers to trends in condition of freshwater habitat.

9

Dams and Mitigation of Their Effects

INTRODUCTION

Dam construction in the Pacific Northwest began late in the 1800s when small irrigation reservoirs were constructed on tributaries of the Snake River in Idaho. Early in the twentieth century, the first hydropower dams were constructed on tributaries of the Columbia, such as the Spokane and Willamette Rivers. During the early 1900s, dam construction moved more rapidly and most of the reservoirs were relatively small (see Figures 3-9 and 3-10). However, beginning in the late 1930s with the initiation of the construction of Bonneville and Grand Coulee, dam construction proceeded at a more rapid pace, as both the number and storage volume of dams in Washington, Oregon, and Idaho increased. During the 45-year period after the authorization of Bonneville (1933) and Grand Coulee (1935) dams, 14 mainstem Columbia River and 13 mainstem Snake River dams were completed.

By the late 1970s, potential sites and public support for major new dams had been virtually exhausted, and the growth phase ended. Dams had been constructed across the migration routes of most Pacific Northwest salmon runs. They range from irrigation diversions with a hydraulic head of only a few feet to dams at Grand Coulee, Dworshak, and Hells Canyon that are several hundred feet high and completely block upstream and downstream passage of anadromous fish. Adult salmon can pass high dams with the aid of trap-and-haul arrangements or even fishways, but the great depths, cross sections, and lengths of the reservoir pools might hinder smolts from finding routes to the sea.

Figure 9-1 shows how the reservoir system has affected the average seasonal

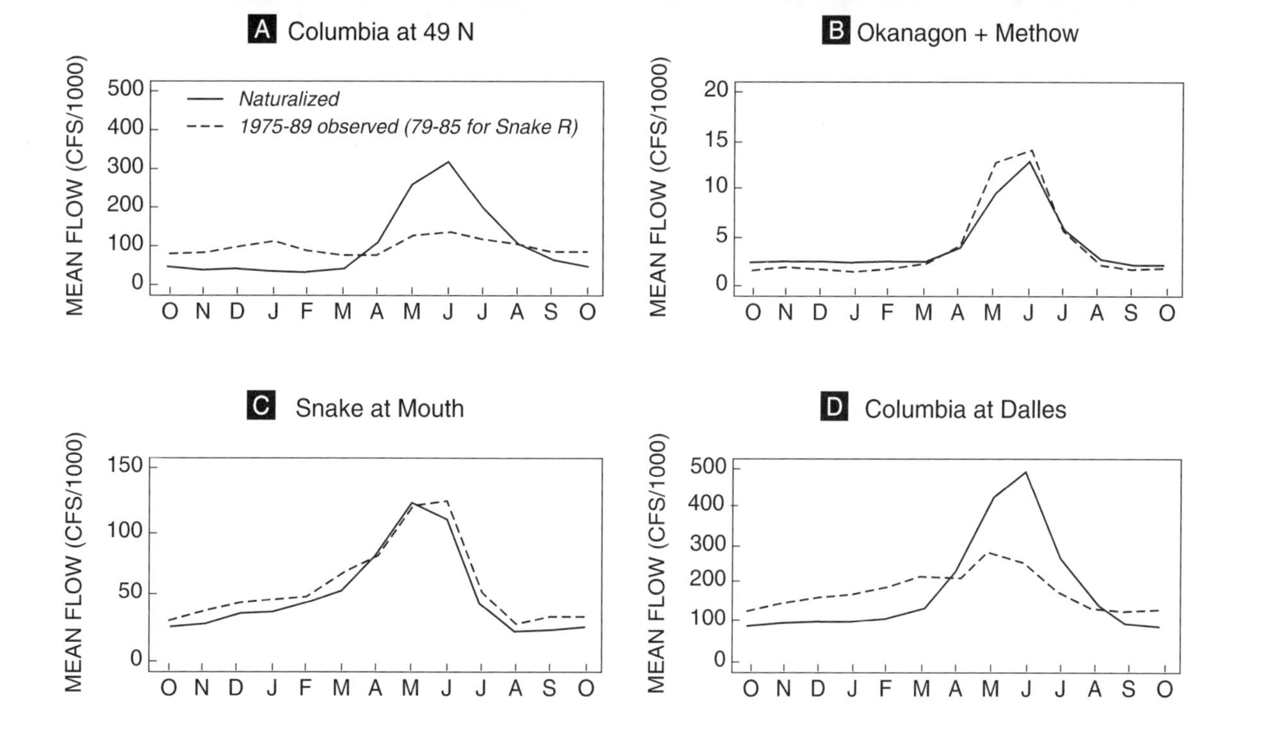

FIGURE 9-1 Effects of reservoir system on Columbia River seasonal discharge at four points: A - Columbia River at the international border reflects Canadian storage, B - sum of Okanogan and Methow rivers reflects regulation by Lake Okanogan and upstream reservoirs in Canada, C - mouth of Snake River reflects effects of all Snake River reservoirs, and D - Columbia River at The Dalles reflects effects of all major storage facilities on the Columbia River system except Willamette and Cowlitz rivers.

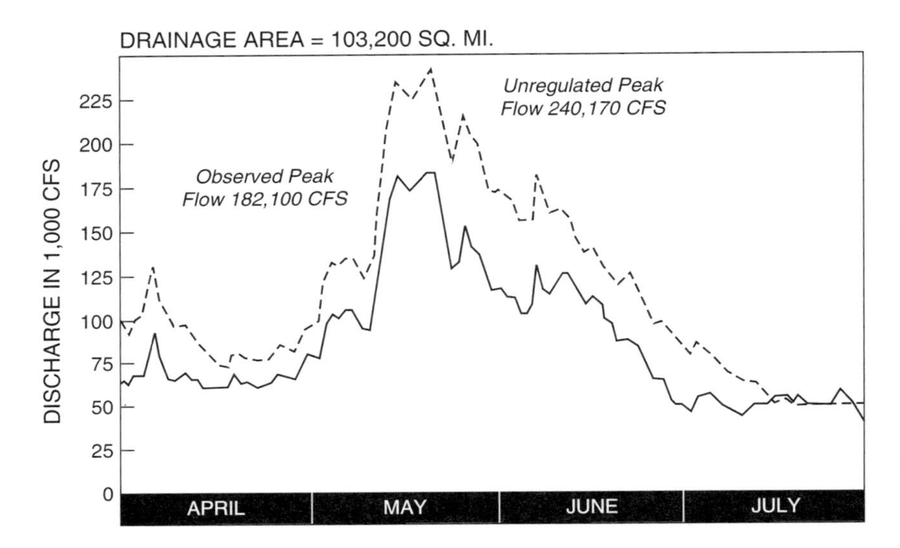

FIGURE 9-2 1993 Snake River hydrograph below Lower Granite Dam. Source: Columbia River Water Report for 1993.

discharge at four points in the Columbia River system: the Columbia River at the international border reflects the effect of Canadian storage, the sum of the Okanogan and Methow rivers reflects regulation by the Lake Okanogan and upstream reservoirs in Canada, the Snake River at its mouth reflects the effects of all Snake River reservoirs, and the Columbia River at The Dalles reflects the effects of all major storage facilities on the Columbia River system except those on the Willamette, Cowlitz, and Lewis rivers. The average seasonal discharge of the Columbia River mainstem has been drastically altered. However, the seasonality of regulated flow of the Snake River has been much less affected (Figure 9-2). That is an important distinction for the discussion that follows because of two common misperceptions: that there has been a major seasonal shift in the mean discharge hydrograph of the Snake River, which Figure 9-2 shows is clearly not the case; and that the reservoir storage in the Snake River is much less than on the mainstem of the Columbia—total storage in the Snake River system, expressed as a fraction of the mean flow, actually is only slightly less than that of the mainstem Columbia. Because there has not been a major shift in the Snake River hydrograph, it is doubtful a priori that the declines in Snake River salmon stocks are due to or reversible by changes in the seasonality of the flow regime of the Snake River alone. These same salmon must traverse the Columbia River, whose seasonal hydrograph has been substantially altered. Even if flow changes are useful in rehabilitation efforts, they are likely to be insufficient without changes in other human interventions in the salmon's life cycle and habitat.

The major difference between the Snake River and the Columbia mainstem is that much of the Snake River storage is used for irrigation rather than hydropower generation. The amount of hydrograph shaping (change in the natural hydrograph) required to meet irrigation requirements is much less than that needed for hydropower because the water-demand peak is in midsummer, typically only 1-2 months after the natural hydrograph peak. However, in contrast with hydropower, part of the water diverted for irrigation is consumptively used—it does not return to the river. In the case of the Snake River, the total consumptive use of water by agriculture—4-5 million acre-feet (MAF) annually—constitutes an appreciable fraction of the natural flow of the river during the months of highest agricultural demand (about 20% of the flow during the period May-September). That is the basis for the argument that agriculture has greater effects on the managed hydrology of the Snake River than does hydropower.

Although the flow regime of regulated rivers usually is less variable over the course of the year than it was before dam construction, water storage in and release from dams can result in large day-to-day or even day-to-night fluctuations in flow and depth. The fluctuations can lead adult salmon to construct their nests in unsuitable places and can strand juveniles. However, intensive studies below Priest Rapids Dam during periods of peaking operation (large diurnal release variations) and load operation (minimal diurnal variations) revealed little or no effect on fall chinook spawning or abundance.

In addition to affecting seasonal hydrographs, the reservoir system has had major effects on flow velocities, water chemistry (especially nitrogen supersaturation downstream of dams), and stream temperatures. Supersaturation with atmospheric gas, chiefly nitrogen, occurs when water is spilled over high dams. Gas is absorbed into the bloodstream of fish during respiration, especially fish that remain close to the surface. When the gas comes out of solution, bubbles form and can subject the fish to a condition similar to the bends suffered by divers. In some years before development of spill deflectors to prevent deep entrainment of spilled water, gas supersaturation caused extensive mortality (Ebel 1969). River managers now coordinate spill at various projects to reduce risk of serious losses. However, supersaturation can still exceed the high-risk levels (125% saturation) in years of high river discharge.

Reservoirs unintentionally provide thermal storage, as well as water storage, so seasonal variations in stream temperature are reduced in much the same way as seasonal variations in streamflow. In general, storage reservoirs tend to increase winter temperatures and reduce downstream summer temperatures and to cause maximum and minimum temperatures to occur later in the year than in the absence of damming. However, water below Bonneville Dam has shown longer and warmer summer conditions over the last 40-50 years (Quinn and Adams, in prep.). Figure 9-3 shows the trend in the date when the spring water temperature on the Columbia mainstem exceeded 15.5° since 1938. The upward trend in spring water temperature is consistent with introduction of storage in upstream

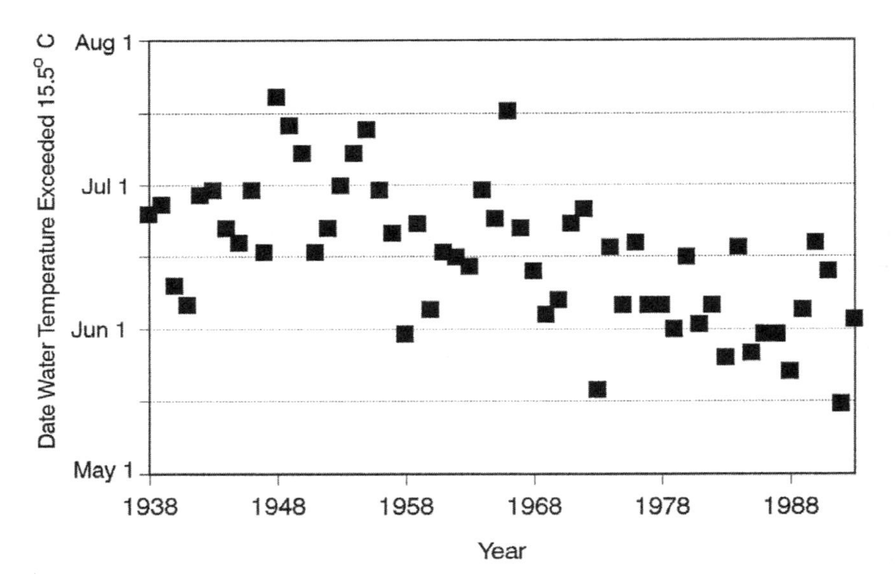

FIGURE 9-3 Spring water temperatures on Columbia mainstem since 1948.

reservoirs, particularly Canadian storage, which came on line in the 1970s (see Appendix E). Although the physical system has not changed since then, the trend apparently continues. That might reflect long-term climatic change or variability; Lettenmaier et al. (1994) found that air temperatures increased significantly in much of the northwestern United States during 1948-1988, especially in late winter and early spring. The effects of long-term changes in water temperature on salmon depend on a complex interaction of early rearing conditions, emergence date, and predator populations.

High dams can inundate substantial amounts of spawning and rearing habitat. Some salmon, notably chinook, spawn in the mainstems of rivers and hence lose usable area when rivers become reservoirs (e.g., John Day, Priest Rapids, Coulee, and Wells dam pools). In addition, juveniles, particularly chinook, might rear in large rivers or feed there during downstream migration (e.g., the Fraser, Columbia, and Sacramento rivers) (Rich 1920, Levy and Northcote 1982, Chapman et al. 1994). The reservoirs might constitute a reduction in desirable habitat. In the Columbia River, however, reservoir rearing might substitute for lost flowing-river habitat and make up for lost quality and quantity of estuarine habitat (Rondorf et al. 1990, Chapman and Witty 1993). The cumulative volume of all Pacific Northwest (Idaho, Oregon, Washington, and northern California) reservoir storage was over 50 MAF by 1980 and currently exceeds 65 MAF (Figure 3-10).

The following discussion focuses on intermediate and large dams. Most smaller dams were built and are operated primarily to generate hydroelectric power. Their effects on salmon populations are discussed in Chapter 3.

EFFECTS OF DAMS ON SALMON

The effect of dams without fish-passage facilities on salmon is clear: the upstream habitat is lost. Such dams block about one-third of the Columbia River watershed to access by anadromous fishes; owing to natural passage barriers, one-third was never accessible. One-third lost to anadromous fish is upstream from Grand Coulee Dam on the Columbia River and the Hell's Canyon complex of dams on the Snake River (Chief Joseph Dam, the reregulating dam for Grand Coulee, also is impassable). Many dams on tributaries are also impassable, such as Mayfield Dam on the Cowlitz River and Round Butte Dam on the Deschutes River (tributaries to the Columbia), Detroit Dam on the North Santiam (tributary to the Willamette), and Dworshak Dam on the North Fork of the Clearwater (tributary to the Snake). The loss of spawning and rearing habitat because of impassable dams is perhaps most acute on the Columbia River system but is by no means restricted to it. Shasta Dam eliminated the upriver runs of salmon on the Sacramento River, and rivers in Puget Sound (e.g., the Skagit River) and the Strait of Juan de Fuca (the Elwha River) have impassable dams as well. On the coast, access to the upper Klamath and Rogue rivers was blocked by Iron Gate and Lost Creek dams, respectively. It is difficult to specify the magnitude of losses attributable to such dams because record-keeping before construction was often poor or nonexistent, and it might not be possible to survey the inundated habitat to estimate potential production.

Not all the impassable dams were large hydroelectric dams. Small splash dams, built to back up water and then float logs downriver, were often impassable and sometimes remained in place long enough to obliterate major salmon runs, e.g., in western Oregon and Washington (Sedell and Luchessa 1982) and in the upper Adams River, British Columbia (Williams 1987). Relatively small irrigation and hydroelectric dams blocked some salmon migrations early in western development—e.g., Black Canyon Dam on the Payette River in Idaho, Grangeville Dam on the south fork of the Clearwater River, and Sunbeam Dam on Idaho's Salmon River. More recently, a common practice at many fish hatcheries has been to block upstream migration at or near the hatchery to aid in collecting returning adults or to isolate adults, possibly carrying diseases, from the hatchery's water supply.

Dam-Related Mortality

Even when dams are constructed with fish ladders for upstream passage of salmon, fish can still be delayed. Turbine discharge flows can disorient salmon and make it difficult for them to find the small attraction flows that lead to the ladder. Ladder designs have evolved greatly since the early 1900s. Early facilities often had excessive in-ladder flow rates and turbulence and lacked sufficient resting areas; salmon either avoided them or found them impassable or too demanding of energy. Flood flows destroyed many wooden fish ladders, e.g., at Condit Dam on the White Salmon River and Grangeville Dam on the south fork of the Clearwater River. Poor concrete quality caused others to fail, such as Sunbeam Dam on the Salmon River. Spillways close to fishway exits tended to pull adults that left the ladders back over the dam and thus caused migration delay (Bjornn and Peery 1992). Delays might not kill fish, but salmon do not feed on the upstream migration and must use stored energy as efficiently as possible to migrate upstream, mature sexually, and spawn successfully (Gilhousen 1980). Adult salmon can be killed if they drop back through turbine intakes, although the rate of loss of fish that drop back is unknown. Counts at successive dams seem to indicate that deaths occur between dams, although it is rare to observe dead salmon there. Poaching might account for some of the loss. Interdam losses have been estimated at up to 25% for the reach from Bonneville Dam to John Day Dam, but current loss estimates are about 4-5% per project there and elsewhere in the Columbia River system (Chapman et al. 1991).

Downstream passage of juveniles through bypass facilities comes at a biological price. These juveniles, hereafter referred to as guided fish, can make contact with deflection-screen surfaces, gatewell walls, the vertical barrier screens in gatewells, the orifice entrance, or portions of the bypass channel or downwell (Figure 9-4). Such encounters can cause impingement, bruising, scale loss (descaling), and stress (Chapman et al. 1991). Because it is the most quantifiable evidence of damage to fishes, descaling is evaluated by biologists quantitatively and is used to indicate facility problems and fish viability.

Stress accompanies passage through bypasses. Fish that hold in currents to resist passing downstream, that contact separators, or that otherwise experience stressful contact (e.g., with equipment) become physiologically stressed. They appear to recover when held for several to 48 hours. However, direct bypass to the river delivers stressed fish to the outfall,[1] where they can become prey for birds and fish, especially northern squawfish (*Ptychocheilus oregonensis*).

Bypass systems also concentrate smolts in a small area. Smolts from the whole width of the Columbia and Snake rivers (which might flow at 250 and 90 thousand cubic feet per second [kcfs], respectively) for example, are gathered into a bypass channel that flows at only 500 cubic feet per second (cfs). If bypassed directly, many thousands of smolts per hour can be delivered in a small volume of water to the dam tailrace, which provides a concentrated stream of

[1]The bypass outfall, the water just delivered from the bypass system to the river, delivers fish to the tailrace or the tailrace edge; the tailrace is the entire river flow just downstream from the dam.

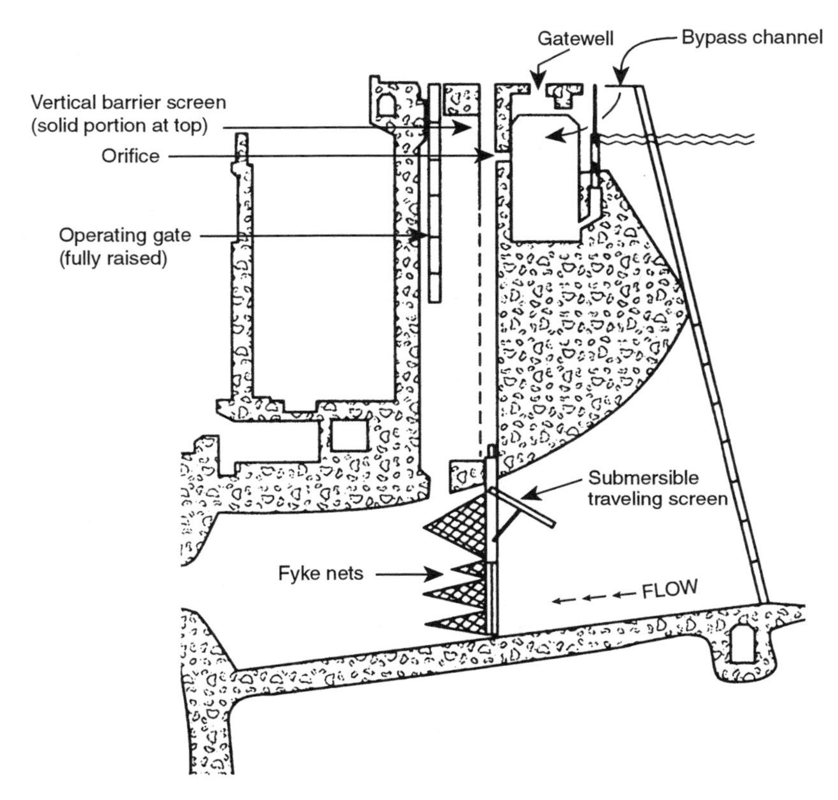

FIGURE 9-4 Cross section of typical dam and bypass system. Fyke nets are used to estimate fish-guidance efficiencies. Source: adapted from USCOE 1993:163.

prey for predators. At the Bonneville second powerhouse, extensive studies of fall chinook passing through turbines and the bypass revealed that the survivals through the two routes were not very different and that predators were keying on the stream of prey from the bypass outfall (Ledgerwood et al. 1991). Similar studies have begun at the Bonneville first powerhouse. At other dams equipped with bypasses, total bypass-related mortality has not been thoroughly investigated.

Some investigators consider bypass-caused deaths to include only those which can be observed (carcasses) in the raceways and sampling facilities incorporated in bypass systems. But those do not provide data on impingement on deflection screens, predation within the gatewell and bypass system, predation caused by the bypass-concentrated stream of prey, stress-related deaths that occur after smolts leave the outfall area, or predation on stressed fish long after they leave the outfall pipe. Carcass counts typically indicate a mortality of 1% in the bypass system (Koski et al. 1985); some studies indicate that bypass-related mortality averages perhaps 5-7% (Matthews et al. 1987).

The location of the outfall is crucial to bypass-related mortality. Expanded evaluations of bypass-related mortality include the first smolt encounters with deflection screens to a point well downstream from the dam, where fish become free of all physical effects of the bypass. Passage through the bypass system is not immediate in most cases. The National Marine Fisheries Service annually evaluates how quickly smolts pass out of the gatewell and through the orifice on the basis of the fraction of smolts that enter the gatewell and leave it within 24 hours. Orifice-passage efficiency of 70% is considered satisfactory; this means that some smolts remain well over 24 h, some are delayed in the bypass channel, and others are delayed beneath the separator (if the system incorporates one).

Seasonal races and individual populations of salmon migrate from rivers to the ocean at specific times of the year. It is generally believed that photoperiod is the primary cue that triggers migration, although flow, temperature, and social interactions can also influence it (Godin 1982). The rate at which smolts migrate downstream depends on both their swimming speed and their orientation to the velocity of the flow. Smolts migrated downstream in the Columbia River more rapidly before dams were constructed than they do now (Raymond 1979, Rieman et al. 1991). It also appears that migration is more rapid in years or times of the year when river flow is greater. However, there is dispute over the validity and interpretation of these data, as discussed in the next section.

There is some evidence that predators, such as northern squawfish, have increased in abundance in the lower Columbia River. The most important fish predators (squawfish) and birds, such as gulls, are native species. There is some predation by nonnative fishes such as walleye (*Stizostedion vitreum*) and smallmouth bass (*Micropterus dolomieui*). Perhaps more important, the reservoirs, tailraces, and bypass outfalls might have improved the river as rearing habitat for these species, and the tailraces and forebays of the dams might lead to an increase in foraging efficiency over that in the undammed river. In addition to possible increases in mortality en route to the ocean that might result from retarded migration, delayed arrival in the estuary or ocean might result in higher mortality or reduced growth.

Storage in the upper Columbia River and Snake River has altered the main Columbia River's hydrograph. Sherwood et al. (1990) analyzed monthly mean flows of the Columbia River and found that large-scale manipulation of the flow cycle began around 1969. Since then, monthly mean flow has varied less. Flow damping has resulted in a reduction in average sediment supply to the estuary. Except for times of major floods, residence time of water in the estuary has increased with decreasing salinity. Detritus and nutrient residence has increased; vertical mixing has decreased. Sherwood et al. (1990) noted that although hydrodynamic changes have probably enhanced the pelagic primary productivity in the estuary, the costs have yet to be evaluated. The changes have enhanced estuarine conditions for detritivorous epibenthic and pelagic copepods; the estuary has been converted to a less-energetic microdetritus-based ecosystem with higher organic sedimentation rates. Sherwood et al. (1990) concluded that

> it is apparent that these changes, and other changes in the fluvial part of the system, have contributed to the dramatic decline in salmon populations. The implications such as have taken place in the Columbia River estuary and watershed need to be incorporated into contemporary estuarine and shorelands management strategies. In particular, proposals for comprehensive hydroelectric and water withdrawal developments, shoreline modifications, and navigation projects should all be evaluated in terms of potential consequences to the estuarine ecosystem and resulting effects on other resources, including fisheries, which depend on a highly coevolved and biologically diverse estuarine environment.

Sherwood et al. (1990) estimated that the estuary of the Columbia River lost 20,000 acres of tidal swamps, 10,000 acres of tidal marshes, and 3,000 acres of tidal flats between 1870 and 1970. They further estimated an 80% reduction in emergent vegetation production and a 15% decline in benthic algal production.

Ebbesmeyer and Tangborn (1993) demonstrated that reservoir storage in the Columbia River had altered the hydrograph by diverting summer flows to winter, which altered coastal sea-surface salinities from California to Alaska. Coastal ocean and estuarine dynamics have changed at various locations along 2,000 km of North Pacific shoreline. The effects of those alterations on trophic dynamics, loss to predators, and migration success are completely unknown. Ebbesmeyer and Tangborn (1993) raised the question of whether homing behavior of salmon might change in response to the altered salinities. If homing were affected, we would expect to see steelhead and salmon from Columbia River hatcheries as strays in coastal rivers of California, Oregon, and Washington. But no evidence of extensive or unusual straying has been found.

Time of Travel

The effect of time of travel on subyearling chinook (i.e., ones that pass their first winter in the sea, or ocean-type chinook) is less clear than the effects of passing through dams. Ocean-type chinook pass their first winter of life (after emerging from the redd) at sea; stream-type chinook spend their first winter of life in the stream before going to sea. Some studies indicate that subyearlings migrate downstream more quickly with higher flows (Rondorf and Miller 1994), but other studies (e.g., Giorgi et al. 1990, Chapman et al. 1994) do not. Subyearling chinook gradually move downstream as they grow—a rearing migration. Rather than the rapid downstream passage of yearling chinook and steelhead, which often reach an average speed of 20 miles/day or more, downstream passage of subyearling chinook is perhaps 2 miles/day. Some subyearling summer-fall chinook do not enter the sea in the first summer of life, as it is commonly supposed; rather, they do not pass McNary Dam until late fall. Growth rates of subyearlings in mainstem Columbia River reservoirs are very high (Chapman et al. 1994). No information exists on whether subyearlings that reach large size in reservoirs of the Columbia River have higher survival rates than small subyearlings that go to sea in July.

The effect of migration speed on smolt survival has been inferred from data on system survival and travel-time data acquired during the 1970s. The accuracy, precision, and relevance of those historical estimates are questionable (Giorgi 1993), and for that reason, the data from the 1970s were abandoned in the early 1980s. Smolt survival through the same river reaches would differ today: the smolt cohort in the Snake River has changed from almost 70% wild in the late 1960s to only 15% wild today (Park 1993). Hatchery steelhead and spring chinook predominate in the main Columbia and Willamette rivers, and spring-summer chinook in the Snake River. Wild fish still compose the majority of fall-run and summer-run chinook of the mid-Columbia and in some tributaries that lie in wilderness areas where hatcheries are not used (e.g., the middle fork of the Salmon River and Chamberlain Creek in Idaho) or are considered refugia for wild gene pools (John Day River in Oregon).

Not only the makeup of the smolt runs, but the river environment itself has been altered. More turbines have been installed in mainstem dams. John Day and other newer reservoirs have matured limnologically. Fish communities have changed (e.g., exotic species have been introduced), and river management has evolved in response to needs for fish conservation. Reach-specific or system survival studies have not been completed that would allow managers to evaluate modern conditions in the river; they are crucially important and should be pursued with vigor and dispatch.

Recently, there has been sponsorship of new research, such as reach-specific survival studies that began in 1993 (Snake River Salmon Recovery Team 1993). If such studies are continued, they would provide steady improvement in the scientific knowledge available for the difficult challenges presented by salmon rehabilitation. For instance, Williams and Matthews (1994) and Steward (1994) critically reviewed the data obtained in the 1970s. They concluded that system and average project mortalities in smolts migrating downstream were overestimated and that the data should not be used to estimate current system and project losses.

The lack of modern survival data that could be used to evaluate survival through various dam-passage routes and through reservoirs has resulted in dissension over the value of high flows for reducing mortality in downstream migrants. Many investigators believe that the older data support a need to provide flows of 80-90 kcfs in the Snake River and 200-220 kcfs in the main Columbia. However, many of the same researchers question the gains in survival that the

Fish Passage Center and Columbia Basin Fish and Wildlife Authority indicate will result from providing flows higher than that.

Survival studies with passive integrated transponder (PIT) tags in 1993 and 1994 indicate survival near 100% through Lower Granite pool and high survival through Little Goose pool. However, estimated mortality "across the concrete" was higher than expected (Iwamoto et al. 1994, Schiewe 1994). Data on interrogation rates of PIT-tagged fish that had been trapped and marked in the lower Salmon and Clearwater rivers also support high survival to the first (Lower Granite) dam (data tabulated in Fish Passage Center 1994). Those preliminary results, if confirmed by additional analysis and studies and augmented by reach-specific survival estimates at downriver dams on the Snake and Columbia rivers, would suggest high priority for mitigation efforts directed at increasing survival at the dam rather than speeding fish through pools. If later work shows that the 1993 and 1994 pool-survival estimates were atypically high, drawdown or flow augmentation could be given higher priority. The committee strongly supports acquisition of additional reach-specific survival information at Snake and Columbia river projects.

Unscreened Diversion Dams

When water is diverted from the river into domestic or agricultural uses, there is always the possibility that downstream-migrating salmon and other fish will be drawn into the diversion channel or pipe or that fish will actively seek it. Many unscreened or inadequately screened diversions have operated on the Columbia River system over the years, and some unscreened or poorly screened diversions remain. An extensive program, funded by BPA, has added new screens and maintained old ones in irrigation-diversion ditches and canals in Columbia River tributaries. No study has balanced the net benefit of screens and comparison to the lost rearing habitat in irrigation canals and ditches downstream from the screens.

Estuarine Dynamics

The estuaries of West Coast rivers, both dammed and undammed, might have been changed to the detriment of salmon. Reservoir storage in the upper Columbia and Snake rivers has altered both the seasonal pattern and the characteristics of extremes of freshwater entering the estuary, as mentioned above. Since large-scale regulation of the flow cycle began about 1969, average sediment supply to the estuary, which is governed by floods, has been greatly reduced (Simenstad et al. 1992). Although changes in discharge have undoubtedly affected estuarine dynamics, many other factors have played roles as well.

MITIGATION OF DAMS' EFFECTS ON SALMON

Before 1980, the Columbia River dams and reservoirs were operated for the primary purposes of irrigation, flood protection, hydropower, navigation, and recreation. As a practical matter, irrigation, flood protection, navigation, and recreation were reasonably compatible with hydropower, and hydropower considerations dominated the operating strategy for the system. The Northwest Power Act of 1980 made fishery protection and enhancement one of the key objectives of reservoir-system operations—in principle, a major change in operating strategy. In this section, seven measures for mitigating dams' effects on salmon are discussed: fish-passage facilities, predator control, transportation, spill, flow augmentation, reservoir drawdown, and dam removal.

Fish-Passage Facilities

Fish ladders to pass salmon upstream of dams were inadequate or absent in the early 1900s. As knowledge developed about fish behavior, response to attraction flows, and in-ladder hydraulics, designs improved. Fish facilities at Rock Island Dam (1933), Bonneville Dam (1938), and McNary Dam (1953) greatly improved on the early designs, although instances of fishes' being delayed by hydropower dams and falling back after passage have been noted (Bjornn and Peery 1992, Dauble and Mueller 1993). Studies continue on adult passage efficiency, in-ladder behavior, fallback, and interdam losses.

On the Columbia River, the first mainstem dams constructed were equipped with fishways that permitted only adult passage. In particular, Rock Island Dam (near Wenatchee) and Bonneville Dam were equipped with fishways when built. Most mainstem Columbia River dams—Bonneville, The Dalles, John Day, McNary, Priest Rapids, Wanapum, Rock Island, Rocky Reach, and Wells—incorporate fishways. Grand Coulee and Chief Joseph dams do not. The four mainstem Snake River dams—Ice Harbor, Lower Monumental, Little Goose, and Lower Granite—also have adult fishways.

Juvenile passage facilities on most mainstem dams in the Columbia River system use deflection screens that project downward into the intakes of turbines and deflect fish upward from the turbine intake into the gatewell (Figure 9-4). Some deflection screens have traveling screen surfaces and are called submersible traveling screens. Recent designs involve bar screens, which are usually constructed of a wedgewire material that is triangular in cross section and presents a flat surface to incident flow. Bar screens are considered most useful where debris and drifting vegetation are less abundant; traveling screens serve best where debris and vegetation would clog a stationary screen. Traveling screens need not operate continuously. Periodic movement of screens can keep them clean and preserve hydraulic configurations in the turbine-intake environment important for fish guidance. Periodic cycling also reduces wear and maintenance requirements.

Deflection-screen designs are based on the premise that downstream-migrating juvenile salmon tend to pass through the turbine intake high in the water column. The screens guide intercepted fish upward into the gatewell. The gatewell's primary purpose is to provide a slot for insertion of a bulkhead to prevent water from passing farther down the turbine intake. Water-surface elevation in the gatewell is the same as that in the reservoir pool. The degree to which deflection screens guide fish up into the gatewell is termed fish-guidance efficiency (FGE). FGE is estimated by dividing the number of fish deflected upward into the gatewell by the sum of that number and the number of deflected fish and smolts trapped in fyke nets arrayed below the deflection screen. For example, if the number of fish deflected upward into the gatewell is the same as the number trapped in fyke nets, the FGE is 50%. FGEs of 70% or greater are considered "acceptable" with agencies, utilities, etc., in discussions about fish passage facilities at dams. Hydroacoustics are often used in connection with FGE studies to determine behavior of fish that encounter or avoid screens in turbine intakes.

A vertical flow is developed in the gatewell by permitting water to escape through a vertical barrier screen (VBS). The VBS prevents smolts from leaving the gatewell with the flow that leaves it. Several feet from the water surface within the gatewell, an orifice, usually about 14 in. in diameter, leads to a bypass channel that passes longitudinally through the dam structure normal to river flow. Makeup water (water to provide a reasonable gallery flow at the upstream end of the bypass gallery, where few orifices have had an opportunity to feed the flow) and water from the orifices provide a substantial flow down the bypass channel. Each turbine in the dam has one or two orifices that permit fish to escape to the bypass channel. Flow through each orifice amounts to 9-12 cfs. Thus, by the time the bypass channel contains all orifice flow and initial makeup water, the flow in the channel can reach perhaps 500 cfs or more.

The bypass channel lies only a few feet below the reservoir elevation. It carries water by a downwell to near-tailrace elevation. Different types of downwell designs have been used. Some lead to raceways where fish are held for later transportation by barge or truck. Others lead to the river for direct bypass of smolts without transportation. Most incorporate fish sampling facilities between the downwell and the raceway or bypass outfall. Where fish are held in raceways for later transport by barge, they must be separated from the large flow produced in the bypass channel and downwell; this is usually accomplished by a separator that consists of screening or wedgewire.

The downstream-migrant facilities described above have been installed or are being installed on Bonneville, The Dalles, John Day, McNary, and the four Snake River dams operated by the Corps of Engineers. On the Columbia River upstream from the mouth of the Snake River, where five dams are operated by public utility districts, the only dam with an operating fish bypass system is Wells

Dam, which, unlike others on the Snake and Columbia rivers, is a hydrocombine design that has the spillways directly over the turbine intakes and therefore cannot have conventional deflection screens in turbine intakes. Wells Dam has a retrofitted spill bypass that encourages juveniles to leave turbine intakes upward and pass downstream in a modest amount of spill from the spill bay.

The remaining middle Columbia dams (Rocky Reach, Rock Island, Wanapum, and Priest Rapids) were built without bypass facilities. For the last 12 years, various studies have been under way to facilitate design of juvenile bypasses. Most options involve some form of deflection screen. The FGEs obtained at various dams with fyke-net studies have been variable, ranging from less than 10% to more than 90%, depending on the dam and species. FGEs with the Priest Rapids and Wanapum prototype screens have been high enough to encourage further work on design and installation. FGEs at Rocky Reach have been very low. Turbine intakes there lie in a cul-de-sac that has the axis of the turbine-intake mouths parallel to the main flow of the Columbia River, rather than normal to flow. A somewhat similar orientation has prevented prompt application of screening technology to The Dalles Dam on the lower Columbia River. The FGEs obtained at dams with deflection screens differ considerably. They have been low at Bonneville Dam, whose bypass facilities were incorporated in the design rather than being retrofitted, and range from at least 70% at John Day and McNary dams to 55% at Lower Granite Dam.

Predator Control

The dynamics of the interaction between squawfish and smolts has received relatively little attention except in the pool of John Day Dam (see Raymond 1979, Rieman et al. 1991), where squawfish predation was estimated to kill millions of downstream migrants each year. Rieman et al. (1991) estimated that 8%, 11%, 7%, 19%, and 61%[2] of salmon that entered John Day pool were killed in the months of April through August respectively; steelhead loss averaged 12% per month in April, May, and June. Those estimates provided a platform for launching a major effort to control squawfish in the Columbia River basin. The program includes bounties and intensive efforts to reduce predator densities. It was designed to reduce the population of large squawfish and is to continue indefinitely. Assumed gains from such control measures are incorporated into smolt-survival models.

Transportation

At Lower Granite, Little Goose, Lower Monumental, and McNary dams, raceways can hold downstream migrants for delivery to barges or trucks for

[2]Because rounding was used, the sum of the percentages is 106%.

transport downstream. The transportation concept relies on "leapfrogging" downriver reservoirs and dams to deliver the juveniles to some point downstream from hydropower projects. In the Columbia River system, transportation programs deliver juveniles to the Bonneville Dam tailrace, largely by barge. Barges control water quality, e.g., by preventing the occurrence of gas supersaturation. Transport conditions can also be monitored in barge containers. In addition, barges can deliver smolts to varied release points and at varied times to confound predators in the lower Columbia River. Data on this point are limited but intriguing. Solazzi et al. (1991) reported that survival to adult capture of hatchery-produced coho smolts was increased by about 50% when smolts were delivered to Tongue Point, near the mouth of the Columbia River, instead of to the Bonneville Dam tailrace, about 185 km upstream. Other work (e.g., Ledgerwood et al. 1991) showed that based on survival from subyearlings to adults, midriver release at the Bonneville tailrace was superior to shoreline release. Experiments with steelhead have begun. Chinook, both yearling and subyearling, should also be tested for time and place of optimum survival after release from barges. However, effects of transportation on homing must be closely monitored (e.g., Solazzi et al. 1991).

State agencies and American Indian tribes currently oppose barging. Some of the objections center around the unnatural migration conditions (see Chapter 5). The efficacy of barging has been questioned because Snake River runs declined while barging was being used as a major mitigation tool. However, the committee notes that many Pacific Northwest runs that have not been barged have also declined. Recent transport-benefit ratios (TBRs) of observed survivals of transported compared with observed survivals of inriver migrants have been 1.6:1.0 in favor of transported fish (1986) and 2.5:1.0 (1989) (Mundy et al. 1994) for yearling chinook at Lower Granite Dam on the Snake River. Transport of yearlings at McNary Dam has yielded a ratio of 1.55:1.00 in favor of transported fish (Townsend and Skalski 1994). For subyearling (ocean-type) chinook, studies at McNary Dam indicate much higher transport-benefit ratios (often over 3.0:1.0) (Park 1993). Transportation might increase disease in transported smolts, but the evidence to date is equivocal (Pascho and Elliott 1989, Pascho et al. 1993). However, the transport-benefit ratios indicated by studies in the 1980s suggest that even if disease transfer occurs, the effects are not sufficient to erase the survival advantage received by transported fish, which avoid passing through turbines at hydroelectric projects. The committee relied on the information available on survival of transported fish in relation to in-river migrants and agrees with the Snake River Salmon Recovery Team (1993) that transportation is the mitigation tool of choice until inriver migration shows higher survival rates than does barging fish.

Spill

For fish-passage purposes, "spill" is water deliberately discharged over dam spillways, rather than discharged through turbines or bypass facilities, with the intent of reducing turbine mortality. Spill is an option that has been considered to improve the survival of juveniles passing dams on the lower Snake and lower and middle Columbia rivers. Schoeneman et al. (1961) found that mortality in chinook that were spilled over McNary Dam (Columbia River) and Big Cliff Dam (Santiam River) was less than 2%. Other studies have indicated similar low loss rates associated with spill.

Like bypass, spill can have a price in terms of gas supersaturation. Gas becomes trapped in bubbles in spilled water that plunges deeply into the stilling basin (the basin of relatively still water where gas bubbles can disperse) downstream from the dam (Ebel and Raymond 1976). Air within the plunging water is dissolved at depth because of increased pressure. It later escapes from the water when turbulence causes the gas to be released into the atmosphere. However, turbulence downstream from tailraces is scant because most dams are close to the pool of the next dam downstream, so supersaturation can persist some distance downstream of the spillway. "Flip-lips" (devices that direct spilled water outward to prevent deep plunging of spill flows, which cause supersaturation) have been installed on most dams. However, when spill is great, the lips become largely useless. An independent panel of experts appointed by the National Marine Fisheries Service (NMFS) on dissolved-gas effects recommended that gas supersaturation be limited to 110% (NMFS Independent Panel on Gas Bubble Disease 1994). Even so, gas supersaturation is often greater than 115% and sometimes as high as 140% (Fish Passage Center weekly reports) below major dams on the Columbia River and Snake River where spill occurs. Gas supersaturation of up to 143% killed many adult salmon in 1968, when all flow was passed through the completed spillway during construction of John Day Dam (Ebel et al. 1975). An emergency at John Day Dam on May 29, 1990, caused all flow to be spilled again, and water in the tailrace was supersaturated up to 145% (Chapman et al. 1991). As a result, dissolved-gas saturation at The Dalles Dam (the next dam downstream of John Day) remained greater than 130% from May 30 through June 2, 1990 (Fish Passage Center weekly reports 1990), and might have been even higher in some parts of The Dalles pool.

Spill is thought to reduce smolt mortality compared with turbine passage as they migrate across dams. However, because of the way it is measured—from the head of the pool to the dam—and because it is not thought to speed fish through reservoirs, it does not affect reservoir mortality. Flow speed, which often directly corresponds with flow volume, can speed the fish through the reservoir, assuming that they are physiologically ready to move downstream. Thus, spill differs from transportation, which can reduce both dam and reservoir mortality of smolts by carrying them around downstream reservoirs.

Flow Augmentation

Of the major effects of the dams on fish, the increase in time of passage through the hydropower projects on the middle Columbia and lower Snake rivers, particularly in the spring, has been identified as a key obstacle to survival of juvenile salmon and steelhead. An attempt to reduce passage time was made through a "water budget," which allocated some upstream storage in the Snake (at Dworshak Reservoir) and in the mainstem Columbia (at Lake Roosevelt, the reservoir of Grand Coulee Dam) to increase spring flows. It was anticipated that most of the additional water released would pass quickly through the downstream run-of-the-river dams (hydropower dams that primarily use riverflow rather than stoed water for power generation; see Appendix F) and therefore would not be available for the generation of power during the peak-demand season. However, the excess power produced by the increased flows at the time of the water-budget releases, although of less value, could be sold to meet demands elsewhere (e.g., in California, by use of the direct-current intertie); this compensated for some extent for the loss due to the water-budget releases (Wood 1993) (see Box 9-1).

The purpose of flow augmentation is to reduce the travel time through the reservoir system at key times in the salmon life cycle (most proposals are for the period April 15-June 14, although some proposals would augment summer discharge as well) to approximate more closely the pre-dam travel times. Flow augmentation was originally termed the "water budget" by the Northwest Power Planning Council (NPPC) in the middle 1980s. The idea was to reserve enough storage at Dworshak Dam's reservoir and Lake Roosevelt (Grand Coulee Dam's reservoir) to meet a flow target of 85 kcfs in the Snake River at Lower Granite in May (about 53% of the mean May flow under pre-dam conditions and about 78% of mean May flow with regulation; see Chapman et al. 1991) and to meet a target of 134 kcfs in the Columbia at Priest Rapids (about 87% of the mean May flow for the period 1971-1993, after further alteration of the natural hydrograph by Canadian storage). The projects on the lower Snake and middle and lower Columbia rivers were to operate near "full pool" (the maximum reservoir water level authorized in the projects' design).

The storage required to meet the streamflow targets was determined by adding the positive differences between the target monthly flows and the critical-period-rule curve for hydropower targets. This resulted initially in a storage requirement of about 4 MAF. NPPC later determined that sufficient unallocated storage capacity was not available on the Snake; of the total reservoir storage of 11.7 MAF on the Snake, only about 2 MAF at Dworshak Dam is federally controlled and thus potentially available. Therefore, the water-budget storage was increased on the Columbia and decreased on the Snake so that the total was 4.65 MAF. The water budget was to be released during the peak smolt outmigration period (April 15-June 14).

Box 9-1
Annual Costs of Fish and Wildlife Investments

The largest annual cost for salmon restoration in the Columbia River is in water. An implicit assumption in design of the system of dams was that there would always be water for fish. Water, however, is not available in the right places, in adequate quantities, or at the right times to meet the requirements of migrating salmon. Fish have no rights to Columbia River water. To implement restoration goals, water has to be purchased from those who control it.

A second assumption in design of the Columbia River system was that substitutions such as artificial propagation and artificial ways of getting up and down river could replace natural production. The second largest annual cost is in bypass facilities and the purchase of hatchery salmon to replace natural runs of salmon adversely affected by dams.

The annual cost of fish and wildlife investments is not a precise or simple number. Costs vary according to rainfall in the region, power markets, water storage from the previous year, the allocation of responsibility for fixed costs, and what gets included as a fish and wildlife expense.

Fish and wildlife program costs usually are discussed together. Fish program costs are greater, but depending on rainfall and replacement electric-power markets, the fish program costs can vary annually by $140 million dollars (Ruff and Fazio 1993, Fazio and Ruff 1994). The Northwest Power Planning Council (NPPC) and Bonneville Power Administration (BPA) report different cost categories and numbers (BPA 1994, NPPC 1994). Each has different incentives in reporting costs. For example, BPA separates NPPC fish and wildlife program costs from the Endangered Species Act (ESA) biological opinion. NPPC sees the ESA program as an extension of the council's fish and wildlife plan.

Total FY 1994 costs for fish and wildlife for both the NPPC and ESA programs add operating expenses, flow revenue impacts and power purchases, interagency transfers for associated projects, related fixed expenses, and foregone revenues. Of the total costs, BPA administrators actually control a relatively small portion.

In assigning costs, BPA makes assumptions about future years with respect to flows, power markets, and other factors. BPA planning does not budget to save money from good market and flow years for helping in bad years. Thus, the drought of the the late 1980s and early 1990s has put severe upward pressure on BPA electricity rates.

The biggest salmon-restoration expense comes from the revenue impacts and purchase of power lost because of altered flows for migration. This category has increased by 300% since FY 1991 and is $150-200 million for FY 1994. Low flows due to drought and poor power markets substantially increase this cost. At very

Wood (1993) argued that the water budget has failed for four reasons. First, the base power flows, which the water budget was designed to augment, are not necessarily maintained during low-flow years on the Snake. Second, the management agencies, notably the U.S. Army Corps of Engineers, view the water budget as "a cooperative arrangement," not a mandate, so water-budget volumes are not guaranteed, especially in low-flow years, when they are most needed.

high flows, hydropower costs drop to zero (Ruff and Fazio 1993, Fazio and Ruff 1994). Many factors go into fixing this amount, including the hydropower costs of the water budget, increased flows, drawdown, and spill. Factors seemingly unrelated to salmon or the region get added. For example, for FY 1994, $8 million was added for reduced capacity of the DC Intertie, which was damaged by the 1994 Los Angeles earthquake.

Interagency transfers and fixed expenses related to hatcheries and passage facilities amounted to $105 million for FY 1994. This was up by two-thirds since 1991. Fixed expenses—which include interest, amortization, and depreciation–were $60 million. Another $40 million was in associated projects, such as operations and maintenance for Lower Snake and Bureau of Reclamation hatcheries, along with Corps of Engineers bypass facilities. Most of the capital expenses for these facilities are funded by borrowing. In FY94, $79 million in principal was added to the debt. A portion of these funds pay for modifications to the dams.

The actual direct FY94 BPA expenses for salmon, resident fish, and wildlife measures amounted to $54 million. This was up one-third from FY91. The salmon portion was 72% of the total. It has increased 17% since FY91.

The total of all these costs was on the order of $350 million, which was about 16% of BPA revenues. The total costs were more than twice what they were in FY91. While it is worthwhile knowing the relative costs of actions to protect salmon, NPPC fish and wildlife program and ESA biological opinion are held out as the major factor affecting "Pacific Northwest electric utility ratepayers" (BPA 1994). The impact of the drought and power markets are seldom mentioned in conjunction with the fish and wildlife costs. The costs in lost power to provide water for irrigation, transportation, recreation, and flood control are not calculated. The primary purpose of the Columbia River system is assumed to be mainly hydropower production and related functions. Fish and wildlife requirements are judged from this perspective.

Cost estimates come from models that simulate operation of the system. The costs allocated to the fish and wildlife program depend on how one sets priorities for water use. Should hatcheries, for example, be charged to the fish and wildlife program or to irrigation or to hydropower? Why should salmon be charged for flaws in the design of the hydropower system. If water flows for fish were given the highest priority, then hydropower generation would have to buy water allocated to fish rather than the reverse. How is the amount of power purchased for fish enhancement known? The selection of interest, amortization, and depreciation rates greatly affects cost estimates. What replacement rates should be used to calculate the forgone revenues? These and many other assumptions, and estimated allocations make the costs frequently quoted highly subject to debate and discussion for the values implied in their calculation.

Third, other operating considerations have higher priority, e.g., for secondary power release and in some cases refill of downstream reservoirs. Fourth, BPA mitigates the cost of the water budget by selling the resulting excess spring power, then buying it back in the summer from other sources, thus causing abnormally low summer flows in the months after the period of spring enhancement.

More recently, the interagency Systems Operation Review (1992) considered a base-case flow-augmentation alternative, in which Dworshak Dam would provide at least an additional 0.30 MAF release in the spring and 0.47 MAF release in addition to 0.43 MAF release from the Upper Snake River Dam (note that these are releases, and not storage allocations). In addition, up to 3 MAF of spring flow augmentation would occur on the Columbia; this would be achieved in part through winter flood-control shifts from the Snake River dams to Grand Coulee Dam in some low-flow years.

The effectiveness of flow-augmentation alternatives has not been demonstrated. On the basis of modeling studies that evaluated the effectiveness of several flow-augmentation alternatives for improvement of juvenile survival, the Systems Operation Review (1992) found that "flow augmentation alternatives produce similar results in all three drainages (Snake, Lower Columbia, Upper Columbia), providing negligible survival benefits for inriver migrants." Flow augmentation might be useful if it provides sufficient water to reduce dam-related mortality (e.g., by spill). However, it is unable to reduce the water-particle travel times through the pools in average-flow years by more than a few days—probably biologically insignificant—beyond the levels already achieved by NPPC's 85-kcfs Lower Granite Dam target. It might well be important that the system operating policy treats the NPPC "targets" of 85 kcfs at Lower Granite Dam and 134 kcfs at Priest Rapids Dam as operating constraints, rather than operating targets; i.e., these targets are given precedence over power production. Flow augmentation should be implemented in such a way that targets are met in all years when storage is available, not just in average and above-average years. Merely focusing on average years is insufficient, as was shown in the years after 1986. In some of those years, spring flows were very low; for example, average May discharge at Lower Granite Dam in 1990 was only 68.2 kcfs. Before Snake River sockeye and chinook salmon were listed under the Endangered Species Act (ESA), the Fish Passage Center allocated the water budget on the basis of numbers of fish passing through the system. It concentrated mitigative efforts on hatchery fish, which tend to move in a relatively narrow period. Wild fish from some tributaries move through the Snake River over a much longer period—some from as early as mid-April, others as late as the end of June (Chapman et al. 1991).

Reservoir Drawdown

A different approach to increasing survival has been suggested in response to the ESA listing of the Snake River sockeye and spring-summer and fall chinook and the general perception among fishery managers that the water budget has failed to stop the decline of Snake River salmon runs. Although the water-budget approach can reduce the time of travel through the reservoir system (in particular, the lower Snake and middle Columbia River reservoirs, all of which are run-of-the-river [see Appendix F]), the effect is at best modest—a few days at mean springtime flows. Average travel time through a reservoir is roughly equal to discharge (volume per unit time) divided by the average cross-sectional area. Therefore, drawdown would reduce travel times by lowering reservoir levels (subject to control of dissolved gases; see the discussion of spill) to decrease the cross-sectional area, hence increasing the water-particle velocities through the run-of-the-river reservoirs. As an extreme case, drawdown could be made to the pre-dam or original river channel.

For example, a drawdown of zero corresponds to full operating pool (essentially the present operating condition), and, for a particular dam, 110 ft of drawdown corresponds to the undammed river channel. Harza and Associates (1994), in reviewing a range of drawdown options considered by the interagency Systems Operation Review (BPA et al. 1992), recommended that only three be considered in detail: natural river, deep drawdown to some level below minimum operating pool, and maximum drawdown (to spill crest). Harza found that

> for the maximum drawdown option [to the] natural river [channel], the maximum travel time advantage (that is, the decrease in travel time through the Lower Snake River reservoir system for the drawn down pools relative to travel time with present pool levels) is 7 days at 100 kcfs, 14 days at 50 kcfs, and 29 days at 25 kcfs. The latter condition (25 kcfs flow) applies less than 2% of the time in the spring migration season (April 15 - June 15), although it is not clear whether this figure includes the effect of the water budget.

The System Operation Review found that the natural river options "decrease travel time (from the mouth of the Salmon River to Bonneville Dam) between 7-10 days, depending on the stock." However, effects are limited to Snake River populations since the action is restricted to the Snake River Basin.

Most of the studies, including Harza's (1994), that have considered a drawdown option have concluded that drawdown to any elevation greater than natural river level is not likely to have biologically significant benefits, in comparison with other alternatives for improving juvenile survival (such as flow augmentation and barging). However, reduction of the reservoir only to minimum operating pool would allow hydropower turbines to continue to operate and would not require reconstruction of fish-passage facilities, such as fish ladders.

In addition to the economic costs associated with elimination of hydropower production, navigation, and recreation during the natural-river drawdown periods, a number of complications accompany drawdown. These include at least the following:

- Loss of spawning habitat for nonsalmon in the reservoir pools and tributaries that directly enter the pools.

- Concentration of predators in the relatively small channel volume.
- Necessity to reconstruct passage facilities for use during the drawdown and refill period, which would otherwise be unusable at reservoir levels below minimum operating pool.
- Increased mortality due to operation of turbines at less than maximum efficiency during the drawdown and refill periods.
- Loss of rearing habitat for subyearling fall chinook.

Direct economic costs are also associated with reconstruction of fish-passage facilities, irrigation withdrawal intakes, and channel stabilization.

The Snake River Salmon Recovery Team (1993) found, on the basis of model predictions, that the natural-river drawdown option would produce the highest inriver survival of yearling migrants in the Snake River basin. This option also would have the potential to increase survival over that predicted for transportation. However, the team added that "the [other] drawdown alternatives are highly uncertain, and even the most optimistic juvenile passage assumptions associated with a four pool drawdown fail to improve survival values of Snake River stock beyond what is achievable with juvenile transportation."

Models that incorporate all life stages of salmon and their responses to various sources of mortality have the potential to be helpful in comparing alternative management and environmental scenarios and as guides to research. The committee has not evaluated any of the models that have been developed on these lines (examples include models produced by the Columbia River Salmon Passage (CRiSP) project (Center for Quantitative Science, University of Washington, undated), but it encourages their development and especially the collection of reliable data for them.

Dam Removal

Although dams are seemingly permanent (albeit recent) features of the Northwest riverine environment, like all artificial structures, they have a finite engineering and economic life expectancy. Structural criteria for dam safety have changed greatly in the 85 or so years since construction of the first high dams in the region, notably to include passage of extreme floods and resistance to earthquakes. Some smaller dams have already undergone significant modifications for these reasons. In addition, dams trap sediment, which significantly reduces their active storage capacity and economic value. Although sedimentation in most Columbia River reservoirs is minor compared to dams on rivers elsewhere, which carry higher sediment loads under natural conditions, the economic life of all reservoirs is ultimately affected by sedimentation. A number of older, low dams elsewhere in the U.S. (notably in the East and Midwest) have been removed because of sedimentation.

Where dams are a significant contributor to the decline of salmon runs, dam removal is an obvious rehabilitation alternative. Like the construction process that created the dams, dam removal would be a major engineering undertaking with major environmental consequences. It would be naive to expect that removal of a dam would allow a stream quickly to revert to its natural state. After removing the structure, in most cases, there would be major and long-lasting downstream effects due to movement of sediment stored behind the structure, and engineered rechannelization of the former reservoir bed would almost certainly be necessary.

The Elwha River Proposal

The Elwha River dams provide a useful case study that gives some idea of the magnitude of the dam removal problem. The Elwha River drains 831 km^2 of the Olympic Mountains, WA, and discharges to the Strait of Juan de Fuca. The Elwha Dam, about 8 km from the mouth of the river, was constructed from 1910 to 1913, and the Glines Canyon Dam, about 21 km from the mouth of the river, was constructed from 1925 to 1927. The entire firm energy production[3] of about 19 MW is presently used at a pulp mill in nearby Port Angeles. Glines Canyon Dam lies in the Olympic National Park, which was created in 1937, as does 83% of the Elwha River drainage. Neither dam has fish-passage facilities, and the lower dam (Elwha) has never had a federal license to operate.

In 1992, Congress authorized the secretary of the interior to acquire the dams and remove them if he determined that their removal was necessary to "the full restoration of the Elwha River ecosystem and native anadromous fisheries" (PL 102-495, Elwha River Ecosystem and Fisheries Restoration Act). A report pursuant to the act (USDI 1994) found that "The removal of the Elwha and Glines Canyon Dams is the only alternative that would result . . . [in meeting the goals of the act]". The report also conducted a preliminary engineering analysis of dam removal alternatives, reviewed briefly here.

The major issues in the Elwha River project are removal of the structures, management of sediment and erosion during and following dam removal, and re-establishing and protecting a channel within the presently inundated area. Although various combinations of dam removal (e.g., only Glines Canyon) and provision of fish passage facilities (at Elwha Dam) were considered, only the option of removal of both dams is reviewed here, because the report concluded that it is the only option that would meet the goals of the Elwha Act.

Removal of the dams is complicated by the necessity to provide a stream channel during the dam removal process, which is estimated to require about 18 months. The channel and temporary discharge structure must be sufficient to

[3]Firm energy production is the amount of energy that could be produced during the worst-case historical conditions (known as the critical period), with all other demands on the system (e.g., irrigation) fixed.

pass floods safely during this period. The options considered for diversion of the river were a) construction of a diversion tunnel, b) construction of a surface diversion channel, c) construction of a low-level diversion through the dam structure; and d) progressive construction of notches through the dam along with top-down removal of the structure. The report did not conclude which option was preferable (although comparative costs were estimated), since the removal options affect sediment management and channel reconstruction as well. Further, site-specific constraints preclude application of some of the options at both sites. For instance, the Elwha Dam was back-filled with large rock and gravel as a result of a structural failure of the dam foundation during the initial filling. Therefore, options c and d are not feasible at this dam. Construction of a surface diversion channel at Glines Canyon Dam is considered infeasible due to construction staging problems and high cost associated with the need to excavate much of the channel in bedrock.

The lowest-cost alternative is a combination of low-level diversion through the dam, or progressive notching of the dam and top-down removal at Glines Canyon, and construction of a surface diversion channel at Elwha Dam. If top-down removal were used at Glines Canyon, it would be accomplished by construction of multiple notches 15.2 m (50 ft) deep by 5.2 m (17 ft) wide and temporary gates, in nine increments of 3.8 m (12.5 ft) each. Dam material above the notch would be removed by barge. At the last step, the dam base at the bottom of the gorge would be removed by cableway or boom crane during summer low-flow conditions. The low-level diversion alternative would make use of a small outlet valve that was installed at the base of the dam during construction. This outlet would be enlarged, and the reservoir level lowered to its level to allow "dry" removal of most of the dam. Removal of the remaining portion of the dam structure would be the same as with the top-down approach. At Elwha Dam, the surface channel alternative would isolate the north spillway structure, which would be removed "dry," then the stream channel would be diverted in a second stage to a channel excavated through the existing north spillway site, and the main concrete dam section as well as power plants and south spillway would be removed.

Perhaps the most difficult aspect of the project would be sediment removal and management. An estimated 8.6 million m^3 of sediment has been deposited behind Glines Canyon Dam in the 70 years since the reservoir was filled, and 2-3 million m^3 has been deposited behind the Elwha Dam (mostly before completion of the upstream dam). A delta has been formed at the head of Lake Mills (formed by Glines Canyon Dam), which is about 21 m deep. The delta consists mostly of coarse materials (small sand and larger); fine sediment is more generally distributed throughout the reservoirs. Three alternatives, and combinations thereof, have been evaluated for sediment management: removal, erosion, and retention. Removal would be accomplished either after draining of the reservoirs by trucking to an upland site or to disposal in salt water, or (before draining of the reservoirs) by pipe transport of dredging spoils as a slurry to salt water or to an upland site.

Both of the removal approaches are comparatively costly, require a long time for completion (as much as 9 years), and would have significant ancillary environmental effects (e.g., to the disposal site). They have the advantage that the final topography of the presently inundated areas could be made to resemble predam conditions closely.

The erosion alternative would allow the river to transport the accumulated material downstream. Resuspension of the materials would be augmented by dredging or other mechanical means. Fine and smaller coarse material would be transported by the river to salt water, and larger materials would likely be redeposited downstream. Although this alternative has the advantage of low cost, it would result in high levels of suspended sediment downstream, and almost certainly would extensively damage downstream fish habitat. Further, it was estimated that approximately 20 years would be required to remove all of the material from Lake Mills. An alternative of dredging the Lake Mills delta material and disposing of the dredged material over Glines Canyon Dam before removal was also investigated. This alternative would effectively store the coarse materials in Lake Aldwell (formed by Elwha Dam), and allow the river to transport the fine sediments downstream. This alternative would reduce the period necessary for stabilization of the channel system with respect to sediment movement and would eliminate the need for an upland disposal site. Because there is insufficient space in Lake Aldwell for disposal of all of the Lake Mills delta material, the Lake Mills bed would not be completely restored to natural topography.

The retention approach would relocate materials deposited in the old (predam) river channel elsewhere within the lake beds, but would otherwise leave the accumulated sediment in its present location. Two retention alternatives were considered. The first would use hydraulic dredging before dam removal to restore the original channel through the Lake Mills delta. The channel would be dredged to the original river-bed elevation. The second alternative would remove less material by dredging, and would allow the river to erode the delta to form a new channel after removal of the dams. However, unlike the erosion alternatives, no attempt would be made to remove sediment away from the original channel. The hydraulic dredging alternative is expected to have the least downstream effect, as much of the fine sediment disturbed during channel excavation would settle out in the reservoirs, rather than being transported downstream. Resettling would be enhanced through use of impermeable silt curtains, and by using the coarser dredged material to form containment cells, that would aid in dewatering of fine sediments. Subsequent to dewatering, the sediment would be graded, compacted, and revegetated.

The estimated total costs of dam removal, sediment management, and revegetation (but excluding the cost of the hydropower loss) was estimated to range from about $70 million to $240 million. The lowest cost alternatives were those that would remove sediment only from the former river channel, use either the notch or low level diversion at Glines Canyon Dam, and a surface diversion channel at Elwha Dam.

Most of the cost of the restoration project (about 75% for the lowest-cost alternative, and well over 90% for the higher-cost alternatives) would be associated with sediment management. Although the least-cost alternatives do not restore the topography of the inundated areas to their original contours, and might result in less spawning habitat in these areas, they have the advantage (in addition to cost) of greatly reducing the movement of fine sediment downstream. Furthermore, only a relatively small amount of the spawning habitat that would be reopened as a result of dam removal is in the inundated area.

The dam removal project would itself be a significant engineering project. In addition to the removal of the dam structure, it would involve (depending on the alternative selected) construction of temporary roads, channels, and diversion dams, as well as extensive dredging and channel work. The greatest uncertainty in performance of the different alternatives (and their cost) is undoubtedly associated with sediment management, and this uncertainty is in turn greatest for alternatives that rely on channel transport of sediment (as opposed to removal or retention). Although there has been some experience with rechannelization via dredging in a manner similar to one of the sediment retention alternatives by British Columbia Hydro in the case of a reconstructed dam, there is essentially no comparable experience with any of the other approaches. In addition, there remains, after project completion, the potential problem of sediment movement due to slope failures induced by channel migration, which might well require permanent channel structures, such as levees, at least for the sediment-retention alternatives.

Applicability of Experience

How applicable is the Elwha River experience to potential removal of dams in the Columbia River system? At this point, the strongest candidates for dam removal in the Columbia system appear to be the run-of-the-river dams in the middle Columbia and lower Snake, which are the targets of current flow-augmentation studies. The most obvious differences between the Elwha River and middle Columbia and Snake dams are the size of the structures and the climate. Annual precipitation in the Elwha River Basin is about 1,700 mm/yr; in the middle Columbia and lower Snake it is in the range 200-400 mm. The humid climate in the Elwha River Basin would aid revegetation efforts; it is quite likely that a native forest could ultimately be restored in the presently inundated areas (albeit with soil enhancement), whether or not the original contours were retained. In the arid middle Columbia and lower Snake region, revegetation (and hence erosion control) in the presently inundated areas would pose a much greater challenge due to the probable necessity to irrigate, at least in the short term. In addition, the channel slopes and river gradients in the case of the much larger middle Columbia and lower Snake River dams are lower than the Elwha, which is a steep mountain stream, especially at Glines Canyon Dam. Therefore, the inundated area and length of inundated channel is much larger for the middle Columbia and lower Snake River dams relative to the Elwha dams, with expected restoration costs for the channel and inundated areas proportionately larger, as well. On the other hand, the amount of sediment deposited behind the middle Columbia and lower Snake dams is likely to be less, because these dams were constructed after upstream dams had already trapped much of the sediment that would otherwise have been deposited. Nonetheless, the costs of removing a middle Columbia or lower Snake River dam would be much larger than those estimated for the Elwha River dams.

Selection of Mitigation Alternatives

Many entities—including the Fish Passage Center, CBFWA, the Idaho Department of Fish and Game, the Oregon Department of Fish and Wildlife, and the U.S. Fish and Wildlife Service—have recommended spill of water at Snake River dams, elimination of transportation, flow increases, and drawdown of the Snake River reservoirs to increase water velocities. Opposing that position have been the Corps of Engineers, NMFS, utilities, and others.

The Snake River Salmon Recovery Team (1993) extensively modeled mitigation alternatives and their effect on the proportion of Snake River smolts that would arrive at a point downstream from Bonneville Dam. The team concluded that no combination of spill, flow augmentation, and drawdown within the limits imposed by present dam structures would produce survivals close to those obtained by transporting smolts from collector dams to the Bonneville tailrace. The team modeled a transport-benefit ratio (TBR) of 2:1. That ratio was based on an average of TBRs obtained in transport tests in 1986 (1.6:1) and 1989 (2.5:1). The study years represented modern river management and community structure. The team found that for all flow regimes, including high discharges (151 kcfs in the Snake and 401 kcfs in the Columbia), a mitigation program using transportation as the main tool was most effective. Only drawdown of the Snake River reservoirs to river grade, which would take years to design and build, potentially offers higher survival than transportation.

No investigator to date has provided the Columbia River region with experimental results that demonstrate higher survival of inriver migrants than transported migrants at any discharge level. Until such experimental data become available, transportation should continue to be used. However, it is essential that managers use an adaptive (experimental) management approach and avoid taking any action that jeopardizes all of the fish in a stream. For example, if some fish in a stream are transported downstream, the action should be designed so its effectiveness can be assessed and compared with other alternatives, such as spill.

10

Fishing

Fishing has long been considered to be a major cause of the declines in salmon abundance of the late nineteenth century and continues to be increasingly restricted. But fishing is only one cause of salmon mortality. Mortality caused by other human economic developments or activities (dams, habitat loss or degradation, pollution, and water diversions) and by natural factors (predators, disease, and environmental variability) together usually exceed fishing mortality and thus influence the rate of fishing that a salmon population can sustain. Commercial, recreational, and treaty fishing, however, generally occur late in the salmon life cycle and can significantly reduce the number of spawning adults and thus have a more direct effect than other sources of mortality on the population size remaining to reproduce. For example, following a major environmental change or catastrophic event, or human-caused mortalities in other stages of the life cycle, reduction of fishing mortality might be the only way to protect the surviving salmon and allow enough of them to spawn to sustain the population. For this reason, and because fishing is often easier to control than other causes of mortality, control of fishing is often used to maintain spawning population sizes.

SALMON FISHERIES IN THE PACIFIC NORTHWEST

Early History

In the 1830s, commercial salmon fishing by non-Indians began on the Columbia River. Not many salmon were taken, because a method of storage for sale had not been perfected. Canning technology arrived at the Sacramento River in the 1860s, but salmon runs there were already in poor condition because of overfishing, hydraulic mining, and habitat destruction. Salmon canning spread to the Columbia River in the middle 1860s and to Puget Sound in the 1870s (Goode and others, 1884-1887, Cobb 1930). Market demand was high, and the Columbia River fleet grew from two gill-net boats in 1866 to more than 1,500 early in the 1880s. By that time, Alaska salmon fisheries also were developing, and their overwhelming influence on the supply of West Coast salmon had begun.

Catches in the Columbia River reached peaks in the 1880s and again in World War I. The number of gill-net boats reached a peak of 2,800 by the middle 1910s (C. L. Smith 1979:108). As exploitation rates grew, catches changed from spring-caught chinook to summer and fall chinook, coho, sockeye, and steelhead, all of which were fished to provide a year-round supply for canning (Cobb 1930, Craig and Hacker 1940). Summer chinook were preferred and by the end of World War I, the summer run was overfished. By 1915, ocean trolling had begun in an attempt to get around the closed season that was introduced to allow salmon escapement from river fishing. The gasoline engine and refrigeration appeared in the early 1900s and made ocean fishing possible.

After World War I, a surplus of Alaska-caught salmon and the Great Depression reduced the salmon fishery, and it stayed reduced until the middle 1940s. After World War II, fishing increased by extending out into the ocean. Recreational angling became important in rivers and the ocean. Along the coast, many began commercial fishing as salmon trollers; later, some changed to trawling for other species. Until the 1970s, ocean trolling took larger and larger shares of the chinook and coho salmon caught.

The expansion of ocean fisheries placed the burden of responsibility for conservation on fishers closer to the spawning grounds, including the American Indians. Regulations increased on these more proximal users, who then argued that fishery-management plans did not treat them fairly. Arguments by Indian fishers culminated in 1974 with Judge Boldt's interpretation of the 1855 Treaty of Medicine Creek. Judge Boldt interpreted the statement that "the right of taking fish . . . is further secured to said Indians, in common with all citizens of the territory" to mean a 50/50 allocation between Indians and non-Indians of the catchable fish passing through the tribes' fishing grounds. His decision expanded the established Indian exemption from state regulation of fishing within the reservation to all usual and accustomed fishing areas of the treaty tribes (Bruun 1982).

Extended jurisdiction under the Magnuson Fishery and Conservation Act of 1976 exerted additional limitations on ocean fisheries. With declining salmon abundance, the ex-vessel value of non-Indian troll salmon fishing fluctuated by a factor of ten, as shown in Figure 10-1 (from PFMC 1995:IV-14). However, these coastal ocean fisheries account for only a small portion of the total catch of salmon.

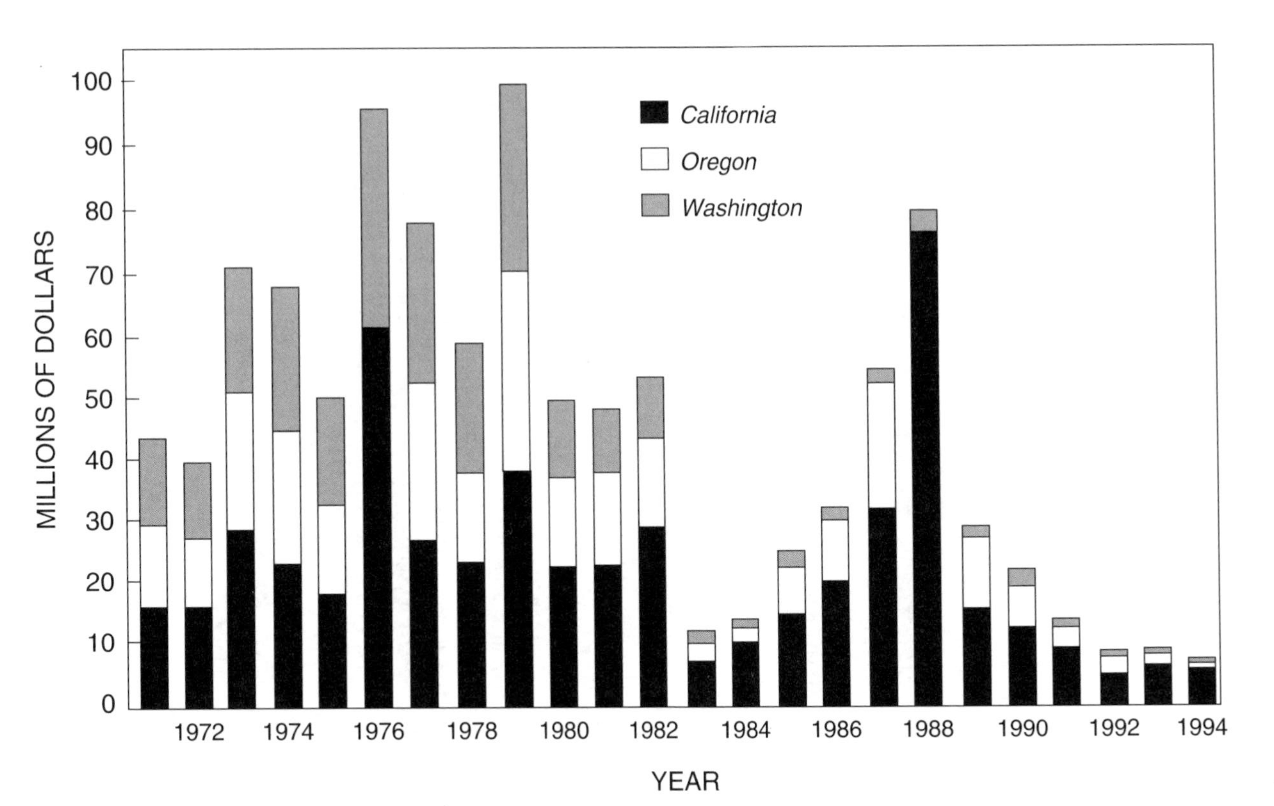

FIGURE 10-1 Exvessel value of nontreaty troll chinook and coho landings (1993 dollars). Periods were chosen to show conditions 15 years ago, in a strong recent year (1988) and in 1994, which was worse than all prior years. Exvessel values are derived from catch of chinook, coho, and pink salmon and reflect annual variation in salmon abundance, allocation between areas and gears, and market prices.

The 1990s

The major groups that fish are treaty Indians; nontreaty commercial fishers; and charter, guided, and general recreational anglers. Treaty fisheries use nets and trolling for ceremonial, subsistence, and commercial purposes. Nontreaty commercial fishers use ocean trolling and nets in the lower Columbia River and Puget Sound. Most recreational charters fish the ocean using hook and line. Anglers fish from jetties, river banks, and private boats in the ocean, estuaries, and rivers.

The 1990s have been poor for all fishing groups in the areas managed by the Pacific Fisheries Management Council (PFMC). Only 53% of ocean trollers registered in the 1990s reported landings, and the average landings by weight in the 1990s were only 43% of those in the 1980s. For the Columbia River gill-net fishery, 1993 was at 57% of the 1979 high of 1,524 licenses. Indian fishers in the 1990s caught only half as much as those in the early 1940s, even though their proportional allocation increased from 5% to 50%. Ocean recreational fishers' effort and catch in the 1990s averaged about 80% of those in the 1980s, although the chinook or coho caught per angler trip remained about the same. For Puget Sound, recreational catches of chinook and coho declined. Both treaty and nontreaty commercial fishers' catches in Puget Sound did not decline as much as those of commercial ocean fishers, probably because catches in the Strait of Juan de Fuca and Puget Sound take five species and rely heavily on sockeye and pink returns to the Fraser River (Table 10-1).

TABLE 10-1 Commercial Net and Troll Salmon Catches in Strait of Juan de Fuca and Puget Sound (Washington Statistical Areas 5 through 13)

	Catch (thousands of fish)				
Year	Chinook	Coho	Pink[a]	Sockeye	Chum
1971-1975	157.9	748.3	2,057.3	409.3	2,196.8
1976-1980	229.6	901.9	3,090.5	702.9	1,373.4
1981-1985	215.5	950.7	3,294.9	756.7	1,837.2
1986	223.9	1,357.1	0.1	1,155.8	2,751.3
1987	213.1	1,782.2	2,069.8	1,301.6	1,946.1
1988	231.4	1,237.5	0.1	1568.7	849.4
1989	253.9	965.4	3,427.7	886.5	2,251.7
1990	250.2	1,067.7	0.3	1,100.5	2,166.7
1991	143.3	598.7	3288.5	1,021.2	1,828.1
1992	136.1	400.0	0.4	1,380.7	616.5
1993	81.01	83.7	2,097.9	1,131.1	2,692.6

[a]Pink salmon runs are strong only in odd years.

Source: Data from PFMC 1994, Table B-37.

Catches and returns to the numerous terminal fisheries (i.e., near the terminus of the salmons' upstream migration to the spawning grounds) and inriver fisheries are highly variable, but in general they are also lower than 10-15 years ago. Coho catches and returns, in particular, are characterized by high annual variability and are now in low abundance. Where terminal catches have been maintained, it has often been because of curtailment of more distant fishing or because of hatchery production.

Allocations between terminal and ocean fisheries within the Pacific Northwest states are complicated by international allocation issues. Canada and the United States have competed for Fraser River sockeye almost from the beginning of commercial fishing. The debate broadened throughout the North Pacific Ocean with the development of ocean troll fisheries and high-seas fishing. The major international issue remaining in the 1990s concerns fishing agreements between Canada and the United States. Fishery and management history is considered below—first for high-seas fisheries involving interactions with Japan and the Soviet Union and second for the Canadian and United States interactions. The concepts underlying salmon management are discussed in detail in Chapter 11.

Fishery-Management Institutions

Since 1977, PFMC has coordinated fishery management in the Pacific Northwest states. PFMC is one of eight regional fishery-management councils created by the Magnuson Fishery Conservation and Management Act of 1976. The regional management councils are innovative institutions made up of governor-appointed citizens with knowledge of fisheries or fish conservation and members of relevant state and federal agencies. PFMC jurisdiction covers ocean waters 3-200 miles from shore and anadromous species throughout the range of their migration. The council recommends rules for salmon management to the secretary of commerce, who has final authority.

PFMC coordinates federal and state actions with respect to salmon management for California, Idaho, Oregon, and Washington. Technical advice comes from a Salmon Technical Team with representatives of the American Indian tribes; Washington, California, and Oregon fishery agencies; the National Marine Fisheries Service; and the U.S. Fish and Wildlife Service. A Klamath River Technical Advisory Team provides guidance on the management of Klamath populations. An industry advisory body, the Salmon Advisory Subpanel, includes representatives of most groups concerned with salmon management—consumers, processors, anglers, charter operators, gillnetters, trollers, and treaty tribes.

International aspects are addressed through the Pacific Salmon Commission (PSC), a bilaterally funded commission created under the Pacific Salmon Treaty. The commission is composed of three advisory panels with equal numbers of Canadian and U.S. members from the fisheries industry, American Indians, and government appointees. There are also several scientific advisory committees and a secretariat. Panels are aligned geographically and functionally. The northern panel addresses fisheries and populations in northern and central British Columbia and southeastern Alaska, including the transboundary rivers in southeastern Alaska. The southern panel addresses issues in southern British Columbia and the Pacific Northwest states, excluding Fraser River sockeye and pinks. The Fraser panel evolved from the former International Pacific Salmon Fisheries Commission and manages Canadian and U.S. fisheries for sockeye and pink salmon in the Strait of Juan de Fuca, outer Puget Sound, and Canadian waters off the Fraser River. The Fraser panel is the only PSC body with any within-season management capacity, and commission staff collect catch, escapement, and biological data.

Puget Sound salmon management is the most complex domestically and internationally. In domestic salmon management, Puget Sound operates under a cooperative arrangement that has evolved since the Boldt decision (see Clark 1985). The Northwest Indian Fisheries Commission and Washington state jointly manage catch in Puget Sound in accord with international agreements reached in PSC.

Commercial fishing in the Columbia River is divided between Oregon and Washington. The Columbia River Compact was created in 1918 to manage Columbia River fisheries. The Washington Department of Fisheries and Wildlife and the Oregon Department of Fish and Wildlife share responsibilities. The compact meets before all major seasons and sets regulations. The *U.S. v Oregon* (Belloni) decisions of 1969 and 1975 require that catch be allocated between treaty and nontreaty fishers. Regulations must meet the requirements of the Columbia River Fish Management Plan, approved by the U.S. District Court in 1988. Idaho too has an interest in how river catches are managed; Idaho has felt that its interests have not been considered in compact decisions.

The Northwest Power Act (1980) requires that the dam-induced losses of salmon in the Columbia Basin be mitigated. The Northwest Power Planning Council (NPPC) determined that Bonneville Power Administration ratepayers are responsible for 8-11 million fish per year above the 2.5 million that were being produced in the late 1970s (Lee 1993a). The act provides monetary resources from the sale of electric power for improving salmon populations. NPPC has set overall goals for salmon restoration in the Columbia Basin (*Strategy for Salmon*, NPPC 1992a). All the region's ratepayers, through NPPC, have an interest in the status of salmon.

Anglers and nontreaty tribes in eastern Oregon, eastern Washington, and Idaho are the farthest upstream consumers of Columbia River salmon spawning in tributaries of the Snake and middle Columbia rivers. They compete most directly against escapement goals for their catch. Because escapement goals have seldom been met, fisheries for these upriver anglers and nontreaty tribes have been affected most severely. Chinook angling in Idaho has been eliminated since the early 1970s, and wild steelhead cannot be kept. Yet, Idaho has had no voice in the Columbia River Compact. Similar problems occur among states in the management of coastal fisheries where individual states' different rules for opening dates, size limits, and gear restrictions have resulted in intraregional conflicts. PFMC has worked to coordinate state rules. The pattern has been for rules to become more time-, location-, and population-specific with the effect of reducing fishing times, locations, and gear.

With the listing of several salmon populations as endangered in the Snake River, the National Marine Fisheries Service takes on an even more important role in salmon management because the endangered-species authority rests with it. Endangered-species listing requires that state and regional planning be coordinated with efforts to restore endangered salmon populations—another layer of review for all salmon-fishing rules established by PFMC.

Fishery-Management Data

Data needed for salmon management are complex, voluminous, and expensive because of the large number of salmon populations, fisheries, communities, and management agencies, as well as the complex, wide-ranging life histories of salmon. Often, several agencies must cooperate to collect data for assessment of a single population. Basic catch, escapement, and economic data are reported annually by several organizations, such as PSC, PFMC, the Pacific States Marine Fisheries Commission, and state and treaty fishery agencies. The data are used in numerous analyses and reports but often are considered inadequate for accurate determination of the biological basis of salmon population dynamics and the social and economic aspects of fishery regulation. Agencies have difficulty in funding the elaborate management process. They lack resources to evaluate appropriate sampling strategies to get accurate catch data, to determine the relative weight of different incentives for fishing, and to estimate costs of and earnings from fishing. As discussed later, the data limitations have had a serious influence on our ability to assess and manage salmon appropriately.

One of the difficult data problems is to quantify nonreported fish deaths associated with salmon and other fisheries. For salmon, these deaths are associated with the release of undersized fish in hook-and-line and some seine fisheries, dropout from gill nets, and incidental catch of juveniles in seine fisheries. Incidental deaths of chinook have been estimated at 30-50% of the reported catch during the middle 1980s (PSC 1987), but estimates for other salmon have not been developed. Trawl fisheries are another source of deaths. The vast majority of salmon caught in trawl fisheries are chinook, followed by chum in some areas and some years. The largest trawl-associated mortality occurs in the Bering Sea but does not influence Pacific Northwest salmon. However, trawl fisheries in the Gulf of Alaska and off Washington, Oregon, and California might influence Pacific Northwest salmon (Table 10-2). The catches are variable, but could be

TABLE 10-2 Estimated Catches of Salmon in Trawl Fisheries

Year	No. Fish Caught in Trawl Fisheries	
	Gulf of Alaska	Washington, Oregon, and California
1977	5,222	14,627
1978	45,603[a]	5,924
1979	21,460[a]	8,666
1980	36,069	8,433
1981	30,860	11,474
1982	6,967[a]	11,798
1983	13,874	5,143
1984	75,846[a]	10,255
1986	20,820	43,790
1987	1,221[a]	13,285
1988	147[a]	16,168
1989	0[a]	9,199
1990	21,085	617
1991	50,873	6,358
1992	26,090	5,099
1993[b]	74,853	8,373

[a]Did not include U.S. domestic trawl catches.
[b]Preliminary estimates.

Source: Data from Low and Berger 1994 (Table 4).

large in some years and are comparable with the catch in past Japanese high-seas fisheries. The incidental deaths today are associated with U.S. trawl fisheries but previously were associated with foreign or joint-venture fisheries.

Human fishers also compete with nonhuman consumers. Sea lions and seals consume many salmon. Marine-mammal populations have been growing rapidly (Beach et al. 1985, Olesiuk et al. 1990) since implementation of the Marine Mammal Protection Act. Sea lions and seals, abundant in British Columbia and the Pacific Northwest, take salmon as far upstream as Bonneville Dam and Willamette Falls in the Columbia basin and at other places of fish concentration, such as Ballard Locks in Seattle (see Box 2-1). Palmisano et al. (1993) estimated that incidental loss due to marine-mammal predation can be as much as 16% of the human salmon catch. Park (1993) claimed that marine mammals including killer whales and birds take more salmon than are caught by humans. Those deaths usually are considered natural deaths but the rate of increase in these predators can be a concern, especially when salmon are at low abundance, or the predators concentrate on depressed salmon populations near and at their spawning grounds.

INTERNATIONAL INFLUENCES

High-Seas Fishing

Japan began exploring offshore salmon fishing in the early 1930s as its access to salmon along the Kamchatka Peninsula became more restricted by the Soviet Union. By 1936, the Japanese fleet in the Bering Sea caught the attention of U.S. fishers in Bristol Bay. Japan agreed to discontinue the expansion in 1938, and World War II soon curtailed such activities. After the war, the Japanese fleet again began to move seaward. Concerns in Canada and the United States led to the 1952 International Convention for the High Seas Fisheries of the North Pacific Ocean. The convention called for Japan to abstain from fishing for salmon east of longitude 175°W to minimize its catch of North American salmon. The abstention line was subject to review by the International North Pacific Fisheries Commission to determine whether another line separated Asian and North American salmon better or divided the catch more equitably. Canada and the United States were to provide evidence that their populations were used fully. All three countries were to cooperate in scientific investigations of the distribution and origin of salmon on the high seas.

A 1956 agreement between Japan and the Soviet Union further restricted Japan's high-seas salmon fishery. The resulting Japan-Soviet Northwest Fisheries Commission established total allowable catches and other regulations in areas west of the abstention line through negotiations that were extended and often difficult. The annual quota for the Japanese catches of salmon declined steadily. The commission also imposed time and area closures and effort limitations to reduce the catches of salmon originating in Asia (Fredin et al. 1977).

Japan's high-seas fishery became increasingly contentious as it expanded and as investigations proceeded, because Asian and North American populations were more mixed than had been expected, the fishery caught immature salmon, and the origins of the large catches were unknown. The abstention line, however, was not changed until the nations unilaterally extended their fishery jurisdictions to 200 miles. The United States declared that its exclusive fishery-management zone extended throughout the migratory range of anadromous species (Magnuson Fishery Conservation and Management Act of 1976)—an assertion that was accepted by Japan and other nations seeking to fish within the U.S. 200-mile fishery zone (Burke 1991). The locations of the U.S. Aleutian Islands extended the new U.S. zone considerably west of the abstention line and north and south of the islands into areas in which Japan had fished traditionally. In 1977, the Soviet Union also declared a 200-mile zone and terminated Japanese salmon fishing there. In 1978, Japan negotiated new agreements with the Soviet Union and with Canada and the United States. On February 15, 1979, a new protocol with Canada and the United States amended the original 1952 convention. The protocol continued scientific studies and cooperation, closed fishing southeast of latitude 56°N and longitude 175°E, limited mothership fleet-days northeast of latitude 56°N, and limited fishing within the U.S. 200-mile fishing zone. A detailed history of the agreements between Japan, Canada, and the United States is in Jackson and Royce (1986). Japan's high-seas salmon fishing was restricted again both in 1985 through a new Japan-Soviet agreement and in 1986 through amendments to the Japan, Canada, and U.S. protocol of 1979. Quota reductions continued through the Japan-Soviet agreements, as did increases in fees. The amended protocol with the United States and Canada phased out the mothership fleet in the central Bering Sea, moved the eastern boundary of the land-based fishery another degree west, and intensified studies in land-based fishing south of latitude 46°N to determine the continent of origin of salmon caught in this area.

By the late 1980s, Japan's high-seas fishing for salmon had been reduced substantially. The number of vessels in its land-based fleet declined from 374 in the early 1970s to 83 in 1991 (Myers et al. 1993). The mothership fleet was reduced from 16 motherships and 460 catcher boats in the 1950s to one mothership with 43 catcher boats in 1988.

A side effect of the many agreements was the displacement of Japanese fishing vessels. The vessels became the first to use drift gill nets to catch neon flying squid (*Ommastrephes bartrami*) in the central North Pacific (Ignell 1991). By the middle 1980s, the landed catch was averaging 258,000 metric tons/year annually and involved over 700 vessels from Japan, Taiwan, and the Republic of Korea. Estimates of the total salmon deaths, including all species and origins, in the Japanese squid drift-net fishery have been 1,614 in 1989 and 141,279 in 1990 (Anon. 1991), 103,895 in 1990 (Yatsu et al. 1993), and 231,000 in 1990 (Pella et al. 1993). The estimate of total salmon deaths in the 1990 Korean fishery was only 4,036 (Pella et al. 1993), and estimates of the Taiwanese catch were not developed, because only two salmon had been observed. Through the late 1980s, world opposition to the use of high-seas drift nets increased. Concerns over the incidental catch of salmon, marine mammals, seabirds, and other fish in squid drift-nets increased public pressure on Japan, Korea, and Taiwan to stop their use.

Forty years of effort to minimize the interception of North American salmon on the high seas culminated in 1992 when Japan notified the Japan-Russia Fisheries Joint Commission of its decision to stop high-seas salmon fishing in 1992 and the United Nations adopted resolution A/RES/46/215 (February 10, 1992) ensuring a "global moratorium on all large-scale pelagic drift-net fishing . . . , by 31 December 1992."

Will the absence of this fishing by Japan aid in the recovery of salmon in the Pacific Northwest states? The answer is probably no, because these fisheries previously caught few salmon originating in the Pacific Northwest. Information regarding this conclusion is summarized below for various species and fisheries.

Japan's salmon fishery never intercepted large numbers of sockeye, pink, and chum originating in the Pacific Northwest, and as early as 1977 the fishing boundaries had eliminated the likelihood of any major interception (Neave et al.

1976, Fredin et al. 1977, Takagi et al. 1981, Harris 1987, 1988, Myers et al. 1993). Adult tagging and scale-pattern analyses indicated that coho from southeastern Alaska and British Columbia were absent or rare south of latitude 50°N between longitude 160°E and 175°W (Harris 1988). However, in 1991, two coho tagged in Washington and Oregon were caught far to the west, indicating that intermingling with Asian and Alaskan populations occurs at least occasionally (Myers et al. 1993). The distribution of chinook in Japan's fishing areas remains uncertain because few tags were recovered from high-seas tagging programs, but high-seas tagging and biological markers have indicated a potential for mixing between Asian and North American chinook in the central North Pacific (Myers et al. 1993). However, interception of large numbers of Pacific Northwest chinook is unlikely, given both the predominance of Kamchatkan and Alaskan chinook in catches even from land-based fishing and the relatively small catches of chinook on the high seas. In addition, catches of coho and chinook by Japan's mothership fishery northwest of latitude 46°N and longitude 175°W and south of latitude 52°N (Figure 10-2) were not consistently large and could not account for declines in populations after the late 1970s.

The effect of distant fisheries on steelhead could be more important because their life history differs from that of the other anadromous salmon. Coded-wire tags have identified steelhead originating in the Pacific Northwest as far west as longitude 163°E and as far south as latitude 41°N (Burgner et al. 1992). Data on

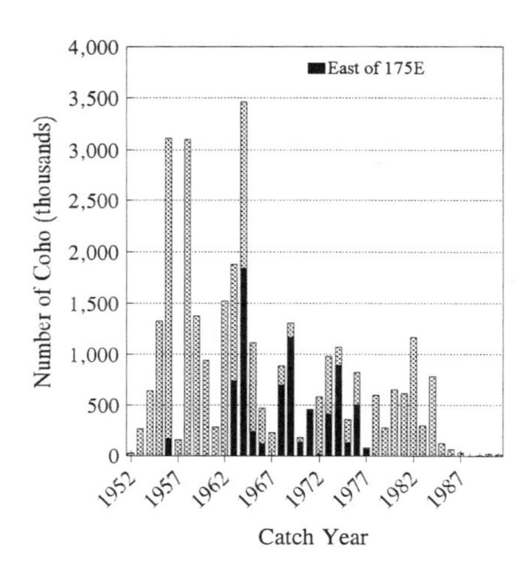

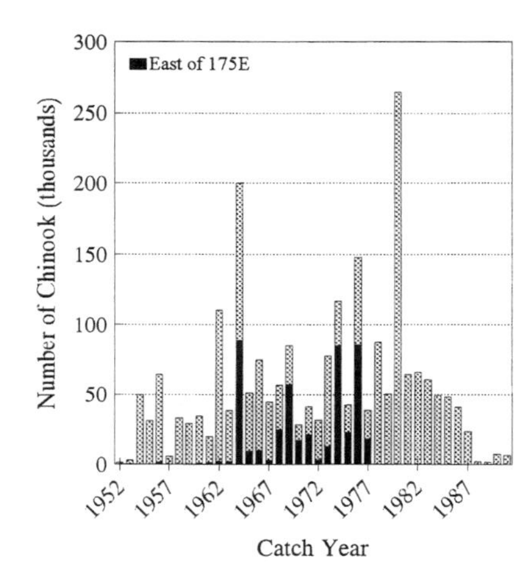

FIGURE 10-2 Japanese mothership fishery catch of coho (left) and chinook (right) in the North Pacific (southwest of 52°N and 175°W), 1952-1991. Total height of bars represents total catch in this region, and black bars indicate catch taken between 175°E and 175°W.

steelhead in Japan's mothership and land-based fisheries before the 1980s were not available but later data indicate that catches were much smaller in the mothership fishery than in the land-based fishery (Burgner et al. 1992). In the land-based fishery, the steelhead catch declined from 23,000 in 1981 to 1,000 in 1991; populations from British Columbia, Washington, Idaho, and Oregon would dominate this catch (Myers et al. 1993). Thus, the land-based fishery, particularly before the 1979 protocol, could have intercepted substantial numbers of steelhead from the Pacific Northwest (Burgner et al. 1992).

Regarding the effect of Japan's high-seas squid fishery on salmon, the Discussion Group on Salmon and Squids (Anon. 1991) concluded that the incidental catch of salmon had no known important biological effect overall—but could if concentrated on specific populations. The latter seems unlikely in the central Pacific, but about half the 1990 incidental salmon deaths occurred in July between longitude 170°E and 172°E at latitude 41°N, and most of them were of coho and chum (Pella et al. 1993). However, earlier information indicates that the fish killed would be predominantly of Asian origin and have virtually no effect on fish originating in the Pacific Northwest. These comments pertain to legal squid drift nets, but a substantial illegal catch of salmon also occurred on the high seas (Pella et al. 1993). The opportunity existed because salmon occur near the northern boundary of the squid fishery. Pella et al. (1993) suggested that at least 10,000 metric tons of salmon (5.5 million fish) was killed illegally in a recent year. With the absence of a drift-net fishery on the high seas, illegal harvest should be reduced greatly, but the issue merits monitoring.

Canadian and U.S. Fishery Interactions

Canada and the United States have a long history of competitive fishing and of fishing agreements. Their relations, although not having the notoriety of the high-seas issues above, have influenced the use and conservation of Pacific Northwest salmon. Shepard (1967), Argue et al. (1987), Roos (1991), and Parsons (1993) provided historical overviews of major events and agreements. Brief discussions of the major fisheries are presented here as background to the discussion of the 1985 Pacific Salmon Treaty between Canada and the United States concerning Pacific salmon.

Fraser River Sockeye

Starting in 1866, Fraser River sockeye provided the base of Canada's developing salmon-canning industry. But by about the turn of the century, catches of Fraser River sockeye by non-Indian trap fishers in the U.S. San Juan Islands rivaled the Canadian catches. This greatly disturbed Canadian fishermen. As early as 1902, reports of the Commissioner for Fisheries of British Columbia reflected the concern (Babcock 1902):

> Unfortunately, there is a divided jurisdiction on the fishing grounds of the Fraser River. The American fishing grounds on Puget Sound must be considered a part of the Fraser River district, as the sockeye captured there were bred in and are endeavoring to return to that river. This divided authority prevents, at least for the present time, the making of suitable protective laws which justly affect the fishery interests on both sides of the lines.

The United States took about 60% of the Fraser sockeye catch until 1935 (Figure 10-3). The period includes the catastrophic rockslides at the Fraser River's Hell's Gate in 1913 and 1914 (Roos 1991). Competitive overfishing and the rockslides severely depressed production. The need for cooperation in managing catches and restoring production was obvious, but almost 35 years of debate passed before a meaningful bilateral agreement was achieved (Roos 1991). A convention between Canada and the United States for the protection, preservation, and extension of the sockeye fishery was signed in 1930 but was not ratified by the United States until 1937 and did not begin regulating harvest until 1946, because additional research was required before implementation. Equal shares of the Fraser River sockeye catch were achieved in convention waters with an average annual U.S. catch of 1.66 million sockeye in 1946-1985 (about 44% of the total Fraser catch in 1946-1977). After 1977, the expansion of Canadian fisheries outside the convention areas and the increasing proportion of sockeye migrating through Johnstone Strait (i.e., exclusively Canadian waters) reduced the U.S. portion to about 25% (Figure 10-4). However, the United States still

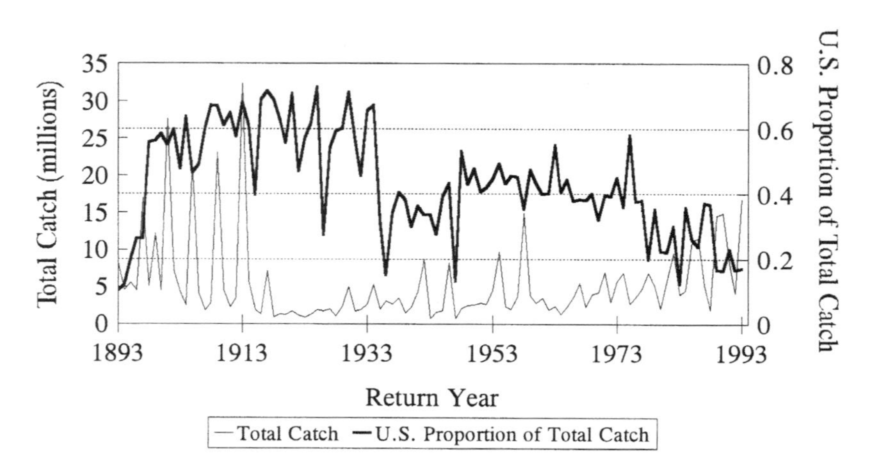

FIGURE 10-3 Historical total catch of Fraser River sockeye salmon, 1893-1993, and proportion of catch taken by fisheries in United States. Total catch includes catch by all gears in all areas reported to have harvested Fraser River sockeye. Source: Data provided by Pacific Salmon Commission.

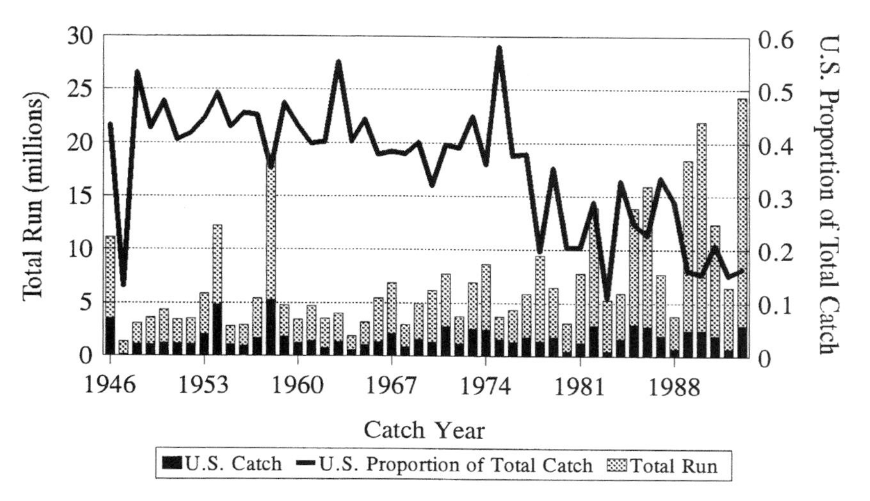

FIGURE 10-4 Total runs of Fraser River sockeye salmon since regulatory control by International Pacific Salmon Fisheries Commission in 1946 and proportion of total catch taken by fisheries in United States. Height of black bar indicates catch taken by United States. Total catch includes catch by all gears in all areas reported to have harvested Fraser River sockeye. Source: Data provided by Pacific Salmon Commission.

receives a substantial catch of Fraser River sockeye; the U.S. average take has been 1.96 million annually from 1986 to 1993 (Figure 10-4).

Fraser River Pinks

Fraser pinks were reduced severely by the Hell's Gate slides (Rounsefell and Kelez 1938, Ricker 1989). The decline was not as evident as for sockeye, because pinks were not in demand by early fisheries, and catches of Fraser pinks were mixed with Puget Sound pinks. Catches in the Fraser River-Puget Sound region declined to 5.3 million pinks in 1939 but recovered after reduced fishing during World War II. As the total catch of pinks increased from 1945 to 1951, the U.S. portion of the catch in convention waters continued at 70% of the total (Roos 1991). In the early 1950s, Canada began expressing interest in establishing a pink salmon agreement to restore their production in the Fraser River. Canada also increased its fisheries at the entrance to the Strait of Juan de Fuca, and by the middle 1950s, the U.S. portion of the pink catch was reduced to about 46%. The Pink Salmon Protocol to the 1936 convention was ratified in 1957. Since 1959, the United States has caught an average of 2.1 million Fraser River pinks (in odd years only), or about 30% of the total (Figure 10-5).

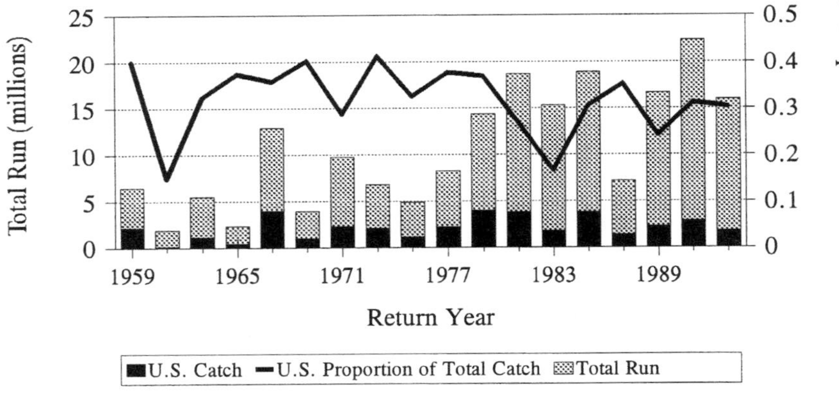

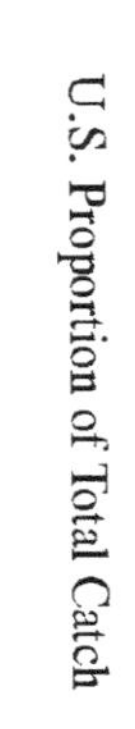

FIGURE 10-5 Total runs of Fraser River pink salmon since regulatory control by International Pacific Salmon Fisheries Commission in 1959 and proportion of total catch taken by fisheries in United States. Height of black bar indicates catch taken by United States Total catch includes catch by all gears in all areas reported to have harvested Fraser River pink salmon. Source: Data provided by Pacific Salmon Commission except for preliminary 1993 spawning escapement value (Canadian Department of Fisheries and Oceans).

Ocean Trolling

Competition between Canada and the United States also occurred (and still occurs) in the ocean troll fisheries. The substantial decline of U.S. chinook catches by ocean trolling in the early 1950s coincided with a rapid increase in chinook catches by trollers in British Columbia (Figure 10-6a). Changes in coho catches between U.S. and Canadian trollers are not as striking, except for the divergence from Washington catches of southeastern Alaska and British Columbia catches after 1976 (Figure 10-6b).

Concerns about ocean troll fisheries are similar to those about high-seas salmon fisheries: trolling catches immature fish in addition to mature fish, deaths include incidental deaths of small, sublegal fish that are hooked and released, and the origins of the fish caught are unknown. U.S. and Canadian chinook and coho mix extensively along the Pacific coast (see Table 10-3, for example), and neither country can take its own production in ocean troll fisheries without incidentally catching the other's production as well. A notable difference from the high-seas situation is that each country intercepts salmon bound for the other country and, until 1978, trollers from both countries often shared fishing grounds (Milne 1964, Argue et al. 1987).

In southern British Columbia, Canada sees the interception of U.S. chinook and coho as its only means to balance U.S. catch of Fraser River sockeye and

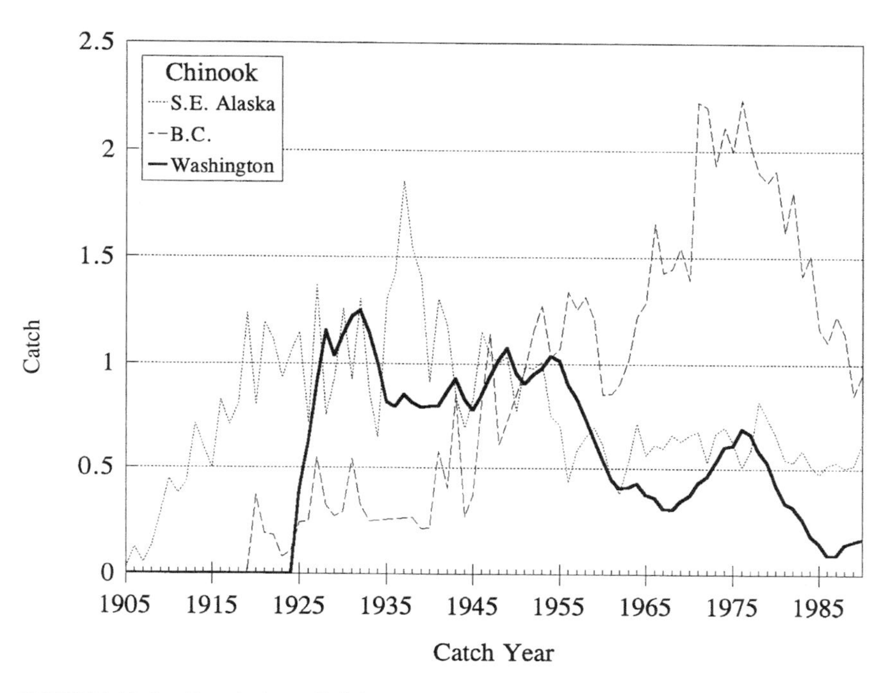

FIGURE 10-6a Trends in troll-fishery catch of chinook salmon by fishers from southeastern Alaska, and British Columbia. For comparison, catch trends are relative to average catches between 1946-1950 within each region. Mean values for these regions were 4.93 million pounds Washington state landed, 457,000 fish landed in southeastern Alaska, and 536,900 fish landed by B.C. fishers.

pinks. Recently, Canadian and U.S. scientists published agreed-on estimates of Canadian interceptions in the troll fishery along the west coast of Vancouver Island (Table 10-4). The data indicate an average annual interception of 236,700 chinook and 1.08 million coho. Interceptions of U.S. salmon also occur in other Canadian fisheries; the above values represent 77% of the total Canadian interceptions of coho (PSC 1994a) but only 50-60% of the total interceptions of chinook (PSC 1993a).

Role of Southeastern Alaska

U.S. fisheries in southeastern Alaska have both a direct and an indirect effect on Pacific Northwest salmon. They not only intercept those chinook from the Pacific Northwest states that migrate far to the north (Appendix I, PSC 1993b, PSC 1994b, Table 10-5), but they also intercept large numbers of Canadian salmon, including sockeye returning to the Fraser River. Alaskan interceptions

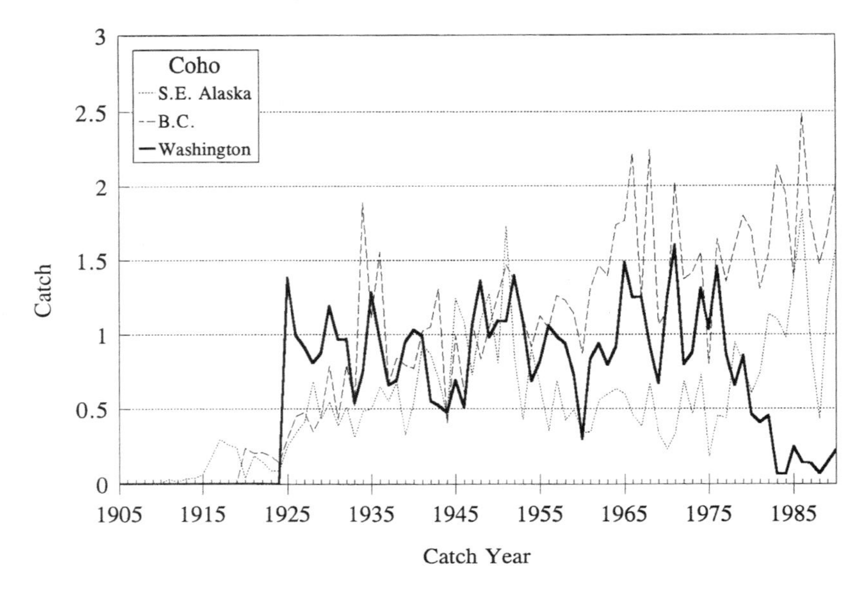

FIGURE 10-6b Trends in troll-fishery catch of coho salmon by fishers from southeastern Alaska, and British Columbia. For comparison, catch trends are relative to average catches between 1946-1950 within each region. Mean values for these regions were 4.27 million pounds Washington state landed, 1.16 million fish landed in southeastern Alaska, and 1.73 million fish landed by B.C. fishermen.

TABLE 10-3 Distribution of Total Fishing Deaths Based on Coded-Wire Tag Recoveries for Fall Chinook

	% of Total Recoveries Averaged from 1988 to 1992	
Recovery Locations	Robertson Creek Fall Chinook[a]	Snake River Fall Chinook[b]
Southeastern Alaska (all gear)	28.0	4.6
West Coast Vancouver Island troll (B.C.)	7.51	5.7
Other B.C. ocean fisheries	15.4	8.7
Pacific Marine Fishery Commission fisheries	0.2	9.0
Total ocean recoveries	51.1	38.0
Terminal catch, Columbia River	—	22.2
Terminal catch, Barkley Sound	17.2	—
Total catch recoveries	68.3	60.3
Freshwater escapement	31.7	39.7

[a]Data on Robertson Creek from Canadian Department of Fisheries and Oceans (B. Riddell, Pacific Biological Station, Nanaimo, B.C.).

[b]Data on Snake River chinook from Columbia River Inter-Tribal Fisheries Commission (Feb. 10, 1994).

TABLE 10-4 Vancouver Island (West Coast) Troll-Fishery Catch[a] and Estimated Percentage of U.S. Populations Caught

	Chinook[b]		Coho[c]	
	Catch%	U.S.	Catch	% U.S.
1984	460,057	64%	2,172,166	50%
1985	354,068	61%	1,389,055	84%
1986	342,063	75%	2,156,833	70%
1987	378,936	89%	1,821,347	48%
1988	408,724	87%	1,595,801	51%
1989	203,695	69%	1,952,009	70%
1990	297,974	59%	1,863,608	46%
1991	202,919	59%	1,889,946	49%
1992	346,741	NA	1,671,822	NA
1993	273,305	NA	948,474	NA

[a] Canadian Department of Fisheries and Oceans statistical areas 21-27 and 121-127.

[b] Estimates of stock proportions for chinook salmon from PSC Report JIC (93)-1 (PSC 1993a).

[c] Estimates of stock proportions for coho from PSC Report TCCOHO (94)-1 (PSC 1994a).

NA = analysis not yet available.

of Canadian sockeye, chinook, and coho are substantially larger than Canada's interception of Alaskan production of the same species (PSC 1993a, PSC 1994a). Interception estimates for pink and chum are under review, but the technical basis for the estimates is weaker than for sockeye, chinook, and coho. Canada believes that the interception imbalance in southeastern Alaska and northern British Columbia has worsened over time and should be redressed (Canadian Discussion Paper, presented to the Pacific Salmon Commission, Feb. 12, 1993). It is a difficult problem for Alaskan managers. Alaskan fisheries have a long history of catching U.S., transboundary (Stikine, Taku, Alsek, Unuk, Whiting, and Chilkat rivers), and Canadian salmon, and their interceptions occur incidentally in fisheries on returning U.S. salmon. Alaskan managers consistently argue that reducing interceptions in these large mixed-population fisheries is impractical. Canadian fishers are equally insistent that to improve management of salmon spawning in northern British Columbia, the Alaskan interceptions should be curtailed. The indirect effect of Alaskan fisheries on Pacific Northwest salmon is simply that in the absence of progress on Alaska-Canada transboundary fisheries, Canada has resisted change in its southern British Columbia fisheries, so pressures are maintained on the entire U.S. delegation. Reducing Canadian interceptions of salmon originating in Pacific Northwest states to levels that will allow these salmon

TABLE 10-5 Annual Distribution of Total Fishing Mortality[a] for Washington and Oregon Chinook Populations Caught in S.E. Alaska for 1985-1993

		Fishing Location (%, all gears included)		
Population	Location	Southeast Alaska	British Columbia	Northwestern United States
Queets Falls	Coastal Washington	37.4	36.4	26.1
Cowlitz Falls	Lower Columbia River	9.4	31.0	59.6
Columbia River fall brights	Middle Columbia River	21.4	31.5	47.1
Hanford Reach wild fall brights[b]	Middle Columbia River	30.3	27.4	42.3
Lewis River fall wild	Lower Columbia River	12.0	31.1	56.9
Lyons Ferry fall brights[c]	Snake River, Middle Columbia River	9.23	8.8	52.0
Willamette Springs	Lower Columbia River (Oregon)	19.0	11.9	69.1
Salmon River Falls	North coastal Oregon	27.6	36.1	36.3

[a]Total fishing mortality includes reported catch and estimated mortalities associated with fishing (e.g., hook-and-release of under-sized fish in troll and recreational fisheries, drop-out from nets, etc.).
[b]Average for years, 1990-1993.
[c]Average for years, 1988-1993.

Source: PSC 1994b.

populations to be sustained will require compromise not only by Canada and Pacific Northwest states, but also by Alaska.

The Pacific Salmon Treaty

The 1985 treaty is the first comprehensive fishery agreement on Pacific salmon between Canada and the United States. Its principles (Article III) follow:

> 1. With respect to stocks subject to this Treaty, each Party shall conduct its fisheries and its salmon enhancement programs so as to:
>   a) prevent overfishing and provide for optimum production; and
>   b) provide for each Party to receive benefits equivalent to the production of salmon originating in its waters.
> 2. In fulfilling their obligations pursuant to paragraph 1, the Parties shall cooperate in management, research and enhancement.

3. In fulfilling their obligations pursuant to paragraph 1, the Parties shall take into account:

a) the desirability in most cases of reducing interceptions;

b) the desirability in most cases of avoiding undue disruptions of existing fisheries; and

c) annual variations in abundance of the stocks.

Article III (1.b) often is referred to as the equity principle; the treaty provides opportunity to conserve and rebuild production of salmon from the Pacific Northwest and for both countries to benefit from their increased production. Annexes in the treaty provide a mechanism to limit interception of Pacific Northwest chinook and coho salmon in ocean troll fisheries, limits to interceptions of chum salmon by fisheries in southern British Columbia and Washington State, continued exploitation of Fraser River sockeye and pink salmon, and a general obligation not to initiate new interception fisheries and not to redirect existing fisheries intentionally to increase interceptions. Total catches of Fraser sockeye and pinks are limited and, unlike the previous Fraser convention, apply to U.S. catch of these populations wherever they are caught inside and outside the area managed by the Fraser panel. Agreements on the conduct of fisheries are contained in chapters of Annex IV and can be revised whenever appropriate, although achieving consensus on changes in the annex has been difficult.

Potential benefits of the treaty for Pacific Northwest salmon are not being fully achieved, as indicated by the recent inabilities of the parties to renegotiate the fishing annexes. Fundamentally, the parties differ in their application of the treaty principles. The United States emphasizes the need to prevent overfishing and to adjust fisheries for conservation. Its approach to the equity principle is that determining national benefits from salmon production will be difficult technically and should be considered over a longer term. Canada agrees with the need for conservation and rebuilding of production but links these with parallel actions to address equity. Without adequate progress toward addressing equity and in light of recent U.S. proposals for fishing annexes to increase U.S. interceptions particularly of Fraser sockeye, renegotiation of the 1994 fishing annexes failed. This is of serious concern for the conservation and rehabilitation of Pacific Northwest salmon. Without a 1994 agreement, ocean troll and recreational fisheries of Washington and Oregon are closed, and Canadian fisheries continue to intercept the same populations without any changes in management plans.

CONCLUSION

Since the nineteenth century, in an effort to maximize catch, salmon fisheries of the Pacific Northwest have exploited a mix of wild-spawning and hatchery-produced salmon. Fishing moved farther into the ocean to catch more and better-quality salmon earlier in their life cycle, but the stream origins of these fish were unknown. Societal pressures pushed catch levels toward those which only the most-productive populations could sustain, but they were often too high for natural populations. Mixed-stock fisheries developed for human convenience, and society watched as local breeding populations of salmon went to extinction or were depressed severely. Fishing impacts and the promotion of regional economic growth combined to alter salmons' environment to their detriment. The existing technocratic model for fishery management, productivity enhancement, and environmental modification has not been able to sustain salmon catches or the diversity of salmon populations. The result has been a major reduction in economic opportunity for fishers. All fishers have without doubt suffered possibly irreparable injury from the status of salmon and the management prescriptions to deal with it. The decline in income is much greater than that in any other major resource industry in the Pacific Northwest, and catches by American Indian fishers are now smaller in numbers of fish than before the Boldt decision.

The committee concludes that fishery management objectives must explicitly recognize the need to conserve and expand the genetic diversity of the salmon resource. To accomplish this, emphasis must be given to minimum sustainable escapements and filling out the dendritic structure of salmon habitats.

A more holistic management approach must recognize the connections between the genetic resource base, habitat, and the resulting salmon production; it must also account for the uncertainty in our scientific advice and for inherent environmental variability. The committee has outlined a process intended to improve the potential sustainability of salmon in the Pacific Northwest. Furthermore, the committee does not believe that the sustainability of Pacific Northwest salmon can be achieved without limiting the interceptions of U.S. salmon in Canada and obtaining the cooperation of Alaska. An effective and cooperative Pacific Salmon Treaty is necessary. The committee does not provide specific recommendations about altering specific fisheries, because there are numerous options and interactions between fisheries. Achieving agreement on changes in fisheries will be difficult and necessitates an effective institutional process.

11

Salmon-Fishery Management Concepts

While Pacific salmon fisheries developed rapidly during their early history, our ability to manage them did not. Much of the basic biological understanding of Pacific salmon and information that could be used to manage salmon fisheries were being developed as the fisheries developed, but their application to management developed much more slowly. In his review of salmon management during the first century of Pacific salmon fisheries, Larkin (1970) suggested that almost from the beginnings of the industry two ideas were implicit in attempts at management: that salmon returned to their home stream to spawn and that catches in each river had to be limited. Those continue to be the biological bases for management, and we continue to struggle with their incorporation into a sustainable management concept.

Papers by McHugh (1970) and Larkin (1970) provided historical perspectives on the development of fishery science and management of Pacific salmon in North America. Initially, scientific investigations consisted largely of descriptive biology and examination of the "home-stream concept." The scientific basis of that concept was debated long after its acceptance in management (see, for example, Jordan 1925, Moulton 1939). But acceptance, coupled with the early recognition that salmon eggs were easily cultured, resulted in hatcheries' becoming the major management activity during the first 50 years of the industry. By the late 1930s, however, management of Pacific salmon was in transition. Larkin (1970:226) reported that "regulations for controlling harvest were inadequate, but insufficient information existed on which to construct better techniques; hatchery practices were fairly advanced but of dubious value; inroads on salmon production as a consequence of the development of other resources were beginning to cause concern." The 1930s began a period of more-quantitative assessment in fishery management (Cushing 1988, McHugh 1970). The quantitative basis of salmon management was provided by Ricker's 1954 seminal paper on stock and recruitment. Since then, management of Pacific salmon fisheries has been premised on his stock-recruitment theory.

STOCK[1] AND RECRUITMENT

Salmon-fishery management assumes that there is surplus production below some upper size of the spawning population. *Surplus* in the case of salmon means that a given number of spawners in an adult generation produces, on average, more progeny than needed to replace the parents and overcome all natural mortality sources from the time fertilized eggs are deposited in the gravel of natal streams, through juvenile and immature life phases, to adulthood. The number of surplus animals varies with the size of the population and the natural mortality rate. Smaller populations tend to have higher productivity than larger populations (i.e., number of progeny returning per adult spawner), and their total production is limited mostly by the number of eggs deposited. In larger populations, production depends more on the interactions between spawners and habitat required for sustaining survival and growth of progeny.

Ricker (1954) noted that factors that become more effective at high densities, called "compensatory" factors by Neave (1953), control or regulate salmon populations. Compensatory mortality factors place more pressure on high-density than on low-density populations. For example, when large numbers of pink salmon reach their spawning grounds, some adults are forced to use less-suitable gravels at stream margins; in crowded conditions, late spawners might even dig out developing embryos deposited by earlier spawners. Those factors decrease the number of progeny produced per female. When chinook or steelhead spawners are less abundant, the resulting fry, fingerlings, and pre-smolts have more access to feeding positions and cover, so they may grow faster and be less vulnerable to predation.

Ricker (1954) termed the relationship between the number of spawners (stock or S) and the production of progeny (recruitment or R), the stock-recruitment function. The term *recruitment* refers to the potential availability of fish to a fishery or to form the next spawning generation. The stock often is referred to as the *escapement*, because these fish escaped capture by a fishery and return to spawn.

Fishery managers have attempted to maximize surplus production (i.e., ani-

[1]The terminological difficulties associated with the word *stock* are discussed in Chapter 4. To permit comparison of the discussion in this chapter with much of the published literature on fisheries, we use the term *stock* here, although we use the term *population* in most of the rest of the report.

mals available for catch) by maintaining the number of spawners at an abundance at which, according to Ricker's stock-recruitment theory, they are likely to produce the largest sustainable catch. Figure 11-1 is an example of a hypothetical Ricker stock-recruitment function. In reality, the function would be fitted statistically through a scatter of data points collected over time. The function represents the average response expected given an escapement under the environmental conditions that existed when the data were collected. If escapements merely replaced themselves in the next generation, those returns would fall along a "replacement line" where R = S (line A in Figure 11-1). However, if the function value R_1 expected for a particular S_1 exceeds the replacement value, then a surplus production ($R_1 - S_1$) could be caught and the population maintained in equilibrium at the same future S and R numbers. Salmon populations can maintain themselves at several levels of abundance, and different salmon populations have different stock-recruitment curves. In Figure 11-1, curve B describes a population with greater productivity than curve C, but one with greater density-dependence at large spawning stocks. Populations with greater productivity can sustain their production at higher exploitation rates.

The S number that, on average, maximizes the catchable number of fish generation after generation is referred to as the optimum escapement, and the associated catch is the maximum surplus reproduction or maximum sustained yield (MSY). The escapement expected to provide MSY is indicated as S_{MSY} in Figure 11-1. It occurs where the slope of the recruitment curve is 1.0, the tangent to the curve parallel to the replacement line. Once S_{MSY} is determined, the rate of exploitation that can be sustained by the population to maintain MSY can also be determined, i.e., $(R_{MSY} - S_{MSY})/R_{MSY}$. In this figure, the surplus production ($R_1 - S_1$) is equal to MSY.

Other stock-recruitment models have been proposed. The Beverton-Holt model (1957) predicts that the number of recruits increases with spawning stock ever more slowly and never exceeds a particular value (asymptote). This model does not turn downward at high S, as with Ricker's model.

Stock-recruitment functions, whether Ricker's or Beverton-Holt's, share several serious limitations for application to salmon management. The principal limitations are related to

- The estimation of the biological production function in a highly variable natural environment.
- Differences between populations and change over time within populations.
- The necessity for accurate data on total fishing mortality by age and population over all fisheries, on number of spawners by age, and on future production.

An individual data point (i.e., the recruitment from a parental spawning

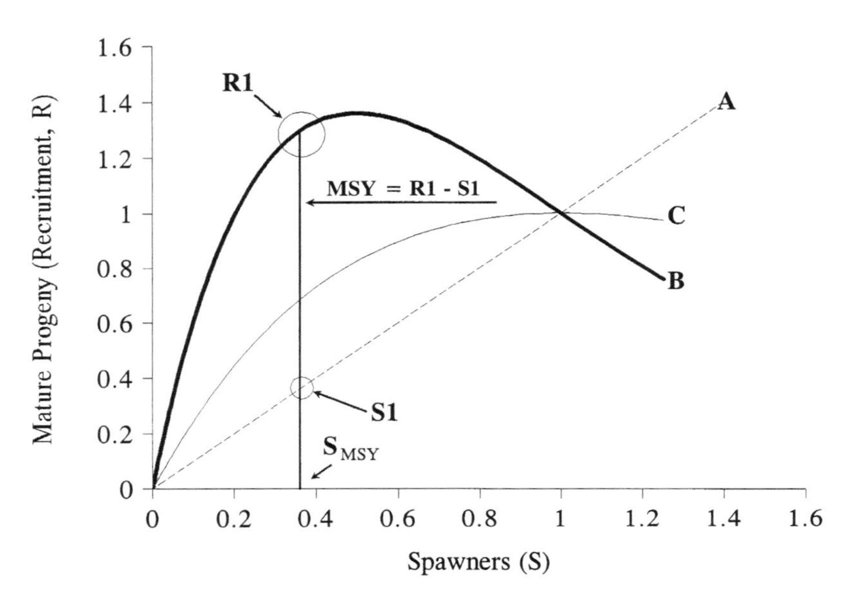

FIGURE 11-1 Hypothetical Ricker stock-recruitment curves relating number of animals reproducing (spawners) and production of mature progeny (recruitment). Other letters explained in text.

stock) reflects biological processes, effects of environmental variability, and random events. Determining an appropriate production function in the presence of this variability requires a long series of data on returns over a wide range of spawning-stock sizes. The uncertainty about a recruitment function is usually high. For example, even in a sockeye population with 41 years of good assessment information, a characteristic recruitment function is not evident (Figure 11-2a). The relationship between spawners and juvenile production in freshwater is more evident (Figure 11-2b), but variability in marine survival weakens both the relationship between spawners and adult returns (Figure 11-2a) and between downstream migrants (smolts) and adult returns (Figure 11-2c). The latter relationship would already account for variation in returns attributable to variation in freshwater survival. Even in the population modeled in Figure 11-2, the estimate of S_{MSY} is uncertain; S_{MSY} = 332,000 with a 90% confidence range between 203,000 and one million spawners. This confidence range was estimated from 1,000 computer simulations of the relationship between adult spawners and adult recruitment. The distribution of the simulation results (Figure 11-3) indicates the uncertainty associated with estimates of the optimal escapement value for this population. Furthermore, the scatter plot of alpha versus beta values (S/R param-

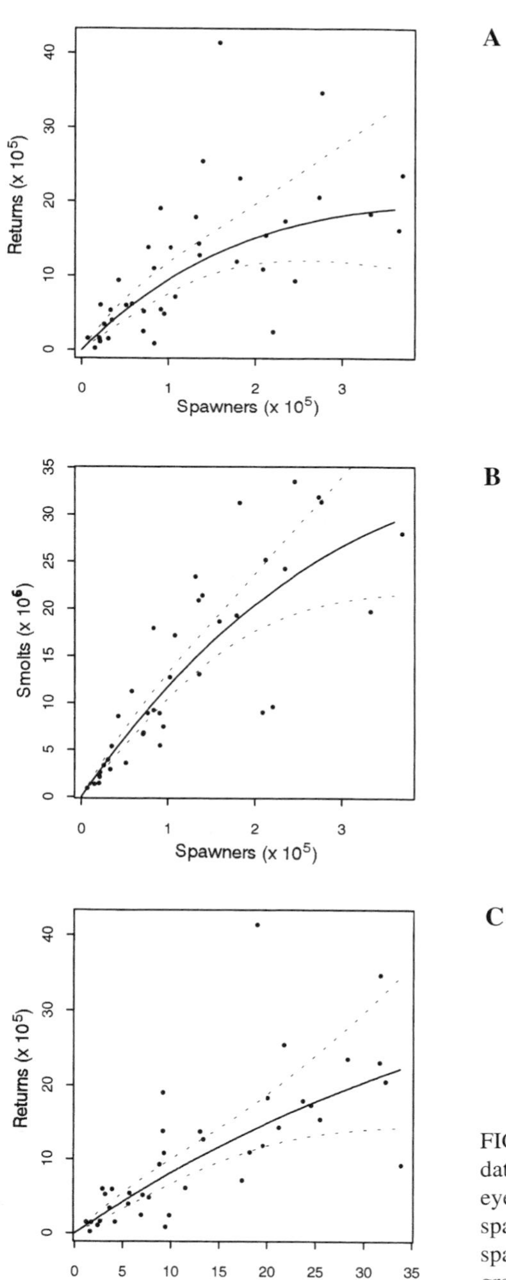

FIGURE 11-2 Ricker stock-recruitment data and functions for Chilko Lake sockeye salmon from Fraser River. A, adult spawners and adult recruitment; B, adult spawners and juvenile downstream migrants (age 1 + smolts); and C, migrants and adult returns.

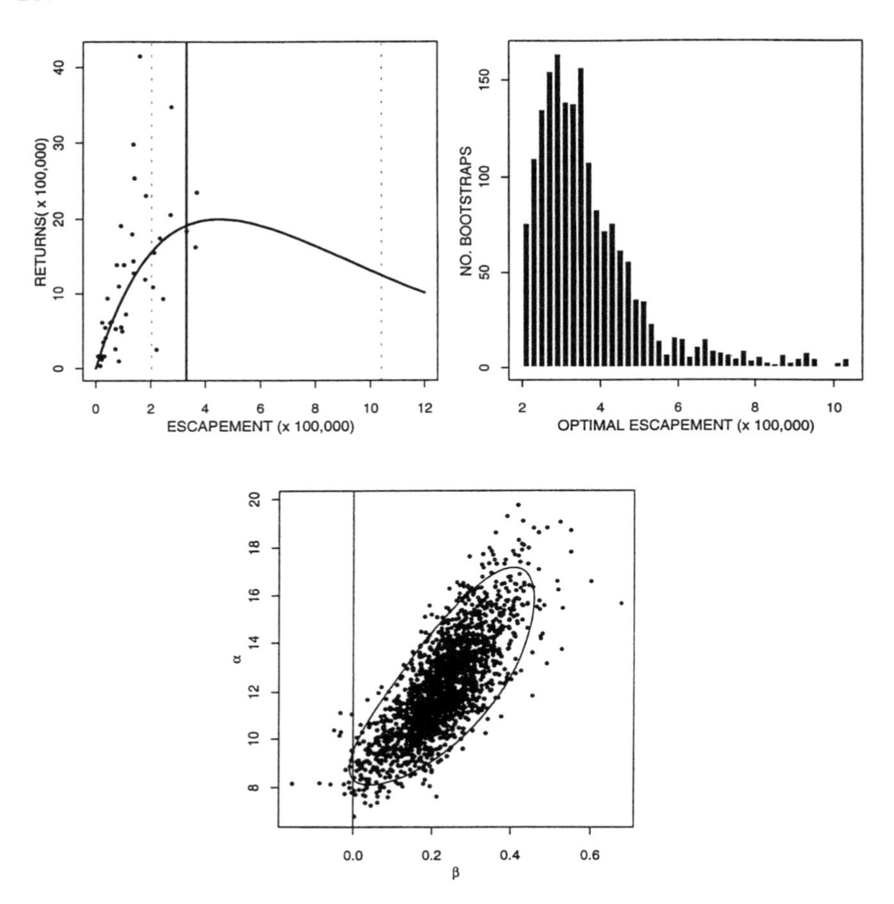

FIGURE 11-3 A, results of 1,000 bootstrap simulations of Chilko sockeye Ricker stock-recruitment function. B, distribution of 90% confidence interval for optimal spawning-stock sizes determined by simulations. C, bivariate scatter plot of Ricker stock-recruitment parameters determined from each simulation.

eters in the Ricker function) indicates that these parameters are correlated (the oval shape of the 90% joint confidence limit indicates correlation). The wide variation in the alpha value is associated with wide variation in beta; this results in a highly uncertain stock-recruitment function for this population. In salmon populations in which recruitment and spawning stock sizes have been monitored, annual variation in the ratio of returns to spawners can vary by a factor of 10. Recently the marine survival rate of chinook salmon released from Robertson Creek Hatchery (on the west coast of Vancouver Island, B.C.) has been shown to vary by a factor of more than 100 (0.1%-13.7% survival to the second year).

Many years of data would assist in accounting for that variability, but long-term data can involve another problem. The function calculated reflects returns per spawner under past environmental conditions. If the environment changes, the stock-recruitment function changes. An obvious example is deterioration of freshwater environments, as evidenced in increased deaths associated with dams, reduction in area available for spawning or rearing because of water abstraction, or sedimentation in spawning gravels. Change in marine survival (see Chapter 2) also can alter the stock-recruitment function. Environmental variability makes questionable how representative any stock-recruitment function will be for current and future environmental situations. Limiting data to periods considered to be more "typical" of existing conditions might be possible, but the resulting decrease in data points would increase uncertainty substantially.

The most common concern about managing for MSY in salmon fisheries is that stock-recruitment functions vary among populations. The MSY for a population is determined by its productivity and sources and magnitudes of density-dependent mortality rates, which reflect the life history of the species and the specific habitat in which the population lives. Stock-recruitment functions are expected to vary, but the paucity of reliable data on population-specific functions makes it hard to account for the differences. An obvious example is the comparison of wild-spawned versus hatchery-reared salmon. A hatchery population can sustain its maximum catch at substantially greater exploitation rates than can a natural population because mortality associated with spawning and freshwater rearing is much lower in a hatchery than in natural systems. Assuming that after release marine mortality sources do not compensate, fewer parents are needed to reproduce the recruitment from a hatchery population (see Chapter 12). Direct comparisons of stock-recruitment functions for hatchery and wild populations (in the same geographic area and period) are rare. One good comparison involves sockeye salmon in the lower Fraser River (Figure 11-4), where an artificial spawning channel in Weaver Creek enhances the fry productivity of that population but later rearing occurs in the natural environment. Two other populations, from Birkenhead and Cultus lakes, are produced naturally and have the same adult run timing as Weaver Creek; all three populations are fished simultaneously. The catchable surplus from Weaver Creek is greater than that in the natural populations. The exploitation rates to sustain these populations at MSY are 0.76 for Weaver Creek, 0.70 for Birkenhead Lake, and 0.62 for Cultus Lake. The spawning channel has increased the productivity of the Weaver Creek sockeye, but fishing to maximize the catch from Weaver Creek would mean overfishing returns to both natural populations.

The hatchery-wild dichotomy presents an extreme example of the "mixed-stock" fishing problem. If fishing responds to apparent abundance without consideration of the stock composition (i.e., the mixture of portions of stock from source populations) or if fishing levels are based on hatchery production, the natural population will be overfished and its production will, on the average,

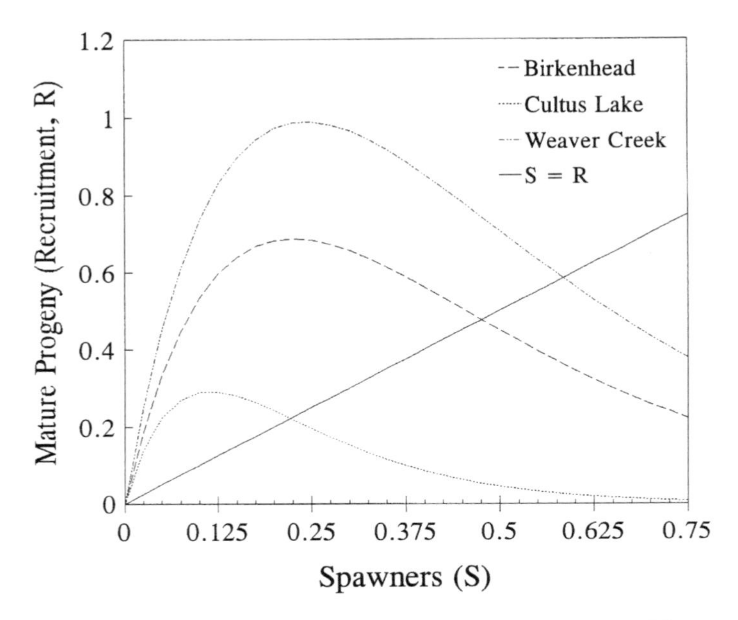

FIGURE 11-4 Ricker stock-recruitment curves for three Fraser River sockeye salmon populations. Weaver Creek population is enhanced but Cultus and Birkenhead populations are both naturally spawning. Source: Data collected 1946-1990 by International Pacific Salmon Fisheries Commission and Canada's Department of Fisheries and Oceans.

decline. Alternatively, if the fishery is managed to sustain the natural population, substantial surplus production will return to the hatchery or could be caught in a single-population, terminal fishery.

The example of mixed-stock fishing represents a much more general problem. Differences in productivity between natural populations cause the same problem, and by-catch of other species in fisheries that are directed at a more productive species is an analogous problem. When fishing occurs on a mixture of populations with different stock-recruitment functions and fishing cannot be regulated at a rate appropriate for each component population, the stage is set for overfishing of the less-productive components (Ricker 1958, 1973; Hilborn 1985). For example, extinction of wild coho salmon in the lower Columbia River has occurred as fishing pressures at sea and in the lower Columbia increased to take hatchery returns; catch levels of 85-95% were directed at the returning fish (Cramer et al. 1991). The less-productive stocks are referred to as "weak stocks," but that term leads to confusion. "Weak" cannot be equated with "small," nor does it imply anything maladaptive, inferior, etc., about animals in the population. The "mixed-stock" (or mixed-population) fishery problem is related to differences in *production rates*, not the relative size of populations.

Apart from natural variability and variation among populations or over time, estimating the S_{MSY} for just one population raises a serious question. Larkin's (1977) discussion of MSY as a management concept identifies the issue of the poor quality of the data available for use in stock-recruitment analysis, and recently the joint U.S.-Canada committee on chinook salmon stated (PSC 1993b:87):

> At present, complete information necessary to determine stock productivity is not available for any individual chinook stock! For a few stocks, enough information has been available to apply stock-recruitment type analyses to estimate productivity parameters, but even these had to involve some major assumptions about age structure in catch and/or escapement and about the error structure of these data. And none include environmental factors, which are known to produce variability in annual production.

To determine stock-recruitment functions is data-intensive, expensive, and statistically nontrivial. Data and cost issues are related to accurate determination of a population's mortality in each fishery and its spawning escapement by age so that production can be related to the parental generation. Salmon tend to be caught in many sequential, mixed-stock fisheries, and their escapement is not determined easily. There are few cases in which this challenge has been met to study salmon population dynamics, and the sensitivity of stock-recruitment analyses to errors in the data is poorly understood. Hilborn and Walters (1992) stated that stock-recruitment analyses can provide "terribly misleading answers" and that (p. 287)

> the types of misleading answers produced by stock and recruitment analysis are almost always the same; the answers mistakenly lead you to believe that recruitment will not decline very much with spawning stock. We think that bad stock-recruitment analyses have been a significant factor leading to over-exploitation and stock collapse for some major fisheries.

Hilborn and Walters reviewed the problems associated with stock-recruitment analyses in greater detail than is appropriate here, but the committee has developed an example of the consequences of such analyses (Box 11-1). The most common outcome of simple stock-recruitment analyses is that the optimum exploitation rate is overestimated and the S_{MSY} underestimated. The consequence of this outcome could be management advice that unintentionally would lead to overfishing and contribute to declining production.

Although MSY concepts have provided the basic paradigm for salmon management since the 1950s, the paradigm has been inadequate, given the fishing pressure and economic development in the Pacific Northwest. Mixed-population fisheries, habitat change, and uncertain assessment advice have all contributed to overfishing and loss of less-productive populations. The committee reiterates

Larkin's caution about the inadequacy of the MSY concept for salmon management (Larkin, 1977:9):

> The foregoing has demonstrated, I hope, that MSY is not attainable for single species and must be compromised: (1) to reduce the risk of catastrophic decline and reduction of genetic variability; and (2) to accommodate the interactions among the species of organisms that comprise aquatic communities.

Given that the limitations of stock-recruitment analyses have been known for many years, why are management strategies based on those models? Part of the answer is that technical improvements in analyses has led to unjustified confidence in abilities to compensate for deficiencies. Much of the answer, however, lies in the socioeconomics of fisheries and fishery management. In the United States and Canada, marine fish are generally viewed as "common property" resources, owned by no one—or by the public—until they are caught. Such a situation is well known to lead to excessive investments in capital and labor and to pressures to overfish resources, particularly when there is open entry (i.e., no limit on the number of people who can fish) (Gordon 1954, Scott 1955, Crutchfield and Pontecorvo 1969). However, salmon fishing in the Pacific Northwest is not now (and has not been for a long while) an open-entry fishery. The states of Washington, Oregon, and Alaska and the province of British Columbia have limited entry into the salmon fishery since the 1960s; this has not prevented overcapacity in boats, gear, and fishing technology, but it has raised greatly the costs of participating in the fishery and reduced overall numbers of people and boats in it. Higher costs of entry and higher investments increase the needs of fishers to pressure regulatory agencies to allow higher catches at the expense of spawning requirements.

The problem has long been recognized (e.g., Wright 1981, Ludwig et al. 1993). Wright stated (p. 38)

> Fishermen make poor management allies due to their perpetual optimism about strengths of the salmon runs and their understandable preoccupation with short-term economic considerations.
>
> There can be little doubt, however, that the salmon fishery lobbyists are currently winning the battle against the spawning-escapement protectors. A team of fishery scientists formed by the Pacific Fishery Management Council concluded that 40% more chinook salmon and coho salmon were needed to meet spawning-escapement requirements, under existing habitat conditions, for the combined areas of California, Oregon, and Washington (PFMC 1978:39).

Similar appraisals can be found in Fraidenburg and Lincoln (1985), Walters and Riddell (1986), and National Research Council (NRC 1994). The remedies suggested most commonly, besides complete but preferably temporary closures of the fisheries (as occurred in 1994), include restructuring managing bodies to remove apparent conflicts of interest (NRC 1994) and privatizing rights of access to salmon stocks through individual transferable quotas or similar devices, perhaps combined with buyouts or other compensations for displaced fishers. A third approach, paradoxically, is to strengthen the involvement of fishers in the management process so that they are encouraged to take more responsibility as stakeholders in either a common property or a privatized fishing situation (cf. Scott 1993). Hanna (in press) suggested that the Pacific Fishery Management Council (PFMC) has already moved a long way toward involving fishers in the management process, at least for other species of fish.

The application of stock-recruitment theory and MSY as the basis of salmon management is complex and of limited applicability. The multitude of populations and habitats, the extent of enhancement programs, and the variability and uncertainty in the data make determining an accurate optimal escapement goal elusive. However, where the necessary data are available, stock-recruitment relations can be clear (see Box 11-2). The definition of the relationship depends on the degree of environmental variability, the causes of density-dependent mortality, and data quality. Given the poor quality of the data available on almost all Pacific salmon populations, we cannot test the stock-recruitment theory rigorously. We have learned that the theory is more applicable in freshwater phases of salmon life history and that environmental variability in the marine habitat ultimately can determine the number of returning adults. Principal lessons are that salmon stock-recruitment relationships are inherently uncertain, that the determination of a specific escapement goal (S_{MSY}) is seldom justified by available data, and that the MSY concept has been inadequate for conserving population diversity or production.

FISHERY MANAGEMENT IN THE FUTURE

The committee explored four general options for managing fisheries to help frame the process of developing a new management paradigm: status quo, no fishing, limited entry, and terminal fisheries.

The Status Quo

One management option is to continue to use the MSY concept while working to improve its predictive powers. The committee has concluded, however, that the MSY concept by itself is inadequate and impractical as a basis for salmon management because the model implies the existence of a continued surplus production, which is fundamentally inconsistent with historical data. In overfished populations, most stock-recruitment data will be from the lower range of escapement numbers. We can adjust for biases in data, but we cannot correct for the absence of data at larger escapements without actual observations. If we estimated S_{MSY} on the basis of historical data and managed perfectly by annually achieving this value, we would learn nothing about the productive potential or dynamics of a population; we would learn more only about natural variability in

Box 11-1
Stock-Recruitment Simulation

Stock-recruitment functions are usually nonlinear, which means that natural environment fluctuations can produce systematically skewed estimates of the long-term response of salmon populations to exploitation. The direction of this error appears to lead to overexploitation, even when statistical procedures generally accepted by fisheries biologists are properly applied. In this example, the committee develops ideas suggested by Hilborn and Walters (1992) to show how advice to management might produce serious errors.

The simulation begins with a "known" stock-recruitment relationship, the values for which are typical for chinook salmon in the Pacific Northwest: $\mathbf{R_t = S_{t-1} \exp (a - bS_{t-1}) \exp\sigma_E^2}$ where a = 1.6, b = 0.2, and sigma (σ_E) = 0.7.

Each brood year was fished at a 75% exploitation rate; this is common for many fall chinook populations but exceeds the 60% rate sustainable at MSY for this stock-recruitment function.

Each simulation was run for 100 years, and data from the last 30 years were collected for stock-recruitment analysis. At the end of each simulation, stock-recruitment parameters (a, b, S_{MSY}) were estimated from the 30 data points. The effects of three known error sources were examined:

Type 1: Environmental variation in recruitment, normally distributed with mean zero and standard deviation σ_E. The value of σ_E was chosen so that production varied by a factor of 2-4.

Type 2: Environmental variation plus observation error in spawning-escapement estimation. The error about S_t was simulated as $\mathbf{S_t * \exp\sigma_s^2}$, where the random normal error has mean zero and standard deviation = 0.57. This value of σ_S was chosen to produce escapement-estimation error of ± 50% about the true S_t.

Type 3: Environmental variation, observation error in spawners, and error in catch estimation. Catch error was simply generated by smoothing the catch (three-point moving average) to simulate assumed age structures or errors in catch allocation between populations.

One thousand simulations of each type were conducted, and frequency distributions for parameters a and b were compared with values of the "known" function. The solid vertical line in the figures represents the parameter values of the "known" function, and the dashed lines encompass 95% of the estimates.

Both a and b have been rescaled for clarity in presentation. The a values are expressed as the expected recruits per spawn ($\sigma = \exp^{a}$), and b as logarithms to spread out the distribution.

The distributions of results for Type 1 to 3 simulations are presented from top to bottom. In each distribution, the most common result of the simulations is represented by the tallest bar. In every case, the most common result is to the right, of the value that we are trying to estimate (i.e., the values in the original, or "known", stock-recruitment function). Furthermore, for the productivity parameter (α), the effects of the error types are compounded as the most common value progressively deviates from the vertical line. Comparing the means of the simulations demonstrates the tendency for errors in stock-recruitment analyses to overestimate productivity and underestimate the number of spawners needed to sustain MSY.

	Recruiters/ spawner	S_{MSY}	Sustainable exploitation rate at MSY
"Known" function	4.95	3.43	0.60
Type 1 results	5.51	1.44	0.63
Type 2 results	6.28	0.98	0.67
Type 3 results	6.49	1.11	0.675

These analyses are clearly very limited and were intended only to demonstrate, under realistic assumptions about error, the potential for misleading information (see Hilborn and Walters 1992). Advice based on these analyses would recommend a sustainable exploitation rate, at MSY, 12.5% greater than could actually be sustained by the population. While we thought we were managing correctly, the population would continue to decline.

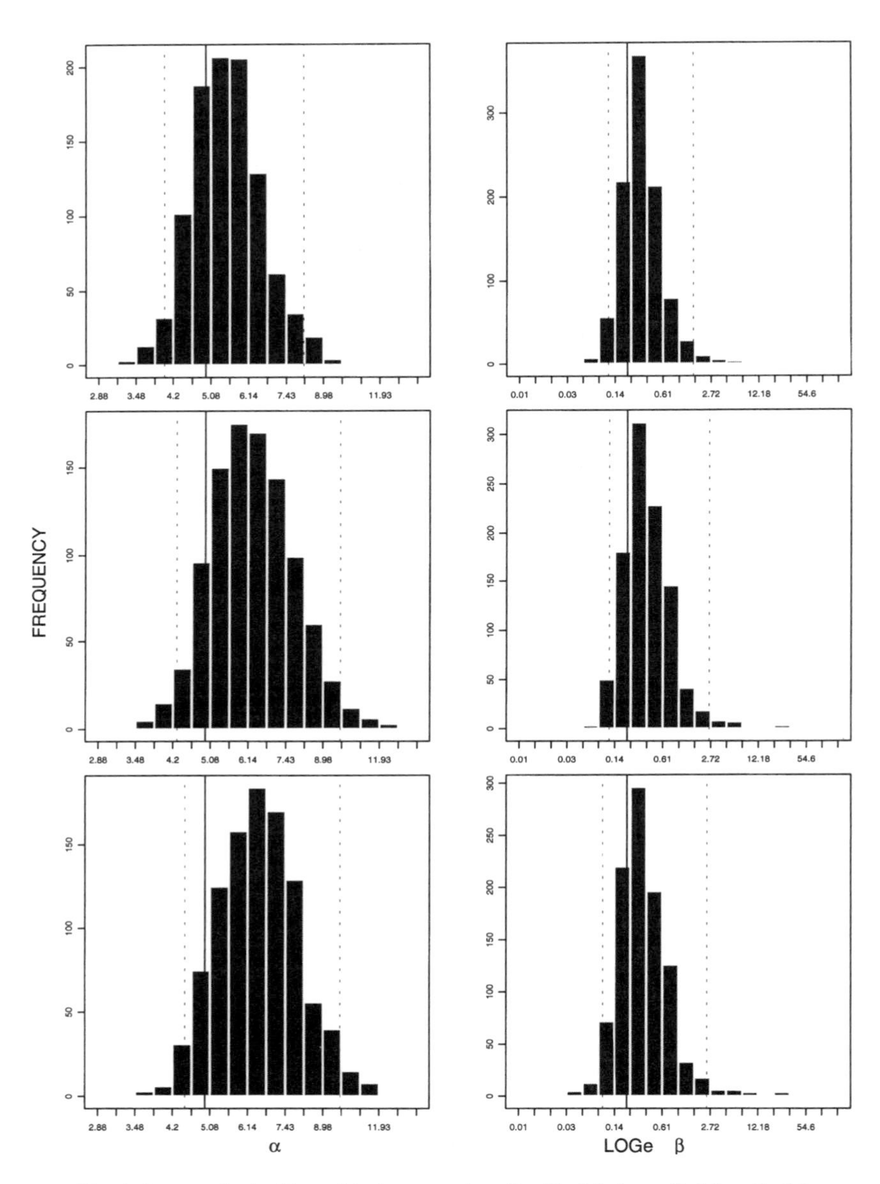

Simulation results for Type 1 to 3 errors described in this box. Solid vertical lines in each figure are true parameter values. Frequency histograms on left show distribution of estimated returns per spawner, and on right, natural logarithm of beta parameter (natural lags used to spread distribution). Results of 1,000 simulations for Type 1 to 3 errors are presented from top to bottom.

Box 11-2

A most striking example of stock-recruitment relations in a salmonid is the detailed study of Elliot (1994) on anadromous brown trout (*Salmo trutta*). The figures below from Elliot's recent book show a clear curvilinear relation between spawning (measured as egg density) and recruitment to later life phases. Note that recruitment is more strongly associated with egg density at the lower ages (R_1 to R_4). The relation becomes much more variable when adult returns (R_5, age 3+ female returns) are related to egg density. The committee readily acknowledges, however, that this population inhabits a spring-fed stream not subjected to extremes of climate or high water velocities. Reprinted by permission of Oxford University Press.

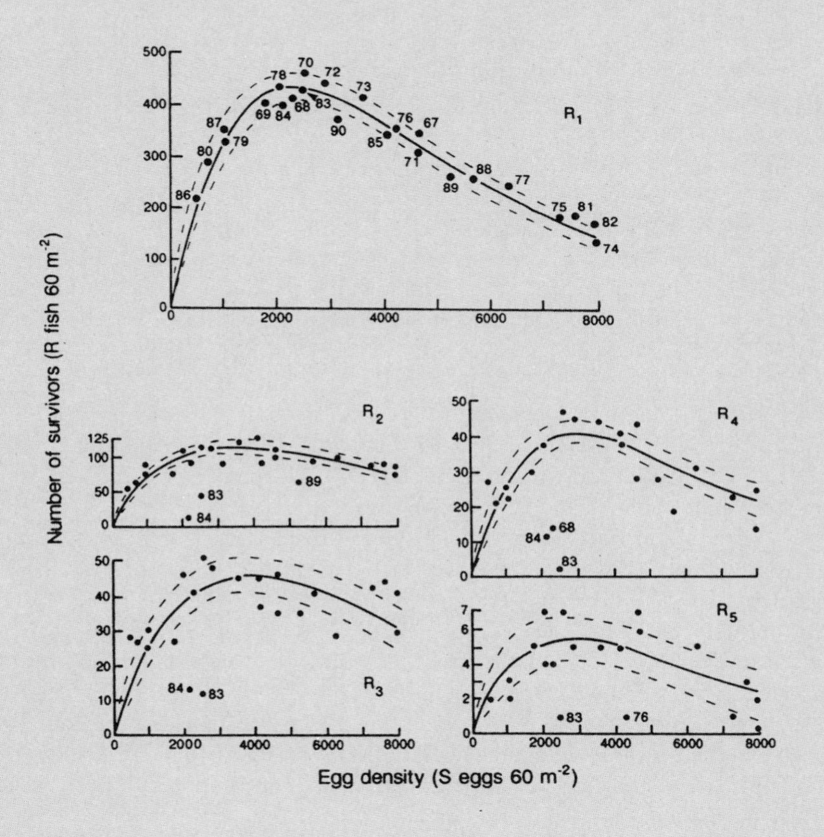

pressed populations have not rebounded. One common argument is that stopping catch certainly cannot hurt and that fishing should not continue when threatened and endangered species are mixed with salmon targeted for catch.

The major problem with the no-fishing option is that the social and economic hardships caused by stopping are substantial, particularly to those who depend on salmon fishing for their livelihood. Fishing cessation usually occurs only when the overall benefits from continued fishing are so reduced that every party gets little from the population. Furthermore, people always expect the factors causing the decline to go away soon and more favorable conditions to return.

Fishers argue that commercial and recreational uses of salmon constitute an important value. They say that fishing is an excellent way of keeping the status of the populations in the public eye. Fishing is a livelihood for professional fishers. Catches are important for economic, subsistence, and ceremonial purposes for American Indians, and the expectation that catches are possible drives the recreational fishery. Eliminating fishing makes salmon less valuable economically.

Eliminating fishing is not a simple issue. Salmon are caught from southern Alaska to the central California coast by netters, trollers, anglers, and charter boats and by both Indian and non-Indian fishers. Should fishing be stopped for Alaska salmon fishers whose runs are generally in much better shape than those returning to the Columbia River? The position of Alaska salmon fishers is that salmon problems in the Pacific Northwest are due to choices made by people of the Pacific Northwest and that Alaskans should not be penalized to fix the region's problems (Pacific Salmon Commission 1993a). Salmon from the Pacific Northwest are also caught in Canadian fisheries, and U.S. treaty and nontreaty fishers catch salmon from Canada. The position of Alaskan salmon fishers has prevented the Pacific Salmon Commission (PSC) from making progress in protecting threatened and endangered salmon runs.

When elimination or control of fishing is discussed, people often think only of commercial fishing, but that is too restricted a view; all fishing kills fish. Is it legally possible to stop treaty fishing? As long as any salmon are available, treaty fishing will continue. Thus, the no-fishing option is complex; it does not imply just a simple decision to close a fishery. Closing one part of the fishery results in another group's getting a larger catch. Treaties give tribes the right to fish in their usual and accustomed places in common with non-Indians. And treaties require the United States to maintain the health of salmon stocks. In a legal sense, treaty fishing would be difficult or impossible to stop; and as long as there is fishing, other fishers will demand fishing opportunities.

If fishing cannot be eliminated legally for some peoples, such as Canadians and treaty fishers, which fishers can be restricted from fishing? Should recreational and charter fishing be stopped? Recreational and charter fishers take relatively few fish and contribute substantially to the economies of coastal communities. One variant of the no-fishing option could be for the United States to stop all ocean recreational, charter, and commercial fishing for salmon in Alaska and the Pacific Northwest. Such a ban would have favorable effects on negotiations with Canada in the PSC. However, it would be fought by all those affected. Another variant of this option would be to close all ocean fisheries in the Pacific Northwest. A complete ban of ocean fishing is close to being realized. It was proposed in 1993 for coho by the PFMC. It was proposed again in 1994 and implemented for coho, leaving only limited fishing periods for chinook. That was not the first time a no-fishing option has been proposed. In 1904, J.P. Babcock, the British Columbia fisheries commissioner, unsuccessfully proposed closing fishing on the Fraser River during 1906 as a conservation measure to build up sockeye stocks. The problem with any partial closure is that, although it might allow some increased escapement, it also redistributes catch among different fishing interests.

Canada and the United States have many points of complementary and cooperative interest that might be negotiated on a smaller, more-specific scale, rather than simple wide-scale closures. One point of complementarity is the catches of Pacific Northwest coho and chinook off the west coast of Vancouver Island and Puget Sound catches of Fraser River sockeye. Another point of complementarity is the possibility of opening ocean fishing areas to a joint fishery of trollers from Alaska, Canada, and the Pacific Northwest.

The Limited-Entry Option

If cessation of fishing is too strong an option, limiting the number of fishers might be helpful. All West Coast commercial salmon fisheries have some form of limit on the number of gill-net and troll licenses. The underlying idea is that the number of fishers should be limited to correspond to the size of catch that can be taken. The objective of the license-limitation program is to restrict fishing capacity to a level closer to the effort that can be maintained. One problem with limited entry is that it has many of the same elements as the status quo. Limiting entry to the degree necessary to produce the needed effect is perceived as a severe step.

A second problem that limited entry does not solve is the natural tendency of fishers to become more effective. New technology, knowledge, and fishing methods make fishers more efficient with the gear that they have. Thus, a limited-entry program must continually reduce the number of fishers in accordance with both resource availability and the capacity of fishing vessels and fishers to catch salmon (Smith and Hanna 1990); a reliable way to do this has not been perfected.

A third problem is that, as with open-access fisheries, successful application of limited entry depends on the ability to calculate accurately the quantity available for fishing. People want a consistent number, but fisheries are inherently variable; no stable number can be given. A safe number would have to be conservative, and fishers would probably complain that it is too low. The MSY mode of management has continually overestimated the amount available for fishing. With management for genetic diversity, as we have been recommending throughout this report, the focus is on achieving spawning escapements. That will mean highly variable catch opportunities for a much smaller fleet of vessels.

The Terminal-Fishery Option

Catching salmon closer to the place where they spawn allows greater separation of hatchery from wild and threatened from nonthreatened populations. A way to achieve that separation is to allow only terminal fisheries. The separation can be even better achieved with live traps. With live-trap, terminal fishing, salmon needing protection can be released if they are identifiable with minimal potential for harm. Because natural mortality in the ocean, after early transitions to ocean life, reduces biomass more slowly than body growth adds biomass to the population, fishing closer to the spawning grounds would increase salmon yields.

Ocean fishers might question the quality of salmon taken in terminal fisheries; the meat of fish caught nearer to their spawning grounds will tend to be less oily and the skin more colored, and they will be less preferred by some consumers. Salmon do deteriorate in quality as they get closer to spawning, but terminal fisheries in estuaries and river mainstems would not necessarily decrease quality and the average size of the fish would be greater. Two advantages of live-trap, terminal fisheries are the potential to separate populations from one another and the ability to set catch rates for what each population can sustain. For example, salmon from threatened populations could be released. The treaty fishery in the middle Columbia would be a place to experiment with terminal fishing. Shifting to a live-trap fishery also has the potential to increase employment. Recreational fishers view set nets as wasteful of the resource and as yielding lower-quality fish. Live traps would improve the perception of Indian fishing on both conservation and quality grounds.

A major problem with the no-fishing option is at least partially solved by adopting terminal fisheries. Alaskan fishers catch salmon destined for Alaska and British Columbia streams, as well as for the Columbia River and the north coastal area. Alaskan ocean fishers question why their opportunity to fish for healthy Alaskan populations should be jeopardized by habitat and hydropower problems in the Pacific Northwest. Canadian fishers who fish mixed U.S., Canadian, and Alaskan populations do not see a reason to limit themselves when the problem is not theirs. They have not built dams on the Fraser River, and in British Columbia the habitat is less altered.

Although the current catch situation—which is unbalanced between areas—will make it politically difficult to restrict fishing to locations of origin, the committee concludes that it is worthwhile and important. Salmon management—especially population-specific management—is likely only practical if catch were allowed only near the point of origin, and in the long run, the salmon and many fishers would benefit once production increased, although which fishers benefit most would involve social factors.

Developing a New Management Paradigm

Given the complexity and scope of the salmon problem, developing a new management concept will be difficult and contentious. The committee starts by identifying several premises based on its experience:

- In Pacific salmon, the presence of many diverse, spatially distributed spawning populations is closely aligned with genetic diversity, maximal use of available habitat, and potential for increasing production from natural spawners.
- The sustainable exploitation rate is a function of a population's productivity determined over all life phases. Catch is only one of numerous mortality sources and cannot be viewed as independent or as an alternative to other sources over which we do not have control. The fishable portion of a return is determined by the brood-year survival to the time of the fishery and the desired spawning-stock size.
- Salmon are a component of ecosystems and they exist in a dynamic evolutionary process. Their production is variable and interconnected with the condition of their communities and habitats.
- Catch is a function of the fishing rate exerted by a fishery and the abundance of salmon recruited to the fishery. A low fishing rate and a high abundance can yield the same catch as higher fishing rate and a lower abundance.
- Productivity varies among populations and over time. The projected return from any population and brood year is highly uncertain. Any management process must acknowledge and account for limitations and uncertainty in assessment information and management capabilities.

Those premises consider only biological aspects of fishery management. But the sustainability of salmon in the Pacific Northwest also is inextricably linked to economic development and societal values. Society in the Northwest has exchanged natural salmon populations for economic development or argued about who was to blame as the resource declined. Figure 11-5, based on data from Matthews and Waples (1991), demonstrates the decline in Snake River spring and summer chinook salmon since the late 1950s. In spite of a progressive decline, major corrective actions were not taken until 1992, when the chinook were listed as threatened under the Endangered Species Act. The greater the decline in the resource, the greater the disruption will have to be to correct the problem. The committee believes that a stronger societal commitment to the biological-resource base must be established if salmon are to be sustained. For

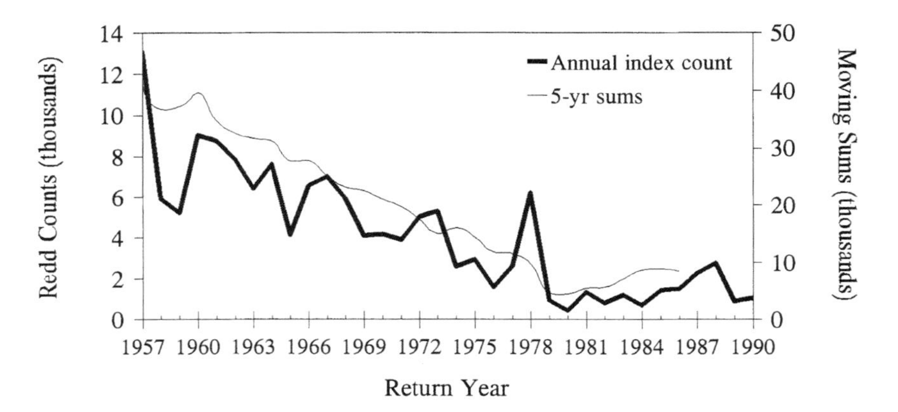

FIGURE 11-5 Trend in spawning-escapement index for Snake River spring and summer chinook salmon. Trend in annual redd counts and 5-year sum (for smoothing) are presented. Data from Matthews and Waples (1991) for Snake system minus Grande Ronde returns.

the fishery-management process to be effective, a strong commitment to the salmon must be an integral part of the process.

A management cycle for fisheries involves four activities: stock assessment to provide the biological advice, development of management plans, conducting the fisheries, and evaluation. The critical elements are sound biological advice, explicit and assessable management objectives (biological, social, economic, etc.), an institutional process for developing management plans, control of fisheries, and accountability in achieving management objectives. We consider those elements below, except for institutional processes and accountability, which are discussed in Chapter 13.

Stock Assessment and Biological Advice

Biological advice is only as sound as the information on which it is based. Advice must recognize limitations and uncertainties in knowledge and in abilities to predict recruitment. For example, the committee suggests that the concept of "optimum escapement" be replaced with a more conservative notion of a minimum sustainable escapement (MSE). An MSE concept avoids a single target escapement value and acknowledges that estimates of S_{MSY} are often biased low and rely on weak historical data. The committee emphasizes that MSE is a minimum and that actual escapements would exceed it and not be scattered about it. The committee's notion was based on protecting against the continued decline in salmon production and on concern about the use of an uninformative, possibly misleading, statistic.

The MSE level could initially be determined from historical stock-recruitment data. When the data are not available, initial escapement levels may be derived from habitat assessments and/or historical escapement trends. This information may then be incorporated in demographic or life history models to determine MSE at a particular level of confidence. In explicitly accounting for uncertainty in achieving the target, the concept of MSE is analogous to minimum viable population size (Shaffer 1981, Simberloff 1988) and population viability analysis (Gilpin and Soulé 1986, Shaffer 1990). It acknowledges that the longer-term sustainability of salmon populations depends on reducing the risk of extinction due to overfishing and stochastic events (environmental and demographic variability) and imprecision in fisheries management. However, in assessing these risks, society must determine the level of security desired for salmon populations over what time period (i.e., how confident do we want to be that an MSE is achieved annually and that a population will exist in 100 or 200 years?).

In many cases, the appropriateness of the initial MSE will be unknown. However, under the MSE concept, populations would generally be at less risk than under the earlier MSY approach, because escapements should exceed the MSE rather than fluctuate around it. MSE would be expected to lead to greater escapements, on the average, than achieved by the conventional application of the MSY concept. In practice, it will be important to allow variability in escapements above MSE in order to examine the productive capacity of populations and habitats, and the appropriateness of the MSE value.

Estimates of MSE should ideally include information about the composition of spawning populations, the maintenance of connections between salmon demes, the role of carcasses as nutrient sources for freshwater ecosystems, intraspecific competition in reproduction, mate selection, and gene flow, but relatively little attention has been given to these factors. The need for levels of escapements that promote competition and fertilization or that maintain niches used by salmon is not well demonstrated with direct research.

In summary, the committee recommends the establishment of minimum safe levels of spawning escapements to reduce the risk of continued loss of salmon populations and production. Actual escapements should always exceed this value, with allowances for assessment error for abundances near this minimum level. Escapements would vary above the minimum depending on the population abundance and sources of mortalities. Escapements substantially above these minima will be needed to maintain salmon productivity (and therefore, sustainable exploitation rates) in many more populations than are presently available. These increased escapements are also likely to have benefits in expanding the number of spawning populations, increasing genetic diversity within populations, and enhancing natural ecological processes.

Management Objectives

The major change in objectives related to the sustainability of salmon must be to broaden the set of biological objectives. That does not imply a priority of biological objectives over socioeconomic objectives, but socioeconomic objectives should complement biological objectives. The committee concludes that the resource base necessary to sustain salmon production consists of genetic diversity (both within and between natural breeding populations) and the habitat used by all life stages of the species. Genetic diversity provides for the continuing evolutionary process and is the biological basis of future salmon production. Therefore, the committee recommends managing for the joint biological objectives of MSE and increased diversity within and between local breeding populations, which will result in increased production in the long run. Increasing the size and number of spawning populations will, on average, increase the abundance of salmon. The committee acknowledges that increasing diversity will require initial reductions in catch because animals must survive to reproduce. However, catch in future years should increase as salmon production increases, even though fisheries probably would be managed at lower catch rates to maintain the diversity within local breeding populations and promote the development of interpopulation diversity.

Figure 11-6 shows what is expected in accordance with MSE. Graphs A, B, and C represent what has occurred commonly in the past. Natural or wild (N) and hatchery (H) populations have been fished simultaneously, but the hatchery population has higher productivity. As total population (N + H) increases, catch often increases to a maximum (Figure 11-6b), but the catch rate (i.e., the portion of the available salmon abundance that is caught) may not be sustainable by N. Consequently, the catch of N + H begins to decrease because of the declining production from N. Eventually, management responds to conservation concerns for N and reduces the catch to conserve N. If that situation is visualized over many natural populations, loss of population diversity can be characterized by Figure 11-6c. Diversity, if measured simply by the existence of spawning populations, would be maintained for a longer period than the catch (N + H). But under increased fishing pressures, the less-productive N will begin to be lost. Diversity would probably stabilize as catch is curtailed to conserve population diversity. Under a management policy to increase interpopulation diversity and achieve minimum escapement levels, the expected outcomes would be increased habitat use by spatially and temporally more diversified salmon populations and an increased catch achieved at a lower, sustainable rate of fishing (Figure 11-6d). The potential cost of this plan is an initially decreased catch of N while diversity is increased. The magnitude of initial loss depends on the specific situation.

A useful analogy of this plan is the idea of salmon runs as a tree. Each stem, branch, and twig on the tree is a potential home for a local breeding population, an isolated reproductive group adapted to the conditions of that particular stem,

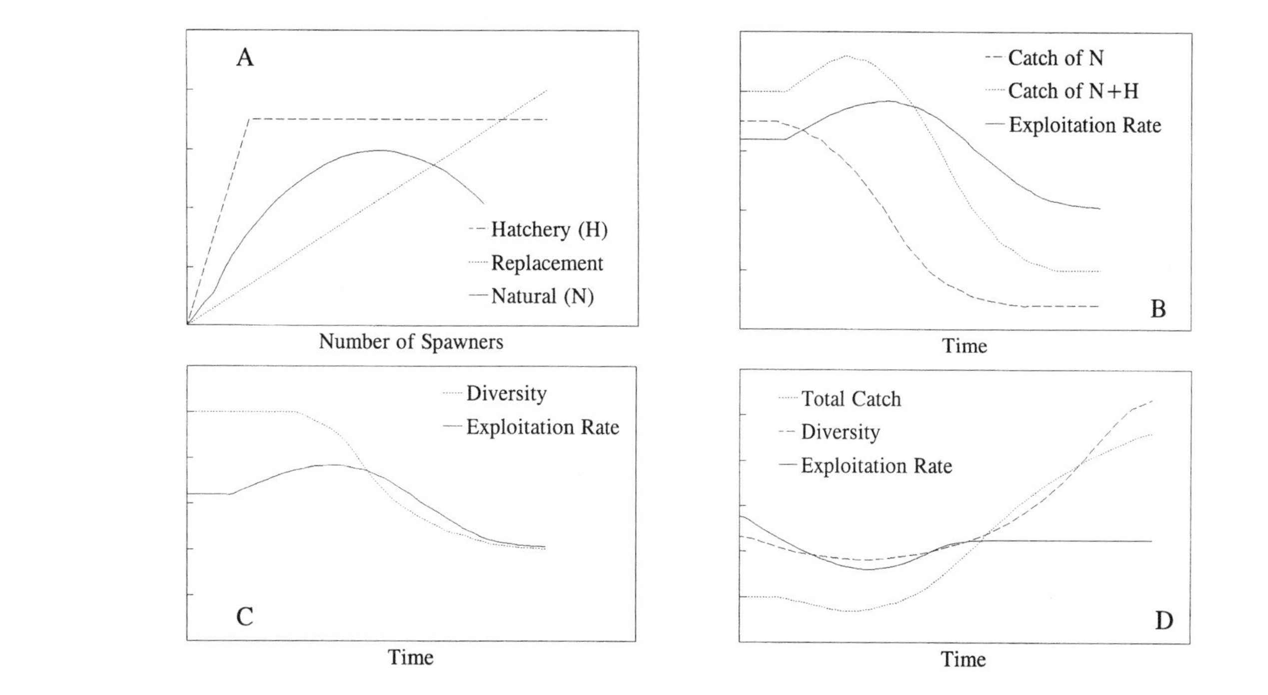

FIGURE 11-6 Schematic portrayal of observed historical trends in salmon populations, catch, and exploitation (plots B and C) and the expected outcome of managing for a minimum safe escapement (MSE) objective and rehabilitation of interdemic diversity (plot D). Plot A is a stock-recruitment curve showing number of recruits as a function of the number of spawners. In plots B-D, the vertical axes represent numbers of fish (for catch curves), numbers of populations (for diversity curves), and percentages (for exploitation rates). The exploitation rate is the total mortality associated with fishing activities, including the landed catch, incidental catch, discards, and hooking mortality. See text for further details.

branch, or twig. Some salmon climb mighty trees like the Columbia and Sacramento with complex branching. Others climb much smaller, less-complex trees like coastal streams. Cutting limbs from the dendritic structure of these salmon trees or placing obstructions on major limbs prevents local breeding populations from filling out the evolutionary potentials offered. That reduces the genetic diversity and viability of the salmon population as a whole and reduces habitat use and the potential production of salmon. A more holistic approach in salmon management would focus attention on filling out the trees' foliage so that viable local breeding populations of salmon inhabit as many branches as possible.

How could those joint biological and socioeconomic objectives be implemented in a management plan? The committee has considered only a general process because details of implementation would involve societal values and decisions. For example, how quickly diversity increases will be associated with how much societal change is acceptable or with the array of economic alternatives in a specific area. A possible process would involve the following:

- Identification of natural populations with the quantitative information needed for a credible population assessment and determination of an MSE and exploitation rate that, on average, would be allowable at this level of spawning-population size. Currently there are few of these "assessment" populations, but the application of a safe escapement level will reduce the risk of misapplication to other populations and should provide reasonable starting points in the plan development. Total fishing mortality would initially be limited to the exploitation rates at the MSE.
- Predictions of available abundance to fisheries. The methods might vary between regions, species, etc., but should account for spawning-population sizes, environmental variation, and interceptions in fisheries outside the management zones. Abundance forecasting also might prove to be highly uncertain, but methods to incorporate in-season information with pre-season estimates (see Noakes 1989) could be useful in controlling fishing impacts.
- Establishment of survey designs for estimating diversity within local breeding populations. The essential need is to measure diversity and how it changes over time. Surveys would be designed to be repeatable annually and to measure quantitatively the spatial and temporal diversity of local breeding populations.
- Conduct of annual evaluations involving quantitatively assessed indicator populations, surveys of the spatial and temporal diversity of local breeding populations within geographic areas, and fishery dynamics. The indicator populations would include natural populations on which accurate stock-recruitment data can be collected and whose dynamics (e.g., freshwater and marine survival rates, productivity, etc. [see Holtby and Scrivener 1989]) can be studied, natural populations that are conducive to repeatable annual estimates of spawning escapement, and hatchery populations whose exploitation rates can be determined. Fish-

ery dynamics are assessed to understand units of effort, relationships between catch and effort, and effort responses to abundance and ultimately to estimate catch levels for a fishery.

• Assessment of progress toward the biological objectives and incorporation of what is learned from evaluations into future management plans. Given the limitations in our knowledge and the inherent variability in the environment, the committee strongly endorses adaptive management (Walters 1986) to achieve sustainability for salmon. For example, the response of natural populations to management changes can be confounded by environmental variability. Experimental designs can be useful in controlling this interaction (see Walters et al. 1988) and in improving detection of changes in diversity over time and under different management plans or fisheries. The use of adaptive management, however, emphasizes the need for effective institutional processes for communication and participation in the development of longer-term management plans.

Control of Fisheries

Meeting the joint management objectives of achieving the MSE and increasing diversity of local breeding populations diversity will not resolve the mixed-population fishing problem or settle allocation debates. Without greater control on fishing impacts, meeting the objectives could even exacerbate these problems. Furthermore, the sequential alignment of fisheries in the Pacific Northwest, from ocean mixed-population fisheries to more terminal fisheries involving fewer populations, could result in inequitable disruption of fisheries. But sequential fisheries also present an opportunity to compensate for fishing impacts among fisheries. Given the complex of fisheries, variations in population size, and the need for social decisions in establishing a fishing plan, the committee felt that it was impractical to comment on any specific fishing options. There are only two general kinds of strategies for meeting the objectives through fishing controls:

• Reducing exploitation rates over all populations in a fishery—fishery-oriented strategies.

• Increasing the specificity (in time, area, gear, species, etc.) of a fishery to avoid or minimize impacts on particular populations—population-oriented strategies.

There are many ways to implement each kind of strategy. Fishery strategies can vary from no fishing through allowing exploitation only in specific fisheries to reducing exploitation rates in all fisheries. Population strategies can divert fishing effort to another time or area, develop a selective fishery for only marked animals (i.e., prohibit retention of unmarked fish), or develop selective fishing gear, such as live traps and fishwheels. Strategies can also be combined to limit exploitation of some populations while maintaining a fishery on others. For example, an ocean troll or recreational fishery might be managed at low exploitation rates that are sustainable by most populations. Terminal fisheries could then be managed to compensate for these ocean-fishery mortalities by either increasing or decreasing further exploitation on a population.

In developing a fishing plan, managers have to balance fishing capacity (number of vessels, effort, market prices, etc.), availability and quality of biological data (on abundance, stock composition, previous fishing impacts, etc.), and societal agreements (allocation requirements, treaty vs. nontreaty, ocean recreational vs. ocean troll, etc.). Each balance has problems. In the Northwest, more people would participate in fisheries if there were more fish. The potential for additional fishing pressure is an important source of uncertainty in how a fleet will respond to a particular fishing plan. The quality of biological data varies among fisheries, but the catch rate is seldom known until after the fishing has ended for the season. Achievement of allocation agreements is uncertain because population-specific fishing mortalities often are unknown or a substantial portion of the allowable catch might be taken in fisheries outside the management region, e.g., in Canada or southeastern Alaska. The most common problem, though, is our limited ability to control in-season fishing impacts, especially on a population-specific level. In the absence of reliable pre-season predictions of population and fishery abundance, fishery managers have developed in-season estimation procedures to monitor abundance and run timing. These procedures normally compare historical test-fishery catches or catch-per-unit effort from specific fisheries, with run-size estimates to develop in-season prediction models. These models frequently also have large uncertainties due to variation in run timing, stock compositions, and environmental conditions; or simply due to measurement error in historical data. In summary, the quality of biological data varies widely between fisheries, and exploitation rates in fisheries are seldom known. This uncertainty places the objective of increasing genetic diversity at risk and argues for the continued application of conservative fishing plans, particularly in the mixed-population ocean fisheries. Fishers should recall, however, that fishing at a lower rate on an increasing population will eventually restore catch levels.

Developing fishing plans for each of the Pacific Northwest regions will necessitate consideration of specific resource problems, distribution of fisheries, and social groups. Choosing a strategy requires establishing priorities and making a number of difficult societal choices. But fishing is only one mortality factor. Fishers can enhance the spawning population by forgoing catch, but salmon also require habitat for long-term sustainability. The control of fishing as a means to approach sustainability in salmon will be only as successful as our ability to address the freshwater-habitat issues. We would also expect greater support from fishers if they could see a successful return on the spawners invested. Presumably, the same would be true of Canada's participation in the Pacific Salmon Treaty.

CONCLUSIONS

Cumulative effects of fishing activities have contributed to depressed production. Fishing must be managed on the basis of total fishing mortalities (catch plus incidental fishing mortalities) and operate at sustainable exploitation rates. Even after a population has recovered, managers and users should *not* expect a return to historic exploitation levels, because those were based on excessive fishing rates. The exploitation levels might be achieved again only if population sizes were rebuilt to their former numbers and survival were good. Large catches could still result from fishing at low sustainable exploitation rates, but on larger abundances of salmon.

12

Hatcheries

Anadromous-salmon hatcheries constitute a primary human intervention in the Pacific Northwest, and in many areas of the Pacific Northwest hatchery fish make up the majority of salmon in rivers and streams (see Box 12-1). In this century, the intended goal of most hatchery programs has been *mitigation*. Mitigation aims to lessen the immediate impact of human actions through definition of a "socially acceptable" altered state (Christie et al. 1987). Hatcheries were expected to lessen the impact of numerous human actions that have dramatically altered freshwater ecosystems of Pacific salmon. For instance, the impact of construction of mainstem dams on the Columbia River was the decline or loss of many upriver populations. The socially acceptable state chosen was large-scale hatchery production in the lower Columbia, below the dams, to substitute for lost upriver populations. Both degraded environments and overfishing have contributed to the dramatic decline in the numbers of mature salmon that escape capture or death and reach natural spawning areas. Hatchery systems have been expected to compensate for this decline in escapements. A number of long-existing Pacific salmon artificial-propagation programs have been touted as successes, but these claims have increasingly been called into question (e.g., Riddell 1993a, Gharrett and Smoker 1993, Bowles 1995, Allendorf and Waples in press).

Most artificial-propagation programs have not undertaken long-term evaluation and documentation of the extent to which intended goals were reached (e.g., increase the catch for a given population, prevent extinction of populations whose spawning grounds were destroyed by dams) and unintended risks were imposed (e.g., adverse genetic or ecological impacts on naturally reproducing fish). For many artificial-propagation facilities, this lack of long-term monitoring makes it

Box 12-1
Recent Examples of Prominence of Hatchery-released Fish

1. Preseason forecasts, based on separate estimates of hatchery-propagated versus natural chinook salmon adults available to net fisheries in Puget Sound (Salmon Technical Team 1993), were that hatchery fish would amount to 73% of available fish in 1992 and 1993; postseason estimates were that hatchery fish comprised 60% in 1990 and 63% in 1991. In some locations, only estimates of hatchery fish are provided; this raises the question of whether any natural fish are present, perhaps in numbers too small to affect estimates of fish "available to net fisheries."

2. By 1987, hatchery-origin fish dominated adult returns in the Columbia River basin, comprising more than 95% of the coho, 70% of the spring chinook, about 80% of the summer chinook, more than 50% of the fall chinook, and about 70% of the steelhead (Columbia Basin Fish and Wildlife Authority, 1990).

3. In the Snake River basin, reliance on hatchery propagation of chinook increased from 0.75 million juveniles released in 1964 to 14.9 million in 1989, but this did not prevent steep declines in numbers of adult returns to the basin (Chapman et al. 1991).

4. About 70% of coho populations in Washington coastal streams were recently classified to be of hatchery origin (WDF et al., 1993).

5. From 1969 to the 1980s, releases of hatchery coho smolts in the Oregon production-index area (California and Oregon coasts and Columbia River) increased from about 10 million to 40-60 million in the 1980s, adult production fluctuated between years but returned to the same level at the end as at the beginning of this interval (Lichatowich and Nicholas in press).

6. Hatchery fish make up about half the overall abundance of steelhead found from Alaska to California but about 70% of steelhead from coastal Oregon and Washington and the Columbia River (Light, 1987).

7. The abundance of hatchery-origin spring chinook on the Rogue River (southern coast of Oregon) rose from a negligible value in 1977 to about that of naturally reproduced fish between 1980 and 1991 while overall productivity of the run dropped from 20,000 adults to below 5,000. Specific data sources and whether "wild" fish values clearly exclude hatchery fish were not reported (Richards and Olsen 1993).

8. From 1970 through 1992, hatchery-propagated fall chinook returning to hatchery facilities made up a relatively small proportion of all adults returning to spawn in the Sacramento River basin but "available data" indicate that most naturally spawning fish were hatchery-produced (Pacific Fishery Management Council 1993, p.II-1); the type of data was not specified.

9. About 57% of the 31,000 adult coho that returned to California streams annually in 1987-1991 were hatchery fish; only 16% had no hatchery ancestry (Brown et al. 1994).

nearly impossible for us to differentiate impacts of hatchery programs from impacts of other human interventions or of natural environmental changes. Since its inception in 1977, even the ambitious Salmon Enhancement Program of British Columbia failed to collect data needed to evaluate its benefits and risks (Hilborn and Winton 1993, Winton and Hilborn 1994). Over the last century, the Pacific Northwest culture passed up the chance to learn adaptively about artificial propagation of anadromous Pacific salmon.

In this chapter, we discuss the effectiveness and effects of hatcheries. In retrospect, it is clear that hatcheries have caused biological and social problems. For example, hatcheries have contributed to the more than 90% reduction in spawning densities of wild coho salmon in the lower Columbia River over the past 30 years (Flagg et al. 1995). Hatcheries were not part of an adaptive management program, so inadequate evidence was gathered to assess their impacts. As in so many other aspects of the salmon problem, there is a dearth of good scientific information about hatcheries' effects. In this chapter, we base our discussions of hatcheries' effects on a synthesis of biological principles and empirical evidence.

Generally, the problems stem from the predominant goals of hatchery operations—increasing run sizes to mitigate other human-induced mortality—and from insufficient incorporation of basic genetic, evolutionary, and ecological principles into hatchery planning, operation, and monitoring. In addition to the direct effects of hatcheries, management practices often associated with hatchery operations but not required by them have also had adverse effects on wild populations. The practices include outplantings (population transfers), overfishing of wild fish in mixed fisheries on hatchery and wild fish, broodstock extraction, and environmental manipulations (examples include the building of weirs and removal of salmon carcasses from streams) associated with the management of hatchery fish in the wild (Campton 1995).

Hatcheries have had some successes, but the ideas behind their operation and the ways that they are managed and operated will need to change if they are to play an important role in maintaining the long-term sustainability of naturally reproducing salmon.

PROBLEMS ASSOCIATED WITH HATCHERY PRACTICES

Traditional approaches of hatchery programs have imposed different types of biological problems on salmon populations, including demographic risks; genetic and evolutionary risks; problems due to the behavior, health status, or physiology of hatchery fish; and ecological problems. One or more of those problems might have affected either the populations that a hatchery program aimed to rebuild, other populations with which they interact, or both (Allendorf and Ryman 1987, Gharrett and Smoker 1991, Kapuscinski 1991, Riddell 1993b, Allendorf and Waples in press). The extent of the effects of any particular hatchery program is hard to document because of a lack of appropriate monitoring over appropriate periods. Given the Pacific Northwest's great diversity of hatchery operations and ecological contexts in which the hatcheries have operated, there probably has been much variability in the degree to which any of the problems has occurred (see Box 12-2). Growing scientific evidence supports the notion that hatchery-caused problems cannot be ignored without further threatening the future of depleted salmon populations.

Demographic Risks

Large-scale releases of hatchery fish have greatly exacerbated the mixed-population fishery problem. Less productive populations in the mixture, often the naturally reproducing ones, are overfished as a consequence of relatively high exploitation rates, which are set in response to the relatively large contribution of hatchery fish to the mixture. Purely because of the demographic problem of excessive mortality of sexually mature adults, wild populations in the mixture are eventually driven to extinction as their escapements drop below replacement levels. Chapter 11 discusses overfishing and mixed-population fishing in greater detail.

Genetic and Evolutionary Risks

Current approaches to artificial propagation can impose four types of genetic risks on hatchery fish and interacting populations (Busack and Currens 1995): extinction due to such genetic problems as inbreeding depression, loss of population identity (between-population genetic variation), loss of within-population genetic variation, and domestication selection. Redesign of hatchery programs can greatly reduce imposition of these risks, particularly the loss of genetic variation between and within populations (Campton 1995).

Population Identity and Within-Population Variability

Numerous studies involving comparisons of hatchery and wild populations have yielded strong evidence that traditional hatchery practices have caused losses of genetic variability between and within anadromous-salmon populations (e.g., Allendorf and Ryman 1987, Riddell 1993a, Allendorf and Waples in press). Only isolated sets of data are available that can be used to estimate roughly the prominence of hatchery-released fish for particular species in particular locations. Box 12-1 contains some examples of these. Release of hatchery fish propagated from nonindigenous broodstock without consideration of existence of local breeding populations and metapopulations (see Chapter 6) was common in the past and was a principal mechanism of loss of genetic variability between populations. Most hatchery programs now avoid this practice. Straying of hatch-

Box 12-2
Weighing Advantages and Disadvantages of Hatcheries—An Example

Empirical information that would allow managers to weigh the risk of domestication due to hatchery rearing against the risk of extinction of a numerically depressed wild population is lacking. That lack constitutes a critical uncertainty for making decisions about whether hatchery fish should be used to rehabilitate a depleted population, particularly when chances of substantial and rapid habitat rehabilitation are low, as is the case for upriver Columbia basin populations whose spawning grounds lie upstream of numerous dams. If managers assume that all descendants of hatchery fish will inherit low fitness in natural environments because of selection imposed by the hatchery environment, they will want to discourage interbreeding of hatchery and wild fish even if all other genetic risks have been minimized in the hatchery population.

That seems to be the assumption underlying the Wild Fish Policy of the Oregon Department of Fisheries and Wildlife (ODFW). The policy (ODFW 1992, Section 635-07-527) limits the number of hatchery fish that can interbreed with wild fish on natural spawning grounds to no more than 50% of the breeding population. Application of the 50% rule requires that other genetic risks of hatchery fish be minimized by use of local broodstock and other measures. The policy increases the risk of demographic extinction of the wild population when the purpose of hatchery releases in the first place is to rebuild the population. If a large number of hatchery-bred adults return, adherence to the 50% rule requires slaughter of many of the returns even if the total number of natural spawners is very low. If the assumption of high domestication risk is wrong, the policy can unnecessarily increase the probability of extinction of the wild population as long as it persists at a small size—this is called demographic stochasticity.

A concrete example of concerns raised by that ODFW policy occurred in the Imnaha River Basin, a tributary of the Snake River situated upstream of eight dams (Berg 1993). Within the Imnaha River Basin, biologists estimate that sufficiently good habitat exists to support at least 3,800 spring chinook. The current population is designated as threatened under the Endangered Species Act. A hatchery program to rebuild the population began in 1982 with wild broodstock from the Imnaha River; each year, about 30% of the hatchery broodstock are of wild origin. To meet the 50% rule, ODFW intended to kill 300 of about 1,000 spring chinook that returned to the Imnaha River in 1993. Objections raised by the Nez Perce Tribe and the Confederated Tribes of the Umatilla Indian Reservation and later formal dispute resolution led ODFW to adopt a variance to its wild-fish policy: 140 of the disputed adults were released into the wild, and 150 were spawned in the hatchery with the intent of releasing their progeny to the wild.

ery fish, another mechanism of loss of between-population variation, is more difficult to prevent. Waples (1991) reviewed evidence suggesting that hatchery fish stray at a higher rate than naturally reproduced fish, but there are few data sets on straying of wild salmon (Quinn 1993).

Within-population genetic diversity will be eroded if the effective population size (N_e) of the hatchery population—a number usually smaller than the number of adults actually mated in the hatchery—is depressed because of insufficient numbers of broodstock and inappropriate mating protocols (Simon 1991, Kapuscinski and Miller 1993). Traditional hatchery broodstock practices have often led to an N_e much smaller than the census population size (Simon et al. 1986, Bartley et al. 1992b). In some hatchery broodstocks, deleterious changes suggestive of inbreeding depression, such as increased incidence of abnormal vertebral columns and missing fins, have been associated with declines in within-population genetic diversity (e.g., Allendorf and Leary 1988). Within-population genetic diversity can also be eroded if artificial selection against a given heritable trait *recurs* breeding season after breeding season as a consequence of some hatchery practices (as explained in the next section). Genes for quantitative traits (i.e., such continuously varying traits as weight, growth rate, and fecundity) usually occur in blocks of polygenes, whose products interact to affect the physiology and fitness of individual fish. Recombination among such polygenes is an important basis of adaptation via natural selection. If repeated artificial selection occurs in a hatchery program, genetic diversity of these polygenes will decline, closing off options for evolution and jeopardizing long-term persistence of the population. For example, culling of smaller fish imposes a selection differential against length, weight, or other size-related traits (see Figure 10 in Kapuscinski and Jacobson 1987). Alleles that influence the size traits are removed from the population (Lacy 1987, Falconer 1989). In addition, alleles that are linked to the alleles for size traits but whose phenotypic effects and fitness value are usually unknown are removed from the population.

Domestication

If fish become domesticated by genetic adaptation to a hatchery, they will have a commensurate decline of fitness in natural environments. Domestication—selection imposed by human actions—can be imposed in two ways. One potential source is nonrandom collection of hatchery broodstock over the duration of a spawning run. Genetic response to this source of selection is well documented. For example, Steward and Bjornn (1990) reviewed numerous observations of hatchery stocks shifting to early run timing (a genetic response) after consecutive collections of hatchery gametes primarily from early-returning adults. A second source of domestication selection is altered selection pressures due to differences between the natural environment and the hatchery culture resulting from physical conditions or operational practices in the hatchery.

Effects of hatchery practices. Historically common hatchery practices, such as size-grading of smolts, differentially affect survival of hatchery fish in natural environments. The risks of such selective practices are often recognized today, and these practices seem to have been stopped or much reduced. Explicit statements about the undesirability of such practices should be included in hatchery policies and guidelines for all hatchery programs that affect wild populations (Kapuscinski 1991, Kapuscinski and Miller 1993).

Disruption of natural sexual selection by artificial matings. The practice of making artificial matings—now the dominant hatchery method—is a serious concern because it disrupts natural patterns of sexual selection with negative implications for fitness of hatchery fish in natural environments. Sexual selection is an important part of the evolution of most vertebrate species. The salmon case is especially striking, in that millions of "trial balloons" (fertilized eggs) for the next generation are produced all at once by the spawners of any generation, followed by the death of the fish that produced them.

Sexual reproduction in salmon, as in most other forms of life, is the exclusive gateway to the genetic future of each of the many populations. The relatively small number of breeding salmon that arrive at a home site to spawn are true survivors. In some manner, throughout their life cycle, they have survived many challenges: predators, scarcity of food, accidents, diseases, and pollutants that might have weakened or killed them. Under natural conditions, sexual selection would normally occur within these few survivors. Sexual selection favors the pairing off and differential reproduction of some individuals at the expense of others. The elaborate structures and colors of spawning salmon—especially males, with humped backs, enlarged and hooked jaws, enlarged canine teeth, and brilliant reds and greens—are used only in the mating process to increase the probability of mating (Quinn and Foote 1994). Breeding competition in Pacific salmon is especially concentrated and intense because most breed only once during one spawning season and then die, allocating all their breeding effort to one massive bout of reproduction.

The most effective breeders in general are few, implying that this process of sexual selection is extremely intense and important evolutionarily. In hatcheries, the whole process is bypassed: humans select adult males and females, strip their eggs and sperm, and then rear the fry. The human breeders have no way of identifying the genetic relatedness of spawners or fish that would be the best natural breeders. Although not all the effects of this inadvertent interference with natural selection are precisely known, it is almost certain that one result is loss of general vigor, adaptation to local environments, and evolutionary fitness. Extrapolating from other examples of human interference in breeding processes in domesticated animals leads to the conclusion that the species' ability to survive unaided in natural conditions must be diminished. For example, substantial differences in potency among male pink salmon have been experimentally demonstrated (Gharrett and Shirley 1985). Such differences—if they have a genetic basis—would provide a mechanism for hatchery processes that bypass natural mating processes to produce genetic changes in breeding ability. Indeed, some direct evidence shows that genetic differences in reproductive behavior have resulted from the bypassing of sexual selection in hatcheries (Fleming and Gross 1993, 1994).

Probability of selection due to physical hatchery conditions. To date, physical and biological conditions in hatchery facilities have differed greatly from those in natural environments. For example, hatchery rearing systems have lacked diversity in habitat structure, cover, diversity in flow and temperature regimes, and exposure of fish to natural prey and predators. The physical conditions of the hatchery rearing environment have the potential to alter selection pressures in one generation, lead to a genetic response (adaptation to the hatchery environment) in descendants, and thus decrease fitness in natural environments. However, the degree to which the potential has been realized is not clear from review of the published literature (Kapuscinski and Miller 1993). Rapid domestication has been demonstrated in captive broodstocks that have been reared in captivity for the entire life cycle over several or many generations (Vincent 1960, Flick and Webster 1962, Flick and Webster 1964, Mason et al. 1967, Moyle 1969, Johnson and Abrahams 1991). In that situation, genetic adaptation to the hatchery environment is greatly enhanced because of the lack of exposure of the population to natural selection in natural environments.

The risk of domestication is less certain in most Pacific salmon hatchery programs, where only part of a fish's life cycle is exposed to hatchery selection pressures and the rest is exposed to a complex suite of selection pressures in freshwater and marine environments, including selection pressure by natural environmental conditions, such as predation in the ocean, and by human-induced environmental conditions, such as dam passage, stream habitat alterations, altered stream fish communities (e.g., Reznick et al. 1990), and possible fishing selection pressure (e.g., Policansky 1993). Interactions among the various sources of natural and anthropogenic selection can be complex and nonlinear and can yield surprising and counterintuitive results. That is even true for a single anthropogenic source of genetic selection: size selectivity of fishing gear. Miller and Kapuscinski (1994) showed that genetic selection on size traits can vary substantially between fishing seasons. No selection, selection against large size, or selection against small size can occur in a given fishing season, depending on the interaction between the relatively constant selectivity curve of the fishing gear and the more variable characteristics of the fish population, such as sex ratios and the frequency distribution of different age classes (because of correlation with size classes). To determine the relative importance of hatchery environments that cause genetic change, compared with other sources of genetic change, and to guide hatchery reforms, the degree and severity of selection due to hatchery conditions would need to be established through research.

Insufficient understanding of the potential of hatchery environments to cause domestication with consequently low fitness in the wild does not justify ignoring the issue in the formulation of rehabilitation strategies. Indeed, the risk of domestication should be explicitly addressed in any rehabilitation strategies that include releases of hatchery fish. Furthermore, questions about domestication should be directly addressed by a research program.

Behavior

The behavior of hatchery-propagated fish differs from that of their wild counterparts. These changes can result from learning or from genetic differences. The hatchery and stream environments differ in many respects, and the foraging, social, and predator-avoidance behavior patterns of fish reared in the different environments are often different (e.g., Moyle 1969, reviewed by Suboski and Templeton 1989, Mesa 1991). The selection regimes posed by hatchery versus stream environments also differ. For example, the behavior of coho salmon surviving a year in the river and in the hatchery will probably differ. Fewer of the hatchery fish will survive the rigors of the stream, and their culling will almost certainly not be random with respect to behavioral traits. The lower mortality in the hatchery might reduce selection pressure against some behavioral traits, but the hatchery environment might also select against different behavioral traits. Hatchery fish could be more aggressive than wild salmon (e.g., Fenderson et al. 1968), and recent studies have shown that both aggression (Riddell and Swain 1991) and boldness when foraging (Johnson and Abrahams 1991) have heritable components.

Thus, hatchery-reared salmon are often more aggressive (and larger) than their wild counterparts for genetic or environmental reasons. Paradoxically, they also generally experience much higher mortality once released from the hatchery (i.e., lower survival from release as smolts to return as adults) than wild fish. Efforts to use hatchery-produced underyearling coho salmon to rebuild populations in Oregon had the worst possible result (Nickelson et al. 1986). The hatchery fish displaced the wild coho that were in the streams and the authors believed that this was the result of competition between the larger hatchery presmolts and the smaller wild juveniles. Other studies have shown fish size to be important in determining the outcome of competitive interactions (Chapman 1962, Mason and Chapman 1965, Chandler and Bjornn 1988). In the Nickelson et al. study (1986), the hatchery-adult returns showed a large shift from late to early spawners; but early hatchery spawners contributed very little successful reproduction, so fewer offspring were in the stocked stream than would have been produced if the initial stocking of larger hatchery fish had not occurred. The lesson from that study is that any efforts to use hatchery fish to rebuild a naturally reproducing population should consider the ecological implications of—and perhaps seek to avoid—changes in time of spawning and size at release for hatchery and naturally reproduced fish.

Fish Health

Although there is much information on the incidence of disease and effects on salmon and steelhead in hatcheries, the effects of disease on released hatchery fish and their interaction with naturally reproduced fish are not well understood (reviewed by Steward and Bjornn 1990). Disease is thought to be directly or indirectly responsible for substantial post-release mortality of hatchery fish. Disease outbreaks are relatively common in hatcheries and are managed with one or more of the following methods: disinfection of influent water if the water source is surface water and if finances permit, separation of carrier fish if they can be identified, therapy, and, if the disease is widespread in the hatchery stock, destruction of all the fish. Steward and Bjornn (1990) suggested that widespread use of chemotherapy to control disease in hatcheries could result in the development of new, drug-resistant strains of viral and bacterial pathogens. Research and development are being actively pursued to improve the sensitivity and cost effectiveness of screening for carrier and diseased fish and to develop vaccines for immunizations against major diseases for which therapy is not effective, such as bacterial kidney disease and infectious hematopoietic necrosis virus.

In spite of comparatively high incidence among some hatchery-fish populations, there is little evidence of transmission of disease from infected hatchery fish to naturally reproduced fish (reviewed by Steward and Bjornn 1990). However, there has not been much research on this question, and most disease-related losses in natural environments would probably go undetected. The ability of hatchery fish to transmit disease probably depends on the ecological characteristics that influence the spread and pathology of the particular disease, the environmental conditions of the particular site, and the abundance and distribution of the released hatchery fish.

Disease concerns interact with genetic factors in some important ways. Inattention to the genetic risks of loss of between- and within-population diversity can seriously compromise inherited modes of combating disease. Substantial evidence has accumulated that parasites (including pathogenic viruses, bacteria, protozoans, helminths, and arthropods) play a critical role in the adaptive evolution and persistence of natural populations of vertebrate species (O'Brien and Evermann 1988). A wide variety of genes that encode or regulate mechanisms for the host to combat infectious disease have recently been discovered in numerous vertebrates, including fish (Stet et al. 1990, Stet and Egberts 1991). These genes are polymorphic within species and populations. Dramatically increased vulnerability to infectious disease has been demonstrated in some natural populations, such as the African cheetah, which have experienced demographic declines and dramatic reductions in overall genetic variability due to human activities. The implication for rehabilitation of anadromous salmon is clear: it is critically important to maintain the remaining genetic diversity within and between populations to conserve diversity for genes that are involved in disease defense.

Differential susceptibility to the myxosporean parasite *Ceratomyxa shasta* has been shown both between species and between populations of the same species of Pacific salmon. Fish originating from Columbia River basin tributaries where the parasite is endemic were less susceptible than fish from river basins that are free of the parasite (Zinn et al. 1977, Buchanan et al. 1983). Furthermore, cross-breeding experiments demonstrated that interpopulation differences are genetically based (Hemmingsen et al. 1986). The results indicate that Columbia basin populations have evolved resistance in response to the natural selection imposed by the endemic *C. shasta*. Differences between populations, correlated with different transferrin genotypes, have also been found in resistance to bacterial kidney disease and vibriosis (Winter et al. 1980, Withler and Evelyn 1990). Other types of evolutionary adaptation to disease agents and other environmental stressors are probably present in the relatively few salmon populations that persist in their native watersheds (e.g., Bower and Margolis 1984). Conservation of the resulting spatial pattern of adaptations within different salmon populations native to different watersheds is essential for rehabilitation of salmon populations and, if rehabilitation is successful, for their long-term sustainability.

Physiology

The physiological state of hatchery fish often is suboptimal, even though they are sometimes larger than naturally produced progeny at the time of release, and this might lead to undesirable impacts on their health, on poststocking survival, and interactions with naturally reproduced fish. Physiological stress in hatchery fish due to crowded rearing conditions, handling, and transportation probably increases their postrelease mortality (reviewed by Steward and Bjornn 1990:45). The contribution of environmental stress to reducing the immune response of salmon is well documented (reviewed by Steward and Bjornn 1990:63).

Incomplete smoltification of some hatchery fish is also a major cause of concern (e.g., Shrimpton and Randall 1992). Research in the Columbia River basin suggests that poorly smolted hatchery fish lack strong downstream migration behavior, so they reside longer in stream habitats, while fully smolted fish tend to migrate downstream with little or no delay (Bradford and Schreck 1989, 1990; Snelling et al. 1991; Snelling and Schreck 1992). In addition to probably reducing adult return rates, longer stream residence is undesirable because it could increase opportunities for hatchery fish to prey on or compete with naturally reproduced fish, particularly if the hatchery fish are larger. The cause of incomplete smoltification of hatchery fish is not well understood but is believed to be related to environmental factors that differ from stream conditions, such as the lack of natural seasonal oscillations in water temperature if the hatchery water supply is a well (Shrimpton and Randall 1992). Research in an adaptive-management context (Walters 1986) is needed to design better hatchery rearing environments so that smoltification problems are prevented.

Ecological Problems

Hatchery programs have given virtually no attention to the ecological context into which fish are released. Yet some ecological factors, such as carrying capacity of proximate and distant stream environments for various juvenile life stages and density-dependent interactions within and between species, can exert substantial control over the fate and impacts of released fish. The ability of hatchery fish to survive and be integrated into natural habitats without imposing unnatural changes on wild populations of salmon and other species depends on the numbers and sizes of fish released, their physiological state, their behavior and health status, and the locations and timing of release.

Another ecological effect to be concerned about is that the use of hatcheries leads to a lack of salmon carcasses in the rivers. In natural situations, the carcasses of the spawning salmon that die provide important nutrients to the river ecosystems (Cederholm et al. 1989, Kline et al. 1990, Kline et al. 1993), including such terrestrial animals as bears and eagles. In addition, at many hatcheries managers prevent upstream migration of naturally reproducing adults although some agencies are beginning to recognize this and will, it is hoped, change it. The carcasses of fish that have been intercepted for breeding purposes are not returned to the river, and this also constitutes a substantial loss of nutrients in the ecosystem. The declines in adult carcasses might have reduced the nutritive capacity of stream environments and triggered low survival rates of naturally reproduced fry and juveniles because of lack of food that would normally result from carcass decomposition. That has been hypothesized as a partial explanation for the depressed natural reproduction of anadromous salmon in the middle fork of the Salmon River in spite of the high quality of the physical habitat in this wilderness area (Scientific Review Group 1993). In addition, a dramatic reduction in the number of redds reduces the number of disturbed redds from which released eggs drift downstream. Such drifting eggs are a source of food for resident trout, char, and sculpins.

Finally, there is a major concern about the effect of hatcheries on the carrying capacity of the rivers and the oceans. In some cases, enormous numbers of fish are released by hatcheries; about 5.5 *billion* smolts of Pacific salmon of all species are released annually from hatcheries around the Pacific rim from California to Russia and Japan; one chinook hatchery in Canada (Robertson Creek) contributed 366,000 chinook to coastwide catches in 1991 alone (Riddell 1993a). Other observations adding to this concern include decreasing body size at maturity and increasing age at maturity of Japanese chum as total returns have increased, suggesting density-dependent rearing limitations in the oceanic environment (Kaeriyama 1989 cited by Riddell 1993a); reduced catch of chinook salmon in the Strait of Georgia, British Columbia when hatchery releases exceeded 8.3 million fish per year (Riddell 1993a); and the suggestion that interannual variability in salmon abundance might increase as releases of hatchery fish increase (McCarl and Rettig 1983, Fagen and Smoker 1989).

ROLES OF HATCHERIES IN THE FUTURE OF SALMON

What roles do hatchery programs have in the rehabilitation of Pacific anadromous salmon? Answering that question implies agreement about the values human place on features of the environments in which Pacific salmon and steelhead occur. Hatcheries are only one human tool among many human and natural tools available to meet various objectives. Just as a carpenter would not invest in expensive tools before knowing what kind of construction was required, fishery managers should not reach for the hatchery tool or any other tool of human intervention before formulating explicit objectives for specific salmon populations and their ecosystems.

Natural aquatic biotic systems are intrinsically dynamic. To rehabilitate salmon populations successfully, human actions must accommodate natural types and rates of change. Emerging findings and ideas in ecology indicate that the long-term persistence and resiliency of natural systems is tied to their dynamic variability. Even if there were no human disruptions of habitat, the abundance of Oregon coastal coho, for example, would still be influenced by natural fluctuations in ocean productivity, with the entire period of fluctuation from low to high productivity lasting around 40 years (reviewed by Lawson 1993). Any human institutions involved in rehabilitation of coastal Oregon coho salmon, whether or not hatchery programs are a component, should take such variations into account. Thus, a period of many years—probably around 40—is needed for completing a single cycle of adaptive management (Lawson 1993).

It is important also not to be blinded by a "salmonocentric" perspective. Although it is critically important to address the needs of all life-history stages of salmon when designing rehabilitation strategies, too narrow a focus on salmon might prevent the attainment of a goal, even if the goal itself is focused on salmon. Rehabilitation strategies designed for freshwater life stages of anadromous salmon must include conservation and, when needed, rehabilitation of important ecosystem linkages. Examples of linkages that might be essential for rehabilitation and long-term sustainability of salmon are food-web interactions among salmon, their predators, and their prey; nutrient cycles, such as the contribution of nutrients by riparian vegetation of a stream; and the influence of diversity in physical structures, such as woody debris, on natural functioning of an entire stream ecosystem and thus on the quality of salmon habitats (chapters 7 and 8).

If there is no societal commitment to prevent and reverse human-induced damage to freshwater habitats, constraints arising from damaged ecosystems are likely to prevent a hatchery from increasing the abundance of adult returns. In the worst case, a hatchery used under such severe conditions could accelerate rather than reverse the decline of populations. Consider, for example, plans for a new hatchery program to rebuild depleted salmon populations in a river basin. Plans are often based on the expectation that more than 0.1% of all released smolts will survive to return as spawning adults (a 0.1% survival rate is the approximate breakeven point at which each adult spawned in the hatchery will be replaced by one adult return). Recently, serious concern has been raised that survival rates of smolts are too low within some habitat-altered basins—between the upstream sites of release and the river mouths—to ever reach the expected 0.1% return rate (e.g., Fast et al. 1991, Currens 1993). If the return rates were lower than 0.1%, the hatchery could "mine" remnant populations of natural spawners, perhaps leading to a decline in their numbers. Regarding natural large-scale controls, natural productivity cycles of the Pacific Ocean will control the maximum possible abundance of hatchery returns by forcing a low value during periods of low productivity and allowing increased values during periods of increased productivity.

The role of hatcheries is affected by society's choice between continual high human inputs to salmon ecosystems and rehabilitation of their natural regenerative capacity. Although many people now agree on the general goal of rebuilding the abundance of anadromous salmon populations, current debate is vigorous about the best actions for reaching this goal. There are essentially two options. The first is substitution—society commits an indefinite and relatively high amount of human input in an attempt to make up for natural regenerative processes that are damaged in salmon ecosystems. As soon as human societies start to attempt to replace natural processes, they are committing themselves to indefinite expenditures. This option therefore involves high costs to society in the form of money, goods, services, and institutional structures. Costs will increase as the ability of human actions to make up for damaged natural regenerative processes decreases and falls short of expectations. Increased control by humans does not protect against surprises in the behavior of a managed system, such as declines in salmon abundance. Such surprises remain possible because of natural phenomena (e.g., oceanic productivity cycles) at a spatial or temporal scale beyond the scale of human controls.

The other option is rehabilitation of the natural regenerative capacity of salmon ecosystems (Christie et al. 1987). This option reduces costs to society but requires a long-term commitment to obtain positive results. It can also require revision of fishery-management goals to reflect more realistic expectations about issues, such as increases in fish abundance that are possible under current versus revised land use and fluctuations in catch levels due to large-scale environmental controls that cause variability in fish abundance. Because this option emphasizes repair of environmental functions and processes, including redirection of human technologies, it is more likely to rebuild salmon populations to a sustainable status.

Choosing between those two broad options for human intervention requires a policy decision. Making such a decision in an informed and democratic manner is extremely difficult, and our nation has an uneven record in addressing imperiled natural resources. Such a decision greatly influences what human interventions will be carried out—including how different technologies will be used—in the name of rebuilding anadromous Pacific salmon populations. Intervention with the hatchery tool will differ in magnitude between the two options. Major hatchery reforms are imperative with either option. In general, this committee favors the rehabilitation option as more likely to be successful over the long term. It is also more consistent with the spirit of several current laws, including the Endangered Species Act.

Under the option of rehabilitation, the role of hatcheries would be much more limited and refined than their historical role because rehabilitation of the natural regenerative capacity of an ecosystem requires congruence of each human intervention with natural structures and processes of genetics, evolution, and ecology. An important building block of natural structures is the pattern of genetic diversity between and within anadromous salmon populations, and an important building block of natural processes is the evolution of salmon populations. Therefore, rehabilitation implies a genetic-conservation goal, as described below.

Hatcheries in the Rehabilitation Option

Two guiding principles should be applied to all uses of hatcheries in light of the goal of rehabilitation. First, a hatchery program should be only one component of a comprehensive rehabilitation strategy designed to remove or substantially reduce the human-induced causes of decline of anadromous-salmon populations. Because human causes of decline differ across the regions of the Pacific Northwest, such comprehensive strategies should be developed by region (see Box 12-3).

The second guiding principle is that all hatchery programs should adopt the genetic-conservation goal of maintaining the genetic resources that exist in naturally spawning and hatchery populations. These remaining genetic resources are the building blocks of the natural "shifting variability system" (Carson 1983) needed for long-term evolution and persistence of salmon populations. All agencies involved in management or rehabilitation of anadromous salmon must recognize that achievement of population-rebuilding goals will be jeopardized without concurrent adoption of a genetic-conservation goal (Riggs 1990, Gharrett and Smoker 1993, Riddell 1993b, Allendorf and Waples in press). As one positive step in this direction, the Northwest Power Planning Council recently revised its 1987 goal of doubling returns in the Columbia River basin to include prevention

Box 12-3
Examples of Hatcheries in Rehabilitation Programs

Native populations of lake trout (*Salvelinus namaycush*) were extirpated from all the Great Lakes except Lake Superior by the 1960s because of overfishing, sea lamprey (*Petromyzon marinus*) predation, and habitat degradation (Marsden and Krueger 1991). A major portion of rehabilitation programs involved release of hatchery-produced fish. Starting in the 1950s, large numbers of hatchery fish were planted in the different Great Lakes, but first indications of successful natural reproduction by these fish were not found until the 1980s. To date, the only unequivocal report of natural reproduction by hatchery adults whose progeny have survived to re-establish a spawning population is in Thunder Bay, Ontario, where indigenous populations were known to be extinct (Johnson and VanAmberg 1994). Other data indicating successful reproduction by hatchery fish come from areas where there had been remnant but scarce populations. The possibility that these remnant populations are responsible for increased natural reproduction cannot be ruled out, because of an observation from one historical spawning reef, Gull Island Shoal of Lake Superior. Although no remnant stock was detected in the 1960s, natural reproduction has since rebounded without any stocking of hatchery fish (Swanson and Swedberg 1980). Hatchery fish have also been observed to reproduce on a number of artificial and natural reefs in other areas, but even intensive sampling for progeny has yielded only fry, so there are questions about their survival to later stages (Jude et al. 1981, Wagner 1981, Nester and Poe 1984, Peck 1986, Marsden et al. 1988, Marsden and Krueger 1991).

Three lessons relevant to Pacific salmon can be drawn from efforts to rehabilitate lake trout populations (C. Krueger, Dept. of Natural Resources, Cornell University, personal communication, September 1994). First, rehabilitation of populations of longer-lived species is very difficult. Although the average lifetime of anadromous Pacific salmon is shorter than that of lake trout, it is still longer than the typical timetables of activities carried out by the organizations that have a stake in salmon ecosystems. Second, rehabilitation requires a social commitment to modify traditional use of natural resources for a long time. Lake trout were on the brink of rehabilitation to self-sustaining reproduction in northern Lake Michigan when management decisions were made that led to increased fishing mortality; this destroyed the capital of reproductively capable adults that had slowly accumulated and were necessary to re-establish a self-sustaining natural population. Third, the potential for rehabilitation can be destroyed if survival in natural environments is sufficiently depressed by habitat damage or introduced species. Compounding the long-standing problem of sea lamprey predation on lake trout is new evidence of failure of year-class recruitment from naturally spawning hatchery fish because of predation by nonnative alewife on lake trout fry (Krueger et al. 1994) and potentially fatal damage by layers of sharp zebra mussel shells to eggs deposited on spawning reefs.

of "further loss of biological diversity [in] anadromous and resident fish populations" (NPPC 1992a:18).

One way that hatcheries could contribute to a comprehensive rehabilitation program is outlined below. A second scenario—use of hatcheries for catch augmentation—is not a method of *biological* rehabilitation, but using hatcheries in this way, while monitoring their performance in a sensibly comprehensive adaptive-management protocol, could reduce or avoid further biological damage.

Temporary Hatcheries

A temporary hatchery can be used to prevent extinction of a severely depleted population or rebuild a depleted population to self-sustaining status while the human causes of its decline are being rehabilitated to the extent feasible (e.g., while freshwater habitat damage is being repaired). Such temporary use supports rehabilitation in two ways. First, human costs of control are reduced because hatchery expenses and hatchery-associated genetic and ecological risks can be terminated when the populations and their habitats are restored so that they can sustain themselves. Second, re-establishment of a self-sustaining anadromous salmon population is a good indicator of improved natural functioning of the overall watershed ecosystem and increases confidence that there will be long-term persistence of various goods and services that humans desire from salmon ecosystems. For example, re-establishment of self-sustaining lake trout populations has been treated as such an ecosystem indicator in the Great Lakes (Edwards et al. 1990). The hatchery facilities required for this scenario, if they are temporary, would be less expensive than current, permanent hatcheries.

Temporary hatcheries could be extended in principle if habitat damage or loss cannot be reversed in a reasonable period and people are not prepared to give up the salmon populations associated with the habitats. To be compatible with the goal of rehabilitation, this extension requires that negative impacts of hatchery fish on other naturally spawning populations be kept negligible. *To be truly compatible with the rehabilitation goal, this type of longer-term holding action should be considered only as a complement to comprehensive efforts to rehabilitate damaged habitat.* If a temporary hatchery is extended as a substitute for serious efforts to reverse initial causes of decline (as was the case in historical mitigation programs), decision-makers will be accepting a future of indefinite and high human input; such a future should be accepted consciously and openly, with full disclosure to stakeholders and thorough weighing of costs and benefits. For instance, hatcheries planned for the Yakima basin (Fast et al. 1991) fit this scenario only if plans also include major efforts to rehabilitate freshwater habitat within the Yakima and its tributaries; recent concerns about high in-basin mortality of released smolts underscore the importance of this point. Otherwise, hatcheries in the Yakima will constitute implementation of the goal of indefinite and high human input.

Catch-Augmentation Hatchery

The decline of Pacific salmon constitutes not only a decline in the biological productive capacity of the Pacific Northwest, but also a decline in human fishing activity. Other human activities have precluded the historically abundant traditional, recreational, and commercial catch of salmon. To restore or rehabilitate those human activities is a valid goal, and it might involve use of a long-term hatchery program to augment catch of some populations above levels that are now supportable by the quantity and quality of available freshwater habitat. This is perhaps the most controversial hatchery scenario in the eyes of many people who advocate environmental protection and genetic conservation. But in the Columbia River basin, some American Indian tribes are concerned that even if upriver populations are successfully rehabilitated to self-sustainability, numbers of returning adults will be so low without augmentation of the runs as to allow little or no catch in treaty-protected freshwater fisheries. To prevent genetic and ecological risks to long-term sustainability of nonaugmented populations and interacting species, this scenario requires feasible methods of separating hatchery fish from naturally spawning fish in freshwater habitats and selective catch of hatchery fish in the fishery (see Chapter 11). These conditions will be hard to meet in many cases and would require changes in fishing methods and management. Such changes require targeted research to develop simpler techniques for identification of hatchery fish and testing of alternative fishing methods for selective catch of hatchery fish, as proposed in Section 5 of the *Strategy for Salmon* (NPPC 1992b).

Regardless of the choice of overall goal (rehabilitation versus high human input) and subordinate hatchery objectives, major hatchery reforms are needed. They are outlined in the recommendations at the end of this chapter.

CONCLUSIONS

Despite some successes, hatchery programs have been partly or entirely responsible for detrimental effects on some wild runs of salmon. The best-documented detrimental effects are loss of wild populations because hatchery fish swamp a mixed fishery, which encourages fishery managers to set exploitation rates that cause overfishing of natural populations, and loss of natural patterns of genetic variability between and within populations. There have been virtually no efforts to test for ecological effects of hatchery programs although scientific knowledge about salmon ecology strongly suggests that detrimental effects are possible.

Hatchery use has not favored conservation of biological diversity. As a result, institutional arrangements in management agencies have not encouraged collaboration between hatchery managers and fisheries and land managers to attempt to protect or enhance the natural functioning of watershed ecosystems, and that natural functioning has deteriorated. Hatcheries can be useful as part of an integrated, comprehensive approach to restoring sustainable runs of salmon, but they are not a panacea and should be used only sparingly and thoughtfully.

The goals, specific objectives, and methods of past hatchery programs were not critically reviewed for scientific validity and practical feasibility. That led to the unrealistic expectation that releases of fish from hatcheries would always lead to increases in salmon abundance. In some cases, there also was the expectation that hatcheries would stabilize fluctuations in salmon abundance. By raising unrealistic expectations (Gale 1987), hatchery programs have discouraged societal acceptance and understanding of variability in salmon abundance as a natural, complex phenomenon influenced by oceanic and other factors that are well beyond human control. The need to make societal expectations more realistic is heightened by recent oceanographic research that suggested that former high survival rates of hatchery fish were influenced more by ocean conditions than by hatchery activities themselves (see Chapter 2).

Overreliance on hatcheries has also discouraged development of institutional arrangements and behaviors that would accommodate natural large-scale fluctuations in salmon abundance. For instance, heavy emphasis on hatcheries as the means of mitigating the effects of dams and overfishing has drained financial and other social resources that could be available for redressing "proximal" human causes of decline (as would be emphasized if the goal were rehabilitation).

Hatchery programs have lacked proper monitoring and evaluation (i.e., there has been no adaptive management). The typical measure of success of a hatchery program has been the number of fish released rather than the number of returning adults, much less the health of nearby wild populations. In addition, cumulative effects of all the hatchery programs in a given region have not been evaluated. Potential effects on the genetic diversity and ecology of wild populations have been ignored, although this is beginning to change. For example, a new Integrated Hatchery Operations Team (IHOT) in the Columbia River basin is an interagency effort involving 11 fisheries comanagement entities and seven cooperating entities. As a promising first step towards improved coordination among and monitoring of different hatchery programs, the IHOT recently issued and the Northwest Power Planning Council approved five general policies. They covered hatchery coordination, hatchery performance standards, fish health, ecological interactions, and genetics. There is also an implementation plan that includes independent audits of all 90 anadromous salmon hatchery facilities in the basin (Integrated Hatchery Operations Team 1995). There has also been insufficient attention to potential effects on long-term sustainability of both hatchery-propagated and naturally reproducing populations.

RECOMMENDATIONS

The approach to hatchery operations should be changed in accordance with the goal of rehabilitation and the ecological and genetic ideas that inform that goal. Whenever hatcheries are used, great care should be taken to minimize their known and potential adverse effects on genetic structure of stocks and on the ecological capacities of streams and the ocean. For example, it is important to prevent blocking wild fish from migrating upstream of hatcheries because this is likely to compromise recovery and sustainability of wild populations. In streams where only hatchery fish are found, managers should test the assumption that blockage is necessary to avoid disease incidence in hatcheries; this should be done by conducting different adaptive management experiments at different hatcheries that use this practice. More distant, upstream wild populations should be conserved even if their presence complicates efforts to keep hatchery fish separated from wild fish. The aim of temporary hatcheries should be to assist recovery and opportunity for genetic expression of wild populations, not to maximize fishing opportunities. Augmentation hatcheries do aim to increase fishing opportunities, but they must do so in ways that will not jeopardize the genetic makeup and long-term persistence of nonaugmented populations. Adoption of this recommendation will almost certainly result in a significant reduction in total output of salmon hatcheries. Some ways of implementing this recommendation follow.

- *The term "supplementation" as a goal of hatchery programs should be abandoned.* The ambiguity of this term has generated confusion about appropriate roles of hatcheries, as illustrated by its many and often incompatible definitions in the published and "gray" literature. Instead, precise thinking and terms are needed to define and select ecosystem-level goals (e.g., rehabilitation of natural regenerative processes versus high human control) that will set the context for any proposed hatchery program. Goals for each hatchery program are needed so that they are compatible with the ecosystem-level goal. Given the ecosystem-level goal of rehabilitation, a compatible hatchery goal might be to assist rehabilitation of a given population to self-sustainability at the carrying capacity of its suite of environments.
- *Hatcheries should be dismantled, revised, or reprogrammed if they interfere with a comprehensive rehabilitation strategy designed to rebuild wild populations of anadromous salmon to sustainability.* It makes the most sense to test the ability of hatcheries to rehabilitate populations whose natural regenerative potential is severely constrained by both short-term and long-term limitations on rehabilitation of freshwater-habitats. Hatcheries should be excluded from re-

gions where the prognosis for freshwater habitat rehabilitation is much higher, as is the case in many watersheds of the Oregon coast (for example Cummins Creek on the central Oregon coast. This recommendation would allow continuation of, for example, the hatchery-supported fishery on chinook at Willamette Falls because it is disrupting neither wild populations or rehabilitation efforts.

- *Hatcheries should be rigorously audited for their ability to prevent demographic, genetic, fish-health, behavioral, physiological, and ecological problems.* Any hatchery that "mines" broodstock from wild (natural) spawning populations should be a candidate for immediate closure or conversion to research. Diseased broodstock should be rigorously culled to minimize disease in progeny.

All hatchery programs should adopt a genetic-conservation goal of maintaining genetic diversity that exists between and within hatchery and naturally spawning populations. All agencies involved in management of anadromous salmon should recognize that achievement of population-rebuilding goals will be jeopardized without concurrent adoption of a genetic-conservation goal.

- *Intentional artificial selection should be discouraged because of four major concerns.* First, intentional selection might reduce within-population genetic variation, which could be decreased already as a result of previously imposed artificial selection, hatchery founder effects, and mating practices. Second, our ability to accomplish the desired selection is uncertain, given the difficulty of determining the intensity of selective forces, the traits on which they are acting, and the uncertainty about how natural selection can reinforce or oppose the imposed artificial selection. Third, counterselection for a given phenotypic distribution, such as later run timing, is highly unlikely to accomplish the goal of re-establishing the original genetic structure. Fourth, our understanding of natural populations is insufficient to predict which phenotypic distributions and underlying genotypes are necessary for optimal fitness over the long term in the environments of anadromous salmon particularly because of their variability and different degrees of human-induced alterations.
- *Genetic and ecological guidelines, based on the most up-to-date information base, are needed for all aspects of hatchery operations.* Scientific peer review and appropriate revision of draft guidelines must occur before they are implemented and evaluated. Incentives are needed for uniform and coordinated application of scientifically sound genetic guidelines in anadromous salmon hatcheries in the five Pacific Northwest regions.
- *Hatchery programs should avoid intentional transplantation of fish and unnatural patterns of straying by adult returns.* This is necessary to prevent disruption of differences in evolved adaptations that exist among populations indigenous to different watersheds.

All hatchery fish should receive identifiable marks. Visible marks, such as clipped fins, have advantages, but other methods, such as passive integrated transponder (PIT) tags and coded-wire tags, are also useful. Marking hatchery fish is important so that managers can distinguish between hatchery and wild runs. Implementation of this recommendation requires careful consideration of possible increases in mortality caused by different marking methods. Support is needed for research in and development of biological and other marks that will minimize handling-related mortality. Research is also needed to find selectively neutral genetic marks that are peculiar to hatchery fish and are clearly inherited by their naturally reproduced descendants but do not influence the fitness of individuals. Recent advances in application of molecular genetic markers to hatchery-bred fish suggest that it is feasible to develop such marks (Doyle et al. 1995).

Decision-making about uses of hatcheries should occur within the context of fully implemented adaptive-management programs that focus on watershed management, not just on the fish themselves. Coordination should be improved among all hatcheries—in release timing, scale of releases, operating practices, and monitoring and evaluation of individual and cumulative hatchery impacts, including a coastwide database on hatchery–wild fish proportions and numbers. Given the differences among the five regions in the status of environmental conditions that affect Pacific salmon populations, a different suite of interventions under adaptive management should be assembled for each region, and each intervention in a region should be treated as an experiment. Hatcheries would be an experimental treatment within adaptive management in some regions but not in others.

13

Institutional Analysis

INTRODUCTION

The long decline of salmon in the Pacific Northwest has been a slow process by human timekeeping: for more than a century, overfishing, habitat destruction and degradation, and substitution of naturally reproducing fish runs with hatchery-produced fish has depleted the genetic diversity and abundances of salmon. It has taken a long time for people to notice. The loss of wild fish was obscured by hatchery production; the dwindling of salmon, especially those east of the Cascade Mountains, was masked by the shifting of catch from the American Indians, who depended upon fish for subsistence, to commercial fishers, who came to depend on fish for economic survival. As salmon abundance declined, the remaining populations have been made more vulnerable to natural fluctuations (Schaefer 1981, Gilpin 1987, Goodman 1987, Hartman and Scrivener 1990).

What has taken a century to do will not be swiftly undone. The problems are of human, institutional origin; remedies, if they are to be found, must entail human, institutional change. If there were once thought to be easy remedies—such as hatcheries or fish ladders or temporary fishing restrictions or large-scale subsidies from the hydropower dams—they have proved to be less durable and more painful than had been hoped. A harder, subtler remedy is deliberate institutional redesign: the recovery team chartered under the Endangered Species Act (ESA) after the listing of Snake River populations called for centralizing authority in the National Marine Fisheries Service (NMFS). The recovery team saw important institutional problems—fragmentation of scientific effort, responsibility, and authority, leading to a lack of accountability. Arguing only that "no one is in charge" (Snake River Salmon Recovery Team 1994:6), the team recommended NMFS, which has a legal mandate to protect the endangered populations. Yet it would be asking too much of this one agency to bring about complex, painful changes in the region's economy. There may be good reasons to centralize and shift the locus of authority, but this response is not adequate for the large, highly diverse natural and social systems that form the habitat of Pacific Northwest salmon.

Institutions do matter. The bitter divisions and economic pain that are the legacy of the northern spotted owl in the Pacific Northwest stand as reminders that the available means of protecting declining species produce strife and that their costs are real. The available institutional incentives might well lead to a continuing drumfire of petitions under the ESA.

It is against this sober specter of conflicts joined and impending that the committee explored the possibilities for institutional change. We have found no easy answers, but constructive possibilities can be identified. Three ideas shape recent discussions of natural-resources policy in the American West: bioregionalism, cooperative management, and adaptive management. Each has a powerful logic rooted in science or reason, each seems socially promising, and each is systematic and thus suggests a strategy of reform. None will be easily adopted comprehensively: each of these ideas forces changes in the "lords of yesterday" (Wilkinson 1992) described in Chapter 5, the extractive approach to federally owned resources that underlies much of the western economy. Because the problems of salmon are intertwined with the routine operation of the economy, the existing institutional structure cannot be easily replaced by one tailored to the needs of salmon. An ecosystem approach must also include the humans within the ecosystem. Each of the three principal organizing concepts thus blends scientific and social questions, bringing diagnostic insights but also raising the stakes for what must be changed if salmon are to be rehabilitated.

The tenor of this appraisal should not be mistaken for pessimism: much can be done to improve salmon management, population by population, habitat by habitat, season by season. A serious and determined effort—which entails the combined actions of parties now bitterly opposed and widely separated in space and interests—is likely, over time, to yield a rehabilitated set of salmon populations and, more important, a fund of human knowledge and a structure of management that seems likely to be able to sustain both fish and fisheries into the indefinite future. But that serious and determined effort will take perseverance and durable support from a people and an economy marked more than anything else by their pace of change over the last century. Rehabilitating the salmon of the Pacific Northwest will take decades, incur direct costs in the billions of dollars (the cost of the Columbia River Basin effort to rebuild salmon populations since 1981 is more than $1 billion, and current costs exceed $200 million per year [Hardy 1993]), and require substantial realignments of property rights and governmental institutions; all this will demand of the leaders of the region both political courage and scientific acumen. These appear at first sight to be immense, perhaps infeasible requirements, until we recall that over the last dozen decades, a regional economy has grown in the Northwest, with an output valued at hundreds of billions of dollars per year, and that novel government institutions, such as the Bonneville Power Administration founded in 1937, and modified property rights, such as constraints on water pollution or the duty to recycle, have become accepted or even forgotten as routine. Much can be done, but it will not be easy, especially at first.

When human responsibility does not match the spatial, temporal, or functional scale of natural phenomena, unsustainable use of resources is likely and will persist as long as the mismatch of scales remains (Lee 1993b). In preindustrial societies, where long-term dependence on renewable resources has been the norm, a wide spectrum of institutions has emerged to link the scale of human endeavor to the scales of nature (Suttles 1968, McCay and Acheson 1987). But since the Renaissance that linkage has not always been visible or necessary in the short run and has often been displaced. Technological change created new resources, extending the human control of nature and making some substances, such as uranium, useful in new ways. As exploration enlarged the reach of markets, natural resources entered world trade: valuable resources—like furs, timber, and salmon—were removed from their native habitats and sold in distant markets at high profit. Unsustainable use was in full swing. But the growing momentum of technological change brought substitutes to market when natural supplies began to dwindle: new fibers for furs, new building materials for timber, and artificially cultured fish for wild salmon. Unsustainable exploitation could proceed to economic extinction, but the markets' warning signals of rising price would simply induce inventors to innovate and consumers to shift to substitutes.

If the march toward extinction of salmon is to be halted, scale mismatches institutionalized over the last century need to be recognized and then reflected in institutional changes that take matters of scale into account. This is not a call for wholesale change: neither removal of mainstem dams nor drastic change in property rights would in itself reopen migration routes for salmon or rebuild natural habitat. But until salmon have migration routes and habitat that can sustain their populations, decline will continue. In discussing bioregionalism, cooperative management, and adaptive management, the committee's purpose has been to use these ideas diagnostically to spell out the predicament faced by salmon today. Only after the diagnosis has been put into perspective can sensible paths to rehabilitation be identified.

BIOREGIONAL GOVERNANCE

The boundaries of property and government, like the less sharply etched patterns of markets, rarely follow the outlines of biology and topography. From an airliner above the Pacific Northwest, a passenger can see forest clearcuts, their rectangular edges slashing across hillsides, with streambeds delineated by protective strips of unlogged trees. Maps show straight-line political borders—cutting across the curvatures of ecosystems and societies—or ones that follow major rivers. Yet rivers are the centers, not perimeters, of biological provinces; rivers are thoroughfares where migration paths merge. And maps show institutionalized error: the border between central Idaho and southwest Montana was drawn on what was thought to be the continental divide, but the divide turns out to be on the higher range to the east, within Montana.

A simple idea for reform is to ground governance in bioregions: to shift the margins of human control and responsibility to match natural boundaries. For salmon, whose genetics are mapped to natal streams, the drainage basin is the logical unit. The idea of managing fish populations as distinct entities, spelled out in Chapter 11, may be feasible only within a geographic template anchored in drainages and migration routes. The idea is simple—and rarely implemented since the arrival of non-Indians. The Olympic National Park in Washington state encompasses a mountain range, a natural province; the Columbia River basin defines the geographic scope of the Northwest's most ambitious plan for salmon (NPPC 1992); the Forest Ecosystem Management Team (FEMAT) identified key watersheds in the range of the northern spotted owl (see Figure 7-4). But John Wesley Powell's dream, a century ago, to organize the settlement of the arid West around its river basins, was ignored in the rush to exploit the region's lands and minerals (Stegner 1954). A map of the Western states, with their straight borders, shows where priorities came to rest. Key watersheds (Figure 7-4), drawn by topography and gravity and populated by spotted owls, must contend with county boundaries (Figure 13-1) drawn by humans with rulers and transits.

Today, the definition of property and state borders has long been settled. More important, trade routes that link different biogeographic provinces are in place: highways, transmission lines for electricity, water diversions, and others. Along these human paths flow commercial transactions in which human values define the relative scarcity of specific resources, including water, salmon, and the farm and forest crops harvested from managed vegetation and animal populations. Any attempt to shift human jurisdictions so as to focus attention on the welfare of salmon must begin with these boundaries and patterns.

Because the migratory range of salmon goes from Alaska to California, from ocean, the biological range of the fish is too large and too diverse to be manageable as a single spatial unit. Most analyses (e.g., Johnson et al. 1991, Moyle and Ellison 1991) and some policies (FEMAT 1993, Washington Forest Practices Board 1993) use the watershed as the unit of planning. There are roughly 1,000 drainages in the Pacific Northwest, counting streams of order 6[1] and higher, and

[1]An order 2 stream is formed by the confluence of two order 1 streams; an order 3 stream is formed by the confluence of two order 2 streams, and so. The higher the order number, the larger the stream or river.

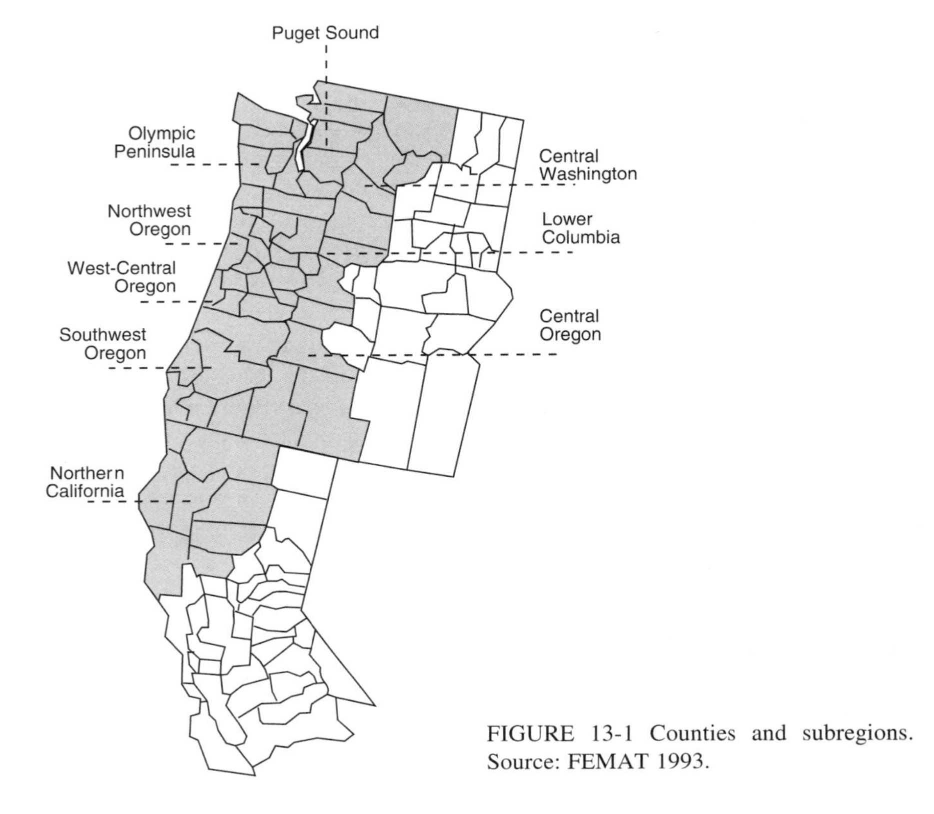

FIGURE 13-1 Counties and subregions. Source: FEMAT 1993.

roughly the same number of local governments in the region, but there is little correspondence between the two sets of boundaries. Any attempt to manage the natal habitats of evolutionarily significant units of the salmon population encounters this fact: human and salmon geographies do not match.

Intermediate-scale alternatives have been tried, with mixed results. The Pacific Northwest River Basins Commission (PNRBC) divided the states of Idaho, Oregon, and Washington into 11 subregions (Figure 13-2), following the contours of major drainages (PNRBC 1969). Beginning in 1987, northwest American Indian tribes and fishery agencies gathered data and formulated plans to manage the Columbia's salmon populations in 31 subbasins (see Figure 13-3), but that biologically coherent approach to the assemblage of populations has been shunted aside in the effort to save the four populations protected under the ESA (Volkman and McConnaha 1993). The PNRBC—whose studies were informative, thoughtful, and in retrospect surprisingly accurate—was ignored by private and public agencies alike; it was disbanded in 1981 as part of a government efficiency drive (Executive Office of the President 1981). Promising analysis has

FIGURE 13-2 Eleven Columbia basin, Puget Sound, and coastal basins that could serve as bioregional planning units. Source: Pacific Northwest River Basins Commission, 1972.

in both cases led to little action, in part because both relied on administrative or planning boundaries that have no visibility or acceptance among local communities. In sum, the Northwest has no socially established boundaries that correspond to bioregions relevant to salmon.

Even if social and ecosystem boundaries were better matched, however, bioregional management would face coupling between bioregions due to the effects of stem, edge, and common mode. *Edge effects* are spatial mismatches: if the land uses in adjacent bioregions are drastically different, there can be spillovers. Heavy timber harvest in one bioregion may affect reseeding in a neighboring one. More generally, species that do well in sharply defined edges are often successful in invading and displacing native species (Moyle and Yoshiyama 1994). *Stem effects* are couplings between regions due to shared biogeography, such as migration routes in the mainstem of a large river system or in ocean areas. The pressure on low-productivity fish populations due to mixed-population fishing is a stem effect, as are the deaths due to dam construction and operation. *Common-mode effects* are those whose spatial scale is larger than that of a bioregion. A national economic recession or a climate change is a common-mode effect. A sense of shared crisis can be heightened when common-mode effects are felt, whereas stem and edge problems tend to sharpen conflicts be-

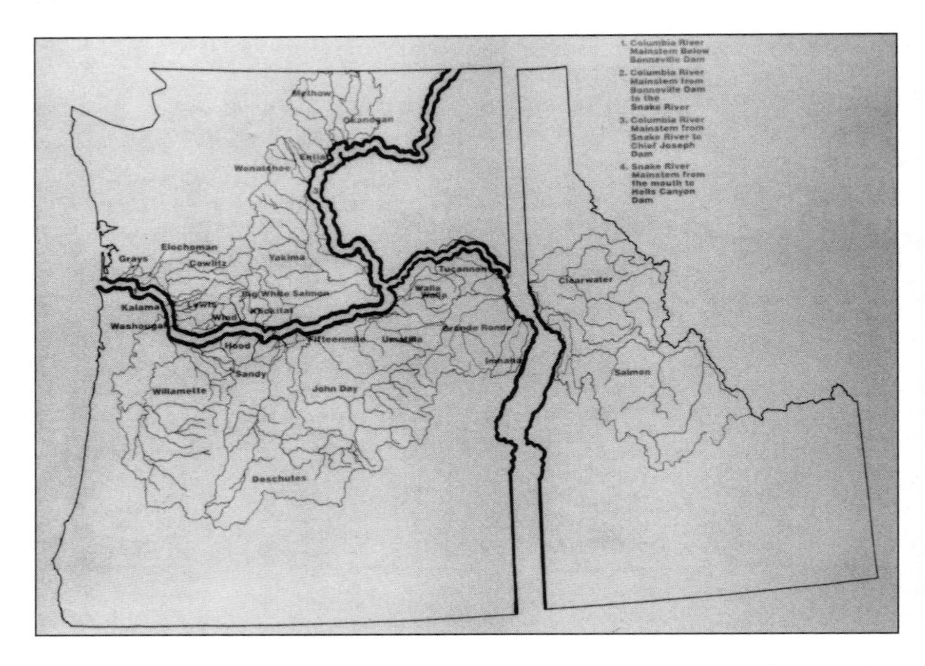

FIGURE 13-3 Subbasins in the Columbia River Basin: Columbia River mainstem below Bonneville Dam, Columbia River mainstem from below Bonneville Dam to the Snake River, Columbia River mainstem from Snake River to Chief Joseph Dam, and Snake River mainstem from the mouth to the Hells Canyon Dam.

tween bioregions. The provision of mainstem habitat for salmon, a stem effect, has proved to be the Achilles' heel of the Columbia River (see Chapter 9). The Northwest Power Planning Council (NPPC) fish and wildlife program did not address the question of how much river flow was biologically necessary in 1982, when the program was formulated. The absence of biological rationale, in turn, made the ESA petitions of 1990 both inevitable and, once filed, unanswerable.

Interdependencies among bioregions might not decrease in intensity with distance in space, time, or function. Some scale mismatches resist localization and thus resist the kind of resolution that the bioregional approach is meant to foster. For that reason, complete decentralization to the bioregional level is inadequate. Stem effects link bioregions that might not be adjacent: the endangered chinook salmon of the Snake River are caught with many abundant fishes in Alaskan waters; as long as ocean fishing continues, it is difficult or impossible to persuade Alaska fishers that they must forgo targeting healthy populations for the benefit of Snake River chinook.

Unless the institutions managing stem habitats are given appropriate resources and jurisdiction, the system will be ungovernable—something that the committee believes has happened in attempts to deal with the ocean fishery. For example, the Pacific Fishery Management Council (PFMC), responsible for ocean-fishing management, has authority over neither ocean nor in-stream habitat. In addition, financial and human resources available through PFMC or other sources to meet salmon-related research needs have declined even as the needs have increased. In the Columbia River Basin, money and information were centralized in the Bonneville Power Administration (BPA) and NPPC; authority was dispersed among fish and wildlife agencies, partly countered by the centralized powers of the U.S. Army Corps of Engineers. That combination, for both historical and functional reasons, has been unable to reach stable resolution; instability continues to undermine salmon.

A critical but elusive need is to handle and to contain conflict. Stem institutions must have real influence because without a comprehensive perspective the fragmentation of an ecosystem by jurisdictional boundaries promotes unsustainable use, as users pursue purposes that often turn out to be incompatible with sustainable management of the whole ecosystem. For conflict resolution to succeed, common goals for the stem must be defined. Yet sustainable management requires a decentralized, fragmented perspective as well, because decisions are carried out by parties whose responsibilities are narrower than the breadth of the analytic tools used by analysts at the stem level. Managing the tension between bioregion and stem entails, paradoxically, the fostering of political competition. The success of decentralized conflict resolution depends in large part on the extent to which various interests see an opportunity to gain from compromise. When the situation is perceived to be a zero-sum game with only win and lose options, the incentive to compromise is undermined (Hanna and Smith 1993). If a common vision can be crafted, however, conflict and the analysis inspired by disputes can link the agendas of bioregions and stem institutions. In the Columbia River Basin, for example, fishery enhancement is funded by hydropower revenues. This arrangement is desirable on economic grounds because those benefiting from electricity help to pay for the damage that its production has caused. The impact on electric-power rates is small, but utilities and large users, such as aluminum refiners, have felt an organizational and political incentive to act as watchdogs (e.g., Carr 1994), seeking to ensure that fishery managers achieve results. Thus a political mechanism reinforces an economic means of recognizing error.

Tension between the centralized knowledge and control of stem institutions and the decentralized experience of the bioregions is perennial. Central planners and governments have a wide-angle, abstract view. Because they must deal with diverse and complex aggregations, there is a constant temptation to oversimplify. To managers at the center, it is imperative to channel the flood of information to manageable proportions. Tradeoffs that make sense at the system level can be incomprehensible or unjust from the local perspective, however.

Stem managers see things in a way that is bound to seem incomplete, unfair, or ignorant to observers at the local level, whose experience is first-hand and whose responsibility and information do not include factors important to overall stability and other system-level properties. Because local knowledge is more detailed, often more persuasive, and sometimes contrary to the inclinations of those responsible for the stem, local actors find themselves in the position of forcing stem managers to attend to matters that are being ignored, to integrate values that are separated by abstract understandings of what is needed, to change the stem manager's grasp of what an action means or the rules by which a decision should be taken.

Those tensions are both inevitable and indispensable when neither stem nor bioregion can see the whole picture—a situation that seems likely to persist, given the difficulties of grasping even parts of the whole today.

Cooperative Management

As conflict over natural resources intensified over the last generation, observers and disputants increasingly turned to cooperative management, "power-sharing in the exercise of resource management between a government agency and a community or organization of stakeholders" (Pinkerton 1992:331). Yet sharing of power has been resisted, in part because the American culture of accountability is rooted in a legal structure in which official government can seek advice but not yield its statutory powers. Although conflict has persisted, cooperative management has a persuasive political claim to representing those with the most at stake, even if it is not a claim that can simply be meshed with the bioregional perspective discussed above.

Given their differing objectives, representatives of user groups, the scientific community, and government agencies should share knowledge, power, and responsibility. In principle, cooperative management brings resource-users and others with stakes in the resource directly into the management process. It assumes that stakeholders have data, understanding, and motivation that can help government officials to assess problems and devise solutions; that cooperation will moderate equity problems that often arise; and that if resource-users are more fully involved, they will be more likely to perceive the management system as legitimate and hence to comply with rules and regulations (McCay 1988 p. 327, Jentoft 1989, Pinkerton 1989).

Cooperative management of salmon rehabilitation implicitly recognizes the central role of equity in fashioning solutions to difficult questions of natural-resource management. If the salmon problem is largely a human problem and if approaches to it are likely to have large social and economic costs, then many of the people affected will demand to be involved trying to solve it. Residents, workers, fishers, those who depend on tourists, and environmentalists have a right to be involved in governance. Their actions have a direct bearing on the welfare of the salmon. Some, such as many residents of coastal communities, have few alternatives when salmon populations decline in abundance or when management rules change. The practicalities of implementation demand that whatever is decided receive support from those who are affected. "People will not support what they cannot understand, and they cannot understand that with which they are not involved" (FEMAT 1993:89). However, as the number of competing interests in a resource increases, the costs, procedural complications, and complex policy tradeoffs of cooperative management are likely to increase as well, often in apparent disproportion to the scope of involvement.

Cooperative management is a general approach to governance that does not dictate the structure of organizations, except that it is probably easier to achieve cooperation among smaller groups of people and in organizations that are less bureaucratized than the ones that characteristically administer complex policies. Cooperative management implies an institutional change, however, a shift in the structure of power that acknowledges the role of various interests—consumers, representatives of different industries, environmentalists—in policy, planning, implementation, and evaluation. Traditional outsiders act in cooperation with each other and with the government agencies that exercise legal mandates in fishery management and have traditionally controlled most of the financial resources and information used in management.

Cooperative management of salmon and water is already institutionalized in the Pacific Northwest. First, beginning with treaty fishing-rights litigation in the late 1960s, the treaty tribes and the state of Washington finally worked out genuine cooperative management for salmon fishing in the Puget Sound in the middle 1980s, and a federal court oversees a similar arrangement in Oregon (Cohen 1986, Pinkerton 1992, Hanna and Smith 1993). Second, the Pacific Fishery Management Council has designed its decision structure to promote active consultation with its salmon advisory committees. Three committees comprising state- and federal-agency scientists, treaty-tribe scientists, university scientists, and representatives of user groups and environmental interests are formal advisers to the council on all decisions related to salmon-fishery management. In addition, numerous public meetings provide forums for public comment on regulatory decisions and management plans (PFMC 1993). Third, in 1990, as the threat of filings under the ESA materialized, the Salmon Summit recognized a wide range of interests as legitimate and redefined the arena for cooperative management to include fisheries, utilities, and the other claimants to multiple use of Pacific Northwest rivers.

The largest attempt to cooperatively manage is the framework established by the Northwest Power Act. By creating a regional planning council, the act brought the governors of the Pacific Northwest states to a decision arena dominated by BPA management, augmenting the presence of the Northwest's congressional delegation, which had played an important but little-noticed role in the creation and management of the regional electric-power system (Lee et al. 1980). The Power Planning Council broadened the political base of the BPA as the governors, through the council, gained authority and influence over this regionally important federal agency.

The council also took its statutory mandate for public involvement seriously, affording opportunities for comment to citizens and organized groups as it developed plans for electric power and for fish and wildlife in the Columbia River Basin. As is often seen in this variant of open government, however, the complexity of the issues and the large geographic scope of the region made it impossible, in a practical sense, for those without substantial resources to stay involved. By the middle 1980s, the "public" with which the council and BPA consulted was a population of organizations, not individuals: electric utilities with direct economic interests in the federal hydropower system; a small set of consumer and civic groups, such as the League of Women Voters, who could field volunteers to follow the council through meetings in four states; and state fishery and wildlife agencies and American Indian tribes with stakes in the council's fish and wildlife program. Indeed, two organizations, the Pacific Northwest Utilities Conference Committee, an organization formed by electric utilities and large industrial consumers of electric power, and the Columbia Basin Fish and Wildlife Authority, a parallel organization created by the tribes and fishery agencies, gained influence within their communities largely because of the magnitude and complexity of the issues raised by the Northwest Power Act. Because the council eschewed any role in managing catch levels, fishing groups played little active role in its deliberations, and their interests were represented by state agencies. Although the Northwest Power Act process falls short of the ideal of "power-sharing in the exercise of resource management" (Pinkerton 1992), it did join the agendas of fisheries and hydropower in a way that has forced conflict into the open and fostered joint action.

Despite the existence of numerous cooperative arrangements on several scales, people remain largely divided, engaged in finger-pointing rather than collaborative problem-solving, and few government agencies operate with formal or legitimized involvement of affected groups or members of the public. Most seem satisfied with the formalities of public hearings, public review and comment on proposed rules and other documents, and the use of advisory committees to supplement the inputs of lobbyists. This is consultative, not cooperative, management and has as yet done little to bring people together to collaborate on one of the biggest challenges imaginable.

The novelty of cooperative management suggests that the assumptions behind it deserve critical examination. As might be expected given both the experimental nature of many cooperative management arrangements and the diversity of social and ecological systems and their contexts, the results of studies of cooperative arrangements for fisheries are mixed (Pinkerton 1989). McCay (1988) found great difficulties in implementing cooperative management in a bay clam rehabilitation project because of conflicting and unrealistic expectations and roles and because of scientific uncertainty, but identified accomplishments as well. Canadian experiments with adaptive management of salmon have also been problematic but show that the cooperative arrangements within which they have taken place have proved to be workable, flexible vehicles (Hilborn and Luedke 1987, Walters et al. 1993). Most cooperative arrangements, such as Washington state's cooperative-management regime for nonfederal forest lands (Halbert and Lee 1990, Pinkerton 1992), have not been designed but are the residue of conflict and negotiation among opposing parties. A multitude of sins, omissions, and inept compromises stemming from such a history can be observed, ranging from adopting lowest-common denominator positions, through resorting to blame as a substitute for analysis, to emphasizing short-term remedies with anticipated early payoffs, such as salmon hatcheries rather than ecosystem management. However, over a decade or more, as Pinkerton (1992) shows for state-forest management, both the agenda of disputing and the institutions within which disputes are conducted can evolve toward durable accommodations. Over a similar temporal scale, PFMC has been able to bring members of the fishing industry directly into the fishery-management process in ways that suggest the value of cooperative arrangements (Hanna in press), and the firm establishment of cooperative management as a working premise in fishery-management practices demonstrates that even bitter divisions can be overcome. For example, Sabatier and Jenkins-Smith (1993) have emphasized the importance of decadal time scales in social dynamics; their studies, in environmental and other policy arenas, document the emergence of issue networks and other stable frameworks for public-policy debate over periods longer than election cycles or budget planning.

Two large, interrelated challenges to cooperative management remain: developing an ecosystem approach to cooperative management and linking the stakes of legitimate stakeholders to biogeography as well as economies. Emerging from conflicts involving government agencies with established spatial jurisdictions, cooperative management arrangements have yet to move beyond the planning stage in dealing with salmon ecosystems as units of management. The shift to an ecosystem perspective is a social as well as scientific realignment, requiring stakeholders to establish legitimate claims and to act at the ecosystem level of biological and spatial organization. Today, stakeholder claims are grounded in economic interests or in universal public interests, such as species conservation. Economic interests might prevail, if resource use continues to yield profits sufficient to influence the governing institutions, but the decline of salmon catches, changes in the economics of electric power, and the maturation of the Pacific Northwest economy all suggest that material interests alone no longer define the governance of environmental quality and natural resources. Indeed, the rise of environmentalism as a politically potent force over the last 25 years, rooted in the suburban populations that harbor electoral and economic power, implies the passing of an extractive policy regime. At the same time, the economic adjustments required to achieve ecological rehabilitation can be large and will sometimes be fiercely resisted; that is the lesson of the spotted owl in the Northwest's ancient forests. Unless social energies among *all* stakeholders are focused at the ecosystem scale, the efforts required to rehabilitate salmon will tax even the most eager participants and hinder widespread adoption of cooperative management as a mode of collective action on behalf of salmon and their ecosystems. There is little evidence that such an ecosystem perspective can or will be reflected in the agendas that characterize conflict and structure action; even the ESA petitions, as Moyle and Yoshiyama (1994) observed, do not draw attention to the landscape units at which ecosystem cooperative management is needed.

The history of the Pacific Northwest, in sum, does not provide clear guidance on the definition of socially coherent bioregions in which ecosystem cooperative management can take root. Although some subregions, such as the coasts of Oregon and Washington, have economic and cultural unity, there have been no successful attempts to govern at the bioregional level. Yet it seems clear that the subregional biological province remains an attractive scale for cooperative management—on that scale, it appears to be possible to make long-range biological and social choices that can be implemented because the underlying social and biological dynamics can readily reinforce the integrity of such spatial units.

Adaptive Management

Because human understanding of nature is imperfect, human interactions with nature should be experimental. Adaptive management applies the concept of experimentation to the design and implementation of natural-resource and environmental policies (Holling 1978, Walters 1986, Hilborn and Winton 1993, Lee 1993a, Walters et al. 1993, Bormann et al. 1994). An adaptive policy is designed from the outset to test clearly formulated hypotheses about the behavior of an ecosystem being changed by human use. In the present instance, the hypotheses are predictions about how salmon or other species will respond to management actions. If the policy succeeds, the hypothesis is affirmed. But if the policy fails, an adaptive design still permits learning so that future decisions can proceed from a better base of understanding.

Adaptive management is highly advantageous when policy-makers face uncertainty, but the adaptive approach is not free: the costs of information-gathering and the political risks of having clearly identified failures are two of the barriers to its use (Volkman and McConnaha 1993). An adaptive approach has been tried in three arenas and proposed in a fourth, all in the Pacific Northwest. Developed initially by a group centered at the University of British Columbia, adaptive management has been used in fisheries as part of Canada's Department of Fisheries and Oceans' sophisticated program of salmon management; two recent case studies (Hilborn and Winton 1993, Walters et al. 1993) led by coinventors of the concept provide appraisals of the successes and limitations of the adaptive process. A second major application began in 1984 in the Columbia

River basin program of the Northwest Power Planning Council (Lee 1993a). A recent retrospective analysis by two senior members of the council staff (Volkman and McConnaha 1993) came to conclusions remarkably similar to those in the studies by Walters et al. and Hilborn and Winton. Third, an abortive attempt to borrow the flexibility implicit in an experimental approach was tried in 1987 in state-forest management in Washington State (Halbert and Lee 1990, Pinkerton 1992). Finally, the idea of adaptive management areas, in which timber harvest would be combined with ecosystem management, is being studied in the USDA Forest Service (USFS) program to manage federal forest lands in the Pacific Northwest (FEMAT 1993).

This body of experience has produced lessons about the practicability of adaptive management and the institutional conditions that affect how experiments on the scale of ecosystems can be conducted:

- Learning takes from decades to as long as a century[2]. Patience, particularly in institutional settings, such as government, that work on much faster cycles, is both necessary and difficult.
- Systematic record-keeping and monitoring are essential if learning is to be possible. But collecting information is expensive and often hard to justify at the outset and during times of budget stringency because the benefits of learning are hard to estimate quantitatively.
- Cooperative management in the design and execution of experiments is indispensable. Experimentation within the context of resource use depends on the collaboration of resource-users.
- Adaptive management does not eliminate political conflict but can affect its character in important, if indirect, ways.

Paradoxically, each of those lessons runs counter to or at cross-purposes with the administrative framework of the Endangered Species Act (ESA). Conserving species and habitats whose biology is poorly understood requires patient observation and analysis that take into account the economic needs of resource-users and that rely on the painstaking fashioning of political consensus. But the urgency that prompts listing of a species as endangered raises the perceived risk of failure and inhibits experimentation even though urgency also makes learning more tangibly valuable.

When one considers the alternative ways to approach biodiversity conserva-

---

[2]The decade-scale variation in the ocean habitats of salmon, discussed in Chapter 3, adds another complicating factor. Although improved understanding of the systematic effects of oceanographic change can illuminate some of the variation in salmon abundance, it is likely that study of particular populations will still need to accumulate information through several ocean fluctuations before reliable patterns can be found.

tion analyzed by Moyle and Yoshiyama (1994), it is apparent that an experimental approach like adaptive management is more likely to work better at the earlier stages of depletion before a procedurally inflexible approach is triggered by severe decline in abundance. It is also among these healthier populations that the risks of experimentation are more likely to be tolerable, and that the long times needed for reliable learning can be afforded. The committee's recommendation to move toward management by population, using such methods as terminal fisheries, is a way to decouple the fate of populations that are endangered, and thus subject to extraordinary protection, from populations that might be able to be managed in a way that can permit more rapid learning. Even with these changes, the move away from hatcheries, which permit far greater control of salmon populations, at much lower monitoring cost, and toward ecosystem management will raise the expense of learning and demand much more precise intervention into the life cycles of specific, identified populations (Volkman and McConnaha 1993). How to carry out experimentation and learning under the more demanding conditions has not been well explored; indeed, experience with artificial propagation suggests that even that presumably easier problem of adaptive management has been formidable (Hilborn and Winton 1993).

FEMAT (1993) suggested that adaptive management be conceptualized as a cycle of planning and adaptation. In contrast with the image of planning as an orderly process, adaptive management is replete with surprises, given the fragmented jurisdictions and conflicting claims that characterize the salmon habitat. Those destabilizing elements, which appear to be inherent in the social dynamics of adaptive processes, underscore the importance of patience, persistence, and a politically grounded determination to make constructive use of inevitable conflicts. Disputes are sure to arise within the spans of space, time, and functional interaction that characterize human behavior; indeed, the conflicts that gave rise to the Northwest Power Act and that have reappeared transformed under the Endangered Species Act are part of a long-term search for an accommodation between humans and the natural environment symbolized by and partly summarized in the plight of salmon.

Lee's (1993a) analysis of adaptive management arrived at a similar set of conditions that influence the feasibility of an experimental approach (see Table 13-1). Today, the last two requirements of adaptive management—institutional stability and an organizational culture that enables experimentation and learning—do not exist. Creating and sustaining them seems difficult or impossible, given the depth of conflict and the fact that open conflict has widened in scope since the American Indian fishing-rights litigation began a quarter-century ago. The problem is one of institutional frailty. Science can identify the problem, but solving it requires political will and steadfast managerial leadership, which science cannot supply. Without a durable commitment to the goal of rehabilitation, with its painful costs, it is unlikely that social conditions under which learning

TABLE 13-1 Institutional Conditions Affecting Adaptive Management

There is a mandate to take action in the face of uncertainty. *But experimentation and learning are at most secondary objectives in large ecosystems. Experimentation that conflicts with primary objectives will often be pushed aside or not proposed.*
Decision makers are aware that they are experimenting. *But experimentation is an open admission that there might be no positive return. More generally, specifying hypotheses to be tested raises the risk of perceived failure.*
Decision-makers care about improving outcomes over biological time scales. *But the costs of monitoring, controls, and replication are substantial, and they will appear especially high at the outset when compared with the costs of unmonitored trial and error. Individual decision-makers rarely stay in office over times of biological significance.*
Preservation of pristine environments is no longer an option, and human intervention cannot produce desired outcomes predictably.
And remedial action crosses jurisdictional boundaries and requires coordinated implementation over long periods.

can take place will be maintained. And without experimental learning, human resuscitation of the declining salmon population will be hit or miss.

The principles of reform, if not its details, can be stated:

- *Bioregional scale.* Jurisdictions and patterns of economic flows should take into account the scale of salmon habitats. Rules and practices in stem habitats, especially rivers and ocean fishing areas, should reflect the need to rehabilitate salmon. In particular, halting the decline of all populations that stakeholders judge to be worth saving must be accorded the highest priority, even at substantial risk to human communities and other, healthier species of fish and wildlife. Intertemporal rules, such as interest or discount rates and lengths of contracts or leases, should take into account that changes in abundances of salmon and in our understanding of the biological needs of managed salmon have undergone substantial shifts on time scales of about a decade. Flexibility to take into account new learning at the decadal time scale is essential.
- *Cooperative management and social diversity.* Stakeholders in salmon and the economic values of their habitats are highly varied and often see themselves as locked in zero-sum struggles, in which the gains of some can come only at the expense of others. Yet the emergence of cooperative management over the last two decades, taking into account the interests of those whose economic claims might be small or unrealized, has demonstrated that arrangements affording compensation to those who yield their claims to exploit a resource can be negotiated and carried out. Assured compensation is, moreover, more conducive to social stability and investment than the pursuit of highly fluctuating and diminishing resources. Acting on the bioregional scale decentralizes authority. This is

valuable especially when it fosters diversity in the approach taken to protection and management of salmon. Diversity is a sensible response to the large uncertainties that remain in our scientific understanding of salmon biology. And diversity in social arrangements and institutions would enable salmon rehabilitation to persist even in the face of common-mode disruptions, such as economic recession and climate change. Decentralization should, however, honor obligations along stem habitats; that has not been done during the decline of the salmon over the last century, and acknowledging and reversing this history is a central challenge to the establishment of cooperative management now.

- *Experimentation and adaptive management.* Because salmon populations are far from their natural equilibria and because the salmon life cycle is comparable in length with many societal rhythms, learning about salmon is obscured by changes among human observers. Persistent, accurate observation is essential to learning, particularly because learning is likely to take decades or longer. Learning by trial and error is inherently slower than deliberate experimentation because errors are often hard to recognize or diagnose correctly, and because trials are not designed for replication or controlled comparison unless an experimental framework is taken into account. Given the already slow pace of learning imposed by environmental fluctuations and the time scale of the salmon life cycle, experimentation should be used as often as it can be afforded, under institutional conditions that provide assurance that experiments can be carried to conclusion. Those conditions have not often been available in the past. An objective of reform should be to make them more readily available in the future, by ensuring funding for ecosystem-scale research and monitoring and by giving scientists a place in the scheme of cooperative management.

Following those principles, the committee would design institutions whose jurisdictions follow major river basins or biologically related drainages, whose decision-making accords power and influence to the socially diverse human groups that inhabit those ecosystems (including appropriate economic incentives), and whose governing procedures and resources enable experimentation and ensure faithful monitoring of ecosystem variables. But such a design will have no life unless those who are governed by it take a role in the processes of design and institutional reform. The recent history of the Pacific Northwest shows a clear biological crisis with environmental and economic impacts that cannot be denied. But there is not yet a social crisis of the kind that, in an open, competitive society, can become the wellspring of institutional change.

A PROPOSAL FOR CONSTRUCTIVE ACTION

The salmon problem is regional in scale: although specific causes and possible remedies vary, salmon populations throughout the Pacific Northwest are in serious decline. The human resources to deal with these declines are substantial but limited. The money and, more important, time to attend to populations that are or will be in trouble over the next decade might well be insufficient to provide for the recovery of all those populations (NRC 1995b); that has been the experience of the ESA (Tobin 1990) even without the extraordinary stresses present in the Northwest. In addition, institutional changes are difficult to achieve and their effects difficult to predict, in large part because institutional responses to change are based on the responses of individuals to the potential changes in their circumstances. For that reason, the committee does not attempt to recommend in detail institutional and procedural changes needed but offers an outline for change as a basis for responding to the decline of salmon populations in a more flexible and more biologically effective fashion than is possible for other species. The details must be worked out by the individuals and institutions of the region.

The committee proposes that the relevant agencies, including the National Marine Fisheries Service, agree on a process to permit the formulation of salmon-recovery plans in advance of listings under the ESA and that the Pacific Northwest states, acting individually and through the Northwest Power Planning Council, provide technical and financial assistance to watershed-level organizations to prepare and implement these preemptive recovery plans. The USFS, Department of the Interior, the Department of Commerce, and the Department of State should also provide technical and financial assistance. The concept of a preemptive recovery plan is described more fully below.

When a salmon population appears to be threatened or endangered, the current institutional structure responds with a federal process. Either in response to a petition or on its own initiative, NMFS, a federal agency, gathers information, hears advocates, and decides whether to list the population as threatened or endangered under the ESA. Listing a species triggers a variety of federally mandated actions, including the preparation of a recovery plan for the population under consideration. Listing a species and the adoption and implementation of the recovery plan can be conflicted, politicized, and delayed because they can involve such actions as designation of critical habitat and the prohibition or reduction of various activities and therefore can bring the risk of serious economic disruption (NRC 1995b). The economic risks are often borne by people different from those who bear the losses due to salmon-population declines. Partly as a result, the seriousness of the risks is often perceived differently. Given the hundreds of populations and metapopulations (networks of populations) into which salmon are divided by their life histories, the widespread decline of many of those populations and metapopulations, the inadequacy of resources of NMFS and the other government parties to the ESA process, and the fact that the ESA process can be triggered by petitioners who can take action without committing substantial resources of their own, the ESA today carries the risk of producing disorder in the Pacific Northwest, whatever the merits of its policy of saving endangered species.

The committee has emphasized three components of a constructive course of action. First, biologically distinct salmon populations spawn in watersheds and migrate to sea along streams that typically flow beyond their natal watershed; as recruits to ocean and inriver fisheries, salmon can be exposed to mixed-population fishing. Action to protect a given population must therefore be organized within the biologically relevant drainage *and* along its migration route in freshwater and at sea. That is, action should be bioregional and include protection or improvement of migration paths.

Second, although it is clear that much of the habitat modification carried out by humans over the last 150 years has been detrimental to salmon, fishery science cannot ensure the recovery of any specific salmon population. Science can identify changes that are likely to be necessary but cannot specify a course of action that is known in advance to be sufficient to guarantee rehabilitation and conservation of a species. There is no correct answer to the question of how much biological diversity and population structure must be maintained or can be lost to provide a long-term future for salmon. Scientific estimates—including the uncertainties associated with them—are only part of the argument: society must decide what degree of biological security would be desirable and affordable if it could be achieved, i.e., the desired probablity of survival or extinction of natural populations over what time and what area, and at what cost. Therefore, a long-term adaptive, experimental approach is both logically compelling and pragmatically indispensable for a large collection of demes like those of Pacific salmon.

Third, acting on the bioregional scale requires cooperation among diverse landowners and water-rights holders. In many biologically relevant regions, the land is controlled by different public and private entities and water flow is shaped by withdrawals and impoundments controlled by landowners and others. There is no single body of law, nor a practical way to consolidate governing powers, sufficient to put each bioregion under the supervision of a single managing entity today. From that practical reality grows the need to rely on cooperative management as a way to act in the face of fragmented control.

The potential for disruption of human activities embodied in the ESA can provide a stimulus for cooperative management if cooperation enables the diverse entities in a bioregion to improve their control of events. Because of the groundwork carried out by generations of fishery biologists and most recently summarized in FEMAT and the subbasin planning process of NPPC, it is possible to identify the geographic outlines of salmon bioregions. And with the resources available within the Columbia basin under the Northwest Power Act, it is possible to formulate and to begin implementation of adaptive management aimed at ensuring the survival of salmon populations. Although the cooperation and resources that can be marshaled in the near future are unlikely to be sufficient to save all the populations that are or might become threatened, it is possible to move beyond the watchful defensiveness of today's uncertain situation.

One way to harness the ESA's potential for disrupting human activities in a biologically constructive fashion is to foster the development of preemptive recovery plans, which incorporate binding contractual commitments from funding sources to undertake adaptive management to rehabilitate a specific salmon population within its bioregion and migration route. The plans would include a set of experimental actions and monitoring methods that ensure the learning of lessons important for salmon management as soon as possible—probably 20 years in many cases—and that lead to a scientifically grounded expectation that the salmon population would increase during that time while preserving its genetic integrity. A recovery plan would need to reflect the commitments undertaken by all the parties that control the various elements of its execution. For that reason, it is likely that cooperative management would characterize plans' development and implementation. The plans would be preemptive in the sense that NMFS would agree that while a recovery plan it has adopted is in operation, the salmon population it covers is protected as much as is possible under the ESA. Protection of populations would be improved by including in the preemptive recovery plan minimum sustainable escapements for salmon populations and approaches for filling out the dendritic habitat of salmon populations.

In sum, the formulation and adoption of a plan would forestall a petition under the ESA to protect the population covered by the plan. No filing on a population would be acted upon by NMFS after a state certified that a recovery plan was being developed and that positive steps were being taken or seriously planned to improve prospects for conservation and rehabilitation of the salmon populations, unless a petitioner demonstrated that a population decline warranting emergency protection superseded the state's certification; this policy would allow time for a plan to be developed but leave the possibility that NMFS could reject a proposed plan (or lack of a plan) and take additional action.

A preemptive recovery plan is attractive because it permits control of the ESA process, but mobilizing the divergent interests in a bioregion, designing an adaptive-management program, and ensuring funding to implement the plan require resources beyond the capability of the inhabitants of most bioregions. Therefore, the active support of state governments is essential. The committee does not believe that any legislation prevents or discourages this from being pursued (although a way would need to be found to allow the government to delay action on a petition to list a species). State governments have already begun to act at the watershed level in natural-resources policies (Oregon Watershed Improvement Council [Krueger 1994] and Washington Watershed Restoration Partnership Program [Harrison 1994]). More important, states exercise legal authority over water rights and existing land-use planning and zoning powers. Finally, the states in the Pacific Northwest already act through NPPC, the agency with the most highly developed base of knowledge relevant to protection and enhancement of salmon on the bioregional scale; it can also direct funding to support recovery-plan implementation within the Columbia River Basin. The activities of the states should be augmented by the USFS and DOI's Bureau of Land Management (BLM), where these agencies are developing adaptive-management areas and watershed-level plans under FEMAT. Assistance will also be needed from the Pacific Fisheries Management Council and the International Pacific Salmon Commission to set fishing restrictions to protect specific populations, and from other relevant agencies that operate beyond state boundaries.

It is unclear whether the support and resources available through federal and state governments will be enough to induce participants on the bioregional scale to create preemptive recovery plans. And it is possible that preemptive plans could seem so attractive that the organizing efforts would strain the resources available from the states and NPPC. If that were the case, it would be necessary to set priorities to provide resources first to bioregions harboring fish populations that are already declining seriously. Regional and state agencies seeking to stretch their resources must take into account, however, the stem effects that link watersheds through migration routes. By focusing on bioregions in trouble today, planning and action could rapidly encumber the limited capacity of the Snake River and other waterways to provide flows for migration or could force drastic, long-term constraints on fishing in an attempt to limit depletions from mixed-population fisheries. Such stem linkages entangle the fates of salmon from distant, apparently unconnected bioregions; that entanglement is clarified but not removed by preemptive recovery plans. Although preemptive recovery plans would have a bioregional focus, it is critical to their success that they take into account the entire life cycle and environment of salmon, including such factors as ocean conditions, international fisheries, large-scale interactions among salmon populations, as well as more obvious but more local factors.

The prospect of constructive action to conserve and rehabilitate at least some salmon populations without the conflict and delays encountered under the present ESA regime leads the committee to advance the idea of preemptive plans. This approach requires no new legislation, and the resources that it calls into play are already available within the region. Although the shifts in incentives proposed are incremental, the bioregion-scale learning that the adaptive plans would produce is likely to lead to more substantial change over the next several decades. The proposed strategy also becomes more compelling as salmon abundances decline, in that bioregions that fail to act are more likely to face petitions under circumstances not of their own choosing. By increasing the incentive to act on behalf of salmon populations, institutional mechanisms can play a role that is more constructive than the protective but conflicted stance of those institutions today.

Preemptive recovery plans seek a better biological result for salmon, but that intent could be undermined by people with divided interests. For example, the 1987 Timber, Fish, and Wildlife agreement in Washington's state-regulated and managed forests had the active support of the environmental community initially, but nongovernment participants found themselves overwhelmed by the volume of regulatory work (Halbert and Lee 1990). It took several years of learning and friction to reach a practicable mode of operation (Pinkerton 1992). Given the mistrust and conflict that are the legacy of the decline of salmon in the Pacific Northwest, it is essential that all parties potentially involved understand that a preemptive recovery plan is a negotiated settlement among parties with different goals (see Lee 1993a:104-113). Indeed, in this respect, the proposal has much in common with habitat conservation plans (HCPs), which are required by Section 10 of the ESA (added in amendments to the act in 1982) of private parties who seek a permit for "incidental take" of listed species (NRC 1995b). Like the preemptive recovery plans proposed here, HCPs are negotiated settlements; but HCP negotiations potentially involve fewer parties and their basis might be slightly narrower—i.e., the right to use some habitat for purposes other than conservation in exchange for setting other habitat aside for conservation.[3]

A recovery plan is an experiment, not a guaranteed recipe for rebuilding salmon populations; there are no guaranteed recipes. The experimental character of adaptive management requires continuing vigilance by all parties involved in implementing a plan. The spatial and temporal scales of human actions aimed at rehabilitating salmon need to match the large range of spatial, temporal, and biological scales of salmon populations. The committee's recommendation builds in these scales of action. Meshing the biological requirements with the institutional rules by which human disagreements are addressed will not be easy. We have proposed, for example, that once a recovery plan covering a 20-year period is adopted only clear evidence that warrants emergency action under the ESA would be accepted as grounds for reopening the plan's contractual obligations. As knowledge accumulates about what salmon need from the habitat, recovery plans already in force might become outmoded, just as the wholesale use of hatcheries to replace lost habitat now seems counterproductive. Yet unless the experimental arrangements embodied in a recovery plan are given time enough to work *and perhaps to fail*, we will never accumulate reliable knowledge of what does and does not make a difference.

It is the duty of government agencies like NMFS, charged with the legal responsibility to protect the nation's interest in endangered species, to persevere in the face of political turbulence so that reliable knowledge of species and ecosystems under pressure can be gained. Independent scientific advice can be of considerable value in testing administrative judgments, as discussed in Chapter 14, but advice informs rather than substitutes for political courage. How much political courage is needed depends heavily on how effective a cooperative management alliance is built during the formulation of the recovery plan. Given the need to act on a bioregional scale so that human actions can take into account the

[3]See Thornton (1991, especially pages 639-652) for a thoughtful discussion of habitat conservation plans under the Endangered Species Act similar in direction to the one we provide here. Regionally based, negotiated approaches to conservation plans were also endorsed by a recent National Research Council report on the Endangered Species Act (NRC 1995b).

needs of salmon throughout their ranges, cooperative management is more likely to succeed when there are social, political, and economic ties that reinforce action on the bioregional scale. The human, social underpinnings of a recovery plan need to be considered as an important variable from the outset (see McCay and Acheson 1987); salmon rehabilitation is social action.

A few subbasins in the Pacific Northwest appear to offer particular promise for implementing the approach described above. Some examples follow; these are based on committee members' experiences, not on a systematic review, and they are not intended to exclude any area not mentioned.

Columbia River System

Yakima River, Washington

Attempts are already under way to rebuild salmon populations in the Yakima basin. Most of the upper Yakima falls within the boundaries of either the national forest system (Mt. Baker-Snoqualmie, Gifford Pinchot) or the Yakima reservation. The lower part of the basin contains large agricultural areas (apples, alfalfa, hops, etc.) and rangeland. There is an effort to develop a basin-management plan for salmon restoration with at least most of the key participants—e.g., tribal, federal, state, and private—involved.

Grande Ronde River, Oregon

The Grande Ronde was one of two subbasins identified as pilot areas for the Oregon Governor's Watershed Restoration Initiative. The USFS and BLM have also spent considerable resources on analysis of habitat in the Grande Ronde River Basin. An interagency restoration group has already been formed; as in the case of the Yakima River Basin, local planning has already begun. Most of the Grande Ronde watershed is in federal ownership although there are some important private landowners along the river. This subbasin contains ESA-listed chinook.

McKenzie River, Oregon

The McKenzie River flows into the Willamette River at Eugene, Oregon, and is the focus of much political attention. The adjacent land includes large areas of USFS and BLM forests and some substantial wilderness areas, but there are sizable tracts of private forest land and agricultural and urban development. Spotted owls abound along the McKenzie River, which might place some constraints on adaptive management. The Pacific Rivers Council (based in Eugene) is trying to form a local basin coalition (Eugene Water and Electric Board and Lane County 1992).

Coastal Streams and Estuaries

Coos Bay, Oregon

The other basin identified for local cooperative planning by the Oregon Governor's Watershed Restoration Initiative (and hence an officially sanctioned mandate to do bioregional adaptive management), Coos Bay is typical of coastal watersheds in Oregon and Washington. Headwaters are in private forests, and lowlands are in agricultural and urban areas. There is considerable political support for restoration here (Cooley and Heikkila 1994).

Rogue River, Oregon

The Rogue River, a large and scenic river in southern Oregon, is unusual among large coastal streams of the region in that it has only one major impoundment (Lost Creek Reservoir), and that one is very near the river's source. The river has been designated a wild and scenic river and is the subject of substantial recreational fishing and tourism. Biologically and physically, the basin—which includes federal, state, and private land—is of great actual and potential value as a salmon-production area.

Willapa Bay, Washington

Willapa Bay is potentially a huge salmon-production area if natural production processes are allowed to operate (Chapter 2). The Willapa Alliance has an effort, though somewhat frail, to get the bioregional planning process started.

Nisqually River, Washington

Near the southern tip of Puget Sound, the Nisqually suffers from most woes except a large urban center. The Fort Lewis military reservation, which owns a large section of the Nisqually, is making serious attempts to conduct environmental planning. The Nisqually River Council, along with the Nisqually Tribe and other landowners, forged a local coalition to develop and implement salmon restoration strategies. This basin could show some substantial gains if attention were paid to restoring floodplain connections and a more natural flow regime and if policies regarding hatcheries and fishing in the area were adapted successfully to rehabilitation.

14

A Scientific Advisory Board to Address the Salmon Problem

GAPS IN KNOWLEDGE

Previous chapters in this report summarize the great deal of knowledge that has been obtained about salmon—clearly enough to substantially improve their prospects for survival if applied wisely. Such information should be used to the fullest extent possible in implementation of projects, watershed planning, and other programs designed to assist the survival and sustainability of salmon. To simply wait for new research, new ideas, and new technology while continuing past practices that have adversely affected anadromous salmon is a mistake. Such a delay only serves to increase the demise of salmon and their ecosystems.

Yet attempts to solve the salmon problem still are hampered by lack of knowledge. The list of central topics that we know too little about is surprisingly long. The topics include, for example, the survival of young fish between dams compared with their survival as they pass through and over dams; the relationship of survival of young fish to the flow rates of water in rivers; the effects on survival of various management practices, including logging, grazing, irrigation, agriculture, and use of hatcheries; the influence of ocean conditions on survival and growth of salmon; the degree to which salmon survival and growth in freshwater and at sea are density-dependent; the effectiveness of transportation and altering flow regimes on dammed rivers in increasing survival of young salmon; the importance of predation in rivers and estuaries in survival of juveniles and adults; the effect of hatcheries on the number of fish returning to spawn, ocean survival, homing ability, and genetic diversity; where and when different populations of salmon migrate in the ocean, and even the status of many populations of salmon; the effectiveness of management institutions; the bioregional relationship between metapopulations of salmon and populations of salmon fishers; the degree of public support for protection of wild salmon versus hatchery programs; the distribution of population and income from fishing by watershed; the relative importance of salmon fishing to community stability; and the relative importance of income from salmon fishing.

Questions directly related to the survival of salmon in various freshwater habitats under various conditions could be answered by focused research in a relatively short time. Indeed, many of them should have been answered by now. Questions concerning long-term changes in environmental conditions in response to climatic and other natural fluctuations and questions concerning complex ecological interactions, especially at sea, are related to several types of research driven by many kinds of motivations in addition to interest in salmon. The committee believes that the greatest short-term gain in ability to protect salmon will come from focused research on freshwater survival especially of young salmon, although knowledge about long-term changes in environmental conditions and complex ecological interactions are also of great importance.

A great deal of money has been spent on research over many years by the Bonneville Power Administration (BPA), the U.S. Army Corps of Engineers (USACE), the National Marine Fisheries Service, and other federal, state, and local agencies (about $70 million a year has been spent recently by BPA and USACE alone). The committee was therefore surprised to find so many gaps in basic scientific knowledge. Because there is much to be learned and money for research is finite, the research money that is available must be spent more effectively, especially given the urgency of the problem.

Research of the kind needed to help improve the salmon situation must be related to policy and management, because informed policy and management decisions are needed. Although basic biological and physical information is needed, by itself it is not enough. Research must not be driven only by short-term politics or even appear to be so motivated, and research decisions must be made in an integrated way, with a coherent view of the general importance and urgency of what needs to be known and with communication among various research organizations and interest groups. In short, there needs to be more objective scientific guidance and planning for research undertaken within the policy context. The difficulties and controversies surrounding management of riverflows in the Columbia River provide a good example of the importance of objective scientific guidance.

SCIENCE, RIVER FLOWS, AND UNCERTAINTY

Chapter 9 demonstrated how important the dynamics of streamflow in the Columbia drainage are to the welfare of migrating juvenile salmon in the spring and summer. Before the dams were built, riverflows east of the Cascade Mountains peaked sharply in the spring because of snowmelt. Salmon are adapted to this seasonal abundance of water. As spring flows have been increasingly impounded by dams and as the water has been forced through hydroelectric turbines, migrating fish have suffered accordingly.

The question of how the Columbia River reservoir system can be operated to benefit fish is mainly an institutional one: How should policy choices be made over time to manage salmon populations while meeting other objectives? If policy choices are to be made, the uncertain relationship between salmon welfare and riverflow must be elucidated and must be compared with the economic welfare of human populations that use the river's waters for multiple purposes.

Since the late 1970s, when the question of whether flows benefit fish arose (when the completion of the Snake River dams enabled river managers to eliminate spills of water over hydropower dams in most years), the problem has been addressed annually as information on natural flows has unfolded with the weather. That process has been frustrating for all participants. Forecasts of seasonal streamflows, on which operating decisions are based, are only modestly accurate before the time of maximum average snow accumulation (about April 1) and are little better than random guesses before about February 1. The inability to forecast streamflows forced decisions to be made that were in reality gambles about what would happen later in the water year. Gambles do not pay off all the time, and the various parties' differing goals fed conflict over how to place bets as data became available.

What has not been widely recognized in the debate over how the reservoir system can and should be operated for multiple purposes (including fish) is the scientific basis of the decisions to be made with uncertain and incomplete information. The concept of burden of proof is familiar: In the absence of unequivocal information, how should a decision be made? The status quo is often presumed to be appropriate unless there is compelling evidence to the contrary. The discretionary judgment of recognized authorities is another basis for making a decision unless there are strong grounds to doubt the process or content of the judgment. The decline of salmon has challenged both those implications that the burden of proof should rest with those seeking change. Instead, statutory, policy, and political changes over the last quarter-century have often shifted the assumptions: as salmon abundances continued to decrease, changes believed to assist the fish were sometimes adopted even in the absence of clear evidence that they would help.

Such shifts in the Columbia River Basin have been at the heart of a program of protection and rebuilding that now costs several hundred million dollars per year, depending on the mountain snowpack and later spring and summer streamflow in any particular year. The costs are due in large part to modifications in dam operations that reduce income from the sale of hydropower. Resource allocations to salmon conservation elsewhere in the Pacific Northwest, where dams are less important in the economics of water, have been proportionately lower. Nonetheless, as costs rose under the press of the Endangered Species Act, electric utilities, users of the Columbia's navigation channels, and others paying or expecting losses from involuntary reallocation of water mounted stiffening resistance. Because changes in the burden of proof have occurred incrementally in response to perceived emergencies, the scientific basis for considering the shifts has been obscured. The committee believes that a scientific perspective can illuminate difficult and contentious choices.

The power of a test is important in judging the value of expensive and uncertain efforts to save declining salmon populations. In a rational process, the most-effective, least-expensive treatments are tried first; when two treatments differ in cost but are of equal effectiveness, the less-expensive treatment should be chosen. Rules like that appear not to have been followed in the Columbia. Instead, a prolonged battle over two approaches has been waged, one in which science has often been subordinated to politics. The first, advocated by fishery managers, has been to increase the flows of the Snake and mainstem Columbia during the juvenile-salmon migration season in an attempt to recreate the conditions in the natural river when migrating smolts took advantage of the spring snowmelt to make a rapid trip to the ocean. The second, championed by the U.S. Army Corps of Engineers and the National Marine Fisheries Service, has been to transport migrating fish; juvenile fish are captured at one or more dams and taken in barges or trucks to below Bonneville, the lowest dam on the Columbia. Transportation is much less expensive because it does not require releasing impounded water in the spring, when power demand is relatively low and the natural supply of water is high because of the snowmelt. The question is whether transportation is biologically effective enough to be used instead of flow-augmentation. Transportation is used now in years of low flow when the available water would be inadequate to carry juveniles to the sea rapidly, and most smolts are transported at Snake River collector dams even when flows are above average, as in 1995.

There is intense controversy about the relative effectiveness of transportation and flow augmentation in other than low-flow years (see Volkman and McConnaha 1993). The cost of releasing water to benefit fish, however, is so high that in all years the amounts of water released fall within the uncertainties of the existing data (Lee 1993a, Chapter 2); that is, it is unclear whether the expensive releases have any biological effect. It would be logical to do experiments to measure the effectiveness of transportation and flow augmentation, but the way the controversy has developed undermines science. The effectiveness of flow augmentation is difficult to appraise because annual flow volume is determined by nature, not humans; in essence, the information comes at the rate of one data point per year. A compilation of studies done in the 1970s constitutes the only data set available (Sims and Ossiander 1981), and it is "sparse and unsatisfactory" (Volkman and McConnaha 1993:1256) in the view of all participants. The effectiveness of transportation is tested by capturing and releasing marked fish. That requires the cooperation of fishery managers, who have too often resisted, fearing that the data would undermine their position in the debate. As a result of the combination of natural and social impediments, scientific understanding has improved little over the last decade as policy has moved toward increasingly expensive flow-augmentation measures of unknown effectiveness.

Given the presence of endangered salmon populations in the Snake River, it might be appropriate to structure the burden of proof to favor possibly beneficial measures without waiting the decades needed to establish the magnitude of the benefits. But it is also important to assess the potential value of the steps that are taken through an assessment of the statistical power of the data available. That would illuminate the chances that transportation or flow augmentation would actually benefit fish, even though the benefits are not yet clearly in view; it would also identify the experiments likely to yield the most useful data. A recent study of the cost-effectiveness of various dam-mitigation options also concluded that studies of reach survival of smolts under several scenarios are essential (Paulsen et al. 1993).

The present committee has not done such an analysis. However, both this committee and the Snake River Salmon Recovery Team (1994) used information that, albeit incomplete, suggests that transportation improves smolts' survival. No other mitigation alternative proposal has data to support it as clearly. The Snake River Recovery Team's judgment that transportation is superior to other proposals to increase the migration speed of fish in the river has aroused controversy; the committee believes that science can clarify the boundaries of argument, in particular by providing better information on travel-time–survival relationships. It is important to clarify the boundaries of dispute so that all can be aware of how the burden of proof might be shifting. If the Recovery Team's judgment were to lead to a new status quo of relying mainly on transportation, then transportation implicitly would become the option to beat in the future. Given the uncertainties that still cloud our understanding of the benefits and costs of all the options, minds should remain open.

In the absence of scientific measurements, however, burdens of proof had shifted onto the other users of river water, and it was inevitable that they would object with increasing vehemence. Gaining reliable knowledge of the relationship between salmon survival and river flows will take a long time; on that, all sides agree. Intense conflict throughout the learning period will inhibit or undermine learning, as already occurred through the 1980s as test fish (fish used for experiments) were refused by the fishery-management agencies. Conflict is inevitable and can be constructive, but institutional arrangements are needed to protect the ability of science to illuminate and focus conflict. The latter point has been poorly understood and often overlooked in practice with consequences that have been expensive and probably biologically destructive.

A SCIENTIFIC ADVISORY BOARD

It will not be simple or inexpensive to establish a sounder basis for research, but there are well-established guidelines for doing it. The research must be—and must be *perceived* to be—objective, focused on the most important problems, scientifically sound, and free of political or policy bias. Few institutions can claim all those qualities on their own; most institutions that fund or conduct research have or are perceived to have vested interests in its outcome. Therefore, it is essential that research into the salmon problem in some way receive input from all relevant interest groups without being controlled or dominated by any of them. This committee recommends the establishment of a qualified, interdisciplinary, highly visible scientific advisory board to advise on the coordination and planning of research activities and to review research results. The board must be adequately staffed and funded, should be as free as possible of political or financial pressures, and should report at a high-enough level to prevent its recommendations from being easily ignored. Its reports should be available to the public. Such a board is necessary, but not sufficient; it will help only if there is a genuine commitment to solving the salmon problem and if the general institutional arrangements are improved.

Although such a board on salmon research would incur costs for travel, lodging, staffing, and other items and would slow some research decisions, we are confident that its advantages would far outweigh its costs. Poor, or badly designed, or inappropriate research is worth little, even if it is done quickly. Therefore, we recommend that establishment of an independent scientific advisory board on Pacific Northwest salmon research be given the highest priority.

Why a Scientific Advisory Board Is Needed

It might not be immediately obvious why an independent scientific advisory board is needed, especially if management agencies pledge themselves to an adaptive management approach, i.e., an approach whose design includes the opportunity to learn from experience and adapt management regimes accordingly (an approach strongly favored by this committee). McAllister and Peterman (1992) and Walters et al. (1993) recently described some of the difficulties of the adaptive-management approach; we describe below a hypothetical situation that illustrates some of the difficulties that a management agency might face and illustrates the potential value of an advisory board.

Suppose that an agency developed a management rule based on an explicit hypothesis, say, the hypothesis that increased river flow increases the survival of salmon smolts. One would expect the agency to emphasize protection of endangered populations while testing the effectiveness of alternatives in a reasoned and systematic fashion, making appropriate use of variations in natural conditions to provide experimental variation.

Such an experiment would take a long time to yield definitive results—perhaps decades, as Walters et al. (1993) described. Even with a clear exposition of the biological and ecological constraints that make such long periods necessary, there is considerable risk that the patience of the agency or others in the region would wear thin, that funding would become tight, or that conditions would change to make the situation seem to be a crisis and that pressure to change the course of action would result. Thus, the frailty of the approach lies not in science, but in the institutional supports for science: will human actors refrain from interfering with this experiment? Given the long periods that are required to learn effectively, the answer to that question must be in doubt.

People who care about the outcome of an experiment will be tempted to intervene in the measuring process to skew the results in their favor. Science has safeguards to account for the self-interest of scientists, chiefly controls and replication. If a scientific finding is valid, it will be replicated under the appropriate conditions. If an explanation is correct, the experimenter will have identified the minimal set of conditions that are necessary to achieve the predicted outcome, and changing one of the conditions will lead to some other result; that is the test of controls. Incorrect science, such as "cold" fusion, fails one or both tests.

The difficulty in the hypothesis concerning salmon survival that our hypothetical agency might consider is that controls and replication are both extremely difficult or impossible to achieve. If the volume of river flow is the dominant independent variable, then in large measure nature determines how much water is available to be used by migrating salmon in a given year. Every river is unique—there is only one Rogue, only one Columbia, so controlled experimentation is in principle impossible. For those reasons, the principal institutional safeguards against the undue influence of human interest on science are weak or unavailable. The best approach to mitigate this difficulty is to separate interest from science, that is, to make the results of the experiment safe from the dictates of policy. And the best mechanism that we can recommend to achieve that is an independent, distinguished, and credible scientific advisory board.

Experiments on mainstem survival—involving different experimental treatments, as they must—are likely to affect the amounts of river flow allocated to fish transport and the fraction of migrating juveniles intercepted for transportation. This kind of adjustment is a common response to new information gained during adaptive experiments—changing the mix of treatments, each supported or criticized by observers in accordance with the accumulating findings. But if the experiments lead to the *discontinuation* of augmented flows and spills for in-river migration or of other methods thought to increase survival (e.g., transportation, fishing regulations, or hatcheries), then one of today's hotly contested options would have been discarded. This outcome would touch off intense conflict. (Argyris and Schon [1978] called this kind of adjustment "double-loop learning"—the attempt to improve performance while allowing the objectives to evolve.) If the political actors expect a threat to any of the mitigation methods, such as in-river migration flows or barge transportation, they might try to derail the experiments. As budget crises, political changes, or other kinds of turbulence come and go, the experiments and a long enough series of their expensive yet crucial data sets seem less and less likely to survive unimpaired.

An independent scientific advisory board could help to insulate the learning process from the defenders of each option for improving survival. Its presence and clear advice would allow the hypothetical agency to obtain an explicit agreement among all parties that inriver flows and spills, transportation, and all other established means of increasing fish survival will continue to be used throughout the period of experimentation unless it becomes clear *to the scientific advisory board* that the experimental results dictate a different course of action. The absence of such an agreement invites interference in the experiment; as time passes, such interference will be ever more harmful because it would contaminate an increasingly long and expensive data set.

Requirements for an Effective Scientific Advisory Board

Vesting authority in an independent scientific panel to redirect adaptive policies invites unaccountable decision-making that could become controversial if the decisions ignore the pleas of those who are seriously affected. Yet, paradoxically, the raison d'etre of the scientific advisory board is to allow scientific learning and analysis to occur independently of the political pressures of interest groups. That leads to two conclusions: the board must be truly independent, comprehensive, interdisciplinary, objective, and balanced; and the board cannot be expected to solve problems by itself. Its job is to increase the efficiency and quality of the science available to inform policy and management decisions. We have pointed out at great length that the salmon problem is to a large degree a human and institutional problem, and we have recommended institutional changes to help solve the problem (Chapter 13). The scientific advisory board must function in the context of institutions and must be seen to do so, or it will not be effective. In addition, many of the experiments needed will be expensive and time-consuming, and demands for different, perhaps less-expensive, experiments will arise. Some will perhaps argue for simultaneous experiments whose cost will exceed the available budget. What is needed is to make the agencies establish common objectives, eliminate duplication, and establish mutually reinforcing long-range sustaining programs.

It is important that the advisory board, without being authoritarian, have some authority, or at least a constituency. We recommend that the advisory board be responsible primarily to the major funders of salmon-related research in the region. Those funders should include the Bonneville Power Administration, the U.S. Army Corps of Engineers, the National Marine Fisheries Service, the U.S. Fish and Wildlife Service, and the U.S. Department of Agriculture. For the advisory board's advice to be effective, those agencies and any others that are willing should agree to give careful consideration to its recommendations concerning research and other science-based activities and to provide public responses to departures from those recommendations. We hope that other researchers and funders of research would also be able to benefit from the advisory board's advice.

If science is to make effective contributions to salmon rehabilitation, experimental designs and their results must be publicly disseminated as soon as possible. The critical times for experimentation are the migration seasons of juvenile and adult salmon. Therefore, to provide enough time to collect and organize data, we encourage scientists, within nine months of the end of the last migration season, to archive a comprehensive data set for each year's studies and make them available to the public at one or more of the region's research universities. We also encourage principal scientists in charge of the experiments to submit a paper or papers reviewing the cumulative experience and scientific findings as often as results warrant; probably at least once every three years. The papers should be submitted to a national peer-reviewed journal of unquestioned reputation.

Membership

The scientific advisory board must have a diverse and distinguished membership that can address all the major scientific issues. In addition to biology, hydrology, fishery science, and engineering, its expertise should include the social sciences, including anthropology, sociology, and economics; and it should include geography and information science. It is critical that members of the advisory board be chosen for their scientific and technical expertise rather than as representatives of a particular agency or organization; to the degree possible, their scientific judgment should be independent of their parent organizations.

There is value in mixing experts who have extensive local knowledge with those who have great expertise but less local knowledge and hence fewer biases. Probably, some advisory board members should come from outside the region. But its membership must be chosen in a way that will give it credibility in the eyes of the people and institutions of the region. It must be independent.

Operation

Several factors are needed for a scientific advisory board to function effectively. Sociologist William Freudenburg analyzed some factors that affect such a board's effectiveness (Freudenburg 1990), and the following advice draws on his analysis. The advisory board must have adequate financial support to operate and to meet often enough to do its job. It must have a professional staff whose primary responsibility is to the board. Borrowing staff from one of the agencies in the region creates the risk that the staff will have divided loyalties and insufficient time to perform their functions. To a large degree, the advisory board must set its own agenda, lest it become bogged down in the details of issues that others deem to be important. Although it must be independent, it must have a mechanism for being responsive to the concerns of people with local knowledge, interest, and concerns. Finally, its reports must be visible and accessible to all. Although the advisory board will not and should not make policy decisions, its scientific advice must be loud and clear enough that it cannot be ignored by accident.

The operation of a successful scientific advisory board will require substantial money for support and logistics and substantial time of its members and staff. But if it is successful in its goals—encouraging thoughtful experimentation, helping other scientists to design and execute research and analyze the results, and helping to create a conduit for information among scientists and between scientists and the public—the investment of money and time will be worthwhile.

15

Conclusions and Recommendations: Toward a Sustainable Future for Salmon

Anadromous salmon in the Pacific Northwest and their habitats have been adversely affected by the region's development, including such factors as forestry; agriculture; grazing; industrial activities; dams; commercial, residential, and recreational development; and fishing. Development and its associated pressures and changes will continue. Considerable action would be needed merely to arrest the decline of salmon and maintain even the current degraded status. Improving the prospects for sustainability of anadromous salmon is complicated and contentious, and it has no simple or single solution. But the Committee on Protection and Management of Pacific Northwest Anadromous Salmonids reached consensus on several important conclusions and recommendations. If the committee's recommendations are adopted, a considerable reallocation of financial and natural resources will follow.

Life-history and migration patterns of salmon complicate their management because, for example, fish hatched in the Columbia River are caught as far away as southeastern Alaska and northern British Columbia. Solutions to the salmon problem must recognize the influence of fishing in Alaska and British Columbia, in addition to that in the Pacific Northwest. Indeed, unless Alaska, British Columbia, and the Pacific Northwest cooperate, solutions to the salmon problem might be impossible. In the absence of such cooperation, any success would entail greater expense for the Pacific Northwest.

In most respects, the salmon problem is a problem of how to match scales of management, governance, fishing, research, and understanding with scales of biology, hydrology, and environmental change in space and time. The salmon traverse a great variety of environments throughout their life cycle, including thousands of miles at sea and up to 1,500 km in rivers. Salmon cross international and state boundaries, and they are important components of several ocean and inland aquatic ecosystems. Salmon catches are governed by local, state, federal, tribal, and international treaties, conventions, agreements, organizations, commissions, and agencies; salmon environments are used for commerce, agriculture, industry, recreation, public safety, and hydropower. Climates that affect salmon fluctuate from year to year and over decades and vary between regions. Human cultures that depend on salmon and the economic factors that affect salmon values have spheres of influence that range from a few to thousands of kilometers and a few to thousands of years. Many of the committee's conclusions and recommendations reflect an attempt to reconcile some of those diverse scales. The committee has developed the conclusions and recommendations that follow within the framework of rehabilitation, rather than degradation, restoration, or substitution.

We present an approach to solving the salmon problem, then a general conclusion, and finally a set of more-specific conclusions and recommendations regarding environmental changes, including habitat changes, both natural and anthropogenic; genetic structure of salmon populations and species and appropriate units to be managed for conservation; fishing and fishery management; hatcheries and other techniques for increasing the number of fish spawned naturally; dams; goals and values; information needs; and institutional and management considerations, including international, federal, state, local, and other jurisdictions.

GENERAL CONCLUSION

Economic development and human population growth without sufficient attention to salmon and their environment have created widespread declines in anadromous salmon abundances in the Pacific Northwest. Although some salmon populations are stable or increasing, the overall pattern is one of decline. Many factors have contributed to salmon declines; it is therefore unlikely that reducing or compensating for only one type of adverse impact will be enough to reverse the decline in any watershed. To rehabilitate salmon populations and their ecosystems, changes in fishing, dam and hatchery operations, and land uses would be required. The degree of change needed in each of these factors will be related to their contribution to the problem in each watershed and to the degree of rehabilitation desired.

Until very recently, the importance and benefits of rehabilitating salmon and their ecosystems have been overridden by the motivation to sustain catch, a reliance on technology, and economic considerations. Weighing the direct and collateral benefits of rehabilitating salmon populations against the dislocations that are sure to occur raises profound questions that should be discussed in ways that allow opportunities for citizens to participate.

The salmon problem took many years to develop, and its solution will require the commitment of time, money, and effort. There is no simple answer to this complex social and biological problem. To achieve long-term protection for a diverse and abundant salmon resource in the Pacific Northwest, two conditions must be met:

- Management must recognize and protect the *genetic diversity* of salmon. It is not enough to focus only on the abundance of salmon: their long-term survival depends on genetic diversity within and between local breeding populations. This diversity and the protection and rehabilitation of salmon habitat are the bases of sustained production of anadromous salmon and of the species' evolutionary futures. Because of their homing behavior and the distribution of their populations and their riverine habitats, salmon populations are dependent on diversity in their genetic makeup and population structure and thus are unusually susceptible to local extinctions (Chapter 6). Attempts to control mortality by fishing and improving environmental conditions and to compensate for mortality with hatchery-produced fish must keep genetic diversity as the highest priority.
- Any solution to the salmon problem must take the effects of growth in human population and economic activity into account. If economic and population growth in the region continue, many of the forces that have reduced salmon runs will continue to make it harder and more expensive to rehabilitate salmon in the Pacific Northwest successfully. The social structures and institutions that have been operating in the Pacific Northwest have proved incapable of ensuring a long-term future for salmon, in large part because they do not operate at the right time and space scales. As described in Chapter 13, differences among watersheds mean that different approaches are likely to be appropriate and effective in different watersheds, even where the goals are the same. This means that institutions must be able to operate at the scale of watersheds; in addition, a coordinating function is needed to make sure that larger perspectives are also considered. Substantial institutional changes would be needed to achieve those goals.

The specific recommendations that follow were made in the context of those two goals. A crucial aspect of the recommendations is the overriding need to focus management goals primarily on genetic diversity rather than on biomass production.

ENVIRONMENTAL CHANGES

Large changes have occurred in salmon habitats, including the ocean. Some changes are natural, others are due to human impacts; some appear to fluctuate, others are more trendlike; some can be directly influenced by human activities, others at present cannot. Rehabilitation must now operate within that context and must acknowledge the inherent uncertainty associated with environmental changes and variability.

Oceanic Conditions

Conclusion. Variations in ocean conditions powerfully influence salmon abundance. Ocean conditions—especially water temperature and currents and associated biological communities—have common effects over wide geographic areas and have become less favorable for salmon in the Pacific Northwest since the late 1970s. Because ocean conditions vary, fish deaths caused by numerous human activities such as fishing could be more damaging just when fish populations are most likely to be depleted by natural conditions.

Recommendation. Fishery management must take the variability in ocean conditions into account. Some might be tempted to attribute all changes in salmon abundance to changes in ocean conditions and to conclude that management related to rivers is therefore unimportant. However, because all human effects on salmon are reductions in the total production that the environment allows, management interventions are *more* important when the ocean environment reduces natural production than when ocean conditions are more favorable. In a situation of such uncontrollable external variation, it would make sense for fishing to take a *fixed and sustainable proportion* of the returning spawners rather than a fixed number, as has been common practice, whether ocean conditions are favorable or unfavorable—as long as the number of returning spawners exceeds a minimal safe threshold based on demographic and genetic considerations. Below that threshold, no fishing should be allowed. Management should attempt to reduce human-caused deaths of fish in rivers and at sea especially when ocean conditions are unfavorable (as measured by estimates of survival rates at sea). Any favorable changes in ocean conditions—which could occur and could increase the productivity of some salmon populations for a time—should be regarded as opportunities for improving management techniques. They should not be regarded as reasons to abandon or reduce rehabilitation efforts, because conditions will change again.

Regional Variation

Conclusion. There is considerable regional variation in the physical, biological, social, cultural, and economic environments of salmon. No unified solution to the salmon problem, management strategy, research strategy, institutional arrangement, or governance structure can be expected to apply to the entire Pacific Northwest.

Recommendation. Any approach to improving the status of salmon popula-

tions must have regional components that, when possible, reflect the bioregions relevant to salmon biology and conservation. Preemptive recovery plans should include management and research strategies, institutional arrangements, and governance structures that are flexible and can be adjusted to fit regional variations.

VALUES AND INSTITUTIONS

Conclusion. Extractive interests have structured regional practices and institutions for the management of natural resources and for the modification of environments for human benefit. Society in the Pacific Northwest is in the midst of assessing values with respect to natural resources and their use. Historically, the region has been governed by an extractive value system. The values were ingrained into the social and political institutions that developed to manage and control resources.

Recommendation. Institutional changes that better reflect societal interests in maintaining biodiversity and the functioning of ecosystems should be sought in light of the conflicts among those interests during a period of change. A broad range of techniques should be used in estimating societal interests, including opinion surveys, focus groups, public participation, and content analysis of written commentary. Because institutional arrangements reflect the commitments of earlier times, continued conflict focused on institutional rules and procedures is to be expected as part of the process of change.

Recommendation. Goals and values should emerge in significant part through cooperative management, so that those most directly involved play an instrumental role in determining how the rehabilitation of salmon takes shape in the places they regard as their own. Efforts to rehabilitate salmon should be accompanied by efforts to communicate with stakeholders and the general public in ways that allow for their evaluation of goals and values of the rehabilitation projects and their participation, where appropriate, in cooperative management.

Recommendation. Interdisciplinary approaches to the salmon problems should be strengthened and should incorporate the expertise required to deal with nonbiological and nonmonetary aspects. Greater effort should be made to use interdisciplinary working groups to evaluate projects, to work on methodologies needed to incorporate monetary and nonmonetary criteria into those evaluations, and to accurately depict and (where appropriate) quantify the value of salmon to the region.

GENETICS AND CONSERVATION

Genetic Resources

Conclusion. Sustained productivity of anadromous salmon in the Pacific Northwest is possible only if the genetic resources that are the basis of such productivity are maintained. We have already lost a substantial portion of the genetic diversity that existed in these salmon species 150 years ago. The possible genetic effects of any actions must be considered when any management decisions are made. The local reproductive population, or deme, is the fundamental biological unit of salmon demography and genetics. An adequate number of returning adults for every local breeding population is needed to ensure persistence of all the reproductive units. The result of regulating fishing on a metapopulation basis and ignoring the reproductive units that make up a metapopulation is the disappearance or extirpation of some of the local breeding populations and the eventual collapse of the metapopulation's production.

Recommendation. Salmon management should be based on the premise that local reproductive populations are genetically different from each other and valuable to the long-term production of salmon. Managing from that perspective will protect habitat and also protect resources for the long term. Efforts should be made to identify and protect remaining native wild populations and their habitats. Minimum sustainable escapements should be established for as many populations as possible. Populations that have unusual genetic adaptations or occupy atypical habitats are of special importance and should be identified and protected. The genetic diversity within existing spawning populations is not replaceable and must be conserved to protect present and future opportunities, including the evolutionary process in salmon. This principle seems self-evident, but risks continue to be imposed on such populations.

Regional Population Structure

Conclusion. The metapopulation model of geographical structure is important for salmon because of the geographical arrangement of salmon into discrete spawning populations (demes) adapted to the environmental conditions in which they reproduce. Local demes of salmon are small enough and exist in variable-enough environments for it to be likely that they will have relatively short persistence times on an evolutionary scale. Although the deme is the functional unit of salmon genetics and demography, the cluster of local populations (the metapopulation) connected by genetic exchange via natural straying is the fundamental unit on an evolutionary time scale. *This conclusion is crucial because it leads to many other conclusions and recommendations about salmon management.* For example, most of this report's conclusions and recommendations about hatcheries, fishing, and habitat rehabilitation are founded on the importance of maintaining appropriate diversity in salmon gene pools and population structure, which has not been adequately recognized by many.

Recommendation. It is important to maintain geographical clusters of demes or metapopulations. The loss of genetic material in small populations is dominated by random events (genetic, demographic, and environmental), so the loss of genetic material in one deme should be largely independent of that in another. In a cluster of demes, a large proportion of the original genetic variation should remain. Thus, small populations might not be viable for long periods in isolation but are important in maintaining genetic variation as parts of a metapopulation.

HABITAT LOSS AND REHABILITATION

Conclusion. Freshwater habitats are critically important to salmon because they constitute the spawning grounds and nurseries in which the genetic makeup of a population is determined. Many human activities—notably forestry, agriculture and grazing, hydropower, and commercial, residential, and recreational development—have contributed to degradation of the riverine and adjacent riparian and near-river habitat and caused loss of habitat of spawning adults and young salmon, and loss of associated components of the ecosystem. So few intact basins or subbasins are in good condition that those few should be viewed as critical salmon refuges and as sources of plants and animals necessary for ecosystem recovery as other watersheds are improved. Part of this source of recovery is provided through the occasional straying of salmon to adjacent streams. Thus, especially if the refuges harbor large populations, this potential for colonizing other suitable habitats is important (Chapter 7). In addition, programs that have relied on artificial habitat and hatchery-production techniques have usually not lived up to expectations and in some cases have actually hastened the decline of wild salmon populations, as described in chapters 7 and 12. Most of the traditional habitat research and current performance standards (e.g., forest-practice rules) have emphasized protecting habitats in headwater stream networks with secondary consideration to lowland systems. In the decline of salmon and their future, more attention needs to be given to rehabilitation of streams subject to county and city planning and land-use authorities. Although rehabilitating habitat will be more difficult in a region experiencing rapid population growth, it is not prudent or appropriate to abandon streams that are degraded, and rehabilitation can be worth the effort. Chapter 7 outlines examples of projects to rehabilitate streams in the Seattle area. All streams providing spawning or rearing habitat can contribute to the long-term survival of salmon populations in river basins.

Recommendation. Riverine-riparian ecosystems and biophysical watershed

processes that support aquatic productivity should have increased protection. Riparian zones are important for the maintenance of aquatic productivity, but insufficient protection has been given to these critical areas in the past. The width of riparian zones requiring protection from harmful human disturbances is usually not known with certainty, but all possible ecological functions should be considered when attempting to define riverine-riparian boundaries. Within the domain of interactions between aquatic and terrestrial environments that characterize the riparian zone, some human activities might occur without major disruption; however, it is critical that the full range of ecological functions be explicitly protected, including all biotic and physical processes that mediate the exchange of energy, water, nutrients, and organic matter between watersheds and their streams. In many cases, the riparian zone in which these exchanges occur may be substantially wider than the narrow border of vegetation often specified in current regulatory language (e.g., state forest-practices acts) for nonfederal forest lands. Riparian zones associated with streams draining rangeland or agricultural or urban areas often lack any regulatory prescription.

Beyond the edge of the riparian zone, it is important that hydrologic processes within watersheds not be altered by human activities to such an extent that patterns of water, sediment, and organic matter inputs to streams degrade aquatic habitat or riparian functions. Human activities resulting in habitat degradation include activities that prevent some important ecological processes (e.g., flooding and groundwater recharge) and activities that alter the rates of other processes (e.g., accelerated erosion). Although land and water will continue to be used in most Pacific Northwest watersheds, recovery of productive salmon habitat will necessitate a concerted effort to rehabilitate the full range of natural conditions in aquatic and riparian ecosystems. To facilitate that recovery, the following six recommendations are offered:

1. Forestry, agricultural, and grazing practices should allow riparian zones to maintain a full range of natural vegetative characteristics, i.e., characteristics occurring in watersheds with natural disturbance regimes. Riparian zones should ideally be wide enough to fulfill all functions necessary for maintaining aquatic productivity.

2. Sediment from all land uses should be reduced to magnitudes appropriate to the geological setting of a river basin. In practical terms, the goal is that human activities should cause no net increase in sediment over natural inputs. Likewise, water temperatures should reflect as closely as possible the normal regime of temperatures throughout the basin.

3. Patterns of water runoff, including surface and subsurface drainage, should match to the greatest extent possible the natural hydrologic pattern for the region in both quantity and quality. Effects of consumptive water uses on both the timing and the quantity of flow should be minimized. Water-management technologies that promote the restoration of natural runoff patterns and water quality

should be strongly encouraged. That will mean implementation of methods to reduce the volumes of water used for irrigation, industrial, and urban uses.

4. Toxic waste products from industrial, mining, agricultural, and urban activities should receive the appropriate treatment before being discharged into any body of water.

5. Habitat reclamation or enhancement activities should emphasize rehabilitation of ecological processes and functions, not artificial creation of habitat. Placement of permanent or semipermanent habitat structures in streams should be discouraged unless it can be clearly demonstrated that no other alternative is available. Existing artificial structures that appear to be impeding natural recovery should be removed.

6. Beneficial long-term effects of natural disturbances, such as flooding, should be preserved or restored whenever possible. Lowland slough and estuarine habitat rehabilitation should receive high priority in coastal regions.

Rehabilitation of riverine-riparian ecosystems will take time. Recovery of important ecological processes, even with appropriate human intervention, may take decades to centuries and will require patience and long-term commitment. Restoration efforts should be coordinated across large areas of the landscape, be accompanied by adaptive management agreements, monitoring and evaluation, and be guided by the results of thorough watershed analyses.

DAMS

Conclusion. Although as many as 90% of young salmon might survive passage over, around, and through any individual major hydropower project on the Columbia-Snake river mainstem, the cumulative reduction in survival caused by passing many projects has adversely affected salmon populations. Partly because salmon do not have rights to water, allocation of water rights usually has not included considerations of their long-term survival. (Of course, the current concern over their survival has included considerations of water availability for them.)

Recommendation. Improve salmon survival rates associated with passing hydropower projects in the Columbia and Snake rivers. The following approaches are recommended:

- Determine existing reach survivals (survival rate of fishes as they pass through a reach or a specified stretch of the river), e.g., by project and project components. On completion of such studies, initiate measures to improve survival, prioritized by the greatest gains obtainable.
- Secure water as need is demonstrated—for example, where changes in annual patterns or total amounts of streamflow are shown to decrease survival—

from water-consumers by subsidizing water conservation by buyout of water rights (which might require legislative changes in transfer of water rights to point of use) and by improved reservoir-system operation, e.g., through improved accuracy of seasonal streamflow forecasts through telemetering, satellite assessments of snow coverage and water content, and other technological means. As long as other rights to public water are considered sacrosanct in any river basin, it will be difficult to respond to demonstrated needs of salmon in a timely manner. All options should be open to consideration, after appropriate evaluation of costs and benefits. For example, flood-control-rule curves should be considered open to change, because in some cases increased seasonal streamflows might need to be balanced against increased risk of downriver flooding.

- Continue downriver transportation of smolts by barge in the Columbia and Snake rivers as long as data indicate that survival in transport exceeds that of inriver migration. It is critical that barging (and any other treatment) be done with experimental controls so that information can continue to accumulate, i.e., enough smolts should continue inriver migration to assess the effectiveness of transportation. And it is essential not to treat all the fish in a river in such a way that failure of a treatment can have catastrophic consequences for the entire population. Careful scientific monitoring is essential for addressing controversies about transportation and other interventions.
- Improve information on the migratory characteristics of salmon in the Columbia-Snake river system. PIT-tag applications should be expanded to enough wild fish and as many hatchery fish as possible to conduct convincing scientific analysis and to separate hatchery from wild fish. The utility of genetic markers that can be safely and quickly detected from fish scales or slivers of fin tissue should be explored. Interrogation facilities (facilities that detect tags) should be set up at all bypasses so that adult returns can be evaluated to compare survival of fish that migrate via bypass, transport, and turbine and spill, and so that reach-specific information can be obtained on tagged smolts. Spawning-ground surveys should be greatly expanded to evaluate homing efficacy in transported and nontransported fish.

Conclusion. The many dams on the Columbia River and its tributaries cumulatively have had large effects on salmon survival. Therefore, the addition of any new major dams in undammed reaches of large rivers in the region (e.g., the Hanford Reach of the Columbia River) would make the situation worse; existing dams should have adequate fish-passage facilities where feasible and appropriate before being relicensed.

Conclusion. Because there has not been a major seasonal shift in the annual Snake River hydrograph, it is doubtful, a priori, that the declines in Snake River salmon populations have resulted from or are reversible by seasonal changes in flow regime alone. Even if flow changes could be helpful in a rehabilitation effort, they are likely to be insufficient without changes in other human interventions in the salmon life cycle and habitat.

Conclusion. Because the Snake River system stores and then diverts substantial quantities of water for consumptive uses, and the volume of water flowing through the system has therefore decreased, beneficial changes in flow regime for salmon can in principle be obtained in a controlled fashion by reallocating human uses of water, including agricultural uses. Whether those changes can be made at lower total social cost than large-scale engineering changes, such as drawdown, would need to be analyzed on a case-by-case basis.

Conclusion. Transportation of smolts to bypass middle Columbia dams might prove better than inriver migration as more data become available on bypass and collection in that region. Because of the stress, injury, post-bypass losses, and delayed arrival of smolts at the ocean resulting from decreased water velocities in reservoirs, the most appropriate use of bypass facilities at most dams might be to collect fish for transportation. Avoidance of mortality at downstream hydropower dams and in reservoirs is an attractive concept. The concept might become even more attractive as means develop to improve survival through release point protocols. Any experiments with transportation should follow the guidelines discussed earlier.

Recommendation. Transport of middle Columbia summer migrants should be investigated. At McNary Dam, upstream from three hydropower projects, transportation of subyearling migrants yielded transport:benefit ratios (observed survivals to adulthood of transported smolts to observed survivals of inriver migrants) of over 3.0:1 in tests in the 1980s.

FISHING AND FISHERY MANAGEMENT

For rehabilitation of salmon populations, the aim of fishery management—as for other management efforts—should be to achieve long-term sustainability based on maintaining genetic diversity. In the recommendations below, the overall goal is to reduce total fishing mortalities (or to increase escapements) to be consistent with the present productivity of salmon and to develop and implement catching technology that ensures minimal mortality in depleted demes.

Too Few Spawners

Conclusion. Not enough fish are being allowed to return to spawn. It is essential to keep in mind that unless enough fish are able to spawn, there will not be enough fish produced to compensate for all the sources of mortality imposed by human activities and to provide sustainable runs of wild salmon. Therefore, a goal of management should be to increase the size and maintain the diversity of spawning populations and to re-establish ecosystem processes.

Recommendation. Escapements should be increased. A shift must be made from focusing on catch to focusing on escapement. Increasing the number of adults that return to spawn (escapement) will enhance opportunities for evolution of genetic diversity through colonization, straying, and competition, and will bolster nutrient input to streams. Management should set new goals of minimum sustainable escapement (MSE), allowing escapements to vary above the MSE, as opposed to managing one fixed escapement. Such a process has been implemented along the Washington coast. "Escapement floors" have been established for each species and system. Returns above the floor are shared by terminal users and spawning escapements. Only larger escapements will allow larger catches. Increased escapements, however, imply reduced catch or increased productivity to sustain a catch in the short term. Over the short term, reduced catch rates can result in a smaller catch. Over the longer term, lower, but sustainable, catch rates applied to increased salmon abundance (resulting from the increases in numbers of spawners) will, on the average, result in increasing catches. As with all aspects of management, an adaptive approach is recommended so that advantage can be taken of experience.

Protection of Genetic Diversity

Conclusion. Long-term survival and production of natural salmon runs depend on maintenance of genetic diversity and metapopulations.

Recommendation. Fishery management should explicitly recognize the need to conserve and expand genetic diversity via natural increases in population sizes. A holistic approach should be taken that recognizes the interdependence of genetics, habitat, and salmon production, and it must account for the uncertainty in scientific knowledge and the inherent variability of biotic and abiotic environmental factors. This is accomplished by never allowing numbers of salmon to decline below the minimum sustainable escapement and by filling out the dendritic structure of salmon metapopulations in a river basin. When escapements *exceed* the MSE, the extra fish should be allocated between escapement and catch. It will often not be possible to maintain all the diverse habitats dictated by the full dendritic structure of the watershed. Land-use planners and managers must be vigilant in preserving as much of the structure as possible (i.e., allowing for connectedness of spawning populations) and then developing approaches for rehabilitating other parts of it.

Strong and Depleted Populations

Conclusion. Past practices of salmon management have not treated strong and depleted populations differently enough, and—more important—salmon management has not been sufficiently based on recognition of the importance of demes. The anadromy of salmon and their tendency to return to their natal streams to spawn results in a population structure in which metapopulations—clusters of demes—are important. Conservation of salmon must take that structure into account to achieve long-term survival of diverse salmon populations.

Recommendation. Management of salmon should be based on the genetic structure of their populations and should allow for separate management regimes for strong and depleted demes and metapopulations whenever possible. In general, the aim is to assure adequate escapements for depleted populations. To achieve this aim, fishing should take place only where the demic identity of the salmon is known and where catching technology can reduce mortality rates in depleted demes. In many cases, that would require fishing to take place in the home-stream estuary or in the river upstream. Inriver gear should be changed to live-catch systems to the greatest possible extent, permitting release of fish of depleted populations or species. Implementing this recommendation initially will require low fishing effort in many areas, especially in the ocean, and will require the cooperation of British Columbia and Alaska, because many salmon that originate in the Pacific Northwest are caught at sea in southeastern Alaska and in British Columbia. Ideally, those fish would be allowed to be caught in terminal fisheries in the Pacific Northwest. If only the ocean fisheries in the Pacific Northwest are closed, more northerly ocean fisheries will still impose a large mortality on a mixture of depleted and strong populations. Because of the diversity of interests and agreements, conventions, and treaties among various parties and nations in the region, and because there are various costs and benefits associated with all methods for rehabilitating salmon populations in the Pacific Northwest, a serious political effort will be needed to achieve the common goal of protecting depleted populations in an acceptable way. Typically, negotiations within the Pacific Salmon Commission (PSC) have been characterized by conflict and stalemate resulting from winner–loser negotiations and the desire of users to avoid disrupting fisheries. The committee notes, however, that management actions have already disrupted some fisheries, especially those in the more terminal areas. Disruption is inevitable during the rehabilitation of Pacific Northwest salmon. An effective Pacific Salmon Treaty is needed, however, to balance changed allocations among fisheries and to balance short-term losses against the long-term gains that will be derived by conserving and rehabilitating the Pacific Northwest salmon. To assist in establishing greater cooperation in the Pacific Salmon Commission, new approaches are encouraged. For example, re-establishing reciprocal fishing agreements would allow each country greater access to its salmon production and might allow financial agreements to aid in fishing-effort reductions through vessel buybacks or the buying out of portions of catch ceilings and vessels to reduce ocean interceptions permanently, or to aid in rehabilitation costs.

HATCHERIES

Role of Hatcheries

Conclusion. The management of hatcheries has had adverse effects on natural salmon populations. Hatcheries can be useful as part of an integrated, comprehensive approach to restoring sustainable runs of salmon, *but by themselves they are not an effective technical solution to the salmon problem.* Hatcheries are not a proven technology for achieving sustained increases in adult production. Indeed, their use often has contributed to damage of wild runs. The current approach to hatchery use—the enhancement of catchable salmon runs—entails a large and continuing input of human energy and money. In addition, such use of hatchery production often results in reduction of already depleted wild runs by further reducing natural populations of salmon (see Chapter 12). In many areas, there is reason to question whether hatcheries can sustain long-term yield, because they can lead to loss of population and genetic diversity and adversely affect natural populations, as discussed in Chapter 12. Therefore, it is unlikely that hatcheries can make up for declines in abundance caused by fishing, habitat loss (including that resulting from dams), etc., over the long term. Hatcheries might be useful as short-term aids to a population in immediate trouble while long-term, sustainable solutions are being developed. Indeed, such a new mission for hatcheries—as a temporary aid in rehabilitating natural populations—could be important in reversing past damage from hatcheries as well as from other causes.

Recommendation. The intent of hatchery operations should be changed from that of making up for losses of juvenile fish production and for increasing catches of adults. They should be viewed instead as part of a bioregional plan for protecting or rebuilding salmon populations and should be used only when they will not cause harm to natural populations. Hatcheries should be considered an experimental treatment in an integrated, regional rebuilding program and they should be evaluated accordingly. Whenever hatcheries are used, great care should be taken to minimize their known and potential adverse effects on genetic structure of metapopulations and on the ecological capacities of streams and the ocean. Special care needs to be taken to avoid transplanting hatchery fish to regions in which naturally spawning fish are genetically different. *The aim of hatcheries should be to assist recovery and opportunity for* *genetic expression of wild populations, not to maximize catch in the near term.* Only when it is clear that hatchery production does not harm wild fish should the use of hatcheries be considered for augmenting catches. Hatcheries should be audited rigorously. Any hatchery that "mines" broodstock from mixed wild and natural escapements to meet its normal operating targets should be a candidate for immediate closure or conversion to research. Diseased broodstocks should be rigorously culled to minimize disease in progeny. It is useful for all hatchery fish to be identifiable. Visible marks, such as finclips, have some advantages, but other methods, such as passive integrated transponder (PIT) tags, coded-wire tags, and genetic markers are also useful. Marking hatchery fish externally is particularly important when fishers and managers need to distinguish between hatchery and wild fish.

Regional Variation in Use of Hatcheries

Conclusion. Current hatchery practices do not operate within a coherent strategy based on the genetic structure of salmon populations. A number of hatcheries operate without appropriate genetic guidance from an explicit conservation policy, although this is beginning to change. Consistency and coordination of practices across hatcheries that affect the same or interacting demes and metapopulations is generally lacking.

Recommendation. Hatcheries should be dismantled, revised, or reprogrammed if they interfere with a comprehensive rehabilitation strategy designed to rebuild natural populations of anadromous salmon sustainably. Hatcheries should be tested for their ability to rehabilitate populations whose natural regenerative potential is constrained severely by both short- and long-term limitations on rehabilitation of freshwater habitats. Hatcheries should be excluded or phased out from regions where the prognosis for freshwater-habitat rehabilitation is much higher, as is the case for many watersheds of the Oregon coast. This recommendation, for example, would allow continuation of the hatchery-supported fishery on chinook at Willamette Falls because it is disrupting neither wild populations nor rehabilitation efforts.

Recommendation. Decision-making about uses of hatcheries should occur within the larger context of the region where the watersheds are located and should include a focus on the whole watershed, rather than only on the fish. Coordination should be improved among all hatcheries—release timing, scale of releases, operating practices, and monitoring and evaluation of individual and cumulative hatchery effects, including a coastwide database on hatchery and wild fish proportions and numbers. Hatcheries should be part of an experimental treatment within an adaptively managed program in some regions but not in others.

Recommendation. All hatchery programs should adopt a genetic-conservation goal of maintaining genetic diversity among and within both hatchery and naturally spawning populations. Agencies involved in management of anadromous salmon should recognize that achievement of population-rebuilding goals will be jeopardized without concurrent adoption of a genetic-conservation goal. Hatchery practices that affect straying—genetic interaction between local wild fish and hatchery-produced fish—should be closely examined for consistency with regional efforts.

INFORMATION NEEDS

Funding Adequacy

Conclusion. Research has been adequately funded but inadequately guided.

Recommendation. An independent, standing scientific advisory board should be established to ensure that the available research dollars are spent most productively to answer the most critical questions as soon as possible. The advisory board would encourage cooperation from other organizations and individuals in the region to help to design and evaluate research and would serve as a conduit for information. It should be composed of experts in relevant disciplines, including natural and social sciences and engineering. The primary funding agencies for salmon research in the region (at least the Bonneville Power Administration, the Army Corps of Engineers, the National Marine Fisheries Service, the Department of the Interior, and the Department of Agriculture) should carefully consider the advice of the advisory board with respect to identification of critical questions, research funding, monitoring, and other science-based decisions concerning salmon. When they do not follow the board's advice, they should provide written justification. The committee's reports should be publicly available.

Adaptive Management

Conclusion. Much of the current uncertainty over the benefits of habitat-improvement projects, hatcheries, and other management and restoration approaches results from lack of scientific monitoring and evaluation. Many habitat programs involving millions of dollars have been undertaken over the last 20 years with little or no monitoring. Even when monitoring has been undertaken, lack of replicates and controls, uneven measurement consistency, and lack of commitment to long-term study have constrained the opportunities to learn from these programs.

Recommendation. Watershed analysis, adaptive management, a careful in-

ventory, and strong regional monitoring programs are needed to provide the context within which management decisions can be made. A systematic evaluation of the condition of Pacific Northwest watersheds and the status of salmon populations must be undertaken. Some states and federal agencies are initiating such efforts. Watershed analysis should encompass multiple land uses throughout river basins. A regional network of reference sites should be established for adaptive-management experimentation similar to the trials now being implemented on federal forest land. Integrative measures of watershed productivity (such as smolt production) must be monitored at many more locations than is the case today. Finally, a clearer picture of the status of salmon populations throughout the region is needed to increase confidence in decisions about how to allocate financial and human resources to solve the salmon problem.

INSTITUTIONS

Conclusion. Continued human population and economic growth threatens the existence of salmon in the Pacific Northwest. In the absence of explicit choices to do otherwise, salmon will continue to decline.

Conclusion. The current set of institutional arrangements contributes to the decline of salmon and cannot halt that decline. Institutional arrangements have a long reach in time, space, and function and are formed and designed on political bases. For the most part, human institutions that affect salmon have taken only incidental account of salmon biology. Because of the character of the social processes by which institutional arrangements emerge and change, rational analysis is necessary but not sufficient for constructive change.

Conclusion. The current set of institutional arrangements is not appropriate to the bioregional requirements of salmon and their ecosystems. A critical institutional need is to link a bioregional (ecosystem) perspective to cooperative management (i.e., joint management by a government agency and a community of stakeholders) as a governing concept. Meeting this need is primarily a political task, not a scientific one.

Conclusion. Political turbulence has thwarted attempts to take a long-term perspective, even though salmon management requires time scales of decades to determine whether a given approach is successful.

Conclusion. Attempts to halt the decline of salmon over the last 30 years have led to institutional reforms in fishing management, funding, habitat conservation, dam operations, and protection of endangered populations. They have not halted the decline but have raised expectations that the decline would be ameliorated.

Recommendation. Because the problems facing salmon are multidimensional, an interdisciplinary approach to solving them is essential. Market mechanisms should be used to help people know the costs of choices, and subsidies that prevent markets from operating effectively should be removed. In addition to the biological and environmental benefits of a sustainable salmon population, effective solutions should improve the operation of market forces and thus reduce some economic factors that now lead people to undervalue salmon and their ecosystems.

Recommendation. Bioregional cooperative management that incorporates stakeholders in governance should make use of local knowledge, provide incentives for long-term learning, and balance local interests against the problems that arise at the edges of bioregions and the requirements to deal with stem effects. Such an approach would reorient and diversify human management in ways that improve the possibilities of sustainability. Bioregional cooperative management is inherently diverse in goals and enables a region to respond to changing conditions with greater resilience. It also provides resilience against political and economic turbulence.

Recommendation. Our limited understanding of salmon and the ecosystems they inhabit requires adaptive management if rehabilitation is to have a chance. Systematic, experimental learning is faster and less expensive than trial-and-error learning, which has proved ineffective within the current institutional arrangements.

Recommendation. All institutional changes should take into account the long time scales of and likelihood of surprise in attempts to rehabilitate salmon and their ecosystems.

Recommendation. Hydropower prices, which internalize the full costs of growth, should be used to provide funding for rehabilitation of salmon and their ecosystems, especially in areas that are affected by hydropower projects.

Recommendation. The institutional framework for fishery management should be unified and streamlined. The committee is reluctant to recommend a detailed model of institutional structure that is most likely to be successful. One reason is that no institutional solution to a similar problem stands out above all others, most have advantages and disadvantages. Another reason is that any institutional structure's success will depend in large degree on having been created by all groups of stakeholders and not imposed from outside. However, three major principles must be adhered to.

1. The institutional structure must allow for a sharing of decision making among all legitimate interests.
2. It must consist of local units small enough to ensure local legitimacy and to respond to local variations in environmental and socioeconomic factors, and it must make use of local knowledge.
3. There must be a mechanism to ensure that the larger-scale environmental and anthropogenic forces behind and consequences of local actions are taken into account, i.e., the interests of the greater region should not be submerged by or sacrificed to local interests.

Our incomplete understanding of salmon, their ecosystems, and the social systems that affect them requires adaptive management, and adaptive management requires a long-term point of view. The following suggestions are made with the three principles in mind.

- Organize a commission for management of each river basin, combining smaller basins into single groups.
- Include American Indian tribes in the process of rehabilitation. Strong populations and terminal fisheries—among the goals of rehabilitation—will benefit everyone in the long run.
- Organize cooperative-management groups to develop more selective fisheries and techniques, such as converting gill-net to live-catch systems and developing techniques appropriate to terminal fisheries.
- The activities of river-basin commissions and preemptive recovery plans must be coordinated with the Northwest Power Planning Council, the Pacific Salmon Commission, the Pacific Marine Fisheries Commission, the National Marine Fisheries Service, and other institutions that have a multibasin focus.

Recommendation. The committee proposes that the relevant agencies in the Pacific Northwest, including the National Marine Fisheries Service, agree on a process to permit the formulation of salmon recovery plans in advance of listings under the Endangered Species Act and that the Pacific Northwest states, acting individually and through the Northwest Power Planning Council, provide technical and financial assistance to watershed-level organizations to prepare and implement these preemptive recovery plans. The U.S. Department of Agriculture, Department of the Interior, Department of Commerce, and Department of State, and Bonneville Power Administration should also provide technical and financial assistance to those efforts. In describing the biological, social, and practical reasons for a constructive course of action, the committee has emphasized three components.

First, because of the unusual biology of salmon populations, action to protect a given population must be organized within the biologically relevant drainage and along its migration route in freshwater and at sea.

Second, although much of the habitat modification carried out by humans over the past 150 years has been detrimental to salmon, research and practice over the past several decades have taught much about habitat needs and improvement. However, science cannot ensure the recovery of any specific salmon population. A long-term adaptive, experimental approach is both logically compelling and pragmatically indispensable for the large collection of demes that constitute Northwest salmon.

Third, acting at the bioregional scale requires cooperation among diverse landowners and water-rights holders. No body of law or practical way to consolidate governing powers is sufficient to put each bioregion under the supervision of a single managing entity today. From that practical reality grows the need to rely on cooperative management as a way to act in the face of fragmented control.

Because of the groundwork carried out by generations of fisheries biologists, and most recently summarized in the Federal Ecosystem Management Assessment Team's (FEMAT) report and the subbasin planning process of the Northwest Power Planning Council, it is possible to identify some of the geographic outlines of salmon bioregions. One way to harness the Endangered Species Act's potential for disrupting human activities in a biologically constructive fashion is to foster the development of *preemptive recovery plans*—which incorporate binding contractual commitments from funding sources—for adaptive management to rehabilitate specific salmon populations within their bioregions and migration routes. The plans would include a set of experimental actions and monitoring methods that ensure that lessons important for salmon management would be learned within several salmon lifetimes (up to 20 years, perhaps even longer) and that lead to a scientifically grounded expectation that the salmon population would increase during that time while preserving its genetic integrity. The recovery plans would need to reflect the commitments undertaken by all the parties that control the various elements of their execution; therefore, cooperative management would probably characterize the plans' development and implementation. The plans would be preemptive in the sense that the National Marine Fisheries Service would agree that, while an adopted recovery plan is in operation, the salmon population it covers is protected as much as is possible under the Endangered Species Act. In sum, the formulation and adoption of a plan would forestall a petition under the Endangered Species Act to protect the population covered by the plan. No filing on a population would be acted upon by the National Marine Fisheries Service for two years after a state certifies that a recovery plan is being developed unless the petitioner demonstrates that a decline warranting emergency protection preceded the state's certification; this policy would allow time for a plan to be developed but set a time limit within which the National Marine Fisheries Service would have to adopt or reject a proposed plan.

The Northwest Power Planning Council, the agency with the most highly developed base of knowledge relevant to protection and enhancement of salmon on the bioregional scale, must be centrally involved in recovery plans for the Columbia River basin. Under current law, the council can also direct funding to support recovery-plan implementation within the Columbia River basin. Outside the Columbia basin, the states should facilitate cooperative management and planning on the bioregional scale. Funds to support those activities can learn from the model of the Timber, Fish, and Wildlife Agreement in Washington state, which combined private, tribal, and state resources. The activities of the states should be augmented by the Bonneville Power Administration, the Forest Service, and the Bureau of Land Management—these agencies are developing adaptive management areas and watershed-level plans under FEMAT. Assistance will also be needed from the Pacific Fisheries Management Council and the International Pacific Salmon Commission to target fishing restrictions to protect specific populations, as well as other relevant agencies that operate beyond state boundaries.

The prospect of constructive action to conserve and rehabilitate at least some salmon populations without the conflict and delays encountered under the present Endangered Species Act regime leads the committee to advance the idea of preemptive plans. This approach requires no new legislation (although a way would need to be found to allow the government to delay action on a petition to list a species), and it calls into play resources that are already available in the region. Although the proposed shifts in incentives are incremental, the bioregion-scale learning that the adaptive plans would produce is likely to lead to more substantial change over the next several decades. The strategy that the committee proposes becomes more compelling as salmon abundances decline, in that bioregions that fail to act are more likely to face petitions under circumstances not of their own choosing. By increasing the incentive to act on behalf of salmon populations, institutional mechanisms can play a role that is more constructive than the protective but conflicted stance of those institutions today.

AN APPROACH TO SOLVING THE SALMON PROBLEM

As described in previous chapters, the salmon problem took many years to develop, and its solution will require the commitment of time, money, and effort. The committee's analyses of the problems and potential solutions lead to the conclusion that there is no "magic bullet." Therefore, like the problem itself, solutions will be complex and often hard to agree on; to be successful, they will need to be based on scientific information, including information provided by social and economic sciences. In addition, to be successful, consensus will be needed about the size of the investments to be made in solving the problem and how the costs should be allocated. This means that solutions will have to be regionally based, just as the salmon problem has regional variations (see Chapter 13).

The committee recommends the following general approach. For each major watershed or river basin, the following should be assessed.

- All the causes of salmon mortality, including their estimated magnitude and the uncertainties associated with the estimates. Factors known to decrease natural production should also be listed.
- Ways of reducing those sources of mortality or compensating for them, their probable effectiveness, and their drawbacks.
- The probable costs of each method of reducing mortality; to be most useful, the estimates should include both market and nonmarket costs. To the degree possible, it is important to identify what groups would bear the major portion of the costs of each method and significant uncertainties in the estimates. (For example, reductions in catch rates would primarily affect fishers and tourists; changes in water use could affect agricultural interests or ratepayers; changes in riparian management could affect forest-products industries or private landowners.)

All the estimates would include substantial uncertainties, due both to lack of knowledge and to fundamental environmental, socioeconomic, and biological uncertainties. Nonetheless, such a process of assessment and evaluation is essential for rational decision making. They will provide a basis for evaluating options—for weighing benefits and costs—and for identifying areas where research is critical. *All the committee's recommendations should be viewed in this context: they need to be considered on a regional basis (i.e., major watersheds) and in a comprehensive framework that includes an analysis of their costs, probable effectiveness, and the ability and willingness of various sectors to bear the costs.*

This will be challenging for several reasons. First, in many cases, the desired information has not been collated or does not exist. Second, considerable time and resources will be needed to perform such analyses even for one watershed. But the most important reason is that estimates of costs and how they might be distributed will require intimate knowledge of each watershed and of people's preferences and habits. These essential estimates should be made with input from the people involved. Nonetheless, the committee believes this approach will lead to improved effectiveness and—if not reduced costs—at least increased cost-effectiveness and reduced controversy.

THE FUTURE

The best approach to establishing a sustainable future for salmon in the Pacific Northwest is to use currently available information to develop workable, comprehensive programs rather than reacting to crises. This report has analyzed many parts of the salmon problem and assessed many options for intervention. However, the effects of more people, more resource consumption, changing economic demands and technologies, and changing societal values are hard to predict. Because the success of programs to improve the long-term prospects for salmon in the Pacific Northwest will depend on the societal and environmental contexts, it is important to develop ways for improving our ability to identify changing contexts and to respond to them. As long as human populations and economic activities continue to increase, so will the challenge of successfully solving the salmon problem.

References

Abell, D.L. (Technical coordinator). 1989. Proceedings of the California Riparian Systems Conference: Protection, Management, and Restoration for the 1990's. U.S. Department of Agriculture, Forest Service, General Technical Report PSW-110. 544 pp.

Agee, J.K. 1993. Fire Ecology of Pacific Northwest Forests. Island Press, Washington, DC. 493 pp.

Alkire, C. 1993. Wild Salmon as Natural Capital. Volume 1, The Living Landscape. The Wilderness Society, Washington, DC.

Allendorf, F.W. 1983. Isolation, gene flow, and genetic differentiation among populations. Pp. 51-65 in C. Schonewald-Cox, S. Chambers, B. MacBryde, and L. Thomas, eds. Genetics and Conservation. Benjamin/Cummings, Menlo Park, CA.

Allendorf, F.W. and N. Ryman. 1987. Genetic management of hatchery stocks. Pp. 141-159 in N. Ryman and F. Utter, eds. Population Genetics and Fishery Management. University of Washington Press, Seattle, WA.

Allendorf, F.W. and R.F. Leary. 1988. Conservation and distribution of genetic variation in a polytypic species, the cutthroat trout. Conserv. Biol. 2:170-184.

Allendorf, F.W. and R.S. Waples. *In press*. Conservation genetics of salmonid fishes. In Conservation Genetics: Case Histories from Nature, edited by J.C. Avise and J.L. Hamrick. Chapman & Hall, New York.

Alm, G. 1959. Connection between maturity, size and age in fishes. Institute of Freshwater Research Drottningholm Reports 40:5-145.

Anderson, J., D. Askren, T. Frever, J. Hayes, A. Lockhart, M. McCann, P. Pulliam, and R. Zabel. 1993. Columbia River Salmon Passage Model, CRiSP.1. Documentation for Version 4. December 1993. Center for Quantitative Science, University of Washington, Seattle, WA.

Andrus, C.W., B.A. Long, and H.A. Froelich. 1988. Woody debris and its contribution to pool formation in a coastal stream 50 years after logging. Can. J. Fish. Aquat. Sci. 45:2080-2086.

Anonymous. 1959. Senate Concurrent Resolution 35, 86th Congress, 1st Session, 105 Cong. Rec. 8026.

Anonymous. 1991. Scientific review of North Pacific high seas driftnet fisheries. Sidney, B.C., June 11-14, 1991. Report for Presentation to the United Nations Pursuant to Resolutions 44/225 and 45/197.

Argue, A.W., C.D. Shepard, M.P. Shepard, and J.S. Argue. 1986. A Compilation of Historical Catches by the British Columbia Commercial Salmon Fishery, 1876 to 1985. Contract report to C. C. Graham, Can. Dept. Fish. Oceans, Vancouver, BC.

Argue, A.W., M.P. Shepard, T.F. Shardlow, and A.D. Anderson. 1987. Review of Salmon Troll Fisheries in Southern British Columbia. Can. Tech. Rep. Fish. Aquat. Sci. 1502: 150 pp.

Argyris, C. and D.A. Schon. 1978. Organizational Learning: A Theory of Action Perspective. Addison-Wesley, Reading, MA.

Ashworth, W. 1995. The Economy of Nature: Rethinking the Connections Between Ecology and Economics. Houghton Mifflin Company, New York.

Babcock, J.P. 1902. Annual Report of the Commissioner of Fisheries (British Columbia) for the year 1901. Victoria, BC.

Backman, T.W.H. and L. Berg. 1992. Managing molecules or saving salmon? The evolutionary significant unit. Wana Chinook Tymoo (Columbia River Inter-Tribal Fish Commission) 2:8-14.

Barber, W.E. and J.N. Taylor. 1990. The importance of goals, objectives, and values in the fisheries management process and organization: A review. No. Am. J. Fish. Manag. 10(4):365-373.

Barnthouse, L.W., A. Anganuzzi, L. Botsford, J. Kitchell, and S. Saila. 1994. Columbia basin salmonid model review; review of Biological Requirements Work Group Report on Analytical methods for determining requirements of listed Snake River salmon relative to survival and recovery. Oak Ridge National Lab., Oak Ridge, TN.

Bartley et al. 1992a (Bartley, D.M., B. Bentley, J. Brodziak, R. Gomulkiewicz, M. Mangel, and G.A.E. Gall). 1992. Geographic variation in population genetic structure of chinook salmon from California and Oregon. Fish. Bull. 90:77-100 (authorship amended per erratum, Fish. Bull. 90(3):iii).

Bartley et al. 1992b (Bartley, D.M. Bagley, G. Gall, and B. Bentley). 1992. Usage of linkage disequilibrium data to estimate effective size of hatchery and natural fish populations. Conserv. Biol. 6:365-375.

Baumgartner, T.R., A. Soutar, and V. Ferreira-Bartrina. 1992. Reconstructions of the history of Pacific sardine and northern anchovy populations over the past two millennia from sediments of the Santa Barbara Basin, California. CalCOFI Rep. 33:24-40.

Beach, R.J., A.C. Gieger, S.J. Jeffries, S.D. Treacy, and B.L. Troutman. 1985. Marine Mammals and Their Interaction with Fisheries of the Columbia River and Adjacent Waters, 1980-1982. Third annual report. Washington Department of Wildlife, Wildlife Management Division, Olympia, WA.

Beacham, T.D. 1986. Type, quantity, and size of food of Pacific salmon (*Oncorhynchus*) in the Strait of Juan de Fuca, British Columbia. Fish. Bull. 84:77-89.

Beamish, R.J. and R. Bouillon. 1993. Pacific salmon production trends in relation to climate. Can. J. Fish. Aquat. Sci. 50:1002-1016.

Beamish, R.J., B.L. Thomson, and G.A McFarlane. 1992. Spiny dogfish predation on chinook and coho salmon and the potential effects on hatchery-produced salmon. Trans. Am. Fish. Soc. 121:444-455.

Beechie, T., E. Beamer, and L. Wasserman. 1994. Estimating coho salmon rearing habitat and smolt production losses in a large river basin, and implications for habitat restoration. No. Am. J. Fish. Manag. 14(4):797-811.

Behnke, R.J. 1992. Native Trout of Western North America. American Fisheries Society Monograph 6. American Fisheries Society, Bethesda, MD.

Bella, David A. 1991. Environmental Strategies and the Protection of Natural Ecosystems. Department of Civil Engineering Working Paper, Corvallis, OR.

Berg, L. 1993. The Imnaha spring chinook story. Oregon's wild fish policy hurts more than helps. Wana Chinook Tymoo (Columbia River Inter-Tribal Fish Commission) 3-4:8-13.

Berg, L. and T.G. Northcote. 1985. Changes in territorial, gill-flaring, and feeding behavior in juvenile coho salmon (*Oncorhynchus kisutch*) following short-term pulses of suspended sediment. Can. J. Fish. Aquat. Sci. 42(8):1410-1417.

Berman, C.H. and T.P. Quinn. 1991. Behavioural thermoregulation and homing by spring chinook salmon, *Oncorhynchus tshawytscha* (Walbaum), in the Yakima River. J. Fish Biol. 39 (3):301-312.

Berry, H. and R. B. Rettig. 1994. Who Should Pay for Salmon Recovery? Compensating for Economic Losses Caused by Recovery of Columbia River Salmon. A Pacific Northwest Extension Publication, PNW 470, April 1994. Oregon State University, Corvallis, OR.

Beschta, R.L. and R.L. Taylor. 1988. Stream temperature increases and land use in a forested Oregon watershed. Water Resour. Bull. 24(1):19-25.

Beschta, R.L. 1991. Stream habitat management for fish in the northwestern United States: the role of riparian vegetation. Am. Fish. Soc. Symp. 10:53-58.

Beschta, R.L., W.S. Platts, and B. Kauffman. 1991. Field Review of Fish Habitat Improvement Projects in the Grande Ronde and John Day River Basins of Eastern Oregon. U.S. Department of Energy, Bonneville Power Administration, Division of Fish and Wildlife, Portland, OR. DOE/BP-21493-1. 53 pp.

Beschta, R.L., J. Griffith, and T.A. Wesche. 1993. Field Review of Fish Habitat Improvement Projects in Central Idaho. U.S. Department of Energy, Bonneville Power Administration, Division of Fish and Wildlife, Portland, OR. DOE/BP-61032-1. 26 pp.

Beschta, R.L., J.R. Boyle, C.C. Chambers, W.P. Gibson, S.V. Gregory, J. Grizzel, J.C. Hagar, J.L. Li, W. C. McComb, T.W. Parzybok, M.L. Reiter, G.H. Taylor, and J.E. Warila. 1995. Cumulative Effects of Forest Practices in Oregon: Literature and Synthesis. Report prepared by Oregon State University, Corvallis, for the Oregon Department of Forestry, Salem, OR.

Beschta, R.L., R.E. Bilby, and G.W. Brown, L.B. Holtby and T.D. Hofstra. 1987. Stream temperature and aquatic habitat: fisheries and forestry implications. Pp. 191-232 in E.O. Salo and T.W. Cundy, eds. Streamside Management: Forestry and Fisheries Interactions. Contributions Number 57, Institute of Forest Resources, University of Washington, Seattle.

Bessey, R.A. 1963. Pacific Northwest Regional Planning—A Review. Bulletin 6 of Washington Department of Conservation, Division of Power Resources.

Beverton, R.J.H., J.G. Cooke, J.B. Csirke, R.W. Doyle, G. Hempel, S.J. Holt, A.D. MacCall, D. Policansky, J. Roughgarden, J.G. Shepherd, M.P. Sissenwine, and P.H. Wiebe. 1984. Dynamics of Single Species: Group Report. Pp. 13-58 in R.M. May, ed., Exploitation of Marine Communities. Dahlem Konferenzen. Springer-Verlag, Berlin.

Beverton, R.J.H. and S.J. Holt. 1957. On the Dynamics of Exploited Fish Populations. Fish Invest. Ser. II, XIX, Her Majesty's Stationery Office, London.

Bilby, R.E. 1988. Interactions between aquatic and terrestrial systems. Pp. 13-29 in K.J. Raedeke, ed. Streamside Management: Riparian Wildlife and Forestry Interactions. Contribution Number 59, Institute of Forest Resources, University of Washington, Seattle, WA.

Bilby, R.E. and P.A. Bisson. 1992. Allochthonous versus autochthonous organic matter contributions to the trophic support of fish populations in clear-cut and old-growth forested streams. Can. J. Fish. Aquat. Sci. 49:540-551.

Bisson, P.A. and G.E. Davis. 1976. Production of juvenile chinook salmon, *Oncorhynchus tshawytscha*, in a heated model stream. USDOC National Oceanic Atmospheric Administration Fish. Bull. 74:763-774.

Bisson, P.A. and R.E. Bilby. 1982. Avoidance of suspended sediment by juvenile coho salmon. No. Am. J. Fish. Manag. 2:371-374.

Bisson, P.A. and J.R. Sedell. 1984. Salmonid populations in streams in clearcut vs. old-growth forests of western Washington. Pp. 121-129 in W.R. Meehan, T.R. Merrell, Jr., and T.A. Hanley, eds. Fish and Wildlife Relationships in Old-Growth Forests. American Institute of Fisheries Research Biologists, Juneau, AK.

Bisson, P.A., R.E. Bilby, M.D. Bryant, C.A. Dolloff, G.B. Grette, R.A. House, M.L. Murphy, K. V. Koski, and J. R. Sedell. 1987. Large woody debris in forested streams in the Pacific Northwest. Pp. 143-190 in E.O. Salo and T.W. Cundy, eds. Streamside Management: Forestry and Fisheries Interactions, Proceedings of a Symposium. University of Washington Institute of Forest Resources, Contribution 57.

Bisson, P.A., J.L. Nielsen, and J.W. Ward. 1988. Summer production of coho salmon stocked in Mount St. Helens streams 3-6 years after the 1980 eruption. Trans. Am. Fish. Soc. 117:322-335.

Bisson, P.A., T.P. Quinn, G.H. Reeves, and S.V. Gregory. 1992. Best management practices, cumulative effects, and long-term trends in fish abundance in Pacific Northwest river systems. Pp. 189-232 in R.B. Naiman, ed. Watershed Management: Balancing Sustainability and Environmental Change. Springer-Verlag, New York.

Bjornn, T.C., D.R. Craddock, and D.R. Corley. 1968. Migration and survival of Redfish Lake Idaho sockeye salmon, *Oncorhynchus nerka*. Trans. Am. Fish. Soc. 97:360-373.

Bjornn, T.C., M.A. Brusven, M.P. Molnau, J.H. Milligan, R.A. Klamt, E. Chacho, and C. Schaye. 1977. Transport of Granitic Sediment in Streams and its effects on Insects and Fish. University of Idaho, Forest, Wildlife, and Range Experiment Station Bulletin 17, Moscow, ID.

Bjornn, T.C. and D.W. Reiser. 1991. Habitat requirements of salmonids in streams. Am. Fish. Soc. Spec. Publ. 19:83-138.

Bjornn, T.C. and C.A. Peery. 1992. A Review of Literature Related to Movements of Adult Salmon and Steelhead Past Dams and Through Reservoirs in the Lower Snake River. Idaho Cooperative Fish and Wildlife Research Unit, U.S. Department of the Interior, Fish and Wildlife Service, Tech. Report 92-1 for U.S. Army Corps of Engineers.

Black, M. 1993. Tragic Remedies: Recounting a Century of Failed Fishery Policy toward California's Sacramento River Salmon and Steelhead. Presented to Colloquium IV, "Conservation Biology of Endangered Pacific Salmonids: Life History, Genetics and Demography," Sept. 8-12, 1993, Bodega Bay, CA. Part of forthcoming California's Last Salmon: The Unnatural Policies of Natural Resource Agencies. University of California Press, Berkeley, CA.

Bledsoe, L.J., D.A. Somerton, and C.M. Lynde. 1989. The Puget Sound runs of salmon: An examination of the changes in run size since 1896. Can. Sp. Publ. Fish. Aquat. Sci. 105:50-61.

BLM (USDI Bureau of Land Management). Undated. 50 Years of Public Land Management 1934-1984. U.S. Department of the Interior, Bureau of Land Management, Washington, DC. 27 pp.

Blumm, M.C. 1981. Hydropower vs. salmon: The struggle of the Pacific Northwest's anadromous fish resources for a peaceful coexistence with the federal Columbia River power system. Environ. Law 11(211):212-300.

Blumm, M. 1990. Anadromous fish law—1979-90. Anadromous Fish Law Memo 50. Extension/ Sea Grant Program. Oregon State University, Corvallis, OR.

Blumm, M.C. and A. Simrin. 1991. The unraveling of the parity promise: Hydropower, salmon, and endangered species in the Columbia Basin. Environ. Law 21: 657-744.

BPA et al. (Bonneville Power Administration, U.S. Army Corps of Engineers, and Bureau of Reclamation). 1992. Screening Analysis, Columbia River System Operation Review. Volume 1, Description and Conclusions. Volume 2, Impact Results. U.S. Department of Energy, Bonneville Power Administration, Portland, OR.

BPA (Bonneville Power Administration). 1994. Bonneville Power Administration Fish and Wildlife Investments. U.S. Department of Energy, Bonneville Power Administration, Portland, OR.

Bonnot, P. 1928. Report on the Seals and Sea Lions of California. Fish Bulletin 14. California Department of Fish and Game, Sacramento.

Booth, D.B. 1991. Urbanization and the natural drainage system—impacts, solutions and prognoses. Northwest Environ. J. 7:93-118.

Bormann, B.T., M.H. Brookes, E.D. Ford, A.R. Kiester, C.D. Oliver, and J.F. Weigand. 1993. Eastside Forest Ecosystem Health Assessment. Volume 5. A Broad, Strategic Framework for Sustainable-Ecosystem Management. U.S. Department of Agriculture, National Forest System, Forest Service Research.

Bormann, B.T., P.G. Cunningham, M.H. Brookes, V.W. Manning and M.W. Collopy. 1994. Adaptive Ecosystem Management in the Pacific Northwest. U.S. Department of Agriculture, Forest Service, Pacific Northwest Research Station, General Technical Report PNW-GTR0341. Portland, OR.

Botkin, D.B. 1990. Discordant Harmonies: A New Ecology for the Twenty-First Century. Oxford University Press, New York.

Bottom, D.L. 1994. To till the water; a history of ideas in fisheries conservation. Draft paper presented at the Symposium on Pacific Salmon and their Ecosystems: Status and Future Options. University of Washington, Seattle.

Boule, M.E. and K.F. Bierly. 1987. History of estuarine wetland development and alteration: what have we wrought? Northwest Environ. J. 3(1):43-61.

Bower, S.M. and L. Margolis. 1984. Distribution of *Cryptobia salmositica*, a hemoflagellate of fishes, in British Columbia and the seasonal pattern of infection in a coastal river. Can. J. Zool. 62(12):2512-2518.

Bowles, E.C. 1995. Supplementation: Panacea or curse for the recovery of declining fish stocks? Pp. 277-283 in H.L. Schramm, Jr. and R.G. Piper, eds. Uses and Effects of Cultured Fishes in Aquatic Ecosystems. Am. Fish. Soc. Symp. 15. American Fisheries Society Bethesda, MD.

Boyd, R. 1990. Demographic history. Pp. 1774-1874 in W. Suttles, ed. Northwest Coast. Smithsonian Institution, Washington, DC.

Bradford, C.S. and C.B. Schreck. 1989. Migratory Characteristics of Spring Chinook Salmon in the Willamette River: 1989 Annual Report. Project DE-A179-88BP92818. U.S. Department of Energy, Bonneville Power Administration. Portland, OR.

Bradford, C.S. and C.B. Schreck. 1990. Migratory Characteristics of Spring Chinook Salmon in the Willamette River: 1990 Annual Report. Project DE-A179-88BP92818. U.S. Department of Energy, Bonneville Power Administration, Portland, OR.

Brannon, E.L. 1972. Mechanisms Controlling Migration of Sockeye Salmon fry. Int. Pac. Salmon Fish. Comm. Bull. 21.

Brannon, E.L. 1987. Mechanisms stabilizing salmonid fry emergence timing. Pp. 120-124 in H.D. Smith, L. Margolis, and C.C. Wood, eds. Sockeye Salmon (*Onchorhynchus nerka*) Population Biology and Future Management. Can. Spec. Publ. Fish Aquat. Sci. 96.

Brett, J.R., J.E. Shelbourn, and C.T. Shoop. 1969. Growth rate and body composition of fingerling sockeye salmon, *Oncorhynchus nerka*, in relation to temperature and ration size. J. Fish. Res. Bd. Can. 26:2362-2394.

Bromley, D.W. 1989. Economic Interests and Institutions: The Conceptual Foundations of Public Policy. Basil Blackwell, New York.

Brodeur, R.D. 1992. Factors relating to variability in feeding intensity of juvenile coho salmon and chinook salmon. Trans. Am. Fish. Soc. 121:104-114.

Brodeur, R.D. and D.M. Ware. 1992. Long-term variability in zooplankton biomass in the subarctic Pacific Ocean. Fish. Oceanogr. 1:32-38.

Brown, B. 1982. Mountain in the Clouds: A Search for Wild Salmon. Simon and Schuster, New York.

Brown, G.W. and J.T. Krygier. 1970. Effects of clear-cutting on stream temperature. Water Resour. Res. 6:1133-1139.

Brown, L.R. P.B. Moyle, and R.M. Yoshiyama. 1994. Historical decline and current status of coho salmon in California. No. Am. J. Fish. Manag. 14:237-261.

Brown, T.G. and G.F. Hartman. 1988. Contribution of seasonally flooded lands and minor tributaries to the production of coho salmon in Carnation Creek, British Columbia. Trans. Am. Fish. Soc. 117:546-551.

Bruun, R. 1982. The Boldt Decision. Legal victory, political defeat. Law & Policy Quart. 4(3):271-298.

Buchanan, D.V., J.E. Sanders, J.L. Zinn, and J.L. Fryer. 1983. Relative susceptibility of four strains of summer steelhead to infection by *Ceratomyxa shasta*. Trans. Am. Fish. Soc. 112:541-543.

Burgner, R.L. 1991. Life history of sockeye salmon (*Onchorhynchus nerka*). Pp. 1-117 in C. Groot and L. Margolis, eds. Pacific Salmon Life Histories. University of British Columbia Press, Vancouver, BC.

Burgner, R.L., J.T. Light, L. Margolis, T. Okazaki, A. Tautz, and S. Ito. 1992. Distribution and origins of steelhead trout (*Oncorhynchus mykiss*) in offshore waters of the North Pacific Ocean. Int. North Pacific Fish. Comm. Bull. 51:92.

Burke, W.T. 1991. Anadromous species and the new International Law of the Sea. Ocean Develop. Intern. Law 22:95-131.

Burns, J.W. 1971. The carrying capacity for juvenile salmonids in some northern California streams. California Fish and Game 57:44-57.

Busack, C. 1990. Yakima/Klickitat Production Project Genetic Risk Assessment. Yakima/Klickitat Production Project Preliminary Design Report, Appendix A. Washington Dept. of Fisheries and Bonneville Power Administration. Division of Fish and Wildlife, Portland, OR. 20 pp.

Busack, C.A. and K.P. Currens. 1995. Genetic risks and hazards in hatchery operations: Fundamental concepts and issues. Pp. 71-80 in H.L. Schramm, Jr. and R.G. Piper, eds. Uses and Effects of Cultured Fishes in Aquatic Ecosystems. Am. Fish. Soc. Symp. 15. American Fisheries Society, Bethesda, MD.

Bustard, D.R. and D.W. Narver. 1975. Aspects of the winter ecology of juvenile coho salmon (*Oncorhynchus kisutch*) and steelhead trout (*Salmo gairdneri*). J. Fish. Res. Bd. Can. 32:667-681.

Callicott, J.B. 1991. Conservation ethics and fisheries management. Fisheries 16(2):22-28.

Campton, D.E. 1995. Genetic effects of hatchery fish on wild populations of Pacific salmon and steelhead: What do we really know? Pp. 337-353 in H.L. Schramm, Jr. and R.G. Piper, eds. Uses and Effects of Cultured Fishes in Aquatic Ecosystems. Am. Fish. Soc. Symp. 15. American Fisheries Society, Bethesda, MD.

Carpenter, S.R., ed. 1988. Complex Interactions in Lake Communities. Springer-Verlag, New York.

Carpenter, S.R. and J.F. Kitchell [eds.] 1993. The Trophic Cascade in Lakes. Cambridge University Press.

Carr, J. 1994. [opinion editorial], *Portland Oregonian*, March 18 1994.

Carson, H.L. 1983. The genetics of the founder effect. Pp. 189-200 in Schonewald-Cox, C.M., S.M. Chambers, B. MacBryde, and L. Thomas, eds. Genetics and Conservation. The Benjamin Cummins Publishing Co., Inc., Menlo Park, CA.

Castle, E.N. 1993. A pluralistic, pragmatic and evolutionary approach to natural resource management. For. Ecol. Manag. 56:279-295.

Cayan, D.R., D.R. McLain, and W.D. Nichols. 1991. Monthly Climatic Time Series Data for the Pacific Ocean and Western Americas. U.S. Geological Survey. Open File Report 91-92. 380 pp.

Cederholm, C.J., L.M. Reid, and E.O. Salo. 1981. Cumulative effects of logging road sediment on salmonid populations in the Clearwater River, Jefferson County, Washington. Pp. 38-74 in Proceedings of a Conference: Salmon Spawning Gravel: A Renewable Resource in the Pacific Northwest? Report 39, State of Washington Water Resource Center, Washington State University, Pullman.

Cederholm, C.J., W.J. Scarlett, and N.P. Peterson. 1988. Low-cost enhancement technique for winter habitat of juvenile coho salmon. No. Am. J. Fish. Manag. 8:438-441.

Cederholm, C.J., D.B. Houston, D.L. Cole, and W.J. Scarlett. 1989. Fate of coho salmon (*Oncorhynchus kisutch*) carcasses in spawning streams. Can. J. Fish. Aquat. Sci., 426:1347-1355.

Center for Quantitative Science, University of Washington. Undated. CRiSP.2: Chinook Salmon Harvesting Model. 4 pp.

Chamberlin, T.W., R.D. Harr, and F.H. Everest. 1991. Timber harvesting, silviculture, and watershed processes. Am. Fish. Soc. Spec. Publ. 19:181-205.

Chandler, G.L. and T.C. Bjornn. 1988. Abundance, growth, and interactions of juvenile steelhead relative to time of emergence. Trans. Am. Fish. Soc. 117:432-443.

Chaney, E., W. Elmore, and W.S. Platts. 1990. Livestock Grazing on Western Riparian Areas. Northwest Resource Information Center, Inc., Eagle, ID. 45 pp.

Chaney, E., W. Elmore, and W.S. Platts. 1993. Managing Change: Livestock Grazing on Western Riparian Areas. Northwest Resource Information Center, Inc., Eagle, ID, and U.S. Environmental Protection Agency, Region 8, Denver, CO. 31 pp.

Chapman, D.W. 1962. Aggressive behavior in juvenile coho salmon as a cause of emigration. J. Fish. Res. Board Can. 19:1047-1080.

Chapman, D.W. 1966. Food and space as regulators of salmonid populations in streams. Am. Nat. 100:345-357.

Chapman, D.W. 1986. Salmon and steelhead abundance in the Columbia River in the nineteenth century. Trans. Am. Fish Soc. 115:662-670.

Chapman, D.W. 1988. Critical review of variables used to define effects of fines in redds of large salmonids. Trans. Am. Fish. Soc. 117:1-21.

Chapman, D.W. and T.C. Bjornn. 1969. Distribution of salmonids in streams, with special reference to food and feeding. Pp. 153-176 in T.G. Northcote, ed. Symposium on Salmon and Trout in Streams. H.R. MacMillan Lectures in Fisheries, Institute of Fisheries, University of British Columbia, Vancouver, BC.

Chapman, D., A. Giorgi, M. Hill, A. Maule, S. McCutcheon, D. Park, W. Platts, K. Pratt, J. Seeb, and F. Utter. 1991. Status of Snake River Chinook Salmon. Report by Don Chapman Consultants, Inc., Boise, ID, to Pacific Northwest Utilities Conference Committee, Portland, OR.

Chapman, D., A. Giorgi, T. Hillman, D. Deppert, M. Erho, S. Hays, C. Peven, B. Suzumoto, and R. Klinge. 1994. Status of Summer/Fall Chinook Salmon in the Mid-Columbia Region. Don Chapman Consultants, Inc., Boise, ID.

Chapman, D.W. and K.L. Witty. 1993. Habitats of Weak Salmon Stocks of the Snake River Basin and Feasible Recovery Measures. Recovery Issues for Threatened and Endangered Snake River Salmon. Technical Report 1 of 11, report of subcontractor to S.P. Cramer and Associates, Inc., Gresham, OR, Bonneville Power Administration Proj. No. 93-013, Contract No. DE-AM79-93BP99654.

Christie, W.J., G.R. Spangler, K.H. Loftus, W.L. Hartmann, P.J. Colby, M.A. Ross, and D.R. Talhelm. 1987. A perspective on Great Lakes fish community rehabilitation. Can. J. Fish. Aquat. Sci. 44 (Suppl. 2):486-499.

Clark, C.W. 1984. Strategies for multispecies management: Objectives and constraints. Pp. 303-312 in R.M. May, ed. Exploitation of Marine Communities. Dahlem Konferenzen. Springer-Verlag, Berlin.

Clark, R.N. and D.R. Gibbons. 1991. Recreation. Am. Fish. Soc. Spec. Publ. 19:459-481.

Clark, W.G. 1985. Fishing a sea of court orders: Puget Sound salmon management 10 years after the Boldt decision. No. Am. J. Fish. Manag. 5:417-434.

Cobb, J.N. 1911. The Salmon Fisheries of the Pacific Coast. Bureau of Fisheries. Document No. 751. Washington, DC.

Cobb, J.N. 1930. The Salmon Fisheries of the Pacific Coast. Bureau of Fisheries Document 1030. Washington, DC.

Cohen, F.G. 1986. Treaties on Trial: The Continuing Controversy Over Northwest Indian Fishing Rights. University of Washington Press, Seattle, WA.

Collins, B.D. and T. Dunne. 1987. Assessing the Effects of Gravel Harvesting on River Morphology and Sediment Transport: A guide for Planners. Report to State of Washington Department of Ecology, Olympia, WA. 45 pp.

Columbia Basin Fish and Wildlife Authority. 1990. Review of the History, Development, and Management of Anadromous Fish Production Facilities in the Columbia River Basin. Compiled by the U.S. Fish and Wildlife Service Office of the Columbia River Coordinator. 52 pp. (Unpublished).

Columbia River System Operation Review. 1993. Power System Coordination: A Guide to the Pacific Northwest Coordination Agreement. U.S. Department of the Army Corps of Engineers, U.S. Department of Energy, Bonneville Power Administration, and U.S. Department of the Interior, Bureau of Reclamation. 46 pp.

Columbia River Water Management Group. 1993. Columbia River Water Management Report, Water Year 1992. U.S. Army Corps of Engineers, North Pacific Division, Portland, OR. 169 pp. + Figures.

Compton, D.E. and F.M. Utter. 1987. Genetic structure of anadromous cutthroat trout (*Salmo clarki clarki*) populations in the Puget Sound area: Evidence for restricted gene flow. Can. J. Fish. Aquat. Sci. 44:573-582.

Cooley, F.E. and P.A Heikkila. 1994. Coquille Watershed Community Action: A Case Study of Salmon Habitat Restoration Efforts. Coquille: Coastal Watersheds Citizen Involvement Project.

Cooper, R. and T.H. Johnson. 1992. Trends in Steelhead (*Oncorhynchus mykiss*) Abundance in Washington and Along the Coast of North America. Washington Department of Wildlife Report 92-20.

Craig, J.A. and R.L. Hacker. 1940. The History and Development of the Fisheries of the Columbia River. Bull. Bur. Fish. 49.

Cramer, S., A. Maule, and D. Chapman. 1991. The Status of Coho Salmon in the Lower Columbia River. Report by Don Chapman Consultants, Inc., to Pacific Northwest Utilities Conference Committee, Portland, OR.

Crenson, M.A. 1971. The Un-Politics of Air Pollution: A Study of Non-Decisionmaking in the Cities. Johns Hopkins Press, Baltimore, MD.

Crutchfield, J.A. and G. Pontecorvo. 1969. The Pacific Salmon Fisheries: A Study of Irrational Conservation. The Johns Hopkins Press, Baltimore, MD.

Currens, K. 1993. Genetic vulnerability of the Yakima Fishery Project. A risk assessment. Contract report for Washington Department of Fisheries. 83 pp.

Cushing, D.H. 1988. The Provident Sea. Cambridge University Press, Cambridge, NY 328 pp.

Dahl, T.E. 1990. Wetland Losses in the United States: 1780s to 1980s. U.S. Department of the Interior, Fish and Wildlife Service, Washington, DC. 13 pp.

Dauble, D.D. and R.P. Mueller. 1993. Factors Affecting the Survival of Upstream Migrant Adult Salmonids in the Columbia River Basin. Recovery Issues for Threatened and Endangered Snake River Salmon Technical Report 9 of 11. Battelle PNW Lab. Report to BPA, Proj. No. 93-026. Task Order No. DE-AT79-93BP00085 to Master Agreement DE-A179-BP6211.

Department of Water Resources. 1988. Dams within the jurisdiction of the state of California. Pp. 17-88 in Depart. Water Resour. Bull., Sacramento, CA.

Donaldson, J.R. 1967. The phosphorus budget of Iliamna Lake, Alaska as related to the cyclic abundance of sockeye salmon. Ph.D. thesis, University of Washington, Seattle, WA.

Doyle, R.W., C. Herbinger, C.T. Taggart, and S. Lochmann. 1995. Use of DNA microsatellite polymorphism to analyze genetic correlations between hatchery and natural fitness. Am. Fish. Soc. Symp. 15:205-211.

Dunne, T. and L.B. Leopold. 1978. Water in Environmental Planning. W.H. Freeman, San Francisco, CA.

Ebbesmeyer, C.C. and C.A. Coomes. 1989. Strong, low frequency (decadal) environmental fluctuations during the 20th century in the North Pacific Ocean, on the Washington Coast, and in Puget Sound. Oceans '89: The Global Ocean. Vol. 1: Fisheries, Global Ocean Studies, Marine Policy and Education, Oceanographic Studies. Pp. 242-246. IEEE Publication Number 89CH2780-5, MTS/IEEE, New York, NY.

Ebbesmeyer, C. and W. Tangborn. 1993. Great Pacific surface salinity trends caused by diverting the Columbia River between seasons. Evans-Hamilton, Inc., Seattle, WA. (Unpublished).

Ebel, W.J. 1969. Supersaturation of nitrogen in the Columbia River and its effect on salmon and steelhead trout. Fish. Bull. 68:1-11.

Ebel, W.J., H.L. Raymond, G.E. Monon, W.E. Farr, and G.K. Tanonaka. 1975. Effect of Atmospheric Gas Supersaturation Caused by Dams on Salmon and Steelhead Trout of the Snake and Columbia Rivers. U.S. Department of Commerce, National Oceanic and Atmospheric Administration, National Marine Fisheries Service, Seattle, WA.

Ebel, W.J. and H.L. Raymond. 1976. Effect of Atmospheric Gas Supersaturation on Salmon and Steelhead Trout of the Snake and Columbia Rivers. U.S. Department of Commerce, National Oceanic and Atmospheric Administration, National Marine Fisheries Service, Marine Fisheries Review 7:1-14.

Eberhardt, L.L. and J.M. Thomas. 1991. Designing environmental field studies. Ecol. Monogr. 61(1):53-73.

Edwards, C.J., R.A. Ryder, and T.R. Marshall. 1990. Using lake trout as a surrogate of ecosystem health for oligotrophic waters of the Great Lakes. J. Great Lakes Res. 16(4):591-608.

Elliott, J.M. 1985. Population dynamics of migratory trout, *Salmo trutta*, in a Lake District stream, 1966-83, and their implications for fisheries management. J. Fish Biol. 27 (Supplement A):35-43.

Elliott, J.M. 1994. Quantitative Ecology and the Brown Trout. Oxford University Press, New York.

Elmore, W. and R.L. Beschta. 1987. Riparian areas: perceptions in management. Rangelands 9(6):260-265.

Elmore, W. 1992. Riparian responses to grazing practices. Pp. 442-457 in R.B. Naiman, ed. Watershed Management: Balancing Sustainability and Environmental Change. Springer-Verlag, New York.

Environment Canada. 1986. Fact Sheet on Wetlands in Canada: A Valuable Resource. Lands Directorate, Environment Canada, Ottawa (Cat. No. En 73-6/86-4E) 8 pp.

Eriksen, A. and L. Townsend. 1940. The Occurrence and Cause of Pollution in Grays Harbor. Washington Pollution Control Commission, Bulletin Number 2, Olympia, WA. 100 pp.

Eugene Water & Electric Board and Lane County. 1992. Scoping Report for an Integrated McKenzie Watershed Program. Lane County Council of Governments, Eugene, OR.

Everest, F.H., R.L. Beschta, J.C. Scrivener, K.V. Koski, J.R. Sedell, and C.J. Cederholm. 1987. Fine sediment and salmonid production: A paradox. Pp. 98-142 in E.O. Salo and T.W. Cundy, eds. Streamside Management: Forestry and Fishery Interactions. Contribution Number 57, Institute of Forest Resources, University of Washington, Seattle.

Executive Office of the President. 1981. Executive Order 12319, September 30 (disbanding regional river basins commissions, including the Pacific Northwest Commission).

Fagen, R. and W.W. Smoker. 1989. How large-capacity hatcheries can alter interannual variability of salmon production. Fish. Res. 8(1):1-12.

Falconer, D.S. 1989. Introduction to Quantitative Genetics. Third Edition. Longman, Harlow, U.K.

Fast, D., J. Hubble, M. Kohn, and B. Watson. 1991. Yakima River Spring Chinook Enhancement Study. Final report to Bonneville Power Administration, DOE/BP-39461-9.

Fausch, K.D., C.L. Hawkes, and M.G. Parsons. 1988. Models that Predict Standing Crop of Stream Fish from Habitat Variables: 1950-85. U.S. Department of Agriculture, Forest Service Gen. Tech. Rep., PNW-213, Pacific Northwest Experimental Station, Portland, OR. 52 pp.

Fazio, J. and J. Ruff. 1994. Power System Impacts of the National Marine Fisheries Service 1994 Biological Opinion. Draft Briefing Paper 94-22, Northwest Power Planning Council, Portland, OR.

Feist, B. 1994. Map of Appropriate Location of Hatcheries in the Columbia River Basin. University of Washington, Seattle.

FEMAT. 1993. Forest Ecosystem Management Assessment Team. A Federal Agency Guide for Pilot Watershed Analysis. Version 1.2. Interagency Working Group, U.S. Department of Agriculture, Forest Service, Portland, OR.

Fenderson, O.C., W.H. Everhart, and K.M. Muth. 1968. Comparative agonistic and feeding behavior of hatchery-reared and wild salmon in aquaria. J. Fish. Res. Board Can. 25:1-14.

Field-Dodgson, M.S. 1987. The effect of salmon redd excavation on stream substrate and benthic community of two salmon spawning streams in Canterbury, New Zealand. Hydrobiologia 154:3-11.

Fish and Game Protector. 1894. First and Second Annual Reports of the Fish and Game Protector to the Governor, 1893-94. Frank C. Baker, State Printer, Salem, OR.

Fish Passage Center. 1994. Annual Report 1993. Annual Report to Bonneville Power Administration, Proj. No. 93-033, Contr. No. DE-FC79-88BP38906. 123 pp. + Appendix.

Fish Passage Center weekly reports. 1990. Portland, OR.

Flagg, T.A., F.W. Waknitz, D.J. Maynard, G.B. Milner, and C.V.W. Mahnken. 1995. The effect of hatcheries on native coho salmon populations in the lower Columbia River. Pp. 366-375 in H.L. Schramm, Jr. and R.G. Piper, eds. Uses and Effects of Cultured Fishes in Aquatic Ecosystems. Am. Fish. Soc. Symp. 15. American Fisheries Society Bethesda, MD.

Fleming, I.A. and M.R. Gross. 1993. Breeding success of hatchery and wild coho salmon *(Oncorhynchus kisutsch)* in competition. Ecol. Appl. 3:230-245.

Fleming, I.A. and M.R. Gross. 1994. Breeding competition in a Pacific salmon (coho: *Oncorhynchus kisutch)*: measures of natural and sexual selection. Evolution 48:637-657.

Flick, W.A. and D.A. Webster. 1962. Problems in sampling wild and domestic stocks of brook trout. Trans. Am. Fish. Soc. 91:140-144.

Flick, W.A. and D.A. Webster. 1964. Comparative first year survival and production in wild and domestic strains of brook trout, *Salvelinus fontinalis*, after stocking. Trans. Am. Fish. Soc. 93:58-69.

Fraidenburg, M.E. and R.H. Lincoln. 1985. Wild chinook salmon management: An international conservation challenge. No. Am. J. Fish. Manag. 5:311-529.

Fraker, M. 1994. California Sea Lions and Steelhead Trout at the Chittenden Locks, Seattle, WA. Marine Mammal Commission, Washington, DC.

Francis, R.C. and S.R. Hare. 1994. Decadal-scale regime shifts in the large marine ecosystems of the Northeast Pacific: A case for historical science. Fish. Oceanogr. 3:279-291.

Francis, R.C. and T.H. Sibley. 1991. Climate change and fisheries: What are the real issues? Northwest Environ. J. 7:295-307.

Fredin, R.A., R.L. Major, R.G. Bakkala, and G.K. Tanonaka. 1977. Pacific Salmon and the High Seas Salmon Fisheries of Japan. Unpubl. processed report, U.S. Department of Commerce, National Oceanic and Atmospheric Administration, National Marine Fisheries Service, Northwest and Alaska Fish. Center, Seattle, WA. 324 pp.

Frenkel, R.E. and J.C. Morlan. 1991. Can we restore our salt marshes? Lessons from the Salmon River, Oregon. Northwest Environ. J. 7:119-135.

Freudenburg, W. 1990. Presentation to the Scientific Advisory Committee of the Department of the Interior's Minerals Management Service, Herndon, VA, an unpublished manuscript.

Frissell, C.A. 1993. Topology of extinction and endangerment of native fishes in the Pacific Northwest and California (U.S.A.). Conserv. Biol. 7:342-354.

Frost, B.W. 1983. Interannual variation of zooplankton standing stock in the open Gulf of Alaska. Pp. 146-157 in W.S. Wooster, ed., From Year to Year: Interannual Variability of the Environment and Fisheries of the Gulf of Alaska and the Eastern Bering Sea. Washington Sea Grant Program, Seattle, WA.

Furniss, M.J., T.D. Roelofs, and C.S. Yee. 1991. Road construction and maintenance. Am. Fish. Soc. Spec. Publ. 19:297-324.

Gale, R.P. 1987. Resource miracles and rising expectations: a challenge to fishery managers. Fisheries (Bethesda) 12(5):8-13.

Gearin, P., R. Pfeifer, S.J.J. Jeffries, R.L. DeLong, and M.A. Johnson. 1988. Results of the 1986-1987 California Sea Lion-Steelhead Trout Predation Control Program at the Hiram M. Chittenden Locks. Northwest and Alaska Fisheries Center Processed Report 88-30. U.S. Department of Commerce, National Oceanic and Atmospheric Administration, National Marine Fisheries Service, Alaska Fisheries Science Center, Seattle.

General Accounting Office. 1988. Public Rangelands: Some Riparian Areas Restored But Widespread Improvement Will Be Slow. U.S. General Accounting Office, Washington, DC. GAO/RCED-88-105. 85 pp.

Gharrett, A.J. and S.M. Shirley. 1985. A genetic examination of spawning methodology in a salmon hatchery. Aquacult. 47:245-256.

Gharrett, A.J. and W.W. Smoker. 1991. Two generations of hybrids between even-and-odd year pink salmon (*Oncorhynchus gorbuscha*): A test for outbreeding depression. Can. J. Fish. Aquat. Sci. 48:1744-1749.

Gharrett, A.J. and W.W. Smoker. 1993. A perspective on the adaptive importance of genetic infrastructure in salmon populations to ocean ranching in Alaska. Fish. Res. 18:45-58.

Gilbert, J., M.-J. Dole-Olivier, P. Marmonier, and P. Vervier. 1990. Surface water-groundwater ecotones. Pp. 199-226 in R.J. Naiman and H. Decamps, eds. The Ecology and Management of Aquatic-Terrestrial Ecotones. United Nations Educational Scientific and Cultural Organization, Paris and Parthenon Publishing Group, Carnforth, United Kingdom.

Gilhousen, B. 1980. Energy Sources and Expenditures in Fraser River Sockeye Salmon During Their Spawning Migration. Int. Pacific Salmon Fish. Comm. Bull. 20. 51 pp.

Gilpin, M.E. 1987. Spatial structure and population vulnerability. Pp. 126-139 in M.E. Soule ed., Viable Populations for Conservation. Cambridge University Press, Cambridge, Great Britain.

Gilpin, M.E. and M.E. Soulé. 1986. Minimum viable populations: Processes of species extinction. Pp. 19-34 in M.E. Soulé, ed. Conservation Biology: The Science of Scarcity and Diversity. Sinauer, Sunderland, MA.

Giorgi, A., D. Miller, and B. Sandford. 1990. Migratory Behavior and Adult Contribution of Summer Outmigrating Subyearling Chinook Salmon in John Day Reservoir. U.S. Department of Commerce, National Oceanic and Atmospheric Administration, National Marine Fisheries Service Final Research Report to BPA, Contr. No. DE-A179-83BP39645.

Giorgi, A. 1993. Flow Augmentation and Reservoir Drawdown: Strategies for the Recovery of Threatened and Endangered Stocks of Salmon in the Snake River Basin. Tech. Report 2 of 11 to BPA, Contract No. DE-AM79-93BP99654. 50 pp.

Gleeson, G.W. 1972. Return of a River: The Willamette River, Oregon. Oregon State University, Water Resources Research Institute WRRI-13.

Godin, J-G. J. 1982. Migrations of salmonid fishes during early life history phases: daily and annual timing. Pp. 22-50 in E.L. Brannon and E.O. Salo, eds. Proceedings of a Symposium on Salmon and Trout Migratory Behavior. University of Washington Press, Seattle.

Gonor, J.J., J.R. Sedell, and P.A. Benner. 1988. What we know about large trees in estuaries, in the sea, and on coastal beaches. Pp. 83-112 in C. Maser, R.F. Tarrant, J.M. Trappe, and J.F. Franklin, eds. From the Forest to the Sea: A Story of Fallen Trees. U.S. Department of Agriculture, Forest Service, Pacific Northwest Research Station, General Technical Report PNW-GTR-229, Portland, OR.

Gonsier, M.J. and R.B. Gardner. 1971. Investigation of Slope Failures in the Idaho Batholith. U.S. Department of Agriculture, Forest Service Research Paper INT-97.

Good, J.W. 1993. Ocean resources. Pp. 110-121 in P.L. Jackson and A.J. Kimerling, eds. Atlas of the Pacific Northwest. Oregon State University Press, Corvallis, OR.

Goode, G.B., and others. 1884-1887. Fisheries and Fishery Industries of the United States. U.S. Bureau of Fisheries, Washington, DC.

Goodman, D. 1987. The demography of chance extinction. Pp. 11-34 in M.E. Soulé, ed. Viable Populations for Conservation. Cambridge University Press, Cambridge, U.K.

Gordon, H.S. 1954. The economic theory of a common property resource: Fish. J. Polit. Econ. 62:124-142.

Gregory, S.V. 1983. Plant-herbivore interactions in stream ecosystems. Pp. 157-189 in G.W. Minshall and J.R. Barnes, eds. Stream Ecology: Application and Testing of General Ecological Theory. Plenum Press, New York.

Gregory, S.V., G.A. Lamberti, D.C. Erman, K.V. Koski, M.L. Murphy, and J.R. Sedell. 1987. Influence of forest practices on aquatic production. Pp. 233-255 in E.O. Salo and T.W. Cundy, eds. Streamside Management: Forestry and Fishery Interactions. Contribution Number 57, Institute of Forest Resources, University of Washington, Seattle, WA,

Gregory, S.V., F.J. Swanson, and W.A. McKee. 1991. An ecosystem perspective of riparian zones. BioScience 40:540-551.

Gresswell, R.E., B.A. Barton, and J.L. Kershner. 1989. Practical Approaches to Riparian Resource Management: An Educational Workshop. U.S. Department of the Interior, Bureau of Land Management, Billings, MT. BLM-MT-PT-89-001-4351. 193 pp.

Grette, G.B. 1985. The abundance and role of large organic debris in juvenile salmonid habitat in streams in second growth and unlogged forests. M.S. thesis, University of Washington, Seattle, WA.

Groot, C. and L. Margolis, eds. 1991. Pacific Salmon Life Histories. University of British Columbia Press, Vancouver, British Columbia. 564 pp.

Groot, C. and T. Quinn. 1987. Homing migration of sockeye salmon, *Oncorhynchus nerka*, to the Fraser River. Fish. Bull. 85:455-469.

Halbert, C.L. and K.M. Lee. 1990. The timber, fish, and wildlife agreement: implementing alternative dispute resolution in Washington State. Northwest Environ. J. 6:139-175.

Hall, J.D. and R.L. Lantz. 1969. Effects of logging on the habitat of coho salmon and cutthroat trout in coastal streams. Pp. 355-375 in T.G. Northcote, ed. Symposium on Salmon and Trout in Streams. H.R. MacMillan Lectures in Fisheries, University of British Columbia, Vancouver, BC.

Hall, J.D. and N.J. Knight. 1981. Natural Variation in Abundance of Salmonid Populations in Streams and its Implications for Designing of Impact Studies. Report EPA-600/S3-81-021, Environmental Protection Agency, Corvallis, OR.

Hanna, S. *In press*. Creating user group vested interest in fishery management Outcomes: A Case Study of a Pacific Fishery Management Council Management Program. In R.M. Meyer, C. Zhang, M.L. Windsor, B. McCay, L. Hushak, and R. Muth, eds. Fisheries Utilization and Policy; Proceedings of the World Fisheries Congress, May 1991; Theme 2 Volume. Oxford & IBH Publishing Co. Pvt. Inc. New Delhi.

Hanna, S. and C.L. Smith. 1993. Resolving allocation conflicts in fisheries management. Soc. Nat. Resour. 6:55-69.

Hanski, I. and M. Gilpin. 1991. Metapopulation dynamics: Brief history and conceptual domain. Biol. J. Linn. Soc. 42:3-16.

Hanson, D.L. and T.F. Waters. 1974. Recovery of standing crop and production rate of a brook trout population in a flood-damaged stream. Trans. Am. Fish. Soc. 103:431-439.

Hardy, R. 1993. Presentation to the committee. University Inn, Seattle, WA. June 23, 1993.

Hare, S.R. and R.C. Francis. 1995. Climate change and salmon production in the Northeast Pacific Ocean. Pp. 357-372 in R.J. Beamish, ed. Climate Change and Northern Fish Populations. Can. Spec. Publ. Fish. Aquat. Sci. 121.

Harmon, M.E., J.F. Franklin, F.J. Swanson, P. Sollins, S.V. Gregory, J.D. Lattin, N.H. Anderson, S.P. Cline, N.G. Aumen, J.R. Sedell, G.W. Lienkaemper, K. Cromack Jr., and K.W. Cummins. 1986. Ecology of coarse woody debris in temperate ecosystems. Pp. 133-302 in A. Macfadyen and E.D. Ford eds. Advances in Ecological Research 15.

Harr, R.D., W.C. Harper, J.T. Krygier, and F.S. Hsieh. 1975. Changes in storm hydrographs after road building and clear-cutting in the Oregon Coast Range. Water Resour. Res. 11:436-444.

Harr, R.D. and F.M. McCorison. 1979. Initial effects of clearcut logging on size and timing of peak flows in a small watershed in western Oregon. Water Resour. Res. 15:90-94.

Harr, R.D. 1986. Effects of clearcutting on rain-on-snow runoff in western Oregon: A new look at old studies. Water Resour. Res. 22:1095-1100.

Harr, R.D., B.A. Coffin, and T.W. Cundy. 1989. Effects of Timber Harvest on Rain-on-Snow Runoff in the Transient Snow Zone of the Washington Cascades. Interim Final Report to the Washington Timber, Fish and Wildlife Sediment, Hydrology, and Mass Wasting Steering Committee for Project 18. U.S. Department of Agriculture, Forest Service, Pacific Northwest Forest and Range Experiment Station, Portland, OR.

Harris, C.K. 1987. Catches of North American sockeye salmon (*Oncorhynchus nerka*) by the Japanese high seas salmon fisheries, 1972-84. Pp. 458-479 in H.D. Smith, L. Margolis, and C.C. Wood, eds., Sockeye Salmon (*Oncorhynchus nerka*) Population Biology and Future Management. Can. Spec. Publ. Fish. Aquat. Sci. 96.

Harris, C.K. 1988. Recent changes in the pattern of catch of North American salmonids by the Japanese high seas salmon fisheries. Pp. 41-65 in W.J. McNeil ed. Salmon Production, Management, and Allocation: Biological, Economic, and Policy Issues. Oregon State University Press, Corvallis, OR.

Harrison, J. 1994. Sharing the watershed: Grande Ronde plan protects endangered salmon and the economy of Northeast Oregon. Northwest Energy News, Fall 1994:15-17.

Hartman, G.F. and J.C. Scrivener. 1990. Impacts of forestry practices on a coastal stream ecosystem, Carnation Creek, British Columbia. Can. Bull. Fish. Aquat. Sci. 223: 148 pp.

Hartt, A.C. and M.B. Dell. 1986. Early oceanic migrations and growth of juvenile Pacific salmon and steelhead trout. International North Pacific Fisheries Commission Bulletin 46, I-IV, 1-105.

Harza and Associates. 1994. Review of Reservoir Drawdown. Final Report. Harza and Associates, to Northwest Power Planning Council, Columbia/Snake River Drawdown Committee, Northwest Power Planning Council, Portland, OR.

Hasler, A.D. 1966. Underwater Guideposts. University of Wisconsin Press, Madison.

Hawkins, C.P., M.L. Murphy, N.H. Anderson, and M.A. Wilzbach. 1983. Density of fish and salamanders in relation to riparian canopy and physical habitat in streams of the northwestern United States. Can. J. Fish. Aquat. Sci. 40:1173-1185.

Hays, S.P. 1959. Conservation and the Gospel of Efficiency: The Progressive Conservation Movement, 1890-1920. Harvard University Press, Cambridge, MA.

Healey, M.C. 1982. Juvenile Pacific salmon in estuaries: The life support system. Pp. 315-341 in V.S. Kennedy, ed. Estuarine Companions. Academic Press, New York.

Healey, M.C. 1991. Life history of chinook salmon (*Oncorhynchus tshawytscha*). Pp. 311-394 in C. Groot and L. Margolis, eds. Pacific Salmon Life Histories. University of British Columbia Press, Vancouver, BC.

Heard, W.R. 1991. Life history of pink salmon (*Oncorhynchus gorbuscha*). Pp. 121-230 in C. Groot and L. Margolis, eds. Pacific Salmon Life Histories. University of British Columbia Press, Vancouver, BC.

Helle, J.H. 1981. Significance of the stock concept in artificial propagation of salmonids in Alaska. Can. J. Fish. Aquat. Sci. 38:1665-1671.

Hellmund, P.C. 1993. A method for ecological greenway design. Pp. 123-160 in D.S. Smith and P.C. Hellmund, eds. Ecology of Greenways. University of Minnesota Press, Minneapolis, MN.

Hemmingsen, A.R., R.A. Holt, and R.D. Ewing. 1986. Susceptibility of progeny from crosses among three stocks of coho salmon to infection from *Ceratomyxa shasta*. Trans. Am. Fish. Soc. 115:492-495.

Hetherington, E.D. 1988. Hydrology and logging in the Carnation Creek watershed—what have we learned? Pp. 11-15 in T.W. Chamberlin, ed. Proceedings of the Workshop: Applying 15 Years of Carnation Creek Results. Pacific Biological Station, Carnation Creek Steering Committee, Nanaimo, BC.

Hewes, G.W. 1947. Aboriginal use of fishery resources in northwestern North America. Ph.D. Dissertation, University of California, Berkeley, CA.

Hicks, B.J., J.D. Hall, P.A. Bisson, and J.R. Sedell. 1991. Response of salmonids to habitat changes. Am. Fish. Soc. Spec. Publ. 19:83-138.

Higgins, P., S. Dobush, and D. Fuller. 1992. Factors in Northern California Threatening Stocks with Extinction. Humbolt Chapter, American Fisheries Society, Arcata, California. 25 pp.

Hilborn, R. 1985. Apparent stock-recruitment relationships in mixed stock fisheries. Can. J. Fish. Aquat. Sci. 42:718-723.

Hilborn, R. 1992. Hatcheries and the future of salmon in the Northwest. Fisheries 17:5-8.

Hilborn, R. and W. Luedke. 1987. Rationalizing the irrational: A case study in user group participation in Pacific salmon management. Can. J. Fish. Aquat. Sci. 44(10):1796-1805.

Hilborn, R. and C.J. Walters. 1992. Quantitative Fisheries Stock Assessment: Choice, Dynamics, and Uncertainty. Chapman and Hall, New York.

Hilborn, R. and J. Winton. 1993. Learning to enhance salmon production: Lessons from the salmonid enhancement program. Can. J. Fish. Aquat. Sci. 50:2043-56.

Hill, M.T., W.S. Platts, and R.L. Beschta. 1991. Ecological and geomorphological concepts for instream and out-of-channel flow requirements. Rivers 2(3):198-210.

Hillman, T.W., J.S. Griffith, and W.S. Platts. 1987. Summer and winter habitat selection by juvenile chinook salmon in a highly sedimented Idaho stream. Trans. Am. Fish. Soc. 116:185-195.

Hittell, J.S. 1882. The Commerce and Industries of the Pacific Coast of North America. A.L. Bancroft & Company, San Francisco, CA. 325 pp.

Holling, C.S., ed. 1978. Adaptive Environmental Assessment and Management. John Wiley & Sons, New York.

Hollowed, A.B. and W.S. Wooster. 1992. Variability of winter ocean conditions and strong year classes of Northeast Pacific groundfish. ICES Mar. Sci. Symposium 195: 433-444.

Holm, B. 1965. Northwest Coast Indian Art: An Analysis of Form. Seattle: Thomas Burke Memorial Washington State Museum Monograph, No. 1, Seattle, WA.

Holtby, L.B. 1988. Effects of logging on stream temperatures in Carnation Creek, British Columbia, and associated impacts on the coho salmon (*Oncorhynchus kisutch*). Can. J. Fish. Aquat. Sci. 45:502-515.

Holtby, L.B. and J.C. Scrivener. 1989. Observed and simulated effects of climatic variability, clear-cut logging and fishing on the numbers of chum salmon (*Oncorhynchus keta*) and coho salmon (*O. kisutch*) returning to Carnation Creek, British Columbia. Pp. 62-81 in C.D. Levings, L.B. Holtby, and M.A. Henderson, eds. Proceedings of the National Workshop on Effects of Habitat Alterations on Salmonid Stocks. Can. Spec. Publ. Fish. Aquat. Sci. 105, Ottawa, Ontario, Canada.

Holtby, L.B., B.C. Andersen, and R.K. Kadowaki. 1990. Importance of smolt size and early ocean growth to interannual variability in marine survival of coho salmo (*Oncorhynchus kisutch*). Can. J. Fish. Aquat. Sci., 47(11): 2181-2194.

House, R.A. and P.L. Boehne. 1986. Effects of instream structures on salmonid habitats and populations in Tobe Creek, Oregon. No. J. Fish. Manag. 6:38-46.

Hume, R.D. 1893. Salmon of the Pacific Coast. Portland, OR.

Hume, R.D. 1908. Solution of the salmon propagation problem. Pacific Fisherman 6(1):25-26.

Hunter, J.G. 1959. Survival and production of pink and chum salmon in a coastal stream. J. Fish. Res. Bd. Can. 16:835-886.

Hyatt, K.D. and J.G. Stockner. 1985. Responses of sockeye salmon (*Oncorhynchus nerka*) to fertilization of British Columbia coastal lakes. Can. J. Fish. Aquat. Sci. 42:320-331.

Hyman, J.B. and K. Wernstedt. 1991. The role of biological and economic analyses in the listing of endangered species. Resources (Resources for the Future), 104:5-9.

Hyman, J.B., K. Wernstedt, and C. Paulsen. 1993. Dollars and Sense Under the Endangered Species Act: Incorporating Diverse Viewpoints in Recovery Planning for Pacific Northwest Salmon. Resources for the Future, Washington, DC.

Ice, G.G. 1985. Catalog of Landslide Inventories for the Northwest. National Council for the Paper Industry for Air and Stream Improvement, NACSI Technical Bulletin 456. 78 pp.

Idaho Department of Lands. 1994. A Draft Cumulative Effects Process for Idaho. Idaho Department of Lands, Boise, ID.

Ignell, S. 1991. The fisheries for neon flying squid (*Ommastrephes bartrami*) in the central north Pacific Ocean. Pp. 97-111 in J.A. Wetherall ed. Biology, Oceanography, and Fisheries of the North Pacific Transition Zone and Subarctic Frontal Zone. NOAA Tech. Rep. NMFS 105.

Integrated Hatchery Operations Team. 1995. Policies and Procedures for Columbia Basin Anadromous Salmonid Hatcheries: Annual Report 1994. U.S. Department of Energy, Bonneville Power Administration, Portland, OR, Project No. 92-043, Contract No. DE-BI79-92BP60629.

Ishida, Y., S. Ito, M. Kaeriyama, S. McKinnell, and K. Nagasawa. 1993. Recent changes in age and size of chum salmon *(Oncorhynchus keta)* in the North Pacific Ocean and possible causes. Can. J. Fish. Aquat. Sci. 50:290-295.

Iwamoto, R.N., W.D. Muir, B.P. Sandford, K.W. McIntyre, D.A. Frost, J.G. Williams, S.G. Smith, and J.R. Skalski. 1994. Survival estimates for the Passage of Juvenile Chinook Salmon Through Snake River Dams and Reservoirs. NMFS annual report prepared for the Bonneville Power Administration, Portland, Oregon. Contract number DE-A179-93BP10891.

Jackson, P.L. and A.J. Kimerling. 1993. Atlas of the Pacific Northwest. Oregon State University Press, Corvallis. 152 pp.

Jackson, R.I. and W.F. Royce. 1986. Ocean Forum: An Interpretative History of the International North Pacific Fisheries Commission. Fishing News Books, Ltd., Farnham. 240 pp.

Jay, T. and B. Matsen. 1994. Reaching Home: Pacific Salmon, Pacific People. Alaska Northwest, Seattle, WA.

Jentoft, S. 1989. Fisheries co-management: delegating government responsibility to fishermen's organizations. Marine Policy, April:137-154.

Johansen, D.O. and C.M. Gates. 1967. Empire of the Columbia—A History of the Pacific Northwest. Harper & Row Publishers, New York. 654 pp.

Johnsen, R.C., L.A. Hawkes, W.W. Smith, G.L. Fredricks, R.D. Martinson, and W.A. Hevlin. 1990. Monitoring of Downstream Salmon and Steelhead at Federal Hydroelectric Facilities—1989. Annual Report, Environmental and Tech. Serv. Div., U.S. Department of Commerce, National Oceanic and Atmospheric Administration, National Marine Fisheries Service, to Bonneville Power Administration, Proj. 84-14, Agreement DE-A179-85BP20733.

Johnson, H.M. 1995. 1995 Annual Report on the United States Seafood Industry. H.M. Johnson & Associates, Bellevue, Washington.

Johnson, J.E. and J. VanAmberg. 1994. Evidence of natural reproduction in western Lake Huron. J. Great Lakes Res.

Johnson, J.I. and M.V. Abrahams. 1991. Interbreeding with domestic strain increases foraging under threat of predation in juvenile steelhead trout (*Oncorhynchus mykiss*): an experimental study. Can. J. Fish. Aquat. Sci. 48:243-247.

Johnson, O.W., T.A. Flagg, D.J. Maynard, G.B. Milner, and F.W. Waknitz. 1991. Status Review for Lower Columbia River Coho Salmon. NOAA (National Oceanic and Atmospheric Administration) Tech. Mem. NMFS (National Marine Fisheries Service), F/NWC-202. Seattle, WA. 94 pp.

Johnson, R.R., C.D. Ziebell, D.R. Patton, P.F. Folliott, R.H. Hamre. 1985. Riparian Ecosystems and Their Management: Reconciling Conflicting Uses. U.S. Department of Agriculture, Forest Service, General Technical Report RM-120. 523 pp.

Johnson, R.W. 1989. Water pollution and the public trust doctrine. Environ. Law 19:485-513.

Johnson, R.W., D. Goeppele, D. Jansen, and R. Paschal. 1992. The public trust doctrine and coastal zone management in Washington State. Washington Law Review 67(3):521-597.

Johnson, S.L. 1984. Freshwater Environmental Problems and Coho Production in Oregon. Oregon Department of Fish & Wildlife, Fish Division, Information Report 84-11, Salem, OR.

Johnson, W.S. 1984. Photoperiod induced delayed maturation of freshwater reared chinook salmon. Aquacult. 43(1-3):279-287.

Jordan, D.S. 1925. Fishes. D. Appleton, and Co., New York. 773 pp.

Jude, D.J., S.A. Klinger, and M.D. Enk. 1981. Evidence of natural reproduction by planted lake trout in Lake Michigan. J. Great Lakes Res. 7:57-61.

Kaczynski, V.W. and J.F. Palmisano. 1993. Oregon's Wild Salmon and Steelhead Trout: A Review of the Impact of Management and Environmental Factors. Oregon Forest Industries Council, Salem, OR. 328 pp.

Kaeriyama, M. 1989. Aspects of salmon ranching in Japan. Physiol. Ecol. Jpn. Spec. Vol. 1:625-638.

Kapuscinski, A.R. and L.D. Jacobson. 1987. Genetic Guidelines for Fisheries Management. Minnesota Sea Grant, St. Paul. 66 pp.

Kapuscinski, A.R. 1991. Genetic Analysis of Policies and Guidelines for Salmon and Steelhead Hatchery Production in the Columbia River Basin. Prepared for the Northwest Power Planning Council [Agreement 90-037], March, 1991. 35 pp. (Unpublished).

Kapuscinski, A.R. and L.M. Miller. 1993. Genetic Hatchery Guidelines for the Yakima/Klickitat Fisheries Project. Contract report for Washington Department of Fisheries. 75 pp. + appendices.

Karr, J.R., L.A. Toth, and D.R. Dudley. 1985. Fish communities of midwestern rivers: A history of degradation. BioScience 35:90-95.

Kauffman, J.B. 1988. The status of riparian habitats in Pacific Northwest forests. Pp. 45-55 in K.J. Raedeke, ed. Streamside Management: Riparian Wildlife and Forestry Interactions. Contribution Number 59, Institute of Forest Resources, University of Washington, Seattle.

Kauffman, J.B., R.L. Beschta, and W.S. Platts. 1993. Fish Habitat Improvement Projects in the Fifteenmile Creek & Trout Creek Basins of Central Oregon: Field Review and Management Recommendations. U.S. Department of Energy, Bonneville Power Administration, Division of Fish and Wildlife, Portland, Oregon. DOE/BP-18955-1. 52 pp.

Kaya, C.M. 1991. Rheotactic differentiation between fluvial and lacustrine populations of Arctic Grayling (*Thymallus arcticus*), and implications for the only remaining indigenous population of fluvial "Montana grayling." Can. J. Fish. Aquat. Sci. 48:53-59.

Kelso, B.W., T.G. Northcote, and C.F. Wehrhahn. 1981. Genetic and environmental aspects of one response to water current by rainbow trout (*Salmo gairdneri*) originating from inlet to outlet streams of two lakes. Can. J. Zool. 59: 2177-2185.

Kimerling, A.J. and P.L. Jackson. 1985. Atlas of the Pacific Northwest. Oregon State University Press, Corvallis, OR. 136 pp.

King County Surface Water Management Division. 1995. Waterways 2000 Acquisition and Stewardship Recommendations: A Report of the King County Open Space Citizen Oversight Committee. Report published February 16, 1995, by King County, Seattle, WA. Unnumbered pages.

Kline, T.C., Jr., J.J. Goering, O.A. Mathisen, P.H. Poe, and P.L. Parker. 1990. Recycling of elements transported upstream by runs of Pacific salmon: I. $\delta^{15}N$ and $\delta^{13}C$ evidence in Sashin Creek, southeastern Alaska. Can. J. Fish. Aquat. Sci. 47:136-144.

Kline, T.C., Jr., J.J. Goering, O.A. Mathisen, P.H. Poe, P.L. Parker and R.S. Scanlan. 1993. Recycling of elements transported upstream by runs of Pacific salmon: II. $\delta^{15}N$ and $\delta^{13}C$ evidence in the Kvichak River watershed, Bristol Bay, Southwestern Alaska. Can. J. Fish. Aquat. Sci. 50:2350-2365.

Konkel, G.W. and J.D. McIntyre. 1987. Trends in Spawning Populations of Pacific Anadromous Salmonids. U.S. Department of the Interior, Fish and Wildlife Service, Fish and Wildlife Technical Report 9, I-III:1-25.

Konopacky, R.C., E.C. Bowles, and P.J. Cenera. 1985. Salmon River Habitat Enhancement. Annual Report 1984. U.S. Department of Energy, Contract DE-A179-84BP14383. Bonneville Power Administration, Portland, OR.

Koski, C.H., S.W. Pettit, J.B. Athearn, and A.L. Heindle. 1985. Fish Transportation Oversight Team Annual Report - FY 1984. Transport Operations on the Snake and Columbia Rivers. National Marine Fisheries Service, NOAA Tech. Memo. NMFS F/NWR-11.

Kraft, M.E. 1972. Effects of controlled flow reduction on a trout stream. J. Fish. Res. Board Can. 29:1405-1411.

Krueger, C.C., D.L. Perkins, E.L. Mills, J.S. DeGisi, and J.E. Marsden. 1994. Alewife predation of lake trout fry in Lake Ontario: blockage of year-class recruitment from spawning of hatchery-origin adults. J. Great Lakes Res.

Krueger, W.C. 1993. Managing areas in mixed ownership. Pp. 307-310 in B. Tellman, H.J. Courtner, M.G. Wallace, L.T. DeBano. Riparian Management: Common Threads and Shared Interests. USDA Forest Service, GTR-RM-226. 419 pp.

Krueger, W.C. 1994. Building consensus for riparian users: Toward the twenty-first century. Nat. Resour. Environ. Issues 1:77-82.

Kusler, J.A. and M.A. Kentula. 1990. Wetland Creation and Restoration: The State of the Science. Island Press, Washington, DC. 594 pp.

Lacy, R.C. 1987. Loss of genetic diversity from managed populations: Interacting effects of drift, mutation, immigration, selection, and population subdivision. Conserv. Biol. 1(2):143-158.

Larkin, P.A. 1970. Management of Pacific salmon of North America. Pp. 223-236 in N.G. Benson, ed. A Century of Fisheries in North America. Am. Fish. Soc. Spec. Publ. 7.

Larkin, P.A. 1972. The stock concept and management of Pacific salmon. Pp. 11-15 in R.C. Simon and P.A. Larkin, eds. The Stock Concept in Pacific Salmon. University of British Columbia Press, Vancouver, BC.

Larkin, P.A. 1977. An epitaph for the concept of maximum sustained yield. Trans. Am. Fish. Soc. 106(1):1-11.

Lawson, P.W. 1993. Cycles in ocean productivity, trends in habitat quality, and the restoration of salmon runs in Oregon. Fisheries (Bethesda) 18(8):6-10.

Ledgerwood, R., E. Dawley, L. Gilbreath, P. Bentley, B. Sandford, and M. Schiewe. 1991. Relative Survival of Subyearling Chinook Salmon That Have Passed Through the Turbines or Bypass System of Bonneville Dam Second Powerhouse, 1990. National Marine Fisheries Service Annual Report to U.S. Army Corps of Engineers. Contr. No. E86900104. 90 pp.

Lee, D.C. and J. Hyman. 1992. The Stochastic Life-Cycle Model: A Tool for Simulating the Population Dynamics of Anadromous Salmonids. U.S. Department of Agriculture, Forest Service, Intermountain Res. Station, Boise, ID; Resources for the Future, Washington, DC.

Lee, K.N. 1991. Unconventional power: Energy efficiency and environmental rehabilitation under the Northwest Power Act. Ann. Rev. Energy Environ. 16:337-64.

Lee, K.N. 1991. Rebuilding confidence: Salmon, science, and law in the Columbia basin. Environ. Law 21:745-805.

Lee, K.N. 1993a. Compass and Gyroscope. Integrating Science and Politics for the Environment. Island Press, Washington, DC.

Lee, K.N. 1993b. Greed, scale mismatch, and learning. Ecol. Appl. 3:560-64.

Lee, K.N. and D.L. Klemka, with M.E. Marts. 1980. Electric Power and the Future of the Pacific Northwest. University of Washington Press, Seattle, WA.

Leopold, A. 1944. Conservation: In whole or in part? Reprinted in S.L. Flader and J.B. Caldicott, eds. 1991. The River of the Mother of God and Other Essays by Aldo Leopold. The University of Wisconsin Press, Madison, WI.

Leopold, L.B., M.G. Wolman, and J.P. Miller. 1964. Fluvial Processes in Geomorphology. W. H. Freeman, San Francisco, CA.

Lettenmaier, D.P., J.R. Wallis, and E.F. Wood. 1994. Hydrometeorological trends in the continental U.S., 1948-88. J. Climatol. 7(4):586-607.

Levings, C.D. 1993. Requirements for genetic data on adaptations to environment and habitats of salmonids. Pp. 49-66 in J.G. Cloud and G.H. Thorgaard, eds. Genetic Conservation of Salmonid Fishes. Plenum, New York.

Levy, D.A. and T.G. Northcote. 1982. Juvenile salmon residency in a marsh area of the Fraser River estuary. Can. J. Fish. Aquat. Sci. 39:270-276.

Li, H.W., C.B. Schreck, C.E. Bond, and E. Rexstad. 1987. Factors influencing changes in fish assemblages of Pacific Northwest streams. Pp. 193-202 in W.J. Matthews and D.C. Heins, eds. Community and Evolutionary Ecology of North American Stream Fishes. University of Oklahoma Press, Norman, OK.

Liao, S. and J.B. Stevens. 1975. Oregon's Commercial Fishermen: Characteristics, Profits, and Incomes in 1972. Oregon State University, Agricultural Experiment Station, Circular of Information, No. 645.

Lichatowich, J. 1992. Managing for sustainable fisheries: Some social, economic, and ethical considerations. Pp. 11-17 in G. Reeves, D. Bottom, and M. Brookes, eds, Ethical Questions for Resource Managers, Pacific Northwest Research Station, U.S. Department of Agriculture, Forest Service, General Technical Report PNW-GTR-288.

Lichatowich, J. and S. Cramer. 1980. Parameter Selection and Sample Sizes in Studies of Anadromous Salmonids. Oregon Department Fish and Wildlife, Info. Report, Fisheries 80-1. 25 pp.

Lichatowich, J.A. 1989. Habitat alteration and changes in abundance of coho (*Oncorhynchus kisutch*) and chinook salmon (*O. tshawytscha*) in Oregon's coastal streams. Pp. 92-99 in C.D. Levings, L.B. Holtby, and M.A. Anderson, eds. Proceedings of the National Workshop on Effects of Habitat Alteration on Salmonid Stocks. Can. Spec. Publ. Fish. Aquat. Sci. 105. 199 pp.

Lichatowich, J.A. and J.W. Nicholas. *In Press*. Oregon's first century of hatchery intervention in salmon production: evolution of the hatchery program, legacy of a utilitarian philosophy and management recommendations. In Proceedings of the Symposium of Biological Interactions of Enhanced Wild Salmonids, June 17-20, 1991. Department of Fisheries and Oceans, Pacific Region, Nanaimo, British Columbia, Canada.

Light, J.T. 1987. Coastwide Abundance of North American Steelhead Trout. University of Washington, Fish. Res. Inst., report to International North Pacific Fisheries Commission by the U.S. National Section.

Lister, D.B., D.E. Marshall, and D.G. Hickey. 1980. Chum salmon survival and production at seven improved groundwater-fed spawning areas. Can. Manuscript Rep. of Fish. Aquat. Sci. 1595.

Lothrop, R.C. 1986. The misplaced role of cost-benefit analysis in Columbia basin fishery mitigation. Environ. Law 16:517-554.

Low, L.L. and J.D. Berger. 1994. Incidental Catch of Salmon in U.S. Groundfish Fisheries in the Bering Sea/Aleutian Islands Region, Gulf of Alaska and the Pacific Coast, 1990-1993. U.S. Department of Commerce, National Marine Fisheries Service, Alaska Fish Sci. Cent., Seattle, WA.

Luchetti, G. and R. Fuerstenberg. 1993. Management of coho salmon habitat in urbanizing landscapes of King County, WA. Pp. 308-317 in L. Berg and P. Delaney, eds. Proceedings of the 1992 Coho Workshop, Nanaimo, British Columbia. North Pacific International Chapter, American Fisheries Society, and Association of Professional Biologists of British Columbia, Vancouver, BC.

Ludwig, D., R. Hilborn, and C.J. Walters. 1993. Uncertainty, resource exploitation, and conservation: Lessons from history. Science 260:17, 36.

Magnuson, J.J. 1991. Fish and fisheries ecology. Ecol. Appl. 1:13-26.

Marsden, J.E. and C.C. Krueger. 1991. Spawning by hatchery-origin lake trout (*Salvelinus namaycush*) in Lake Ontario: data from egg collections, substrate analysis, and diver observations. Can. J. Fish. Aquat. Sci. 48:2377-2384.

Marsden, J.E., C.C. Krueger, and C.P. Schneider. 1988. Evidence of natural reproduction by stocked lake trout in Lake Ontario. J. Great Lakes Res. 14:3-8.

Mason, J.C. and D.W. Chapman. 1965. Significance of early emergence, environmental rearing capacity, and behavioral ecology of juvenile coho salmon in stream channels. J. Fish. Res. Board Can. 22:173-190.

Mason, J.W., O.M. Brynilson, and P.E. Degurse. 1967. Comparative survival of wild and domestic strains of brook trout in streams. Trans. Am. Fish. Soc. 96:313-319.

Mathur, D., W.H. Bason, E.J. Purdy, Jr., and C.A. Silver. 1985. A critique of the Instream Flow Incremental Methodology. Can. J. Fish. Aquat. Sci. 42:825-831.

Matthews, G. and R. Waples. 1991. Status Review for Snake River Spring and Summer Chinook Salmon. NOAA Tech. Memo. NMFS F/NWC-200, Northwest Fisheries Center, National Marine Fisheries Service, Seattle, WA.

Matthews, G.M., D.L. Park, J.R. Harmon, C.S. McCutcheon, and A.J. Novotny. 1987. Evaluation of Transportation of Juvenile Salmonids and Related Research on the Columbia and Snake Rivers—1986. National Marine Fisheries Service, Coastal Zone and Estuarine Studies, Montlake, WA.

McAllister, M.K. and R.M. Peterman. 1992. Experimental design in the management of fisheries: a review. No. Am. J. Fish. Manag. 12:1-18.

McCarl, B.A. and R.B. Rettig. 1983. Influence of hatchery smolt releases on adult salmon production and its variability. Can. J. Fish. Aquat. Sci. 40(11):1880-1886.

McCay, B.J. 1988. Muddling through the clam beds: cooperative management of New Jersey's hard clam spawner sanctuaries. J. Shellfish Res. 7(2):327-340.

McCay, B.J. and J.M. Acheson, eds. 1987. The Question of the Commons: The Culture and Ecology of Communal Resources. University of Arizona Press, Tucson.

McHugh, J.L. 1970. Trends in fishery research. Pp. 26-56 in N.G. Benson, ed. A Century of Fisheries in North America. Am. Fish. Soc. Spec. Publ. 7.

McNabb, D.H. and F.J. Swanson. 1990. Effects of fire on soil erosion. Pp. 159-176 in J.D. Walstad, S.R. Radosevich, and D.V. Sandberg, eds. Natural and Prescribed Fire in Pacific Northwest Forests. Oregon State University Press, Corvallis, OR. 317 pp.

McNeil, W.J. 1964. Redd superposition and egg capacity of pink salmon spawning beds. J. Fish. Res. Board Can. 21:1385-1396.

McPhail, J.D. and Lindsey, C.C. 1986. Zoogeography of the freshwater fishes of Cascadia (the Columbia River systems and rivers north to the Stikine). Pp. 615-637 in C.H. Hocutt and E.O. Wiley, eds. The Zoogeography of North American Freshwater Fishes. Wiley, New York.

Meehan, W.R. and T.C. Bjornn. 1991. Influences of forest and rangeland management of salmonid fishes and their habitats. Am. Fish. Soc. Spec. Publ. 19:47-82.

Meffe, G.K. 1992. Techno-arrogance and halfway technologies: Salmon hatcheries on the Pacific Coast of North America. Conserv. Biol. 6(3):350-354.

Merriam, C.H. 1901. Food of sea lions. Science N.S. 13:777-779.

Mesa, M.G. 1991. Variation in feeding, aggression, and position choice between hatchery and wild cutthroat trout in an artificial stream. Trans. Am. Fish. Soc. 120:723-727.

Miles, E.L. 1987. Science, Politics and International Ocean Management. Policy Papers in International Affairs no. 33. University of California Institute of International Studies, Berkeley, CA.

Miller, L.M. and A.R. Kapuscinski. 1994. Estimation of selection differentials from fish scales: A step towards evaluating genetic alteration of fish size in exploited populations. Can. J. Fish. Aquat. Sci. 51:774-783.

Miller, R.J. and E.L. Brannon. 1982. The origin and development of life history patterns in Pacific salmonids. Pp. 296-309 in E.L. Brannon and E.O. Salo, eds. Proceedings of the Salmon and Trout Migratory Behavior Symposium. University of Washington, School of Fisheries, Seattle, WA.

Milne, D.J. 1964. Sizes and Ages of Chinook, *Oncorhynchus tshawytscha*, and Coho, *O. kisutch*, Salmon in the British Columbia Troll Fisheries (1952-1959) and the Fraser River Gill-Net Fishery (1956-1959). Fish. Res. Board Can. MS Rep. Ser. 776. 42 pp.

Morgan, J.R. 1987. Large marine ecosystems; an emerging concept of regional management. Environ. 29(10):4-34.

Moulton, F.R., editor. 1939. The Migration and Conservation of Salmon. Publication of the American Association for the Advancement of Science. Number 8.

Moyle, P.B. 1969. Comparative behavior of young brook trout of domestic and wild origin. Prog. Fish. Cult. 31(1):51-56.

Moyle, P.B. 1976. Inland Fishes of California. University of California Press, Berkeley, California 405 pp.

Moyle, P.B. and B. Herbold. 1987. Life-history patterns and community structure in stream fishes of western north America: comparisons with eastern North America and Europe. Pp. 25-32 in W.J. Matthews and D.C. Heins, eds. Community and Evolutionary Ecology of North American Stream Fishes. University of Oklahoma Press, Norman, OK.

Moyle, P.B. and J.P. Ellison. 1991. A conservation-oriented classification system for the inland waters of California. Calif. Fish Game 77:161-180.

Moyle, P.B. and R.M. Yoshiyama. 1994. Protection of aquatic biodiversity in California: A five-tiered approach. Fisheries 19(2):6-18.

Muckleston, K.W. 1993. Water resources. Pp. 71-80 in P.L. Jackson and A.J. Kimerling, eds. Atlas of the Pacific Northwest. Oregon State University Press, Corvallis, OR.

Mullan, J.W. 1990. Status of chinook salmon stocks in the mid-Columbia. Pp. 45-55 in D. Park, Convenor. Status and Future of Spring Chinook Salmon in the Columbia River Basin—Conservation and Enhancement. NOAA Tech. Memo. NMFSa F/NWC-187, Northwest Fisheries Center, Montlake, WA.

Mullan, J.W., K.R. Williams, G. Rhodus, T.W. Hillman, and J.D. McIntyre. 1992. Production and habitat of salmonids in Mid-Columbia River Tributary Streams. U.S. Fish and Wildlife Service Monograph I, U.S. Government Printing Office, 1993-695-600, Washington, DC. 489 pp.

Mundie, J.H. 1969. Ecological implications of the diet of juvenile coho in streams. Pp. 135-152 in T.G. Northcote, ed. Symposium on Salmon and Trout in Streams. H.R. MacMillan Lectures in Fisheries. Institute of Fisheries, University of British Columbia, Vancouver, BC.

Mundy, P.R., D. Neeley, C.R. Steward, T.P. Quinn, B.A. Barton, R.N. Williams, D. Goodman, R.R. Whitney, M.W. Erho, Jr., and L.W. Botsford. 1994. Transportation of Juvenile Salmonids From Hydroelectric Projects in the Columbia River Basin: An Independent Peer Review. Final Report to U.S. Department of the Interior, Fish and Wildlife Service, Portland, OR.

Murphy, M.L. 1985. Die-offs of pre-spawn adult pink and chum salmon in southeastern Alaska. No. Am. J. Fish. Manag. 5:302-308.

Murphy, M.L. and W.R. Meehan. 1991. Stream ecosystems. Am. Fish. Soc. Spec. Publ. 19:17-46.

Mutz, K.M. and L.C. Lee (technical coordinators). 1987. Wetland and Riparian Ecosystems of the American West. Proceedings of the Society of Wetland Scientists' Eighth Annual Meeting, Seattle, WA.

Myers, K.W., C.K. Harris, Y. Ishida, L. Margolis, and M. Ogura. 1993. Review of the Japanese Landbased Driftnet Salmon Fishery in the Western North Pacific Ocean and the Continent of Origin of Salmonids in This Area. I.N.P.F.C. Bulletin 52. 86 pp.

Naiman, R.J., T.J. Beechie, L.E. Benda, P.A. Bisson, L.H. MacDonald, M.D. O'Connor, C. Oliver, P. Olson, and E.A. Steel. 1992. Fundamental elements of ecologically healthy watersheds in the Pacific Northwest Coastal Ecoregion. Pp. 127-188 in R.J. Naiman, ed. Watershed Management: Balancing Sustainability and Environmental Change. Springer-Verlag, NY.

Narver, D.W. 1971. Effects of logging debris on fish production. Pp. 100-111 in J.T. Krygier and J.D. Hall, eds. Proceedings of the Symposium on Forest Land Uses and Stream Environment. October 19-20, 1970. Oregon State University, Corvallis, OR.

NRC (National Research Council). 1989. Irrigation-Induced Water Quality Problems: What Can Be Learned from the San Joaquin Valley Experience. National Academy Press, Washington, DC.

NRC (National Research Council). 1992a. Restoration of Aquatic Ecosystems: Science, Technology, and Public Policy. National Academy Press, Washington, DC.

NRC (National Research Council). 1992b. Assessment of the U.S. Outer Continental Shelf Environmental Studies Program. III. Social and Economic Studies. National Academy Press, Washington, DC.

NRC (National Research Council). 1994. Improving the Management of U.S. Marine Fisheries. National Academy Press, Washington, DC.

NRC (National Research Council). 1995a. Wetlands: Characteristics and Boundaries. National Academy Press, Washington, DC.

NRC (National Research Council). 1995b. Science and the Endangered Species Act. National Academy Press, Washington, DC.

Neave, F. 1953. Principles affecting the size of pink and chum salmon populations in British Columbia. J. Fish. Res. Board Can. 9:450-491.

Neave, F., T. Yonemori, and R. Bakkala. 1976. Distribution and Origin of Chum Salmon in Offshore Waters of the North Pacific Ocean. I.N.P.F.C. Bulletin 35. 79 pp.

Nehlsen, W., J.E. Williams, and J.A. Lichatowich. 1991. Pacific salmon at the crossroads: stocks at risk from California, Oregon, Idaho and Washington. Fisheries 16:4-21.

Nelson, K. and M. Soulé. 1987. Genetical conservation of exploited fishes. Pp. 345-368 in N. Ryman and F. Utter, eds. Population Genetics and Fishery Management. University of Washington Press, Seattle.

Nelson, R.L., M.L. McHenry, and W.S. Platts. 1991. Mining. Am. Fish. Soc. Spec. Publ. 19:425-457.

Nester, R.T. and T.P. Poe. 1984. First evidence of successful natural reproduction of planted lake trout in Lake Huron. No. Am. J. Fish. Manag. 4:126-128.

Netboy, A. 1980. The Columbia River Salmon and Steelhead Trout: Their Fight for Survival. University of Washington Press, Seattle.

Netboy, A. 1986. The damming of the Columbia River: The failure of bio-engineering. Pp. 33-48 in E. Goldsmith and N. Hilyard, eds. The Social and Environmental Effects of Large Dams—Volume 2: Case Studies. Wadebridge Ecological Centre, United Kingdom.

Nichols, D. 1990. An Inventory of Water Diversions in Oregon Needing Fish Screens. Habitat Conservation Division, Oregon Department of Fish and Wildlife, Portland, OR. 27 pp. + appendices.

Nickelson, T.E., M.F. Solazzi, and S.L. Johnson. 1986. Use of hatchery coho salmon (*Oncorhynchus kisutch*) presmolts to rebuild wild populations in Oregon coastal streams. Can. J. Fish. Aquat. Sci. 43:2443-2449.

Nickelson T.E., J.W. Nicholas, A.M. McGie, R.B. Lindsayh, D.L. Bottom, F.J. Kaiser, and S.E. Jacobs. 1992. Status of Anadromous Salmonids in Oregon Coastal Basins. Oregon Department of Fish and Wildlife, Portland, OR. 83 pp.

Nickelson, T.E. and R.E. Hafele. 1978. Streamflow Requirements of Salmonids. Federal Aid Progress Reports, Oregon Department of Fish and Wildlife, Portland, OR, AFS-62. 26 pp.

Niemi, G.J., P. DeVore, N. Detenbeck, D. Taylor, A. Lima, J. Pastor, J.D. Yount, and R.J. Naiman. 1990. Overview of case studies on recovery of aquatic systems from disturbance. Environ. Manag. 14:571-587.

NMFS (National Marine Fisheries Service). 1995. Proposed Recovery Plan for Snake River Salmon. U.S. Department of Commerce, National Oceanic and Atmospheric Administration, National Marine Fisheries Service, Northwest Region.

NMFS Panel on Gas Bubble Disease, June 28, 1994.

NMFS and WDFW (National Marine Fisheries Service and Washington Department of Fish and Wildlife). 1995. Environmental Assessment on Protecting Winter-Run Steelhead from Predation by California Sea Lions in the Lake Washington Ship Canal, Seattle, Washington. U.S. Department of Commerce, National Oceanic and Atmospheric Administration, National Marine Fisheries Service Northwest Regional Office, Seattle; and Washington Department of Fish and Wildlife, Olympia, WA.

Noakes, D.J. 1989. A nonparametric approach to generating inseason forecasts of salmon returns. Can. J. Fish. Aquat. Sci. 46:2046-2055.

Noggle, C.C. 1978. Behavioral, physiological and lethal effects of suspended sediment on juvenile salmonids. M.S. Thesis, University of Washington, Seattle.

Norgaard, R. 1994. Development Betrayed: The End of Progress and a Coevolutionary Revisioning of the Future. Routledge, New York.

Northcote, T.G. 1978. Migratory strategies and production in freshwater fishes. Pp. 326-359 in S. D. Gerking, ed. Ecology of Freshwater Fish Production. Blackwell Scientific Publications, Oxford, England.

Northcote, T.G. and P.A. Larkin. 1989. The Fraser River: A Major Salmonid Production System. Proceedings of the International Large River Symposium. Can. Spec. Publ. Fish. Aquat. Sci. 106.

NPPC (Northwest Power Planning Council). 1982. Columbia River Basin Fish and Wildlife Program. Northwest Power Planning Council, Portland, OR.

NPPC (Northwest Power Planning Council). 1986. Compilation of Information on Salmon and Steelhead Losses in the Columbia River Basin. Appendix D of the 1987 Columbia River Basin Fish and Wildlife Program. Portland, OR. 252 pp.

NPPC (Northwest Power Planning Council). 1987. Columbia River Basin Fish and Wildlife Program. Northwest Power Planning Council, Portland, OR.

NPPC (Northwest Power Planning Council). 1991. Columbia River Basin Fish and Wildlife Program. Northwest Power Planning Council, Portland, OR.

NPPC (Northwest Power Planning Council). 1992a. Strategy for Salmon. Vol. I. Northwest Power Planning Council, Portland, OR. 43 pp. Document #92-21.

NPPC (Northwest Power Planning Council). 1992b. Strategy for Salmon. Vol. II. Northwest Power Planning Council, Portland, OR. 98 pp. Document #92-21A.

NPPC (Northwest Power Planning Council). 1992c. System Planning Model documentation update, Version 5.15. November, 1992, Northwest Power Planning Council, Portland, OR. 29 pp.

NPPC (Northwest Power Planning Council). 1994. Elements of Bonneville's Fish and Wildlife Revenue Impacts 1991-94. Portland, OR.

O'Brien S.J. and J.F. Evermann. 1988. Interactive influence of infectious disease and genetic diversity in natural populations. Trends Ecol. Evol. 3:254-259.

Olesiuk, P.F. 1993. Annual prey consumption by harbor seals (*Phoca vitulina*) in the Strait of Georgia, British Columbia. Fish. Bull. 91:491-515.

Olesiuk, P.F., M.A. Bigg, G.M. Ellis, S.J. Crockford, and R.J. Wigen. 1990. An assessment of the feeding habits of harbor seals (*Phoca vitulina*) in the Strait of Georgia, British Columbia, based on scat analysis. Can. Tech. Rep. Fish. Aquat. Sci. No. 1730, 143 pp.

ODFW and WDF (Oregon Department of Fish and Wildlife and Washington Department of Fisheries). 1991. Status Report. Columbia River Fish Runs & Fisheries, 1960-1990. Portland, OR; Olympia, WA.

Oregon Fish and Game Commission. 1919. Biennial Report of the Fish and Game Commission of the State of Oregon, 1919. 32 pp.

Pacific Fishing. 1994. StatsPack '94: Salmon. Pp. 74-79 in Pacific Fishing 1994 Yearbook, March 1994.

Pella, J., R. Rumbaugh, L. Simon, M. Dahlberg, S. Pennoyer, and M. Rose. 1993. Incidental and illegal catches of salmonids in the north Pacific drift-net fisheries. Pp. 325-358 in J. Ito, W. Shaw, and R.L. Burgner, eds. Symposium on Biology, Distribution, and Stock Assessment of Species Caught in the High Seas Driftnet Fisheries in the North Pacific Ocean. I.N.P.F.C. Bulletin 53 (III).

PFMC (Pacific Fisheries Management Council). 1978. Freshwater Habitat, Salmon Produced, and Escapements for Natural Spawning Along the Pacific Coast of the United States. Pacific Fishery Management Council, Portland, OR.

PFMC (Pacific Fishery Management Council). 1979. Freshwater Habitat, Salmon Produced, and Escapements for Natural Spawning Along the Pacific Coast of the United States. Report of the Anadromous Salmonid Environmental Task Force, Portland, OR.

PFMC (Pacific Fishery Management Council). 1993. Review of the 1992 Ocean Salmon Fisheries. Pacific Fishery Management Council, Portland, OR.

PFMC (Pacific Fishery Management Council). 1994. Review of 1993 Ocean Salmon Fisheries. Pacific Fishery Management Council, Portland, OR.

PFMC (Pacific Fishery Management Council). 1995. Review of 1994 Ocean Salmon Fisheries. Pacific Fishery Management Council, Portland, OR.

Pacific Northwest River Basins Commission (PNRBC). 1969. Columbia-North Pacific Regional Comprehensive Framework Study. 16 Appendices. Pacific Northwest River Basins Commission, Vancouver, WA.

Pacific Northwest River Basins Commission. 1971. Columbia-North Pacific Region Comprehensive Framework Study of Water and Related Lands. Appendix IX, Irrigation. Pacific Northwest River Basins Commission, Vancouver, WA. 343 pp.

Pacific Northwest River Basins Commission. 1972. Columbia-North Pacific Region, Comprehensive Framework Study of Water and Related Lands. Vancouver, WA. 373 pp.

Palmisano, J.F., R.H. Ellis, and V.W. Kaczynski. 1993. The Impact of Environmental and Management Factors on Washington's Wild Anadromous Salmon and Trout. Washington Forest Protection Association and Washington Department of Natural Resources, Olympia, WA. 371 pp.

Park, D.L. 1993. Transportation as a Means of Increasing Wild Juvenile Salmon Survival. Recovery Issues for Threatened and Endangered Snake River Salmon. U.S. Department of Energy, Bonneville Power Administration, Division of Fish and Wildlife, Technical Report 4 of 11.

Subcontract of Don Chapman Consultants to S. P. Cramer & Assoc., Inc., for BPA, Proj. No. 93-013, Contr. No. DE-AM79-93BP99654.

Parsons, L.S. 1993. Management of marine fisheries in Canada. Can. Bull. Fish. Aquat. Sci. 225. 763 pp.

Pascho, R.J. and D.G. Elliott. 1989. Juvenile fish transportation: Impact of Bacterial Kidney Disease on Survival of Spring/Summer Chinook Salmon Stocks. U.S. Department of the Interior, Fish and Wildlife Service Annual Report 1988, Contract E6880047, U.S. Army Corps of Engineers, Walla Walla District.

Pascho, R.J., D.G. Elliott and S. Archord. 1993. Monitoring of the in-river migration of smolts from two groups of spring chinook salmon, *Oncorhynchus tshawytscha* (Walbaum), with different profiles of *Renibacterium salmoninarum* infection. Aquacult. Fish. Manag. 24:163-169.

Pascual, M.A. and T.P. Quinn. 1994. Geographical patterns of straying of fall chinook salmon, *Oncorhynchus tshawytscha* (Walbaum), from Columbia River (USA) hatcheries. Aquacult. Fish. Manag. 25 (Supplement 2):17-30.

Paulsen, C.M., J.B. Hyman, and K. Wernstedt. 1993. Above Bonneville Passage and Propagation Cost Effectiveness Analysis. Resources for the Future, Washington DC.

Pearce, D.W. 1993. Economic Values and the Natural World. The MIT Press: Cambridge, MA.

Pearce, D.W. and R.K. Turner. 1990. Economics of Natural Resources and the Environment. Johns Hopkins Press, Baltimore, MD.

Pearcy, W. G. 1992. Ocean Ecology of North Pacific Salmonids. Washington Sea Grant Program, University of Washington Press, Seattle. 179 pp.

Pease, J.R. 1993. Land use and ownership. Pp. 31-39 in P.L. Jackson and A.J. Kimerling, eds. Atlas of the Pacific Northwest. Oregon State University Press, Corvallis, OR. 152 pp.

Peck, J.W. 1986. Dynamics of reproduction by hatchery lake trout on a man-made spawning reef. J. Great Lakes Res. 12:293-303.

Pentec Environmental Inc. 1991. Methods for testing effectiveness of Washington Forest Practices Rules and Regulations with regard to sediment and turbidity. Washington Department of Natural Resources, Forest Regulation and Assistance, Olympia, WA. 100 pp.

Peterson, N.P. 1982. Immigration of juvenile coho salmon (*Oncorhynchus kisutch*) into riverine ponds. Can. J. Fish. Aquat. Sci. 39:1308-1310.

Peterson, N.P. and L.M. Reid. 1984. Wall-base channels: their evolution, distribution, and use by juvenile coho salmon in the Clearwater River, Washington. Pp. 215-226 in J.M. Walton and D.B. Houston, eds. Proceedings of the Olympic Wild Fish Conference, March 23-25, 1983, Port Angeles, WA.

Phinney, L.A. and P. Bucknell. 1975. A Catalog of Washington Streams and Salmon Utilization, Volume 2, Coastal Region. Washington Department of Fisheries, Olympia, Washington.

Pinay, G., H. Decamps, E. Chauvet, and E. Fustec. 1990. Functions of ecotones in fluvial systems. Pp. 141-170 in R.J. Naiman and H. Decamps, eds. The Ecology and Management of Aquatic-Terrestrial Ecotones. United Nations Educational Scientific and Cultural Organization, Paris and Parthenon Publishing Group, Carnforth, United Kingdom.

Pinkerton, E.W., ed. 1989. Cooperative Management of Local Fisheries: New Directions for Improved Management and Community Development. University of British Columbia Press, Vancouver.

Pinkerton, E.W. 1992. Translating legal rights into management practice: Overcoming barriers to the exercise of co-management. Human Organization 51:330-41.

Pinkerton, E.W. 1993. Co-management efforts as social movements. Alternatives 19(3):34-38.

Platt, J. and D. Dompier 1990. "A history of state and federal fish management in the Columbia and Snake River basins," CRITFC News (Columbia River Inter-Tribal Fish Commission) July, 9-11.

Platts, W.S. 1991. Livestock grazing. Am. Fish. Soc. Spec. Publ. 19:389-423.

Platts, W.S. and W.F. Megahan. 1975. Time trends in riverbed sediment composition in salmon and steelhead spawning areas: South Fork Salmon River, Idaho. Trans. No. Am. Wildl. Nat. Resour. Conf. Washington, DC. 40:229-239.

Poff, N.L. and J.V. Ward. 1990. Physical habitat template of lotic ecosystems: recovery in the context of historical pattern of spatial heterogeneity. Environ. Manag. 14:629-645.

Policansky, D. 1983. Size, age and demography of metamorphosis and sexual maturation in fishes. Am. Zool. 23:57-63.

Policansky, D. 1993. Fishing as a cause of evolution in fishes. Pp. 2-18 in T.K. Stokes, J.M. McGlade, and R. Law, eds. The Exploitation of Evolving Resources. Lecture Notes in Biomathematics Volume 99. Springer-Verlag, Berlin.

Pratt, K. and D. Chapman. 1989. Progress Toward the Run Doubling Goal of the Northwest Power Planning Council. Don Chapman Consultants, Inc., Report to Pacific Northwest Utilities Conference Committee, Portland, OR.

PSC (Pacific Salmon Commission). 1987. Joint Chinook Technical Committee Report. PSC Report TCCHINOOK (87)-5. Pacific Salmon Commission, Vancouver, BC, Canada.

PSC (Pacific Salmon Commission). 1993a. Third Report on the Parties' estimates of Salmon Interception 1980-1991. PSC Report JIC (93)-1. 25 pp. + 6 appendices. Pacific Salmon Commission, Vancouver, BC, Canada.

PSC (Pacific Salmon Commission). 1993b. Joint Chinook Technical Committee 1992 Annual Report. PSC Report TCCHINOOK (93)-2. 103 pp. + 13 appendices. Pacific Salmon Commission, Vancouver, BC, Canada.

PSC (Pacific Salmon Commission). 1994a. Interim Estimates of Coho Stock Compositions for 1984-1991 Southern Area Fisheries and for 1987-1991 Northern Panel Area Fisheries. PSC Report TCCOHO (94)-1. 25 pp. + 3 appendices. Pacific Salmon Commission Vancouver, BC, Canada.

PSC (Pacific Salmon Commission). 1994b. Joint Chinook Technical Committee 1993 Annual Report. Appendix F. PSC Report TCCHINOOK (93)-1. 121 pp. + 13 appendices. Pacific Salmon Commission, Vancouver, BC, Canada.

Pyle, R.M. 1986. Wintergreen: Rambles in a Ravaged Land. Charles Scribner's Sons, New York.

Quinn, T.P. 1985. Homing and the evolution of sockeye salmon (*Oncorhynchus nerka*). Contrib. Mar. Sci. 27:353-366.

Quinn, T.P. 1990. Current controversies in the study of salmon homing. Ethol. Ecol. Evol. 2:49-63.

Quinn, T.P. 1993. A review of homing and straying of wild and hatchery-produced salmon. Fish. Res. 18:29-44.

Quinn, T.P. 1994. Anthropogenic influences on fish populations of the Georgia Basin. Part I. Salmonids. Pp. 219-229 in R.C.H. Wilson, R.J. Beamish, F. Aitkens and J. Bell, eds. Review of the Marine Environment and Biota of Strait of Georgia, Puget Sound and Juan de Fuca Strait. Can. Tech. Rep. Fish. Aquat. Sci.

Quinn, T.P. and D.J. Adams. *In press*. Long-term changes in the temperature and flow regimes of the Columbia River, and the migratory timing of American shad and sockeye salmon. Ecology.

Quinn, T.P. and A.H. Dittman. 1992. Fishes. Pp. 145-211 in F. Papi, ed. Animal Homing. Chapman and Hall, London.

Quinn, T.P. and C.J. Foote. 1994. The effects of body size and sexual dimorphism on the reproductive behaviour of sockeye salmon, *Oncorhynchus nerka*. Anim. Behav. 48: 751-761.

Quinn, T.P., R.S. Nemeth, and D.O. McIsaac. 1991. Homing and straying patterns of fall chinook salmon in the lower Columbia River. Trans. Am. Fish. Soc. 120:150-156.

Quinn, T.P. and M.J. Unwin. 1993. Variation in life history patterns among New Zealand chinook salmon (*Oncorhynchus tshawytscha*) populations. Can. J. Fish. Aquat. Sci. 50:1414-1421.

Radtke, H. 1992. Economic Contribution of Salmon to Oregon's Coastal Communities. Paper Prepared for Governor's Coastal Salmon Initiative Conference.

Raleigh, R.F. 1971. Innate control of migrations of salmon and trout fry from natal gravels to rearing areas. Ecol. 52(2):291-297.

Raymond, H.L. 1979. Effects of dams and impoundments on migrations of juvenile chinook salmon and steelhead from the Snake River, 1966-1976. Trans. Am. Fish. Soc. 108:505-529.

Reeves, G.H., F.H. Everest, and J.D. Hall. 1987. Interactions between the redside shiner (*Richardsonius balteatus*) and the steelhead trout (*Salmo gairdneri*) in western Oregon: The influence of water temperature. Can. J. Fish. Aquat. Sci. 44:1602-1613.

Reeves, G.H., F.H. Everest, and T.E. Nickelson. 1989. Identification of Physical Habitats Limiting the Production of Coho Salmon in Western Oregon and Washington. U.S. Department of Agriculture, Forest Service, General Technical Report PNW-GTR-245, Pacific Northwest Research Station, Portland, OR.

Reeves, G.H., F.H. Everest, and J.R. Sedell. 1993. Diversity of juvenile anadromous salmonid assemblages in coastal Oregon basins with different levels of timber harvest. Trans. Am. Fish. Soc. 122:309-317.

Reeves, G.H. and J.R. Sedell. 1992. An ecosystem approach to the conservation and management of freshwater habitat for anadromous salmonids in the Pacific Northwest. U.S. Department of Agriculture, Forest Service, Pacific Northwest Research Station, Corvallis, OR. Pp. 408-415 in R.E. McCabe, ed. Trans. No. Amer. Wildl. and Nat. Resour. Conf. 57; Crossroads of Conservation: 500 years after Columbus, Charlotte, NC, March 27-April 1, 1992. Wildlife Management Institute, Washington, DC.

Reice, S.R., R.C. Wissmar, and R.J. Naiman. 1990. Disturbance regimes, resilience, and recovery of animal communities and habitats in lotic ecosystems. Environ. Manag. 14:647-659.

Reimers, P.E. 1973. The Length of Residence of Juvenile Fall Chinook Salmon in Sixes River, Oregon. Oregon Fish Commission Research Report 4(2), Oregon Department of Fish and Wildlife, Portland, OR. 43 pp.

Reisenbichler, R.R. 1988. Relation between distance transferred from natal stream and recovery rate for hatchery coho salmon. No. Am. J. Fish. Manag. 8:172-174.

Reisner, M. 1986. Cadillac Desert. Viking Penguin, Inc., New York. 582 pp.

Reznick, D.A., H. Bryga, and J.A. Endler. 1990. Experimentally induced life-history evolution in a natural population. Nature 346:357-359.

Rice, R.M. 1992. The science and politics of BMPs in forestry: California experiences. Pp. 385-400 in R.J. Naiman, ed. Watershed Management: Balancing Sustainability and Environmental Change. Springer-Verlag, New York.

Rich, W.H. 1920. Early history and seaward migration of chinook salmon in the Columbia and Sacramento rivers. Bull. U.S. Bur. Fish. 37:1-74 (Docket number 887).

Rich, W.H. 1939. Local Populations and Migration in Relation to the Conservation of Pacific Salmon in the Western States and Alaska. Am. Assoc. Adv. Sci. Publ. 8:45-50.

Rich, B.A., R.J. Scully, and C.E. Petrosky. 1992. Idaho Habitat/Natural Production Monitoring, Part I General Monitoring Subproject. Annual Report 1990, Idaho Department of Fish and Game, BPA Contr. No. DE-A179-84BP13381, Proj. 83-7.

Richards, J. and D. Olsen. 1993. Columbia River Salmon Production Compared to Other West Coast Production Areas. Phase I Analysis for the U.S. Army Corps of Engineers.

Ricker, W.E. 1954. Stock and recruitment. J. Fish. Res. Bd. Can. 11:559-623.

Ricker, W. E. 1958. Maximum sustained yields from fluctuating environments and mixed stocks. J. Fish. Res. Bd. Can. 15:991-1006.

Ricker, W.E. 1972. Hereditary and environmental factors affecting certain salmonid populations. Pp. 19-160 in R.C. Simon and P.A. Larkin, eds. The Stock Concept in Pacific Salmon. University of British Columbia, Vancouver.

Ricker, W.E. 1973. Two mechanisms that make it impossible to maintain peak-period yields from stocks of Pacific salmon and other fishes. J. Fish. Res. Bd. Can. 30:1275-1286.

Ricker, W.E. 1987. Effects of the fishery and of obstacles to migration on the abundance of Fraser River sockeye salmon (*Oncorhynchus nerka*). Can. Tech. Rep. Fish. Aquat. Sci. 1522, Ottawa, Ontario, Canada.

Ricker, W.E. 1989. History and present status of the odd-year pink salmon runs of the Fraser River region. Can. Tech. Rep. Fish. Aquat. Sci. 1702. 37 pp.

Riddell, B.E. 1993a. Salmonid enhancement: lessons from the past and a role for the future. Pp. 338-355 in D. Mills, ed. Salmon in the Sea and New Enhancement Strategies. Blackwell Scientific Publications, Ltd., Oxford, U.K.

Riddell, B.E. 1993b. Spatial organization of Pacific salmon: What to conserve? Pp. 23-41 in J.G. Cloud and G.H. Thorgaard, eds. Genetic Conservation of Salmonid Fishes. Plenum Press, New York.

Riddell, B.E. and D.P. Swain. 1991. Competition between hatchery and wild coho salmon (*Oncorhynchus kisutch*): genetic variation for agonistic behavior in newly-emerged wild fry. Aquacult. 98:161-172.

Rieman, B.E., R.C. Beamesderfer, S. Vigg, and R.P. Poe. 1991. Estimated loss of juvenile salmonids to northern squawfish, walleyes, and smallmouth bass in John Day Reservoir, Columbia River. Trans. Am. Fish. Soc. 120:448-458.

Riggs, L.A. 1990. Principles for Genetic Conservation and Production Quality. Results of a Scientific and Technical Clarification and Revision. Prepared for The Northwest Power Planning Council, [Contract No. C90-005], March 1990. 20 pp. + appendices. (Unpublished).

Rittel, H.W.J. and M.M. Webber. 1973. Dilemmas in a general theory of planning. Policy Sciences 4(1):155-160.

Robinson, W.L. 1987. An industrial forester's perspective on streamside management. Pp. 289-294 in E.O. Salo and T.W. Cundy, eds. Streamside Management: Forestry and Fishery Interactions. Contribution No. 57, Institute of Forest Resources, University of Washington, Seattle.

Rogers, D.E. and G.T. Ruggerone. 1993. Factors affecting marine growth of Bristol Bay sockeye salmon. Fish. Res. 18:89-103.

Rondorf, D.W., G.A. Gray, and R.B. Fairley. 1990. Feeding ecology of subyearling chinook salmon in riverine and reservoir habitats of the Columbia River. Trans. Am. Fish. Soc. 119:16-24.

Rondorf, D.W. and W.H. Miller. 1994. Identification of the Spawning, Rearing and Migratory Requirements of Fall Chinook Salmon in the Columbia River Basin. Annual Report 1992, BPA, Proj. No. 92-029, Contr. No. DE-A179-91BP21708-2. 210 pp.

Roos, J.F. 1991. Restoring Fraser River Salmon: A History of the International Pacific Salmon Fisheries Commission, 1937-1985. Pacific Salmon Commission, Vancouver, BC. 438 pp.

Rosenberg, A.A., M.J. Fogarty, M.P. Sissenwine, J.R. Beddington, J.G. Shepherd. 1993. Achieving sustainable use of renewable resources. Science 262:828-829.

Rosenfeld, C.L. 1993. Landforms and geology. Pp. 40-47 in P.L. Jackson and A.J. Kimerling, eds. Atlas of the Pacific Northwest. Oregon State University Press, Corvallis, OR.

Rothschild, B. 1983. On the allocation of fisheries stocks. Pp. 85-91 in J.W. Reintjes, ed. Improving Multiple Use of Coastal and Marine Resources. American Fisheries Society, Bethesda, MD.

Rounsefell, G.A. and G.B. Kelez. 1938. The salmon and salmon fisheries of Swiftsure Bank, Puget Sound and the Fraser River. Bur. of Fish. Bull. No. 27.

Ruff, J.D. and J.F. Fazio. 1993. Hydropower Costs of Columbia River Salmon Recovery. Northwest Power Planning Council, Portland, OR.

Sabatier, P.A. and H.C. Jenkins-Smith. 1993. Policy Change and Learning. An Advocacy Coalition Approach. Westview Press, Boulder, CO.

Safina, C. 1994. Where have all the fishes gone? Issues in Science and Technology, Spring 1994:37-43.

Sagoff, M. 1992. Has Nature a Good of Its Own? Pp. 57-71 in R. Costanza, B.G. Norton, and B.D. Haskell, eds. Ecosystem Health. Island Press, Washington, DC.

Salmon Technical Team. 1993. Preseason Report I, Stock Abundance Analysis for 1993 Ocean Salmon Fisheries, February 1993. Pacific Fishery Management Council, Portland, OR.

Salo, E.O. 1991. Life history of chum salmon (*Oncorhynchus keta*). Pp. 231-310 in C. Groot and L. Margolis, eds. Pacific Salmon Life Histories. University of British Columbia Press, Vancouver, BC, Canada.

Salo, E.O. and T.W. Cundy. 1987. Streamside Management: Forestry and Fishery Interactions. Contribution No. 57, Institute of Forest Resources, University of Washington, Seattle, WA.

Sandercock, F.K. 1991. Life history of coho salmon (*Oncorhynchus kisutch*). Pp. 397-445 in C. Groot and L. Margolis, eds. Pacific Salmon Life Histories. University of British Columbia Press, Vancouver, BC, Canada.

Scarnecchia, D.L. 1981. Effect of streamflow and upwelling on yield of wild coho salmon (*Oncorhynchus kisutch*) in Oregon. Can. J. Fish. Aquat. Sci. 38:471-475.

Schaefer, M.L. 1981. Minimum population sizes for species conservation. BioScience 31(2): 131-134.

Schaller, H., C. Petrosky, E. Weber, and T. Cooney. 1992. Snake River Spring/Summer Chinook Life-Cycle Simulation Model for Recovery and Rebuilding Plan Evaluation. Oregon Department of Fish and Wildlife, Portland, OR; Idaho Department of Fish and Game, Boise, ID; Columbia River Inter-Tribal Fish Commission, Portland, OR; and Washington Department of Fisheries, Olympia, WA. 16 pp. + appendices.

Schiewe, M. 1994. Preliminary Survival Estimates for Passage of Juvenile Salmonids Through Lower Granite Reservoir and Lower Granite Dam. NMFS Memo from Schiewe, through U. Varanasi to J.G. Smith, dated August 5, 1994.

Schlosser, I.J. 1991. Stream fish ecology: A landscape perspective. BioScience 41:704-712.

Schoeneman, D., R. Pressey, and C. Junge. 1961. Mortalities of downstream migrant salmon at McNary Dam. Trans. Am. Fish. Soc. 90:58-72.

Scientific Review Group. 1993. Critical Uncertainties in the Fish and Wildlife Program. SRG Report Number 93-3, 27 January 1993. Bonneville Power Administration, Portland, OR. (Unpublished.)

Scott, A. 1955. The fishery: The objectives of sole ownership. J. Polit. Econ. 63:116-124.

Scott, A. 1993. Obstacles to fishery self-governance. Mar. Resour. Econ. 8:187-199.

Scott, W.B. and E.J. Crossman. 1973. Freshwater Fishes of Canada. Bull. Fish. Res. Bd. Can. 184.

Scudder, G.G.E. 1989. The adaptive significance of marginal populations: A general perspective. Pp. 180-185 in C.D. Levings, L.B. Holtby, and M.A. Henderson, eds. Proceedings of the National Workshop on Effects of Habitat Alteration on Salmonid Stocks. Can. Spec. Publ. Fish. Aquat. Sci. 105.

Sedell, J.R., R.J. Naiman, K.W. Cummins, G.W. Minshall, and R.L. Vannote. 1978. Transport of particulate organic material in streams as a function of physical processes. Internationale Vereinigung fur Theoretische und Angewandte Limnologie Verhandlungen 20:1366-1375.

Sedell, J.R. and K.J. Luchessa. 1982. Using the historical record as an aid to salmonid habitat enhancement. Pp. 210-223 in N.B. Armantrout, ed. Acquisition and Utilization of Aquatic Habitat Inventory Information. Proceedings of a symposium held October 28-30, 1981, Portland, OR. The Hague Publishing, Billings, MT.

Sedell, J.R., J.E. Yuska, and R.W. Speaker. 1984. Habitats and salmonid distribution in pristine, sediment-rich river valley systems: S. Fork Hoh and Queets Rivers, Olympic National Park. Pp. 33-46 in W.R. Meehan, T.R. Merrell, Jr., and T.A. Hanley, eds. Fish and Wildlife Relationships in Old-Growth Forests. Am. Inst. Fish. Res. Biol., Juneau, AK.

Sedell, J.R. and J.L. Froggatt. 1984. Importance of streamside forests to large rivers: the isolation of the Willamette River, Oregon, U.S.A., from its floodplain by snagging and streamside forest removal. Internationale Vereinigung fur Theoretische und Angewandte Limnologie Verhandlungen 22:1828-1834.

Sedell, J.R., P.A. Bisson, F.J. Swanson, and S.V. Gregory. 1988. What we know about large trees that fall into streams and rivers. Pp. 47-81 in C. Maser, R.F. Tarrant, J.M. Trappe, and J.F. Franklin, eds. From the Forest to the Sea: A Story of Fallen Trees. U.S. Department of Agriculture, Forest Service, Pacific Northwest Research Station, General Technical Report PNW-GTR-229, Portland, OR.

Sedell, J.R., G.H. Reeves, F.R. Hauer, J.A. Stanford, and C.P. Hawkins. 1990. Role of refugia in recovery from disturbance: Modern fragmented and disconnected river systems. Environ. Manag. 14:711-724.

Sedell, J.R., and F.H. Everest. 1991. Historic Changes in Pool Habitat for Columbia River Basin Salmon Under Study for TES listing. U.S. Department of Agriculture, Forest Service, Pacific Northwest Research Station, Draft General Technical Report, Portland, OR.

Sedell, J.R., and R.L. Beschta. 1991. Bringing back the "Bio" in bioengineering. Am. Fish. Soc. Symp. 10:160-175.

Seehorn, M.E. 1985. Stream Habitat Improvement Handbook. U.S. Department of Agriculture, Forest Service, Southern Region Technical Publication R8-T-7, Atlanta, GA.

Seehorn, M.E. 1992. Stream Habitat Improvement Handbook. U.S. Department of Agriculture, Forest Service, Southern Region, Atlanta, GA.

Seiler, D. 1989. Differential survival of Grays Harbor basin anadromous salmonids: Water quality implications. Pp. 123-135 in C.D. Levings, L.B. Holtby, and M.A. Henderson, eds. Proceedings of the National Workshop on Effects of Habitat Alteration on Salmonid Stocks. Can. Spec. Publ. Fish. Aquat. Sci. 105.

Servisi, J.A. 1989. Protecting Fraser River salmon (*Oncorhynchus spp.*) from wastewaters: An assessment. Pp. 136-153 in C.D. Levings, L.B. Holtby, and M.A. Henderson, eds. Proceedings of the National Workshop on Effects of Habitat Alteration on Salmonid Stocks. Can. Spec. Publ. Fish. and Aquat. Sci. 105.

Shaffer, M.L. 1981. Minimum population sizes for species conservation. BioScience 31:131-134.

Shaffer, M.L. 1990. Population viability analysis. Conserv. Biol. 4:39-40.

Shapiro and Associates, Inc. 1980. Humbolt Bay Wetlands Review and Baylands Analysis. U.S. Army Corps of Engineers, San Francisco, CA. 3 volumes.

Shapovalov, L. and A.C. Taft. 1954. The life history of the steelhead rainbow trout (*Salmo gairdneri*) and silver salmon (*Oncorhynchus kisutch*) with special reference to Waddell Creek, California, and recommendations regarding their management. Calif. Dept. Fish Game Fish Bull. 98. 375 pp.

Shepard, M.P. 1967. History of international competition for Pacific salmon. Fisheries Research Board Can. Manuscript Report 921. 15 pp.

Sherwood, C.R., D.A. Jay, R.B. Harvey, and C.A. Simenstad. 1990. Historical changes in the Columbia River estuary. Prog. Oceanogr. 25:299-352.

Shirvell, C.S. 1989. Habitat models and their predictive capability to infer habitat effects on stock size. Pp. 173-179 in C.D. Levings, L.B. Holtby, and M.A. Anderson, eds. Proceedings of the National Workshop on Effects of Habitat Alteration on Salmonid Stocks. Can. Spec. Publ. Fish. Aquat. Sci. 105. 199 pp.

Shrimpton, J.M. and D.J. Randall. 1992. Smolting and survival in wild and hatchery coho salmon. World Aquacult. 23(4):51-54.

Sibert. J.R. 1979. Detritus and juvenile salmon production in the Nanaimo estuary: II. Meiofauna available as food to juvenile chum salmon (*Oncorhynchus keta*). J. Fish. Res. Bd. Can. 36:497-503.

Sidle, R.C., A.J. Pearce, and C.L. O'Loughlin. 1985. Hillslope Stability and Land Use. Water Resour. Monogr. 11.

Simberloff, D. 1988. The contribution of population and community biology to conservation science. Ann. Rev. Ecol. Syst. 19: 473-511.

Simenstad, C.A., K.L. Fresh, and E.O. Salo. 1982. The role of Puget Sound and Washington coastal estuaries in the life history of Pacific salmon: An unappreciated function. Pp. 343-364 in V.S. Kennedy, ed. Estuarine Comparisons. Academic Press, Toronto, Ontario, Canada. 709 pp.

Simenstad, D.A., D.A. Jay, and C.R. Sherwood. 1992. Impacts of watershed management on land-margin ecosystems: the Columbia River Estuary. Pp. 266-306 in R.J. Naiman, ed. Watershed Management: Balancing Sustainability and Environmental Change. Springer-Verlag, New York.

Simon, R.C., J.D. McIntyre, and A.R. Hemmingsen. 1986. Family size and effective population size in a hatchery stock of coho salmon (*Oncorhynchus kisutch*). Can. J. Fish. Aquat. Sci. 43:2432-2442.

Simon, R.C. 1991. Management techniques to minimize the loss of genetic variability in hatchery fish populations. Am. Fish. Soc. Symp. 10:487-494.

Sims, C. and F. Ossiander. 1981. Migrations of Juvenile Chinook Salmon and Steelhead Trout in the Snake River From 1973 to 1979. NMFS Final Report of Research to U.S. Army Corps of Engineers, Contr. No. DACW68-78-C-0038.

Sissenwine, M.P. 1984. Why do fish populations vary? Pp. 59-94 in R.M. May, ed. Exploitation of Marine Communities. Dahlem Konferenzen. Springer-Verlag, Berlin.

Sissenwine, M.P. and A.A. Rosenberg. 1993. Marine fisheries at a critical juncture. Fisheries 18(10):6-14.

Smith, C.L. 1973. Participation in Willamette Valley Environmental Decisions. Oregon State University, Water Resour. Res. Inst. No. 15.

Smith, C.L. 1979. Salmon Fishers of the Columbia. Oregon State University Press, Corvallis, OR.

Smith, C.L. 1981. Satisfaction bonus from salmon fishing: Implications for economic evaluation. Land Econ. 57(2):181-194.

Smith, C.L. 1994. Connecting cultural and biological diversity in restoring Northwest salmon. Fisheries 19(2):20-26.

Smith, C.L. and S. Hanna. 1990. Measuring fleet capacity and capacity utilization. Can. J. Fish. Aquat. Sci. 47:2085-2091.

Smith, C.L. and B.S. Steel. 1995. Values for valuing salmon. In Pacific Salmon and Their Ecosystems: Status and Future Options. Chapman Hall, New York.

Smith, F.J. 1973. Understanding and Using Marine Economic Data Sheets. Oregon State University Sea Grant Program, Corvallis.

Smith, F.J. 1979. What Are Salmon Worth? Oregon State University Sea Grant, SG48, Corvallis.

Smith, G.R. and R.F. Stearley. 1989. The classification and scientific names of rainbow and cutthroat trouts. Fisheries (Bethesda) 14:4-10.

Smith, P.E. and H.G. Moser. 1988. CALCOFI times series: An overview of fishes. Calif. Coop. Oceanic Fish Invest. Rep. 29(0):66-78. National Marine Fisheries Service, La Jolla, CA.

Smoker, W.A. 1955. Effects of streamflow on silver salmon production in western Washington. Ph.D. Dissertation, University of Washington, Seattle.

Smoker, W.W., A.J. Gharrett, and M.S. Stekoll. *In press*. Genetic variation in timing of anadromous migration within a spawning season in a population of pink salmon. In B.E. Riddell, ed. Proceedings on the International Symposium on Biological Interactions of Enhanced and Wild Salmonids, June 5-6, 1991, Nanaimo. Can. Spec. Publ. Fish. Aquat. Sci.

Snake River Salmon Recovery Team. 1993. Snake River Salmon Recovery Plan Recommendations. Draft. National Marine Fisheries Service.

Snake River Salmon Recovery Team. 1994. Snake River salmon Recovery Plan Recommendations. Final Report. National Marine Fisheries Service.

Snelling, J.C., C.B. Schreck, and C.S. Bradford. 1991. Migratory Characteristics of Spring Chinook Salmon in the Willamette River: 1991 Annual Report. Project DE-A179-88BP92818. U.S. Department of Energy, Bonneville Power Administration, Portland, OR.

Snelling, J.C. and C.B. Schreck. 1992. Migratory Characteristics of Spring Chinook Salmon in the Willamette River: 1992 Annual Report. Project DE-A179-88BP92818. U.S. Department of Energy, Bonneville Power Administration, Portland, OR.

Solazzi, M.F., T.E. Nickelson, and S.L. Johnson. 1991. Survival, contribution, and return of hatchery coho salmon (*Oncorhynchus kisutch*) released into freshwater, estuarine, and marine environments. Can. J. Fish. Aquat. Sci. 48:248-253.

Solley, W.B., R.R. Pierce, and H.A. Perlman. 1993. Estimated Use of Water in the United States in 1990. U.S. Department of the Interior, Geological Survey Circular 1081. 76 pp.

Sousa, W.P. 1984. The role of disturbance in natural communities. Ann. Rev. Ecol. Syst. 15:353-391.

Spaulding, W.M., Jr. and R.D. Ogden. 1968. Effects of Surface Mining on the Fish and Wildlife Resources of the United States. U.S. Department of the Interior, Fish and Wildlife Service, Resource Publication 68, Washington, DC.

Spencer, C.N., B.R. McClelland, and J.A. Stanford. 1991. Shrimp stocking, salmon collapse, and eagle displacement. BioScience 41:14-21.

Stahlberg, M. 1993. The Register-Guard, (Eugene, OR.), Aug. 3, 1993 P. 1D.

Stanford, J.A. and J.V. Ward. 1992. Management of aquatic resources in large catchments: recognizing interactions between ecosystem connectivity and environmental disturbance. Pp. 91-124 in R.B. Naiman, ed. Watershed Management: Balancing Sustainability and Environmental Change. Springer-Verlag, New York.

Starbird, E.A. and L.J. Georgia. 1972. A river restored: Oregon's Willamette. Nat. Geogr. 141(6):816-835.

Stearns, S.C. 1976. Life-history tactics: A review of the ideas. Quart. Rev. Biol. 51:3-47.

Stearns, S.C. 1989. The evolutionary significance of phenotypic plasticity. BioScience 39: 436-446.

Steel, B., P. List, and B. Shindler. 1992. Oregon State University Survey of Natural Resource and Forestry Issues: Comparing the Responses of the 1991 National and Oregon Public Surveys.

Steel, B., P. List, and B. Shindler. 1993. Conflicting Values about Federal Forests: A Composition of National and Oregon Publics. Society and Natural Resources.

Steel, B. and M. Brunson. 1993. Western Rangelands Survey: Comparing the Responses of the 1993 National and Oregon Public Surveys. Washington State University at Vancouver.

Stegner, W. 1954. (1982 edition.) Beyond the Hundredth Meridian. John Wesley Powell and the Second Opening of the West. University of Nebraska Press, Lincoln.

Stet, R.J.M. and E. Egberts. 1991. The histocompatibility system in teleostean fishes: From multiple histocompatibility loci to a major histocompatibility complex. Fish Shellfish Immunol. 1(1):1-16.

Stet, R.J.M., P. Kaastrup, E. Egberts, and W.B. Van Muiswinkel. 1990. Characterization of new immunogenetic markers using carp alloantisera; Evidence for the presence of major histocompatibility complex (MHC) molecules. Aquacult. 85(1-4):119-124. Elsevier Science Publ., Amsterdam (Netherlands).

Stevens, J.B. 1972. The economic performance of Oregon's commercial fisherman in 1972. Marine Fisheries Review paper 1261:17-22.

Steward, C.R. and T.C. Bjornn. 1990. Supplementation of salmon and steelhead stocks with hatchery fish: A synthesis of published literature. Part 2. In W.H. Miller (ed.). Analysis of Salmon and Steelhead Supplementation, Parts 1-3. Technical Report 90-1, U.S. Department of Energy, Bonneville Power Administration, Portland, OR.

Steward, C. 1994. Assessment of the Flow-Survival Relationship Obtained by Sims and Ossiander (1981)a for Snake River Spring/Summer Chinook Salmon Smolts. Final Report, Proj. No. 93-013, Contr. No. DE-AM79-93BP99954, Task Order DE-AT79-93BP00121.

Stewart, H. 1979. Looking at Indian Art of the Northwest Coast. University of Washington Press, Seattle, WA.

Stone, L. 1897. The Artificial Propagation of Salmon on the Pacific Coast of the United States. Bulletin of the Fish Commission 16.

Stuehrenberg, L., G. Swan, and P. Ocker. 1994. Migrational Characteristics of Adult Spring, Summer, and Fall Chinook Salmon Passing Through Reservoirs and Dams of the Mid-Columbia River. Draft final report, funded by mid-Columbia River public utility districts and National Marine Fisheries Service, Northwest Fisheries Science Center, Seattle, WA.

Suboski, M.D. and J.J. Templeton. 1989. Life skills training for hatchery fish: Social learning and survival. Fish. Res. 7:343-352.

Sullivan, K., T.E. Lisle, C.A. Dolloff, G.E. Grant, and L.M. Reid. 1987. Stream channels: The link between forests and fishes. Pp. 39-97 in E.O. Salo and T.W. Cundy, eds. Streamside Management: Forestry and Fishery Interactions. Contribution No. 57, Institute of Forest Resources, University of Washington, Seattle, WA.

Suttles, W. 1968. Coping with abundance: Subsistence on the Northwest coast, Pp. 56-68 in R.B. Lee and I. DeVore, eds. Man the Hunter. Aldine, New York. Reprinted in Coast Salish Essays. University of Washington Press, Seattle, WA, 1987.

Suzumoto, B.K. 1992. Willapa Fisheries Enhancement Project. Report prepared for the Willapa Alliance. Published by Ecotrust, Portland, OR. 73 pp.

Swanson, B.L. and D.V. Swedberg. 1980. Decline and recovery of the Lake Superior Gull Island Reeflake trout (*Salvelinus namaycush*) population and the role of sea lamprey (*Petromyzon marinus*) predation. Presented at the Sea Lamprey Int. Symp., Northern Michigan Univ., Marquette, MI, USA, July 30 -August 8, 1979. Can. J. Fish. Aquat. Sci. 37(11):2074-2080.

Swanson, F.J., L.E. Benda, S.H. Duncan, G.E. Grant, W.F. Megahan, L.M. Reid and R.R. Ziemer. 1987. Mass failures and other processes of sediment production in Pacific Northwest forest landscapes. Pp. 9-38 in E.O. Salo and T.W. Cundy, eds. Streamside Management: Forestry and Fishery Interactions. College of Forest Resources, University of Washington, Institute for Forestry Research & Control. No. 57. 471 pp.

Swanston, D.N. 1991. Natural Processes. Am. Fish. Soc. Spec. Publ. 19:139-179.

Swanston, D.N. and F.J. Swanson. 1976. Timber harvesting, mass erosion, and steepland forest geomorphology in the Pacific Northwest. Pp. 199-221 in D.R. Coats, ed. Geomorphology and Engineering. Dowden, Hutchinson, and Ross, Inc., Stroudsburg, PA.

Systems Operation Review (SOR). 1992. Columbia River System Operation Review. Screening Analysis, Vol. 1; Description and Conclusions. August, 1992. 230 pp. U.S. Department of Energy, Bonneville Power Administration; U.S. Army Corps of Engineers, No. Pac. Div.; U.S. Bureau of Reclamation. 1992.

Systems Operation Review (SOR). 1992. U.S. Department of Energy, BPA; U.S. Army Corps of Engineers, No. Pac. Div.; U.S. Bureau of Reclamation. 1992. Columbia River System Operation Review. Screening Analysis, Vol. 2. Impact results. June, 1992. 363 pp.

Takagi, K., K.V. Aro, A.C. Hartt, and M.B. Dell. 1981. Distribution and origin of pink salmon (*Oncorhynchus gorbuscha*) in offshore waters of the North Pacific Ocean. International No. Pac. Fish. Comm. Bull. 40. 195 pp.

Taylor, E.B. 1990. Phenotypic correlates of life history variation in juvenile chinook salmon, *Oncorhynchus tshawytscha*. J. Anim. Ecol. 59:455-468.

Taylor, E.G. 1991. A review of local adaptation in Salmonidae, with particular reference to Pacific and Atlantic salmon. Aquacult. 98:185-207.

Thomas, D.W. 1983. Changes in Columbia River Estuary Habitat Types of the Past Century. Columbia River Estuary Study Taskforce (CREST), Astoria, OR. 51 pp.

Thomas, J.W., M.G. Raphael, R.G. Anthony, E.D. Forsman, A.G. Gunderson, R.S. Holthausen, B.G. Marcot, G.H. Reeves, J.R. Sedell, and D.M. Solis. 1993. Viability Assessments and Management Considerations for Species Associated with Late-Successional and Old-Growth Forests of the Pacific Northwest. Research Report, U.S. Department of Agriculture, Forest Service, Portland, OR. 530 pp.

Thompson, P.B., R.J. Mathews, and E.O. van Ravensway. 1994. Ethics, Public Policy, and Agriculture. Macmillan Publishing Company, New York.

Thompson, W.F. 1951. An Outline for Salmon Research in Alaska. University of Washington, Fisheries Research Institute Circular 18, Seattle, WA.

Thompson, W.F. 1959. An approach to the population dynamics of the Pacific red salmon. Trans. Am. Fish. Soc. 88:206-209.

Thompson, W.F. 1965. Fishing treaties and the salmon of the North Pacific. Science 150:1786-1789.

Thornton, R.D. 1991. Searching for consensus and predictability: Habitat planning under the Endangered Species Act of 1973. Environ. Law 21:605-606.

Thorpe, J.E. 1993. Impacts of fishing on genetic structure of salmonid populations. Pp. 67-80 in J.G. Cloud and G.H. Thorgaard, eds. Genetic Conservation of Salmonid Fishes. Plenum, New York.

Tobin, R.J. 1990. The Expendable Future. U.S. Politics and the Protection of Biological Diversity. Duke University Press, Durham, NC.

Townsend, R.L. and J.R. Skalski. 1994. A Comparison of Statistical Methods of Estimating Treatment-Control Ratios (Transportation Benefit Ratios) Based on Spring Chinook Salmon on the Columbia River. Report of Center for Quantitative Science, School of Fisheries, University of Washington, to BPA, Proj. No. 91-051, Task No. DE-AT79-91BP16570, Contr. No. DE-A179-87BP35885.

Trenberth, K.E. 1990. Recent observed interdecadal climate changes in the northern hemisphere. Bull. Am. Meterol. Soc. 71:988-993.

Triska, F.J., V.C. Kennedy, R.J. Avazino, G.W. Zellweger, and K.E. Bencala. 1989. Retention and transport of nutrients in a third order stream: Hyporheic processes. Ecology 70:1893-1905.

Trotter, P.C. 1989. Coastal cutthroat trout: A life history compendium. Trans. Am. Fish. Soc. 118:463-473.

Trotter, P.C., P.A. Bisson, and B.R. Fransen. 1993. Status and plight of the searun cutthroat trout. Pp. 203-212 in J.G. Cloud and G.H. Thorgaard, eds. Genetic Conservation of Salmonid Fishes. Plenum Press, New York.

Truman, D.B. 1971. The Governmental Press. Second edition. Knopf, New York.

Tschaplinski, P.J. and G.F. Hartman. 1983. Winter distribution of juvenile coho salmon (*Oncorhynchus kisutch*) before and after logging in Carnation Creek, British Columbia, and some implications for overwinter survival. Can. J. Fish. Aquat. Sci. 40:452-461.

UNESCO (United Nations Educational, Scientific, and Cultural Organization). 1992. Oceanic Interdecadal Climate Variability. IOC Technical Series 40. UNESCO. 40 pp.

USCOE (U.S. Army Corps of Engineers). 1993. North Pacific Division. Seventh Progress Report: Fish Passage Development and Evaluation Program 1984-1990. July 1993. U.S. Army Corps of Engineers. North Pacific Division. Environmental Resources Division. Portland, OR. 403 pp.

USCOE et al. (U.S. Army Corps of Engineers, Bonneville Power Administration, and U.S. Bureau of Reclamation). 1992. Columbia River Salmon Flow Measures Options Analysis/EIS. U.S. Army Corps of Engineers, Walla Walla District, Walla Walla, WA.

USDC (U.S. Department of Commerce) Bureau of the Census. 1975. Historical Statistics of the United States, Colonial Times to 1970.

USDC (U.S. Department of Commerce) Bureau of the Census. 1995. Statistical Abstracts of the United States.

USDI (U.S. Department of the Interior) Bureau of Land Management. Undated-a. Taylor Grazing Act in Oregon 1934-1984. USDI Bureau of Land Management, Washington, DC.

USDI (U.S. Department of the Interior) Bureau of Land Management. Undated-b. 50 Years of Public Land Management 1934-1984. USDI Bureau of Land Management, Washington, D.C. 27 pp.

USDI (U.S. Department of the Interior) National Marine Fisheries Service, and Lower Elwha-Skallam Tribe. 1994. The Elwha Report - Restoration of the Elwha River Ecosystem and Native Anadromous Fisheries. A Report to Congress. Port Angeles, WA.

U.S. Superintendent of Indian Affairs, Oregon Territory. 1855. June 11, 1855.

Utter, F.M., J.E. Seeb, and L.W. Seeb. 1993. Complementary uses of ecological and biochemical genetic data in identifying and conserving salmon populations. Fish. Res. 18:59-76.

Utter, F.M., R.S. Waples, and D.J. Teel. 1992. Genetic isolation of previously indistinguishable chinook salmon populations of the Snake and Klamath rivers: limitations of negative data. Fish. Bull. 90:770-777.

VanCleve, R. and R.G. Ting. 1960. Conditions of salmon stocks in the John Day, Umatilla, Walla Walla, Grande Ronde, and Imnaha as reported by various fisheries agencies. 83 pp. + appendices. (Unpublished)

Van Hyning, J. 1968. Factors affecting the abundance of fall chinook salmon in the Columbia River. Ph.D. Dissertation, Oregon State University, Corvallis, OR.

Vannote, R.L., G.W. Minshall, K.W. Cummins, J.R. Sedell, and C.E. Cushing. 1980. The river continuum concept. Can. J. Fish. Aquat. Sci. 37:130-137.

Veldhuisen, C. 1990. Coarse woody debris in streams of the Drift Creek Basin, OR. MS thesis, Oregon State University, Corvallis, OR. 109 pp.

Vincent, R.E. 1960. Some influences of domestication upon three stocks of brook trout (*Salvelinus fontinalis* Mitchill). Trans. Am. Fish. Soc. 89:35-52.

Volkman, J.M. and W. E. McConnaha. 1993. Through a glass, darkly: Columbia River salmon, the Endangered Species Act, and adaptive management. Environ. Law 23:1249-72.

Wagner, W.C. 1981. Reproduction of planted lake trout in Lake Michigan. No. Am. J. Fish. Manag. 1:159-164.

Walker, D.E., Jr. 1967. Mutual Cross-Utilization of Economic Resources in the Plateau: An Example from Aboriginal Nez Perce Fishing Practices. Washington State University Laboratory of Anthropology, Report of Investigations No. 41. Pullman, WA.

Walstad, J.D., S.R. Radosevitch, and D.V. Sandberg (eds). 1990. Natural and Prescribed Fire in Pacific Northwest Forests. Oregon State University Pres, Corvallis, OR. 317 pp.

Walters, C.J. 1986. Adaptive Management of Renewable Resources. Macmillan, New York.

Walters, C.J., J.S. Collie, and T. Webb. 1988. Experimental designs for estimating transient responses to management disturbances. Can. J. Fish. Aquat. Sci. 45:530-538.

Walters, C.J., R.D. Goruk, and D. Radford. 1993. Rivers Inlet sockeye salmon: an experiment in adaptive management. No. Am. J. Fish. Manag. 13:253-262.

Walters, C.J. and B. Riddell. 1986. Multiple objectives in salmon management: The chinook sport fishery in the Strait of Georgia, B.C. Northwest Environ. J. 2:1-15.

Waples, R.S. *In press.* Evolutionarily significant units and the conservation of biological diversity under the Endangered Species Act. In Evolution and the Aquatic Ecosystem, J.L. Nielsen and D.A. Powers, eds. American Fisheries Society, Bethesda, Md.

Waples, R.S. 1991. Genetic interactions between hatchery and wild salmonids: lessons from the Pacific Northwest. Can. J. Fish. Aquat. Sci. 48 (Suppl. 1):124-133.

Waples, Robin S. 1991. Definition of "Species" Under the Endangered Species Act: Application to Pacific Salmon. NOAA Technical Memorandum NMFS F/NWC-194. U.S. Department of Commerce, National Oceanic and Atmospheric Administration, National Marine Fisheries Service.

Ward, J.V. and J.A. Stanford. 1989. Riverine ecosystems: the influence of man on catch dynamics and fish ecology. Pp. 56-64 in D.P. Dodge, ed. Proceedings of the International Large River Symposium (LARS). Can. J. Fish. Aquat. Sci. Spec. Publ. 106. 629 pp.

Warren, C.E. 1971. Biology and Water Pollution Control. W.B. Saunders, Philadelphia.

WDFW and ODFW (Washington Department of Fish and Wildlife and Oregon Department of Fish and Wildlife). 1994. Status Report. Columbia Fish Runs and Fisheries 1938-93. Olympia, WA. 271 pp.

WDF et al. (Washington Department of Fisheries, Washington Department of Wildlife, and Western Washington Treaty Indian Tribes). 1993. 1992 Washington State Salmon and Steelhead Stock Inventory (SASSI). Washington Department of Fisheries, Olympia, WA. 212 pp.

WDF and ODFW (Washington Department of Fisheries and Oregon Department of Fish and Wildlife). 1992. Status Report. Columbia River Fish Runs and Fisheries 1938-91. Olympia, WA 224 pp.

Washington Department of Wildlife. 1992. 1991-92 Steelhead Sport Catch Summary. Olympia, WA.

Washington Forest Practices Board. 1993. Standard Methodology for Conducting Watershed Analysis. Version 2.0. Washington Department of Natural Resources, Olympia, WA.

Washington State Department of Ecology. 1981. Inventory of Dams in the State of Washington. WDOE-81-18. 142 pp.

Waters, T.F. 1983. Replacement of brook trout by brown trout over 15 years in a Minnesota stream: Production and abundance. Trans. Am. Fish. Soc. 112:137-146.

Welsh, T.L., S.V. Gebhards, H.E. Metsker, and R.V. Corning. 1965. Inventory of Idaho Streams Containing Anadromous Fish Including Recommendations for Improving Production of Salmon and Steelhead. Part I - Snake, Salmon, Weiser, Payette, and Boise River drainages. State of Idaho, Department of Fish and Game. 174 pp.

Wendler, H.O. 1966. Regulation of commercial fishing gear and seasons on the Columbia River from 1859 to 1963, Washington Department of Fisheries, Fisheries Research Papers, 2(4):19-31.

Whale, R.J. and R.Z. Smith. 1979. A Historical and Descriptive Account of Pacific Coast Anadromous Salmonid Rearing Facilities and a Summary of Their Releases by Region, 1960-76. U.S. Department of Commerce, National Oceanic and Atmospheric Administration, Tech. Rep. NMFS SS RF-736. 40 pp.

White, R.J., R. Karr, and W. Nehlsen. 1995. Better roles for fish stocking in aquatic resource management. Pp. 527-547 in H.L. Schramm, Jr. and R.G. Piper, eds. Uses and Effects of Cultured Fishes in Aquatic Ecosystems. Am. Fish. Soc. Symp. 15, American Fisheries Society, Bethesda, MD.

Wildavsky, A. 1979. Speaking Truth to Power. Little Brown, Boston, MA.

Wilderness Society. 1993. The Living Landscape. Volume 2. Pacific Salmon on Federal Lands. The Wilderness Society, Bolle Center for Forest Ecosystem Management, Seattle, WA. 87 pp + appendices.

Wilkinson, C.F. 1992. Crossing the Next Meridian. Island Press, Washington, DC. 376 pp.

Wilkinson, C.F. 1980. The public trust doctrine in public land law. U.C. Davis Law Rev. 14(2):269-316.

Wilkinson, C.F. and D.K. Conner. 1983. The law of the Pacific salmon fishery: Conservation and allocation of a transboundary common property resource. Kansas Law Rev. 32:17-109.

Williams, C. 1980. Bridge of the Gods, Mountains of Fire: A Return to the Columbia Gorge. Friends of the Earth, Elephant Mountain Arts, New York.

Williams, J. and G. Matthews. 1994. A review of flow/survival relationships for juvenile salmonids in the Columbia River basin. NMFS, Northwest Fish Sci. Lab., MS submitted to Fish. Bull.

Williams, J.E., J.A. Lichatowich, and W. Nehlsen. 1992. Declining salmon and steelhead populations: New endangered species concerns for the west. Endangered Species Update 9:1-8.

Williams, I.V. 1987. Attempts to re-establish sockeye salmon (*Oncorhynchus nerka*) populations in the Upper Adams River, British Columbia, 1949-84. Pp. 235-242 in H.D. Smith, L. Margolis, and C.C. Wood, eds. Sockeye Salmon (*Oncorhynchus Nerka*) Population Biology and Future Management. Can. Spec. Publ. Fish. Aquat. Sci. 96.

Willingham, W.F. 1992. Northwest Passages: A History of the Seattle District U.S. Army Corps of Engineers, 1896-1920. U.S. Army Corps of Engineers, Seattle, WA.

Wilson, E.O. 1992. The Diversity of Life. Harvard University Press, Cambridge, MA.

Wilson, K. and G.E.B. Morren, Jr. 1990. Systems Approaches for Improvement in Agriculture & Resource Management. Macmillan, New York.

Wilson, P. 1994. Spring FLUSH (Fish Leaving Under Several Hypotheses) Version 4.5. Draft Documentation, July 16, 1994, Columbia Basin Fish and Wildlife Authority.

Wilzbach, M.A. 1985. Relative roles of food abundance and cover in determining the habitat distribution of stream-dwelling cutthroat trout (*Salmo clarki*). Can. J. Fish. Aquat. Sci. 42:1668-1672.

Winner, L. 1977. Autonomous Technology. MIT Press, Cambridge, Massachusetts.

Winter, G.W., C.B. Schreck, and J.D. McIntyre. 1980. Resistance of different stocks and transferrin genotypes of coho salmon, *Oncorhynchus kisutch*, and steelhead trout, *Salmo gairdneri*, to bacterial kidney disease and vibriosis. Fish. Bull. 77:795-802.

Winton, J. and R. Hilborn. 1994. Lessons from supplementation of chinook salmon in British Columbia. No. Am. J. Fish. Manag. 14(1):1-13.

Wissmar, R.C. and F.J. Swanson. 1990. Landscape disturbances and lotic ecosystems. Pp. 65-89 in R.J. Naiman and H. Decamps, eds. The Ecology and Management of Aquatic-Terrestrial Ecotones. UNESCO, Paris, and Parthenon Publishing group, Carnforth, United Kingdom.

Withler, R.E. and T.P.T. Evelyn. 1990. Genetic variation in resistance to bacterial kidney disease within and between two strains of coho salmon from British Columbia Canada. Trans. Am. Fish. Soc. 119 (6):1003-1009.

Wood, C.A. 1993. Implementation and evaluation of the water budget. Fisheries 18(11):6-17.

Wooster, W.S. 1992. King Crab Dethroned. Pp. 15-30 in M.H. Glantz, ed. Climate Variability, Climate Change, and Fisheries. Cambridge University Press, Cambridge, England, UK; New York.

Wright, S. 1940. Breeding structure of populations in relation to speciation. Am. Nat. 74:232-248.

Wright, S. 1951. The genetical structure of populations. Annals of Eugenics 15:323-354.

Wright, S. 1969. Evolution and the Genetics of Populations. Vol. 2 The Theory of Gene Frequencies. University of Chicago Press, Chicago, IL.

Wright, S. 1981. Contemporary Pacific salmon fisheries management. No. Am. J. Fish. Manag. 1:29-40.

WRRI. 1995a. Gravel Disturbance Impacts on Salmon Habitat and Stream Health. Volume I: Summary report. Report prepared by Water Resources Research Institute, Oregon State University, Corvallis, for Oregon Division of State Lands, Salem, OR. 52 pp.

WRRI. 1995b. Gravel Disturbance Impacts on Salmon Habitat and Stream Health. Volume II: Technical background report. Report prepared by Water Resources Research Institute, Oregon State University, Corvallis, for Oregon Division of State Lands, Salem, OR. 228 pp.

Wurtsbaugh, W.A. and G.E. Davis. 1977. Effects of temperature and ration level on the growth and food conversion efficiency of Salmo gairdneri Richardson. J. Fish Biol. 11:87-98.

Wydoski, R.S. and R.R. Whitney. 1979. Inland Fishes of Washington. University of Washington Press, Seattle, WA.

Yatsu, A., K. Hiramatsu and S. Hayase. 1993. Outline of the Japanese squid driftnet fishery with notes on the by-catch. Pp. 5-24 in J. Ito, W. Shaw, and R.L. Burgner (eds.), 1993. Symposium on Biology, Distribution, and Stock Assessment of Species Caught in the High Seas Driftnet Fisheries in the North Pacific Ocean. I.N.P.F.C. Bulletin 53 (III).

Ziemer, R.R. 1981. Storm flow response to road building and partial cutting in small streams of northern California. Water Resour. Res. 17:907-917.

Zinn, J.L., K.A. Johnson, J.E. Sanders, and J.L. Fryer. 1977. Susceptibility of salmonid species and hatchery strains of chinook salmon (*Oncorhynchus tshawytscha*) to infections by *Ceratomyxa shasta*. J. Fish. Res. Board Can. 34:933-936.

Appendixes

APPENDIX

A

Biographical Information on Committee Members and Staff

CHAIR

John J. Magnuson is a professor of zoology and director of the Trout Lake Biological Station and director of the Center for Limnology, University of Wisconsin-Madison. Formerly, Dr. Magnuson was chief, Tuna Behavior Program, Biological Laboratory, U.S. Bureau of Commercial Fisheries, U.S. Fish and Wildlife Service; and chairman, oceanography and limnology graduate program, University of Wisconsin-Madison. He has been a member of the affiliated graduate faculty, University of Hawaii; director, Ecology Program, National Science Foundation; member, Advisory Committee on Marine Resources Research, Food and Agricultural Organization of the United Nations, Ocean Policy Committee; chairman, Fisheries Task Group, Science Advisory Committee to Great Lakes Fishery Commission; and chairman, NRC Committee on Sea Turtle Conservation, NRC Committee on Fisheries, and NRC Committee to Review Atlantic Bluefin Tuna. Dr. Magnuson is a member of the American Fisheries Society (president, 1981-82), American Society of Ichthyologists and Herpetologists; Animal Behavior Society; Ecological Society of America; American Institute of Biological Sciences. Areas of expertise include behavioral ecology of fishes; locomotion of scombrids; distributional ecology of fishes and macroinvertebrates in ocean fronts or gradients; comparative studies of factors determining community structure of fishes in lakes; ecology of Great Lakes; long-term ecological research on northern lake ecosystems; general ecology; fish and wildlife sciences. He has a BS and MS from the University of Minnesota and a PhD in Zoology from the University of British Columbia. Address: Center for Limnology, University of Wisconsin, Madison WI 53706.

MEMBERS

Fred W. Allendorf is a professor of zoology, University of Montana. Formerly, he was the program director, Population Biology and Physiological Ecology, National Science Foundation; visiting scientist, Department of Genetics, University of California, Davis; NATO fellow, Genetics Research Unit, University of Nottingham, England; and lector, Department of Genetics and Ecology, Aarhus University, Denmark. He has served on panels on Conservation and Restoration Biology, International Programs, and Population Biology and Physiological Ecology of the National Science Foundation; as council member of the American Genetic Association; member, Genetics Nomenclature Committee; and chair, DNA Markers Subcommittee, American Fisheries Society. Dr. Allendorf is a member of the Society for the Study of Evolution, American Society of Naturalists, the Genetics Society of America, Society for Conservation Biology, American Association for the Advancement of Science, American Society of Ichthyologists and Herpetologists, American Fisheries Society, Sigma Xi, American Genetic Association, Desert Fishes Council, Ecological Society of America, and the Montana Native Plant Society. His research interests include evolutionary genetics of populations and conservation biology. He has a BS, zoology, Pennsylvania State University; MS, fisheries, University of Washington; and PhD, genetics and fisheries, University of Washington. Address: Division of Biological Sciences, University of Montana, Missoula MT 59812.

Robert L. Beschta is professor of forest hydrology, College of Forestry and Department of Forest Engineering, Oregon State University. Formerly, he was acting department head, forest engineering, Oregon State University; instructor and research associate, University of Arizona; forest hydrologist and research forester, U.S. Forest Service. His areas of expertise are riparian and watershed management; hydrology of wetlands, rangelands, riparian areas, and forests; precipitation and runoff of mountain slopes; the effects of vegetation on the hydrology of riparian areas; channel rehabilitation and morphology; water quality monitoring; peak-flow simulation models; and slope stability. Dr. Beschta has a BS, forest management, Colorado State University; MS, forest hydrology, Utah State University; and PhD, watershed management, University of Arizona. Address: Forest Engineering Department, 213 Peavy Hall, Oregon State University, Corvallis OR 97331-5706.

Peter A. Bisson is an aquatic biologist at Weyerhaeuser Company, Tacoma, Washington. Formerly, he was a research assistant, fisheries and wildlife, Oregon State University. Dr. Bisson is a member of the American Fisheries Society (currently Western Division vice-president), associate editor for salmonids, *Transactions of the American Fisheries Society*, and a member of the American Society of Limnology and Oceanography, Ecological Society of America, and North American Benthological Society. His research expertise includes the structure and function of stream ecosystems, fish production, population dynamics and community structure; analysis of environmental data bases; environmental impacts of land-use practices and non-point source pollution; zoogeography and systematics of freshwater fishes. Dr. Bisson has a BA, environmental biology, University of California, Santa Barbara; MS, and PhD, fisheries and wildlife, Oregon State University. Address: Weyerhaeuser Company, Technology Center WTC-1A5, Tacoma WA 98477. [After November 1995, Dr. Bisson will be with the U.S. Forest Service in Olympia, Washington.]

Hampton L. Carson is a professor emeritus of genetics and molecular biology, University of Hawaii, Manoa, where he was professor. Formerly, he was a professor at Washington University (St. Louis) and instructor, zoology, University of Pennsylvania. He was a member of the Wheelock Expedition, Labrador; professor, biology, University of Sao Paulo in 1951 and 1977; and Fulbright research scholar, University of Melbourne. Dr. Carson is a member of the National Academy of Sciences, the Society for Study of Evolution (president, 1971), American Society of Naturalists (president, 1973), Genetics Society of America (president, 1982), American Association for the Advancement of Science, and American Academy of Arts and Sciences. His interests are genetic changes in populations that lead to the origin of species, species formation on oceanic islands as paradigm, chromosomes and evolution, evolution of parthenogenesis, sexual behavior and genetic shifts in populations. He has a AB and PhD, Zoology, University of Pennsylvania. Address: Department of Genetics and Molecular Biology, John A. Burns School of Medicine, University of Hawaii at Manoa, 1960 East-West Road, Honolulu, HI 96822.

Donald W. Chapman is a consulting biologist and president of Don Chapman Consultants, Inc., and is an adjunct professor, Idaho State University, Pocatello. Formerly, Dr. Chapman was visiting professor, Montana State University; inland fishery biologist, FAO, Cartagena, Colombia; stock assessment specialist, FAO, Kigoma, Tanzania; visiting associate professor of limnology, University of Wisconsin, Madison; leader, Idaho Cooperative Fishery Unit and professor, University of Idaho; director of research, Oregon Fish Commission; associate professor, Oregon State University; coordinator of Alsea Watershed Study, and Exec. Secretary of Water Resources Research Institute; assistant professor, Oregon State University and Coordinator, Alsea Watershed Study. He is a member of the American Fisheries Society, associate editor for salmonids, *Transactions of the American Fisheries Society*; member, National Marine Fisheries Service Endangered Species Act Technical Advisory Committee. Dr. Chapman's research interests are catch and stock assessment, anadromous fish passage problems, population productivity in salmon and steelhead trout, habitat evaluation in salmon and steelhead spawning and rearing areas, fishery resource management, and best management practices for land use. He has a BS, forest management, MS, fisheries, and PhD, fisheries, Oregon State University. Address: Don Chapman Consultants, Inc., 3653 Rickenbacker, Suite 200, Boise ID 83705.

Susan S. Hanna is associate professor, agricultural and resource economics, Oregon State University. Her current research projects include an economic assessment of alternative management approaches in the west coast multispecies groundfish fishery, analysis of exvessel markets for groundfish, and an evaluation of salmonid predator-control programs on the Columbia River. Dr. Hanna is a member of the Scientific and Statistical Committee of the Pacific Fishery Management Council, the Scientific Committee of the Outer Continental Shelf Advisory Board, U.S. Department of the Interior, and the Executive Committee of the International Institute of Fisheries Economics and Trade. Her research interests include fisheries economics, fisheries management, economics of common property resources, and ecological economics. She has a BA, sociology and MS, agricultural and resource economics, University of Maine; Ph.D., agricultural and resource economics, Oregon State University. Address: Department of Agricultural and Resource Economics, Oregon State University, Corvallis, OR 97331-3601.

Anne R. D. Kapuscinski is professor of fisheries and conservation biology and an extension specialist (aquaculture), University of Minnesota. Formerly, she was fish genetics research assistant, aquaculture instructor, Oregon State University. Dr. Kapuscinski has served as outside reviewer, Northwest Power Planning Council, hatchery genetic policies and various genetic conservation planning documents; team leader, Northwest Power Planning Council Genetics Workshop: Sustainability of Anadromous Salmon and Trout Populations; outside reviewer, Bonneville Power Administration, comments on hatchery fish questions related to Endangered Species Act petitions for Pacific salmon stocks; chair, symposium on conservation of fisheries genetic resources, American Fisheries Society meeting; Northwest Power Planning Council, workshops on genetic production principles for Columbia River Basin fisheries management; co-chair, Fisheries Genetics Workshop, Midwest Fish and Wildlife Conference; American Fisheries Society North Central Technical Committee on Fish Genetics. Areas of research expertise include quantitative and molecular genetics of fish; aquatic biotechnology risk assessment and management; aquaculture; and genetic conservation. She has a BA, biology, Swarthmore College; MS and PhD, fisheries, Oregon State University. Address: Department of Fisheries and Wildlife, 130 Hodson Hall, University of Minnesota, St. Paul MN 55108.

Kai N. Lee is professor and director, Center of Environmental Studies, Williams College. Formerly, at the University of Washington, he was assistant professor, environmental studies, political science, and Program in Social Management of Technology; adjunct associate professor, marine studies; adjunct professor, public affairs; adjunct associate professor, fisheries. Dr. Lee was visiting professor, Institute of Economic Research, Kyoto University. He was also a White House Fellow and Assistant to the Secretary of Defense; member, vice-chair (1985-86), and chairman (1984-85, 1986-87), fish and wildlife committee, Northwest Power Planning Council. At the University of California, Berkeley, he was a postdoctoral fellow, Institute of Governmental Studies (1971-73) and assistant research social scientist, Institute of International Studies. He was a member of the NRC Board on Environmental Studies and Toxicology and is now a member of the NRC's Board on Sustainable Development. He served as a member of the board of directors of Friends of the Earth USA, a nongovernmental organization advocating environmental protection, since 1993. His areas of expertise are environmental education, adaptive management and sustainable development, and public policy and politics. Dr. Lee has an AB, physics, Columbia University, and PhD, physics, Princeton University. Address: Center for Environmental Studies, Kellogg House, Williams College, P.O. Box 632, Williamstown, MA 01267.

Dennis P. Lettenmaier is professor, Department of Civil Engineering, University of Washington. Dr. Lettenmaier is a member of the American Geophysical Union, American Meteorological Society, American Society of Civil Engineers, American Water Resources Association, Western Snow Conference, and U.S. National Committee for International Union of Geodesy and Geophysics. His areas of research expertise are global hydrology, hydrological forecasting, and water resource system modeling. Dr. Lettenmaier has a BS, mechanical engineering, University of Washington; MS, mechanical and environmental engineering, The George Washington University; and PhD, civil engineering, University of Washington. Address: Department of Civil Engineering, FX-10, University of Washington, Seattle, WA 98195.

Bonnie J. McCay is professor, Department of Human Ecology, Cook College, and Department of Anthropology, Faculty of Arts and Sciences, Rutgers the State University. She was visiting scientist, Department of Applied Behavioral Sciences, College of Agricultural and Environmental Sciences, University of California, Davis. She is a member of the Agriculture, Food & Human Values Society (council member, 1988-89); fellow, American Anthropological Association; fellow, American Association for the Advancement of Science; American Fisheries Society (various positions in the socioeconomics section); Anthropology Study Group for Agrarian Systems; Canadian Sociology and Anthropology Association; Columbia University Seminar on Cultural Evolution and Ecological

Systems; International Association for the Advancement of Appropriate Technology for Developing Countries; International Association for the Study of Common Property. Her research interests are culture and common property; social science issues in fisheries management; and ecological anthropology. Dr. McCay has a BA, Portland State University; MPH and PhD, anthropology, Columbia University. Address: Department of Human Ecology, Cook College, Rutgers, the State University, P.O. Box 231, New Brunswick NJ 08903.

Gordon M. MacNabb is president and chief executive officer, PRECARN Associates, Inc., and president, G.M. MacNabb Consultants. Formerly, he was chairman, Columbia River Treaty, Permanent Engineering Board; associate to the principal, Queen's University; president, Natural Sciences and Engineering Research Council of Canada (1978-86); chairman, Energy Supplies Allocation Board (1979-80); president, Uranium Canada, Ltd. (1975-85); and deputy minister, senior assistant deputy minister (energy), and assistant deputy minister (energy), Energy, Mines, and Resources, Canada. Mr. MacNabb is a member, Professional Engineers of Ontario, and vice-president and fellow, Canadian Academy of Engineering. His areas of expertise are hydroelectric and river basin planning, including the Columbia River; engineering issues, Columbia River Treaty; and treaty negotiations, St. John River, Maine and New Brunswick; and operating entities, Columbia River. He has a BS, Civil Engineering, Queen's University, and 11 honorary degrees. Address: PRECARN Associates, 30 Colonnade Road, Suite 300, Nepean, Ontario, Canada K2E 7J6.

Thomas P. Quinn is an associate professor, fisheries, University of Washington. Formerly, he was research associate, oceanography, University of British Columbia. He is a member of the American Fisheries Society, American Society of Ichthyologists and Herpetologists, Animal Behavior Society, and the American Institute of Fishery Research Biologists. Dr. Quinn's main areas of expertise are the migration, homing, and spawning behavior of Pacific salmon. He has a BA, biology, Swarthmore College; MS, University of Washington; and PhD, Fisheries, University of Washington. Address: School of Fisheries, University of Washington, Seattle, WA 98195.

Brian E. Riddell is section head, Salmon Stock Assessment and Enhancement Evaluation, Department of Fisheries and Oceans, Pacific Biological Station, Nanaimo, B.C., Canada. Other experience at the Pacific Biological Station includes head, Salmon Stock Assessment Program; head, Salmon Populations Section; head, Salmon Genetics Unit, head, International Salmon Unit. Dr. Riddell was the first chairman (1985-89) and member of the steering committee for the Salmon Sub-Committee of the Pacific Stock Assessment Review Committee; associate editor for genetics, *Transactions of the American Fisheries Society*; and chairman, organizing committee, International Symposium on Interaction between Enhanced and Wild Salmonids (1989-92). His areas of expertise include population biology and genetics of Pacific salmon, including conservation genetics of small natural populations in exploited landscapes and the impact of intensive culture on enhanced stocks and local natural populations; behavioral genetics and developmental genetics in salmonids; population dynamics and fishing mortality estimates on Pacific salmon, particularly chinook salmon; international fishery issues. Dr. Riddell has a BS, marine biology, University of Guelph, and PhD, ecology/population biology, McGill University. Address: Department of Fisheries and Oceans, Biological Science Branch, Pacific Biological Station, Nanaimo, British Columbia, Canada V9R 5K6.

Earl E. Werner is professor, biology, College of Literature, Science, and the Arts, University of Michigan, Ann Arbor. He has served as vice-president of the Ecological Society of America and editor, *Ecology and Ecological Monographs*. Formerly, he was assistant professor, University of Iowa and associate professor and professor, zoology, Michigan State University, Kellogg Biological Station. He is a member of the American Association for the Advancement of Science; American Society of Naturalists; Ecological Society of America; British Ecological Society; International Behavioral Ecology Society; International Society of Theoretical and Applied Limnology. His areas of expertise are community ecology, population biology, and behavioral ecology. He has an AB, zoology, Columbia University, and PhD, zoology-ecology, Michigan State University. Address: Biology Department, University of Michigan, Ann Arbor.

STAFF

David Policansky is associate director of the Board on Environmental Studies and Toxicology at the National Research Council. Formerly, he taught and did research at the University of Chicago, the University of Massachusetts at Boston, and the Grey Herbarium of Harvard University. He was visiting scientist at the National Marine Fisheries Service Northeast Fisheries Center. He is a member of the Ecological Society of America, the American Fisheries Society, and the advisory councils to the University of Alaska's School of Fisheries and Ocean Sciences and the University of British Columbia's Fisheries Centre. He was a member of the editorial board of *BioScience*. His interests include genetics, evolution, and ecology, particularly the effects of fishing on fish populations, ecological risk assessment, and natural resource management. He has directed approximately 15 projects at the National Research Council dealing with natural resources and ecological risk assessment, most recently on the Endangered Species Act. He has a BA, biology, from Stanford University and an MS and PhD, biology, from the University of Oregon. Address: Board on Environmental Studies and Toxicology, National Research Council, 2101 Constitution Ave. NW., Washington, DC 20418.

Tania Williams is research associate in the Board on Environmental Studies at the National Research Council. Formerly, she was a consultant at Wilson, Hill Associates, business manager at Kitty Hawk Investment Corporation, telecommunications specialist at Telecom Specialists, Inc., and research assistant at Alex. Brown & Sons, Inc. At the National Research Council, she has helped manage projects on Alaskan outer continental shelf environmental information and wetlands characterization, and she manages a project reviewing the U.S. Fish and Wildlife Service's Biomonitoring of Environmental Status and Trends Program, and the U.S. National Committee on Scientific Problems of the Environment. Her interests include natural resources, botany, and science policy. She has a BS, psychology, Allegheny College and has studied at Ealing College in London and the University of Baltimore. Address: Board on Environmental Studies and Toxicology, National Research Council, 2101 Constitution Ave. NW., Washington, DC 20418.

Adriénne Davis is senior program assistant in the Board on Environmental Studies and Toxicology. Formerly, she was a legal clerk-typist at the U.S. Patent and Trademark office. At the National Research Council, she has worked at the Toxicology Information Center and on a variety of projects, including DNA forensics, environmental research, and the Endangered Species Act. Her interests are information technology and management and education. She has a BS, business education, University of Maryland and MA, computers in education and training, Trinity College. Address: Board on Environmental Studies and Toxicology, National Research Council, 2101 Constitution Ave. NW., Washington, DC 20418.

APPENDIX
B

Meeting Dates and Locations

December 7-8, 1992
Washington, D.C.
National Academy of Sciences

February 1-3, 1993
Portland, Oregon
Holiday Inn, Lloyd Center

June 23-25, 1993
Seattle, Washington
University Inn

September 8-10, 1993
Portland, Oregon
Portland Marriott

December 13-14, 1993
Irvine, California
Arnold and Mabel Beckman Center

February 27 - March 1, 1994
Irvine, California
Arnold and Mabel Beckman Center

June 16-17, 1994
Seattle, Washington
Battelle Conference Center

APPENDIX

C

Acknowledgements

The committee would like to thank the following for contributing to its deliberations:

Rick Applegate, Fish & Wildlife Division, NPPC
Bill Bakke, Oregon Trout
Don Bevan, Endangered Species Act Snake River Salmon Recovery Team
Gerald Bouck, Fish and Wildlife Division, Bonneville Power Administration
Doug DeHart, Oregon Department of Fish and Wildlife
John Donaldson, Columbia Basin Fish & Wildlife Authority
Bob Francis, University of Washington
Bill Frank, Chair, Northwest Indian Fisheries Commission
James Geisinger, Northwest Forestry Association
Eugene Green, Sr., Columbia River InterTribal Fish Commission
Randall Hardy, Bonneville Power Administration
Gordon Haugen, USDA Forest Service, PAC Fish Coordinator
Arleigh Isley, Cattlemen's Association
Ralph Johnson, University of Washington
James Lichatowich, fisheries consultant
Irene and Kent Martin, Salmon for All
Peter Moyle, University of California, Davis
Phil Mundy, private consultant
Willa Nehlsen, Pacific Rivers Council
Evelyn Pinkerton, University of British Columbia
Jay Rasmussen, Oregon Coastal Zone Management Association

Henry Regier, University of Toronto
Roger Schiewe, Bonneville Power Administration, Dittmer Center
James Sedell, USDA Forest Service
J. Gary Smith, NMFS, Northwest Region, Deputy Director
Glen Spain, Pacific Coast Federation of Fishermen's Associations
Anne Squier, Office of Governor, Oregon
Michael Tillman, NMFS, Office of Protected Resources, Director
Frank Warrens, Pacific Fisheries Management Council
Warren Wooster, University of Washington
Al Wright, Pacific Northwest Utilities Conference Committee

The committee would also like to thank those that briefed it at its public sessions:

Portland, March 1, 1993

Lionel Boyer, Shoshone-Bannock Tribe
Steve Culley, FINS
John Davenport, forestry consultant
Art Goddard, Canadian Consulate General
Lon Peter, Public Power Council
Carl Schreck, Oregon State University

Seattle, June 24, 1993

Michael Anderson, The Wilderness Society
Mike Carr, Oregon Trout
Darryll Olsen, Northwest Irrigation Utilities
Ray White, consultant

Portland, September 9, 1993

James Baker, Sierra Club
Dan Diggs, U.S. Fish and Wildlife Service
Victor Kaczynski, consultant
Tom Marlin, Coalition for Anadromous Salmon and Steelhead Habitat
John Palmisano, consultant
Larry Riggs, Genetic Resource Consultants
Dan Rohlf, Northwest Environmental Defense Center and for American Rivers Council
Gary Spackman, Idaho Department of Water Resources

February 12-15, 1994 (writing session)

Janet Fischer and the staff of the University of Wisconsin's Trout Lake Laboratory, for their support at a writing meeting held there.

APPENDIX

D

Major Landforms and Their Rivers

MOUNTAIN RANGES

A series of mountain ranges, such as the Olympic and the Coast Range, extend from northern Washington's Olympic Peninsula southward along the Washington and Oregon coast to California. Rivers flowing from the Olympic Mountains drain to both coastal and Puget Sound basins. The Olympic Mountains receive the most precipitation (up to 400 cm/yr) of any mountain range in the Pacific Northwest (Muckleston, 1993), and the rivers draining the western slopes are large relative to their watershed areas. South of Gray's Harbor, Washington, the Coast Range consists of relatively low-elevation, highly erodible mountains of moderate relief extending southward to the Coos River, Oregon. Rivers originating in the Coast Range and flowing to the ocean tend to be short, but some larger rivers, such as the Chehalis in Washington and the Umpqua in Oregon, cross the mountains from the Willamette-Puget Lowland. In southern Oregon and northern California, the Klamath-Siskyiou Mountains are characterized by steep topography. The headwaters of the Rogue and Klamath rivers are in the southern Cascade Mountains and cut across the coastal mountains. Farther south in California, the Coast Range again consists of low-elevation mountains dominated by erosive sedimentary deposits. Further south, precipitation is much lower along the northern California coast, with annual rainfall often less than 100 cm/yr. The Sacramento River passes through these mountains, draining the southern Cascades and western Sierra Nevada.

Willamette-Puget Lowland

This broad valley separates the Olympics and Coast Range from the Cascade Mountains. Rivers on the western side of the Willamette-Puget Lowland drain coastal mountains, and those on the eastern side drain the Cascades. In general, the eastern river basins are larger and more-important salmon-producing systems than the smaller western rivers.

Cascade Mountains

These mountains in Washington, Oregon, and northern California and with the coastal mountains above support most of the existing Pacific salmon populations. Annual precipitation of 200 cm/year is common in the North; river basins west of the Cascade crest produce two-thirds of the total runoff for the Pacific Northwest region although they drain less than one-fourth of the area (Muckleston 1993). The backbone of the Cascades is a series of intermittently active and extinct volcanoes extending from Glacier Peak and Mt. Baker in northern Washington to Mt. Lassen in northern California. Volcanic activity has has effects on salmon. Most recently, the 1980 eruption of Mt. St. Helens caused a major debris flow that dammed the mouths of several Toutle River tributaries and formed three new lakes. Many Cascade volcanoes experience eruptions every 100-1,000 years. Short-term effects of such events on salmon have been devastating, but the long-term consequences have included the creation of many new and productive spawning and rearing areas.

A series of large river basins drain the northwestern Cascades to Puget Sound. Some, such as the Skagit River's, support all seven Pacific salmon and are the only U.S. drainages in the Pacific Northwest to do so. Two large, lower Columbia River tributaries, the Cowlitz and Lewis, drain much of the western Cascades in southern Washington. In western Oregon, steelhead and chinook are produced in the Clakamas, Santiam, McKenzie, and upper Willamette rivers, which drain the western Cascades to the Willamette Valley. Farther south, the Cascades are drained by the Umpqua, Rogue, and Klamath. Some headwaters of the Sacramento River originate in the southernmost part of the Cascades.

Along the eastern slopes of the Cascades, precipitation dwindles rapidly, and major river basins become less numerous. The northern Cascade Range in Washington contains several important salmon-producing tributaries of the upper Columbia River. The Yakima River subbasin in central Washington flows from the eastern Cascades and constitutes one of the most important drainages for salmon in the middle Columbia. In Oregon, the eastern Cascade slopes are drained mostly by the Deschutes River and to a small extent by the Klamath River in the South. Precipitation in those river basins ranges from greater than 200 cm/yr at some high elevations less than 20 cm/yr at lower elevations (Jackson and Kimerling, 1993).

Columbia Intermountain Region

The three major divisions of this region are the Columbia River Basin, the Central Mountains (which include the Blue, Wallowa, and Ochoco mountains), and the high lava plains of central Oregon and the Snake River (Rosenfeld, 1993). Fewer salmon occur in this region, but historical populations included large runs of chinook, coho, sockeye, and steelhead. The region receives little precipitation, often less than 10 cm/yr in the plains and only 40-80 cm/yr in the Central Mountains (Jackson and Kimerling, 1993). Major river basins are few. The upper Columbia River has several salmon-producing tributaries fed mainly by snowmelt from the northern Rocky Mountains. Two important tributaries of the Snake River, the Clearwater River and portions of the Salmon River, drain eastern Washington and northern Idaho. Several major Columbia and Snake River subbasins—including the John Day, Grande Ronde, Malheur, and Owyhee rivers—flow from the Central Mountains. Parts of the high lava plains (Harney basin and upper Snake River) have historically been inaccessible to salmon.

Northern Rocky Mountains

These mountains, which extend into Canada, contain headwaters of the Fraser, and Columbia, and Snake rivers in the United States. and include part of eastern British Columbia, northeastern Washington, northern and central Idaho, and western Montana. The geology is dominated by granitic rocks (Rosenfeld, 1993). They can be highly erosive. Tributaries of the Thompson River, a major subbasin of the Fraser River in British Columbia, support substantial populations of sockeye and steelhead. Sockeye were historically abundant in the Salmon River Mountains of central Idaho but are now nearly extinct there. Chinook and steelhead are the only other salmon returning to rivers in the northern Rocky Mountains of Washington and Idaho.

Great Basin

These southernmost areas of eastern Oregon, southern Idaho, and northeastern California contain internally draining basins with no outlets to the Pacific Ocean. Thus, these river basins support no anadromous salmon.

APPENDIX E

International Treaty Considerations in Operation of the Columbia River System

In September 1943, the U.S. Senate Committee on Commerce requested that the Corps of Engineers undertake a comprehensive survey of the Columbia River basin in the United States. This was succeeded in March 1944 by a reference by the Canadian and U.S. governments to the International Joint Commission (IJC) calling for a determination as to "whether a greater use than is now being made of the waters of the Columbia River system would be feasible and advantageous." The urgency of such an analysis was underlined by the disastrous flooding in the basin in 1948, especially in the vicinity of Portland. The need for Canadian involvement is evident: about 30% of the Columbia River flow originates in the 15% of the river basin in that country. The need for Canadian storage for effective flood control, as well as for optimum regulation of the river for power generation, was obvious.

Extensive studies of the river system, especially the portion within Canada, were carried out under the auspices of the IJC during the 1950s. However, it was not until 1959 that the commission made its report to the two governments on the "principles" for cooperative use of Columbia River storage in Canada. Negotiations on a Columbia River Treaty began in 1960, and final approval of a treaty and protocol was obtained in January 1964.

Under the terms of the treaty, Canada has constructed three storage facilities on the mainstem and tributary of the Columbia and has committed to operate 15.5 MAF of that storage under the operating terms of the treaty. The treaty also authorized the construction of the Libby Dam on the Kootenai River in Montana, whose reservoir floods into Canada and provides an additional 5 MAF of storage.

Hence, the treaty has permitted the construction and operation of about 20.5 MAF of usable storage—55% of the total storage within the Columbia basin.

During the negotiations of the IJC principles and the treaty itself, additional power and flood control were the major objectives. The tenor of the discussion during this period is well illustrated by the following paragraph from the introductory pages of the IJC "Principles":

> The principle [sic] benefits in the downstream country from cooperative use of storage of waters within the Columbia River System are improvements in hydroelectric power production and prevention of flood damage. Although other benefits would also be realized from such cooperative use, the outlook at this time is that their value would be so small in comparison to the power and flood control values that formulation of principles for their determination and apportionment would not be warranted.

Similarly, the Columbia River Treaty itself is an international agreement that sets forth the obligations and the sharing of benefits from the operation for power and flood control. The fourth paragraph of the preamble of the document reflects the total focus on those benefits:

> Recognizing that the greatest benefit to each country can be secured by cooperative measures for hydroelectric power generation and flood control, which will make possible other benefits as well. . . .

It is apparent from this brief history of events that neither the benefits from other uses, such as navigation, nor the effects of the treaty on the fish population of the region form part of the international agreements that were reached. Studies might well have been carried out by each of the two parties involved concerning the fishery effects of projects or of the overall operating regime, but no remedial requirements are contained in the treaty documents, other than the maintenance of minimum flow levels at various projects. The basic obligations are as follows:

For Canada:

> To provide 15.5 million acre feet of usable storage at the Mica, Arrow Lakes and Duncan Lake sites

and to operate that storage in accordance with

> operating plans designed to achieve optimum power generation downstream in the United States.

For the United States:

> To maintain and operate the hydroelectric facilities
> . . . in the United States of America in a manner that makes the most effective

> use of the improvement in streamflow resulting from operation of the Canadian storage for hydroelectric power generation in the United States of America power system

and to discharge that obligation

> by reflecting in the determination of downstream power benefits to which Canada is entitled the assumption that the facilities . . . were maintained and operated in accordance therewith.

APPENDIX

F

Reservoir-System Operation

The motivation for construction and operation of a system of reservoirs is to smooth the natural variability of streamflows. If one considers the extreme case of an ephemeral stream (for instance, a small tributary of the lower Snake or middle Columbia River) in periods of no flow, it is clearly impossible to extract water for so-called beneficial uses, such as water supply. During periods of storm flow, withdrawals might be unnecessary or impossible; in fact, flood damages can result from the high flows. In such situations, upstream reservoir storage might serve the dual purpose of flood protection and water supply.

In the Columbia River Basin, only a few streams, which contribute minimally to the flow of the major tributaries, are ephemeral (these streams are generally in the middle and lower Snake River drainage and the middle Columbia). However, the natural flow of all streams is variable both from one year to another and from one season to another. Geographically, the within-year variability tends to be highest for streams that drain high-elevation areas, where much of the winter precipitation is stored as snow and runs off in the spring and summer.

Between-year variability of major Columbia River tributaries tends to be lowest for streams close to the Pacific Coast and generally declines from south to north. It also tends to be lower for high-elevation streams than low-elevation streams and decreases with increasing drainage area. Contrary to common perception, the interannual variability of the Columbia River, which is dominated by high-elevation runoff from the Cascade Mountains and the interior of British Columbia, is among the lowest of major world rivers. Also, expressed as runoff per unit area (or depth), the flow of the Columbia River is much higher than that of other major rivers of the western United States, such as the Colorado and Missouri rivers. That has important implications for reservoir operation, as discussed below.

Reservoirs can, and usually are, operated for a variety of purposes, including water supply (agricultural, municipal, and industrial), hydropower generation, cooling of thermal electric-power plants, navigation, recreation, and fishery protection and enhancement. Reservoirs can be of two types: storage or run-of-the-river. The difference between the two depends on the size of active storage capacity (the water stored behind the dam that can be controlled, i.e., total storage minus dead storage) relative to the mean annual or seasonal flow of the river. If the total storage capacity is equivalent to only a small part of the mean annual flow, such as the flow of a few days, the reservoir is run-of-the-river. Otherwise, it is considered a storage reservoir. In the Columbia River Basin, there are 36 major dams, of which nine (Libby, Hungry Horse, Kerr, Albeni Falls, Grand Coulee, and Dworshak in the United States and Mica, Duncan, and Arrow in Canada) have reservoirs with over 1 MAF of usable storage. In the case of the U.S. reservoirs, most (including Grand Coulee, the largest) were originally authorized for agricultural water supply. In the Columbia River Basin, the federal Columbia Basin Reclamation Project provides irrigation water to about 500,000 acres, primarily of Columbia River water diverted from Roosevelt Lake (formed by Grand Coulee Dam), some of which returns to the river above McNary Dam. In the Snake River Basin, the Minidoka Project irrigates over 1 million acres in the upper Snake River plain. This water is stored by a system of six reservoirs, of which American Falls is the largest.

Notwithstanding their primary purpose (water supply), operation of the U.S. mainstem Columbia River reservoirs is based mainly on hydropower demand. Although the amount of water diverted for agricultural water supply increased substantially during the 1950s and 1960s, it has since dropped somewhat and remains a relatively small fraction of the mean flow of the river at Grand Coulee Dam. In addition, the peak agricultural water demand is only slightly out of phase with the natural hydrograph of the river (which peaks in mid-June on the average).

The basic objective of the hydropower operation policy is to maximize the amount of "firm" power that can be generated by the reservoir system. Firm power is the amount of power that could have been generated by the present power and storage system every year in a specified historical period. (Today, 1928-1958 is specified by the Columbia River Treaty.) The firm power is determined by hydrologic conditions in a 42.5-month critical period—from September 1, 1928, to February 29, 1932. A set of rule curves specify the amount of water that would be released from each of the storage reservoirs during each month of the critical period. The rule curves are the basis for operation of the reservoir system; each year is assumed to be the first year of the critical period, and the storage reservoirs are not drawn below the rule curve corresponding to the first year of the critical period unless the reservoirs did not refill the previous spring, in which case the second-year rule curve is used, and so on.

Within the confines of the rule curves, releases are made to maximize hydropower generation at the time of the year when it is most highly valued and to meet flood-control requirements. In the Pacific Northwest, the seasonal peak electric power demand is in the winter for space-heating. From the standpoint of electric-power generation, then, the ideal system operation would result in a "shaped" hydrograph due to the reservoirs that peaked in midwinter. That is almost precisely out of phase with the natural hydrograph, which peaks in early summer. Although the regulated Columbia River hydrograph peaks in the summer, rather than winter, as would be the "ideal" case, the regulated hydrograph is clearly much less peaked than the natural one and has the effect of greatly increasing winter flows (by storing spring and summer runoff for later release) relative to the natural hydrograph.

In the Columbia system, flood-control objectives are more or less compatible with hydropower generation; for flood-control purposes, some of the storage capacity needs to be available in the spring to store peak runoffs. Winter releases for hydropower generation have the effect of providing for the required storage during this period.

The above considerations dictate some of the characteristics of the reservoir-system operation. Hydropower generation is proportional to the product of rate of release (discharge) and the elevation difference between the tailwater and the reservoir level (head). Most major dams have several turbines, each of which has an efficiency that depends on head and discharge. Generally, turbine efficiency is highest at discharges near the maximum. There are therefore some operational considerations in ensuring that each unit is operating close to its maximum efficiency. However, because run-of-the-river reservoirs have no appreciable storage, they are essentially always operated at maximum head, except for minor fluctuations due to turbine-efficiency considerations, as noted above, and diurnal fluctuations in hydropower demand.

Index

G

H

I

J

K

L

O

P

R

S

T

U

V

W

Y

Z